Statistics

An Introduction to Numerical Reasoning

Statistics

An Introduction to Numerical Reasoning

PILOT EDITION

Ray A. Waller

HOLDEN-DAY, INC.
San Francisco London Dusseldorf
Johannesburg Panama Singapore Sydney

This book is a PILOT EDITION published in a format that can be revised and updated quickly in order to incorporate new developments in both subject matter and computer technology. Both the publishers and authors are anxious to receive your comments and suggestions for improvements. These should be forwarded directly to Holden-Day, Inc., 500 Sansome Street, San Francisco, CA. 94111.

STATISTICS: AN INTRODUCTION TO NUMERICAL REASONING
Pilot Edition

Library of Congress Catalog Card Number: 78-60357
ISBN: 0-8162-9314-7
Printed in the United States of America

1234567890 809

Production Coordination: Michael Bass & Associates
Text Design: Michael Bass
Cover Design: Robert Barringer
Editing: Johanna Peterson

Preface

This book contains the basic concepts and fundamentals for a first course in statistics. The necessary mathematics prerequisite is a working knowledge of algebra. However, to facilitate easy reference to and review of the pertinent results from algebra, Appendix A reviews the rules of exponents, the use of logarithms, the binomial theorem, and some definitions for specific functions.

The study of statistics that you are about to begin can aptly be called a *premeditated utilization of numbers*. That statement is made because *statistics* in a general sense is the use of numerical data for decision making. Thus, a study of statistics requires that certain techniques and procedures be mastered. Obviously, decisions can be and are made without the use of statistics, yet most if not all decisions can be made by using statistics. Furthermore, the use of statistics in our decision making process tends to replace part of our subjectivity with objectivity. That is, the analysis and interpretation of numerical measurements become a natural part of decision making.

In an oversimplified form, we can describe the statistical input to decision making as follows: There is some phenomenon of interest about which certain decisions are to be made. It is frequently convenient to view the decisions as questions to be answered. Information can be gathered concerning those questions through the use of an experiment. Thus, the study begins with the statement of specific questions concerning some topic of interest and continues by completing an experiment (data collection procedure) which yields numerical data that can be analyzed and used as input information for the decision making process.

Many elementary statistics textbooks begin by assuming that the data (numerical measurements) are available, and they concentrate on the analysis of that data. Yet much is to be learned by the statistics student while collecting data. So this textbook describes study situations (experiments) that can be performed without involving elaborate equipment or large investments of either time or money. The student is urged to conduct as many experiments as possible and to subject the results to any appropriate analyses presented in this text.

Since many of the basic statistical techniques are the same for students in diverse disciplines such as accounting, agriculture, biology, business, chemistry, ecology, education, engineering, home economics, management, milling, psychology, sociology, and zoology, the text frequently uses hypothetical rather than realistic examples and exercises to show methods of analysis. Such data illustrate the necessary steps for each technique without devoting unnecessary attention to tedious arithmetical calculations. Furthermore, the data collected by conducting one or more experiments in some field of personal interest provides a personalized study of real data, which offers more realism than the analysis of theoretical data presented in a text. So again each student is urged to ask questions of personal interest and to conduct experiments for collecting data that can be subjected to the various statistical techniques as they are developed. Only then can a true appreciation of the power of numbers in decision making be gained.

The simultaneous treatment of classical and nonparametric statistical techniques is another feature of the text. In Chapters 10 and 11, the text presents the classical methods for testing statistical hypotheses as well as some corresponding nonparametric techniques for similar situations. The parallel presentation offers a better opportunity to compare and contrast the differences and similarities of the procedures. The same approach is used in the discussion of correlation in Chapter 12.

The starred sections contain supplementary material and may be omitted without breaking the continuity of the study. Also, proofs and mathematical developments are enclosed in boxes to indicate that they may, but need not, be included as an integral part of the text materials.

Further flexibility is available to the instructor in that the text includes ample material to allow for the teaching of a variety of courses. For example, it is possible to present a course that covers the basic ideas of statistical analysis without giving attention to the sampling situations or derivations by using only the examples and exercises provided. On the other hand, it is possible to slant the course heavily toward the use of the suggested sampling situations and rely on the students' own data for examples and exercises. Similarly, different sections of the book may be selected to provide courses for students from different curricula. Some examples of selected material for different courses of either one-semester or one- to two-quarters length are as follows:

General introductory statistics: Chapters 1, 2, 3 and 4 (except starred sections), Sections 5.1, 5.2, 5.4, 6.1, 6.2, Chapters 7 and 8, Sections 9.1, 9.2, 9.4, 10.1, 10.2, 12.1, 12.2 (point estimation portion only) and 12.3.

Statistics for undergraduate mathematics majors: Chapters 1-8 including derivations and proofs, Sections 9.1, 9.2, 10.1, and 11.1.

Statistics for majors in the social sciences: Chapters 1, 2, 3 (except starred sections), Sections 4.1-4.5, 5.1, 5.4, 6.1, 6.2, Chapter 7, Sections 8.1, 8.2 (binomial model portion only), 9.1-9.4, 12.1, 12.2 (point estimation portion only) and 12.3.

Statistics for majors in the biological sciences: Chapters 1, 2, 3, 4 (except starred sections), Sections 5.1, 5.4, 6.1, 6.2, Chapters 7, 8, Sections 9.1, 9.2, 9.4, 10.1, 11.1, 12.1, 12.2 (point estimation portion only) and 12.3.

Statistics and probability for fourth-year high school mathematics students: Chapters 1, 2, 3, Sections 4.1-4.4, 5.1, 5.2, 6.1, 6.2, 6.3, 7.2 (binomial models portion only) and Chapter 8.

The foregoing suggested choices of material are only some examples of a variety of possible courses that can be presented from this text. Of course, an instructor of an introductory course in statistics may wish to organize a class around other sets of selected topics. The only restriction in doing so is that the basic concepts in the chapters are interrelated according to the following diagram.

```
                                  / 9
1 — 2 — 3 — 4 — 5 — 6 — 7 — 8 <
                                  \ 10 < 11
                                        \ 12
```

Numerous individuals are due thanks for their assistance and support during the development of the manuscript. In particular, I offer my thanks to Dr. L. S. Laws, Dr. R. L. Iman, Dr. W. J. Conover, Mr. R. E. Thomas and Dr. R. A. Sundheim for reading portions of the manuscript and suggesting improvements; to Professor E. S. Pearson for permission to use Tables 1, 8 and 18 from his book *Biometrika Tables for Statisticians*; to Dr. H. C. Fryer for permission to use Tables IV, V and VI from his book *Concepts and Methods of Experimental Statistics*; and to Mrs. Vicki Claar and Mrs. Carolyn Waller for typing early drafts of the manuscript. Also, I am indebted to the Literary Executor of the late Sir Ronald A. Fisher, F.R.S., to Dr. Frank Yates, F.R.S., and to the publishers, Oliver and Boyd, Edinburgh, for permission to reprint Table III from Fisher and Yates' book, *Statistical Tables for Biological, Agricultural and Medical Research*. I gratefully acknowledge the assistance of those individuals named above and other colleagues and students too numerous to mention who in a variety of ways have improved the final copy. Also, I welcome reactions to and corrections of the book.

Contents

CHAPTER 1

An Introduction to Numerical Reasoning

1.1 STATISTICS - SOME DEFINITIONS AND REMARKS

What is statistics? That question is usually among the first asked by a student when told his curriculum requires one or more courses in statistics. Thus, whatever statistics is, we must suspect that statisticians (those practitioners of statistics) are poor public relations persons. Otherwise we could assume as common knowledge at least a definition of a science (or an art, as some prefer) that has blossomed and matured in this century. Since that is not a valid assumption, we begin by giving some definitions and examples of the word, statistics.

> Statistics is the science of *collecting*, *summarizing*, *analyzing*, and *interpreting* numerical data.
> Statistics are the *numbers* being studied.
> Statistics are the *measurements* of a beauty contestant.
> Statistics are the annual report *figures* for a specific company.

Thus, when we ask for either a definition or an example of statistics, we often get a wide range of responses, some of which are more informative than others.

This simply illustrates that statistics is many things to many people. Statistical techniques have been and are widely used in research of all types. The purpose of this book is to present the basic techniques of statistics, while indicating to some degree the scope of statistical applications.

The importance of statistical techniques is indicated by the following quote attributed to a nonstatistician. The English novelist, historian, and sociologist, H. G. Wells (1866-1946), is quoted as saying, "Statistical thinking will one day be as necessary for efficient citizenship as the ability to read and write." We have not perhaps arrived at that day yet, but we do see uses of statistical thinking about us each day. Average numbers are presented as being representative of groups of measurements. For example, a student's GPA (grade point average) is used to represent all of the student's grades. Also, the likelihood of rain occurring during a specified time period is reported as the probability of measurable precipitation.

A further directive to use numbers in decision-making was given by the English scientist, Sir Francis Galton (1822-1911), when he said, "Whenever you can, count." This might be paraphrased to say that whenever you can, measure. That is, use numerical measurements whenever possible in your decision-making.

Still considering the meaning of statistics, we now present two working definitions of the word. First, *statistics is* the science for analyzing numerical data. Second, *statistics are* the numerical data themselves, as well as functions of those data. The wide implications of those definitions become more apparent as we progress in our study of statistics.

These definitions make it appear (and rightly so) as if a study of statistics is a study of numbers. Thus, a student might ask the following questions: Why study statistics? Who needs to study statistics? How can a bunch of numbers provide any useful information? To such questions the reply could be as follows. We study numerical measurements to gain knowledge of various phenomena that exist about us and are of interest to us. Anyone who is making decisions that are, or can be, dependent on (or influenced by) numerical measurements needs statistics. He may be able to make decisions without statistics, but he can likely do better with statistics. Statistics offers techniques introducing numerical evidence into our decision-making process. That is, statistical methods for preparing, presenting, and interpreting numerical evidence are useful in decision-making, both as a means for availing all pertinent facts for evaluation by the decision-maker when he is making his decision and a means by which the decision can be justified to others.

To be specific in our discussion, we need some basic

concepts. In particular, we need to indentify the following items: The set of questions for which answers are desired (the decisions to be made), the variable to be measured and/or studied to obtain data, the source of the numerical observations (measurements), the method or technique for collecting data, and the uses to be made of the data.

Toward that end, specific definitions and statistical terminology are presented in the following sections so that a familiarity with the language of statistics can be gained. Certainly a knowledge of the basic terms is fundamental to studying statistics.

1.2 POPULATIONS AND QUESTIONS

We first define a *population* as that set of individuals to be studied. We use the word, individuals, in a very broad sense to include elements of such sets as the following:

P1: all students who are enrolled in Basic Statistics (BS) at MYU (My University) for the fall term of this academic year
P2: all rats used in a maze study in the MYU psychology department on November 10th of this year
P3: all loaves of whole wheat bread baked by the Local Biscuit, Bun, and Bread Bakery (LBBBB) on December 21st of this year
P4: all wheat plots used this year by the MYU Agriculture Experiment Station in the fall planting season
P5: all 100-watt light bulbs produced by the Bright As Day (BAD) Light Company during the month of October
P6: all cans of green beans canned by the Sealem Tight Company on March 7th
P7: all exchanges of Good As Gold (GAG) common stock during the transactions of last February.

We note that the symbol, Pi, is used here to identify the seven populations for easy reference in the subsequent development. In that notation the P denotes population and the i denotes the number of a particular population. A similar notation is used throughout this chapter for illustrating the basic terminology.

Once we have unambiguously defined our population (i.e., once our population is a well-defined set of objects) so that we have a designated set of elements (individuals) for study, our next step is to identify questions that we would like to investigate and eventually answer. For example, we might ask the following questions about the preceding populations. We attach numbers to the questions, symbolized by Q's, to correspond to those on the P's.

Q1: How well do statistics students at MYU perform on the Principles of Basic Statistics examination?
Q2: What is the average time required for a rat to complete the maze?
Q3: What is the average weight of a loaf of bread?
Q4: Do all varieties of wheat exhibit the same yields?
Q5: What is the average length of life for a BAD 100-watt light bulb?
What percent of the light bulbs is defective?
Q6: What percent of cans is properly sealed?
Q7: How many shares of GAG common stock are traded per day on the average? What is the average closing price per share for February?

Those are only a few of the possible questions that we can ask concerning the populations identified as P1, P2, ..., P7. The reader should try to list at least one other question of interest for some of the populations.

EXERCISES

1. (a) Identify three populations (distributions) from your major subject area or an area of interest to you which have properties that could be studied by the use of numerical measurements.
(b) For each of the chosen populations in part (a), state one or more questions that you want to answer.

2. Explore statistics in everyday experiences. Look for uses of statistics in newspapers, news magazines (e.g., *Newsweek*, *Time*, *U.S. News and World Report*), stories, and advertising. Watch for the use of statistics in television news reports and commercials. Ask questions about what you observe. Who collected the numbers? How were the numbers obtained? Do the numbers represent the group(s) for which they are being presented?

1.3 DATA COLLECTION AND EXPERIMENTS

After defining a population to be studied and listing a set of questions concerning that population, the next step is to specify methods of techniques for collecting numerical data that provide helpful information in answering those questions. One of the first questions *data collectors* must face is whether to collect data from *all* elements in the population (i.e., study the population by means of *census*) or from only a *part* of the population (i.e., view the population by means of a *sample*). If we are to base decisions on measurements taken from a *sample* (some subset of all the elements in the population), it is best that some consideration be given to sampling techniques before proceeding with an investigation. The following problem illustrates some of the issues we face when beginning an investigation.

PROBLEM: Consider as your population those students enrolled for at least 9 credit hours of work at MYU. Suppose you are going to select a sample (subset) of these students to study the presidential preference of students at MYU. How many students would you select for your sample? How will you make your selection of students? Where will the selected students be contacted? How will you phrase your questions? What responses do you anticipate? How will you analyze your data? That is, how will you incorporate information from your sample into your decision-making process?

Many of the questions we face when conducting a statistical investigation are included in the preceding problem. Thus, our objective is to systematically approach the questions raised for that problem. We continue to give basic definitions in the present chapter and reserve questions of sample selection; data reduction, summarization, and presentation; and statistical analyses for subsequent discussions. However, it is worthwhile to give some thought to the preceding problem and to attempt to answer the questions posed. By thinking about these questions, other questions as to how the study should be conducted will most likely be encountered. It is the aim of this text to both promote the discovery of and create an appreciation for the use of statistics by raising and answering such questions.

The details of sample selection are discussed in Chapter 2, so we continue now as if we have some set of individuals (either a population or a sample) from which we seek *observations* (numerical measurements) for input into our decision-making process. The means by which we solicit and obtain information from individuals within the set (either a population or a sample) under study is called an *experiment*. Note that we use the term experiment in a very general sense. That is, any operation, such as a laboratory experiment, a field study, a door-to-door survey, an opinion poll, or any other method of measurement aimed at the collection of data is called an experiment. In statistics an experiment is defined as a repeatable process for which the result of any one trial is uncertain. That is, we require that the conditions and methods of data acquisition can be specified and repeated. However, there is always an element of uncertainty concerning which particular observation will be realized on a given trial. For example, in an opinion survey, all respondents should be confronted with the same questionnaire, but we should not be able to predict with certainty the response of any one person without examination. Thus, we try to organize experiments so that all of the observations are secured under nearly identical conditions.

The result for any trial of an experiment is called an *outcome*, while the set of all possible outcomes is called the *sample space* for the experiment. In the following examples, we use the symbols E, O, and S_E, respectively, to illustrate some experiments, outcomes, and experimental sample spaces for the preceding populations and questions.

E1: Each student enrolled in Basic Statistics (BS) during the fall term takes the Basic Statistics examination on December 15th. The exam consists of 50 questions.

O_1: Mr. Sharp answers 45 questions correctly.

S_{E1}: {0, 1, 2, ..., 50}. Here we assume the outcome for any trial of the experiment to be the number of correct answers.

E2: Each rat is introduced to the maze to see whether or not it can successfully complete the maze in 20 minutes or less.

O_2: Rat #2 successfully completes the maze.

S_{E2}: {successful, unsuccessful}. The sample space represents the possibilities for each rat.

E3: Each loaf of whole wheat bread baked on December 21st by the LBBBB is weighed.

O_3: Loaf #1 weighs 17.3 ounces.

S_{E3}: $\{w | 10 \leq w \leq 20\}$. (Explanation: The symbolism is read as the set of all values, w, such that w is greater than or equal to 10, but less than or equal to 20. The w denotes the weight of a loaf in ounces.)

E4: The wheat of MYU experimental plots of wheat is harvested. The yield is recorded.

O_4: Plot #5 produces 75 bushels.

S_{E4}: $\{y | 40 \leq y \leq 80\}$. We suppose that all

yields, y, are between 40 and 80 bushels, inclusively.

E5: (1) A group of 100-watt light bulbs produced by the BAD Light Company is tested to determine life-time.

$O_{5(1)}$: Light bulb #10 burns for 783 hours.

$S_{E5(1)}$: $\{t \mid 0 \leq t \leq 1000\}$. Some light bulbs are defective and never burn, while others burn for 1,000 hours.

(2) A group of 100-watt light bulbs are inspected and classified as either defective or nondefective.

$O_{5(2)}$: Light bulb #50 is defective.

$S_{E5(2)}$: {defective, nondefective}.

E6: On March 7th, 1000 cans of green beans packed by the Sealem Tight Company are tested and classified as sealed or not sealed.

O_6: Can #98 is sealed.

S_{E6}: {sealed, not sealed}.

E7: (1) The number of shares of GAG stock traded is recorded for each of the 18 business days in February.

$O_{7(1)}$: 557 shares of GAG are traded on February 9th.

$S_{E7(1)}$: $\{n \mid 0 \leq n, n \text{ is an integer}\}$.

(2) The closing price per share of GAG stock is recorded for each of the 18 business days in February.

$O_{7(2)}$: The closing price for GAG on February 15 is \$3.17.

$S_{7(2)}$: $\{p \mid 2.75 \leq p \leq 3.53\}$.

As indicated by the foregoing examples, the outcomes for experiments may be either qualitative (e.g., defective or nondefective) or *quantitative* (e.g., some number which denotes price, weight, or time). Likewise, the sample space for an experiment may be either a set of qualities or a set of numbers produced by the experiment. Classes of outcomes may be useful for describing the results of an experiment, but for the use in most statistical analyses, *numerical measurements* are needed. Thus, most experiments are designed to either produce quantitive measurements directly or to assign numerical (quantitive) measurements to qualitative observations. The way that we designate which numbers are to be observed for or assigned to the outcomes of an experiment is presented in the next section.

1.4 RANDOM VARIABLES

By inspecting the outcomes and sample spaces for the foregoing experiments, we find that the outcomes vary when the experiment is repeated. It is that variation within the population being studied that introduces uncertainty to our study. To accommodate that uncertainty and provide the numerical observations needed for statistical analyses, we define *random variables*. A *random variable* is a rule (function) that assigns a *real number* to each outcome of an experiment. That assignment may be a classification code for a qualitative outcome, such as 1 for a defective item and 0 for a nondefective item, or an actual numerical measurement for a quantitative outcome, such as the weight of a loaf of bread.

The set of points produced by evaluating a random variable for all outcomes of an experiment is called the *sample space of that random variable.* We note that the sample space of a random variable is always a set of numbers. In fact, the sample of numerical observations for any statistical analysis is a subset of the sample space. Furthermore, note that we have used "sample space" to denote both the set of possible outcomes for an experiment and the set of possible values of a random variable. The two sample spaces may coincide when the outcome of the experiment is a numerical measurement on the specified random variable. The following examples of random variables, V, and their sample spaces, S, for the experiments in Section 1.3 illustrate the basic concepts.

V_1: *percentile* rank on a BS examination. The random variable, V_1, assigns a percentile rank to each possible raw score.
S_1: $\{0, 1, 2, \ldots, 100\}$.

V_2: *time* (minutes) required by any rat that completes the maze in less that 20 minutes.
S_2: $\{t | 0 < t \leq 20\}$.

V_3: *weight* (ounces) of a loaf of whole wheat bread.
S_3: $\{w | 10 \leq w \leq 20\}$.

V_4: *yield* (bushels per plot) of each plot.
S_4: $\{y | 40 < y \leq 80\}$.

V_5: (1) *length* (hours) of life in a light bulb
(2) *percentage* of defectives in the test group

S_5: (1) $\{t \mid 0 < t \leq 1000\}$
(2) $\{p \mid 0 \leq p \leq 100\}$.

V_6: *percentage* of cans that are improperly sealed

S_6: $\{p \mid 0 \leq p \leq 100\}$.

V_7: (1) *number* of shares traded
(2) *price* in dollars per share at closing time

S_7: (1) $\{n \mid 0 \leq n,\ n \text{ is an integer}\}$
(2) $\{p \mid \$2.75 \leq p \leq \$3.37\}$.

Further applications and illustrations of the preceding concepts are supplied by the following exercises.

EXERCISES

1. Identify two populations from any subject of interest, and list one or more questions for which answers are desired. (Results from Exercise 1 on page 4 may be used.)
2. Identify some random variables that can be measured to gain information and help answer questions concerning the populations in Exercise 1.
3. Outline one or more experiments that could be used to measure the variables indicated in Exercise 2.
4. Give the sample space for each variable in Exercise 2.
5. Take one particular use of statistics that you have observed for Exercise 2 on page 4. What population is studied? What questions are posed and answered? What variables are measured? Have the measurements been obtained by an experiment? How would you devise a study using the same questions?

1.5 TYPES OF STATISTICS

We have now defined and illustrated some of the concepts encountered when beginning a statistical investigation of a problem. So let us proceed by assuming that an experiment has produced a set of observations (outcomes) that are to be statistically analyzed. The use to be made of the results from an analysis determines whether we are working in an area of *descriptive statistics* or *statistical inference.*

If the purpose of our study is to *describe* and report the measurements that we have collected, then we use the techniques of *descriptive statistics* to summarize and present this data. This is true when the observations under study are obtained by conducting a census of the population or when we wish to restrict interpretation to the observations in a particular sample (that subset of the population which has yielded the particular set of numerical measurements available for analysis).

On the other hand, if the purpose for studying the available data (sample) is to investigate a larger set (namely, the population) of individuals, then the techniques we use are those of *statistical inference*. That is, we use those statistical techniques that allow us to discuss the behavior of a variable in a population by studying its behavior in a representative subset (sample) of that population. The meaning of a representative sample is discussed in Chapter 2.

Thus, we can concisely say that descriptive statistics is that branch of statistics concerned with summarizing and reducing a set of data for presentation and interpretation, while statistical inference is concerned with using sample evidence to make generalizations about the larger population from which the sample is obtained.

1.6 NOTATION

A very important part of an introductory course in statistics is that of becoming aware of and familiar with the notation and vocabulary to be used. We have to this point presented numerous definitions for terms used by statisticians, and more terms are defined as our study progresses. Another part of the language of statistics is the symbolism used to represent the measurements made on various variables and the rules of summation used to summarize the values of a variable observed in a random sample.

It is common to let a capital Roman letter denote a *variable* being studied and to let the lower case of the same letter denote the *observed value* of the variable associated with a particular element (an outcome) in the sample. Usually a subscript is used on the lower-case letter to indicate which element yielded that value of the variable. For example, let T denote the time (in minutes) required for a rat to complete the given maze. Then $t_1 = 3.7$, $t_2 = 5.2$, $t_3 = 10.9$, and $t_4 = 3.8$ denotes the values of the variable, T, produced by the first, second, third, and fourth rats to try the maze. As a second example, let W denote the weight (in ounces) of a loaf of whole wheat bread. Then $w_1 = 12.1$, $w_2 = 11.9$, and $w_3 = 12.2$ denote the respective weights of the first, second, and third loaves of bread weighed.

After selecting a random sample of size n from some population and measuring a designated variable, say X, for each of the selected elements, we have available a set of data that may be conveniently denoted by $(x_1, x_2, \ldots, x_n)$. Several standard statistical techniques from both the area of descriptive statistics and the area of statistical inference require the sum of the observations, the sum of squares of the observations, or various other functions of the data that involve the addition of a number of quantities. Thus, it is worthwhile to define a symbol for adding several numbers and to present some rules for using the summation symbol, Σ (Greek sigma). First, we define the sum of $x_1, x_2, x_3, \ldots, x_n$ as follows:

$$\sum_{i=1}^{n} x_i = x_1 + x_2 + x_3 + \ldots + x_n. \qquad (1.1)$$

The symbolism on the left side of Eq. (1.1) is read as: Sum all the values of x_i as i ranges from 1, 2, ..., n. The subscript, i, is called an *index* of summation and changes for the different terms in the sum.

EXAMPLE 1: Suppose $x_1 = 3$, $x_2 = 7$, $x_3 = 5$, $x_4 = 8$, $x_5 = 6$. Then,

$$\sum_{i=1}^{5} x_i = 3 + 7 + 5 + 8 + 6 = 29, \quad \sum_{i=1}^{3} x_i = 3 + 7 + 5 = 15,$$

$$\sum_{i=3}^{4} x_i = 5 + 8 = 13, \quad \sum_{i=1}^{5} x_i^2 = 9 + 49 + 25 + 64 + 36 = 183.$$

To work effectively with the summation symbol, three rules are needed.

RULE 1: The sum of the sum (or difference) of two variables is equal to the sum (or difference) of the respective sums. In symbols,

$$\sum_{i=1}^{n} (x_i + y_i) = \sum_{i=1}^{n} x_i + \sum_{i=1}^{n} y_i,$$

$$\sum_{i=1}^{n} (x_i - y_i) = \sum_{i=1}^{n} x_i - \sum_{i=1}^{n} y_i. \qquad (1.2)$$

Note that *Rule* 1 merely states that changing the order of terms in an addition expression does not change the result. That is,

$$\sum_{i=1}^{n} (x_i + y_i) = (x_1 + y_1) + (x_2 + y_2) = + \ldots + (x_n + y_n),$$

$$= (x_1 + x_2 + \ldots + x_n) + (y_1 + y_2 + \ldots + y_n),$$

$$= \Sigma x_i + \Sigma y_i.$$

RULE 2: Any constant (elements with values that do not change as the index changes) can be factored outside Σ. In symbols,

$$\sum_{i=1}^{n} ax_i = a \sum_{i=1}^{n} x_i. \qquad (1.3)$$

Note that *Rule 2* states that multiplying each element of a sum by a given number, a, is equivalent to multiplying the sum of all elements by a. That is,

$$\sum_{i=1}^{n} ax_i = ax_1 + ax_2 + \ldots + ax_n = a(x_1 + x_2 + \ldots + x_n),$$

$$= a \sum_{i=1}^{n} x_i.$$

RULE 3: The sum of a constant as the index goes from 1 to n is equal to n times the constant. In symbols,

$$\sum_{i=1}^{n} a = na.$$

Note that *Rule 3* states that adding a constant n times is equivalent to multiplying the constant by n. That is,

$$\sum_{i=1}^{n} a = a + a + \ldots + a = a(1 + 1 + \ldots + 1) = na.$$

EXAMPLE 2: Given that

$$\sum_{i=1}^{10} x_i^2 = 75$$

and

$$\sum_{i=1}^{10} x_i = 15,$$

use Eqs. (1.2) - (1.4) to find

$$\sum_{i=1}^{10} (x_i^2 - 2x_i + 4).$$

SOLUTION:

$$\sum_{i=1}^{10} (x_i^2 - 2x_i + 4) = \sum_{i=1}^{10} x_i^2 - \sum_{i=1}^{10} 2x_i + \sum_{i=1}^{10} 4 \text{ by Eq. (1.2)},$$

$$= \sum_{i=1}^{10} x_i^2 - 2\sum_{i=1}^{10} x_i + 4(10) \text{ by Eqs. (1.3) and (1.4)},$$

$$= 75 - 2(15) + 40 \qquad \text{by substitution},$$

$$= 85.$$

The limits of summation (i.e., the set of values over which the index ranges) are often understood and thus need not be written for every step of the solution for a given exercise. For example, when the summation symbol is used in reference to a sample of size n, we use Σ and

$$\sum_{i=1}^{n}$$

interchangeably. Also, the sum of all observations, x, in a given set, say A, is sometimes needed for statistical analyses. Therefore, it is expedient to use

$$\sum_{x \varepsilon A} x$$

to denote the sum of all x that are elements of the set A.

EXAMPLE 3: Suppose 6 seniors at MYU are asked to indicate the number of times they have visited the student health center during their student careers and the resulting x values are contained in A = (2, 5, 7, 3, 4, 9). Verify that

$$\sum_{x \varepsilon A} (x - 5)^2 = \sum_{x \varepsilon A} x^2 - 10 \sum_{x \varepsilon A} x + 150.$$

SOLUTION:

$$\sum_{x \in A} (x - 5)^2 = (-3)^2 + 0^2 + 2^2 + (-2)^2 + (-1)^2 + 4^2$$
$$= 34,$$

$$\sum_{x \in A} x^2 = 4 + 25 + 49 + 9 + 16 + 81$$
$$= 184,$$

$$\sum_{x \in A} x = 2 + 5 + 7 + 3 + 4 + 9$$
$$= 30.$$

Since 34 = 184 - 10(30) + 150, the result is verified for the elements of set A.

EXAMPLE 4: Show that

$$\sum_{i=1}^{n} (x_i - c)^2 = \sum_{i=1}^{n} x_i^2 - 2c \sum_{i=1}^{n} x_i + c^2 n$$

for any constant, c.

SOLUTION:

$$(x_i - c)^2 = x_i^2 - 2cx_i + c^2 \qquad \text{by multiplication,}$$

$$\Sigma(x_i - c)^2 = \Sigma x_i^2 - \Sigma 2cx_i + \Sigma c^2 \qquad \text{by Eq. (1.2),}$$

$$= \Sigma x_i^2 - 2c\Sigma x_i + c^2 n \qquad \text{by Eqs. (1.3) and (1.4).}$$

Note that *Example 3* is a special case of *Example 4*, where n = 6 and c = 5.

The preceding notation for variables and realizations (observations) of variables and the rules for using the summation symbol, Σ, are a basic part of the statistical language and must be mastered in order to understand statistics. The student should recognize that the rules are not new but are, in fact, only a statement in symbols of how arithmetic calculations are completed.

EXERCISES

1. Use the definition of Σ to express the following sums:

   (a) $x_2 + x_3 + x_4 + x_5$

   (b) $x_1^2 + 2x_2^2 + 3x_3^2$

   (c) $(x_1 - 1)^2 + (x_2 - 4)^2 + (x_3 - 9)^2 + (x_4 - 16)^2$

   (d) $x_1 + 2x_2^2 + 3x_3^3 + \ldots + nx_n^n.$

2. Use the definition of Σ to expand the following sums:

   (a) $\sum_{i=1}^{5} i^2 x_i$ (b) $\sum_{i=3}^{6} (x_i + 2i)$ (c) $\sum_{i=1}^{3} (x_i + 2y_i).$

3. Let $x_1 = 2, x_2 = 10, x_3 = 5, x_4 = 7, x_5 = 6.$

   (a) Find

   $$\sum_{i=1}^{5} x_i, \quad \sum_{i=1}^{5} x_i^2, \quad \sum_{i=3}^{5} (x_i + 1)^2.$$

   (b) Find

   $$\sum_{i=1}^{5} (x_i - 6)^2$$

   directly and by the result in Example 4.

   (c) Find

   $$\sum_{i=1}^{5} x_i/5, \quad \sum_{i=1}^{5} (x_i - 6)^2/4.$$

4. Let $x_1 = 3, x_2 = 1, x_3 = 5, y_1 = 4, y_2 = 2, y_3 = 6.$

   (a) Find

   $$\sum_{i=1}^{3} x_i, \quad \sum_{i=1}^{3} y_i, \quad \sum_{i=1}^{3} x_i y_i.$$

   (b) Verify that

   $$\sum_{i=1}^{3} (x_i - 3)(y_i - 4) = \sum_{i=1}^{3} x_i y_i - 36.$$

(c) Find

$$\sum_{i=1}^{3} (x_i + y_i)^2.$$

*1.7 HISTORY

We conclude this chapter by presenting some historical background for statistics and probability. Statistical inference techniques are quite modern in that most of them have been developed, promoted, and accepted for use only as recently as the late 1890's. However, the use of numbers for descriptive purposes is probably quite old. It seems quite likely that early man observed counts of animals, people, weapons, caves, etc., for use in decisions affecting his daily life. We know that in Biblical times the number of animals and servants a man owned was an indication of his worth. Thus, numbers have been used in the form of descriptive statistics for a long time.

On the other hand, history of statistical inference can be said to begin with the history of probability because statistical inference is basically the utilization of probability ideas in data analysis. That is, probability is the tool we use to bridge the gap between the descriptive statistics of a sample and the corresponding characteristics of the population from which the sample is observed. Thus, our outline of statistical history includes both the early thoughts and examples of descriptive statistics, as well as some of the early work in probability.

Descriptive statistics began in the area that we now call vital statistics. Those statistics pertaining to number of births, number of deaths, percentage of male births, causes of death, etc., are the earliest form of descriptive statistics we find on record. Newman [1]* presents part of a small book, *Natural and Political Observations made upon the Bills of Mortality*, published by an English army captain, John Graunt, in 1662. Graunt is proclaimed by Newman as the founder of statistics for that work in which he analyzes the birth and death statistics for London from 1604 to 1661. We note that the value of his work was immediately recognized and resulted in his being named an original member of the Royal Society in 1662. To illustrate the type of statistics reported for the year 1632, we give some of the data as recorded in Newman's book [1].

First, we find a rather extensive list of the number of deaths due to various causes during 1632, including the following: abortive and stillborn, 445; affrighted,

*Numbers in brackets here and throughout the text refer to the references cited at the end of each chapter.

l; aged, 628; consumption, 1,797; cut of the stone, 5; grief, 11; kill'd by several accidents, 46; lunatique, 5; plague, 8; rising of the lights, 98; suddenly, 62; and teeth, 470. Newman [1, p. 1425] provides a complete list of those statistics.

Second, Graunt presents a discussion of the apparent difference between numbers of males and females. Those differences appear in both the number of burials and the number of christenings. Both sources of data indicate a small excess of males. His records indicate approximately 52% males and 48% females in the London population during the 1600's. Those figures differ only slightly from the later figures (51.7% males, 48.3% females) reported by Fisz [2, p. 4] for percentages of male and female births in Poland from 1927 to 1932.

The work by Graunt in 1662 certainly represents a major step forward for descriptive statistics. It is an approach (possibly the first) where a mass of numbers (data) is assembled, summarized, presented, and interpreted for the reader. His primary objective was to present the data available to him, as well as an interpretation, in such a way that it would be useful information. His work was expanded by William Petty (1623-1687) and enlarged further by Edmond Halley (of Halley's comet) in 1693 when Halley published what can be called the first life insurance table (Newman [1], p. 1437).

From its beginning in the 1600's, statistics was a technique for reporting political science and state affairs data. Statistics continued to be basically a method for describing and presenting numbers until the late 19th century. Even though the marriage of statistics and probability did not produce the useful techniques of statistical inference until the late 19th century, the shift in emphasis away from descriptive statistics probably began with the work of the mathematician, Abraham De Moivre (1667-1754), as he used life tables and probabilities to determine annuity values.

The pioneering work of De Moivre was followed by statistical applications to astronomy made by two mathematicians and astronomers, Pierre Simon de Laplace (1749-1827) and Karl Friedrick Gauss (1777-1855). Another real stimulus to statistics was provided by Adolphe Quetelet (1796-1874), a Belgian astronomer and mathematician, who arranged for a census in 1829 and used the data obtained to study criminal tendencies. By studying the census data, he was able to predict with astonishing accuracy the number, types, etc., of crimes from year to year. (We note that those predictions by Quetelet were forms of statistical inference.)

Following next was Francis Galton (1822-1911), a cousin of Charles Darwin, who treated heredity as a purely statistical matter. He collected genetic data on sweet peas, moths, hounds, and humans. Even though his basic and pioneering research did not in itself produce many results because of some incorrect assumptions he made, Galton's methodology contributed much toward the application of statistics to biology. In fact, Karl Pearson (1857-1936) used Galton's work to make major contributions in the area of correlation analysis (see Chapter 12) while teaching at University College in London.

Then out of England arose another true giant in the field of statistics. Sir Ronald Alymer Fisher is credited by Mood [3] as having been responsible for at least half of all essential and important contributions to the theory of statistics as it was known midway through this century. Fisher published his first paper in 1912, his first book in 1925 (which was to appear in 13 editions), and never ceased writing and contributing to statistics until his death in 1962.

From the beginning efforts of those pioneers, statistics has grown into a discipline in which more than 20,000 individuals are employed in the United States alone. Present-day statisticians perform many and varied tasks, which include assisting researchers in agriculture, biology, education, psychology, sociology, etc., with the analysis of their data. This is not surprising in that many activities of plants and animals (even men) are studied by means of numerical observations, which in turn are dependent on chance events. Furthermore, most studies rely on sample information rather than on census data, thus presenting an even stronger case for the use of statistics. In particular, we need that portion of statistics called statistical inference, which depends heavily on probability. Therefore, we conclude this section with an outline of probability history as it relates to statistics.

Ore[4] prefers to date the birth of probability with Cardano (1501-1576), who in 1520 wrote a book analyzing games of chance. That work was not published until 1663, and by that time the state of probability had advanced beyond the thinking of Cardano. Others believe (see Weaver[5]) that the real beginning of probability might be as early as 1150 A.D. when Rabbi Ben Ezra used permutations and combinations for astrological purposes. In this regard, we can reason that ever since man began gambling, he has had some ideas or thoughts about probability. When did gambling begin? We read (Matthew 27:35) that Jesus' garments were divided by casting lots (i.e., by chance). Even before that, Canaan was divided by lots (Joshua 14:2,18:6). Thus, man has been using elements of chance in his decision-making for a long period of time. Hence, we might reason that the seeds of probability were sown long ago. In fact, gambling was labeled as undesirable around 100 A.D. by the Code of Manu, which advised, "Let the King prohibit gambling and betting in his kingdom, for these are the vices that destroy the kingdoms of princes."

Regardless of the particular beginnings of probability, it is significant to note that Kramer [6] believes that prior to 1600 *at most* six problems are recorded which in some way indicated probability. One of these early problems can be stated as follows: How many different outcomes can result from the experiment of rolling 3 dice? That question was discussed in Venice about 1477 in a commentary of *Divine Comedy* by Dante. This preceded by several years the 15-page handbook for gamblers (*Liber de Ludo Aleae*) in which Cardano analyzed various games of chance.

In 1654 one of the first attempts was made to solve a problem which today would be called the theory of probability. It was then that a French gentleman, Chevalier de Mere, who at least occasionally visited the gaming rooms, posed the following problem to Blaise Pascal, 17th-century philosopher, mathematician, and religious fanatic: The house bets even money that at least one 6 will appear when a fair die is rolled 4 times. de Mere believed that the bet was slightly favorable to the house and asked Pascal, who at the age of 19 invented the first calculating machine, to assist him in checking his suspicion. Pascal solved the problem through joint correspondence with Pierre de Fermat, 17th-century mathematician, which he initiated by a statement that may be translated as follows: Mister de Mere is very bright, but he is no mathematician and that, as you know, is a very grave defect. There is evidence that Pascal and Fermat jointly attacked other problems in the area of probability. For example, Pedoe [8] suggests that they studied the problems of justly dividing the stakes in a two-person game where a given number of points, say X, wins all, but the game has to be terminated before either player accumulates X points.

Another interesting correspondence involving probability took place in 1693 between the English diarist, Samuel Pepys, and the English mathematician, Isaac Newton. The problem posed by Pepys and solved by Newton is as follows: Individuals A, B, and C are going to roll dice according to the following plan. A rolls 6 dice and wants to observe at least one 6; B rolls 12 dice and wants to observe at least two 6's; C rolls 18 dice and wants to observe at least three 6's. If all dice are fair (balanced), which individual has the best chance of realizing the desired outcome? A full discussion of that problem is provided, along with some of the letters from the Pepys-Newton correspondence, in Pedoe [8] and Schell [9].

Beginning with the analysis of games of chance, ideas about probability grew from the contributions of men such as Jacob Bernoulli (1654-1705), a Swiss mathematician. In fact, he was one of the eight mathematicians from three generations of the Bernoulli family (see Bell [7], 132). His work, *Ars Conjectandi*, was published posthumously in 1713. That work discussed probability and its application to insurance, statistics, and heredity.

Another giant step in probability, as it influenced statistics, came in the work of the previously mentioned mathematician, De Moivre. In 1716 De Moivre published a book, *The Doctrine of Chance*, but his major contribution was contained in a brief paper that appeared in 1733. That paper studied the limit of binomial type probability and actually presented what we today study as the normal distribution (see Chapter 5). De Moivre was also a statistician in spirit in that he used observations to predict the time of his death. As his health began to fail, he noticed that his need for sleep increased by about 10 minutes each day. Using that as a basis, he correctly predicted that he would die on the day after he had slept 23 3/4 hours.

The previously mentioned Frenchman, Laplace, published *Theorie Analytique des Probabilities* in 1812. That book launched probability theory and set it sailing on its way as a discipline to be studied in its own right, as well as one to serve as the basis for all of statistical inference. He worked with the normal probability curve and used that curve to study problems in astronomy. We might add that Laplace certainly never doubted the worth of probability, as he is reported to have said, "It is remarkable that a science which began with the consideration of games of chance should have become the most important object of human knowledge."

Further work with the normal probability curve was conducted by the previously cited German mathematician and astronomer, Karl Friedrick Gauss, who used the normal curve to conduct studies in astronomy. In recognition of his contributions to using and understanding the curve, it is sometimes called the Gaussian curve.

Following the work of Gauss, we find the ideas of statistics and probability growing in many directions as a result of the research efforts of capable and talented individuals. There were those people who concentrated on the theory of probability for the sake of probability. That branch of the tree was and still is deeply rooted in higher mathematics. Another group of researchers devoted their efforts toward developing a methodology of data analysis. Those efforts have borne fruit in the sense that many useful techniques for summarizing and reducing data are now available. Further techniques have been developed for making statistical inference about unknown characteristics of the population under study. Thus, contributions have been made by researchers who have successfully blended the theory of probability with numerical measurements to analyze and interpret various phenomena. We call such individuals statisticians, and as shown in the preceding, brief, historical outline, the science (or art) that they practice is the result of many years of concentrated research efforts.

Additional technical references about the history of probability and statistics can be found in recent issues of the journal, *Biometrika*, as well as in the book by Pearson and Kendall [10].

1.8 SUMMARY

This first chapter presents basic ideas about the nature of statistics, defines fundamental concepts and terms needed for statistical analyses, and provides a brief history of statistics and probability. The idea that statistical methodology involves summarizing, analyzing, and interpreting numerical observations for decision-making is proposed. The purpose of this text is to present that methodology and demonstrate its wide range of applications. To further illustrate the need for statistics in a variety of fields, each student can reference some of the journals listed in the following exercises or consult the essays in Tanur [11], the examples in Wallis and Roberts [12], and the discussions in Kirk [13]. These references, in addition to most of the others cited in the historical discussion, are written for appreciation by novices in statistics as well as for those who have more experience in manipulating (analyzing) numbers.

1.9 REFERENCES

[1] Newman, J. R. 1960. *The world of mathematics:* vol. 3, New York: Simon and Schuster, pp. 1416-1531.

[2] Fisz, M. 1963. *Probability theory and mathematical statistics*. New York: Wiley, p. 4.

[3] Mood, A. M. 1950. *Introduction to the theory of statistics*. New York: McGraw-Hill, p. 282.

[4] Ore, O. 1953. *Cardano: the gambling scholar*. Princeton: Princeton Univ. Press.

[5] Weaver, W. 1963. *Lady luck: the theory of probability*. Garden City, N. Y.: Doubleday.

[6] Kramer, E. E. 1964. *The main stream of mathematics*. Greenwich, Conn.: Fawcett, p. 194.

[7] Bell, E.T. 1937. *Men of mathematics*. New York: Simon and Schuster.

[8] Pedoe, D. 1963. *The gentle art of mathematics.* New York: MacMillan, pp. 36, 43-48.

[9] Schell, E. D. 1960. Samuel Pepys, Issac Newton, and probability. *The American Statistician.* 14: 27-29.

[10] Pearson, E. S., and Kendall, M. G. 1970. *Studies in the history of probability and statistics.* New York: Hafner.

[11] Tanur, J. M., ed. 1972. *Statistics: A guide to the unknown.* San Francisco: Holden-Day.

[12] Wallis, W. A., and Roberts, H. V. 1965. *The nature of statistics.* Glencoe, Ill. The Free Press of Glencoe.

[13] Kirk, R. E., ed. 1972 *Statistical issues: a reader for the behavioral sciences.* Monterey, Calif.: Brooks-Cole.

1.10 REVIEW EXERCISES

1. Consult one or more of the following sources to see the quantity and type of data available in various areas of statistical use:

Agricultural Statistics
Automobile Facts & Figures
The Budget of the U. S.
Congressional Record
Construction Review
Life Insurance Fact Book
Minerals Yearbook
Petroleum Facts & Figures
U. S. Census of Population
U. S. Census of Housing.

2. Consult one or more issues of at least *five* of the journals listed below to answer the following questions.

(a) Does the journal publish articles involving statistics?
(b) What types of data are presented for the reader's consideration?
(c) How are experimental results presented?

Accounting Review
American Journal of Int'l. Law
Antarctic Journal
Appraisal Journal
Behavior Science Notes
Brewer's Digest
Business Economics
Demography
Ecology
Federal Reserve Bulletin
Food Research Inst. Studies
Genetics
Historical Journal
Human Biology
Human Genetics
Hydrology
Improving College & Univ. Teaching
Issues in Criminology
Journal of the Am. Med. Assoc.
Journal of Biosocial Science
Journal of Business World Affairs

Journal of Economic Theory
Journal of Exptl. Biology
Journal of Exptl. Psychology
Journal of Finance
Journal of Marine Research
Journal of Marriage & the Family
Liberal Education
Management Science
Monthly Labor Rev.
Operations Research
Personnel Psychology
Philosophy
Physical Anthropology
Political Studies
Population Studies
Religious Education
Review of Politics
Sociology & Social Research
Sociometry
Transportation Journal
Western Folklore
World Agric. Situation.

3. Consult the latest census data available to determine the following:

   (a) The population of your state.
   (b) The number of births in your state (or county) during recent years.
   (c) The percentage of male births for those reported in part (b).
   (d) The number of U. S. citizens over age 85 at the time of the last census.

4. Consult *Men of Mathematics* by Bell [7], and prepare a report on one of the early statisticians, e. g., Pascal, Bernoulli, Laplace, Gauss.

5. Refer to *World of Mathematics* by Newman [1] to prepare a report on Graunt's work in statistics.

6. As a class activity, write to the following source to secure copies of "Careers in Statistics" for general information about what statistics is and what statisticians do.
   Executive Director, The American Statistical Association, 810 18th Street, N. W., Washington, D. C. 20006.

CHAPTER 2

Probability and Sampling

2.1 INTRODUCTORY REMARKS

Sample selection is basic to much of statistical analysis. In many experimental situations, it is either impossible or impractical to secure and study data from every element of a population to answer questions concerning some phenomena. Thus, sampling is necessary in order to proceed with an investigation. Even in those cases where the acquisition of census data (data for the full population) is a possibility, we are frequently forced to work with sample data for economic reasons. As examples of the foregoing ideas, consider the following situations: (1) The nature of a test may be such that the product being tested is destroyed (destructive testing). For example, to show a profit and remain in business, the BAD Light Company must determine the average life-time of its 100-watt bulbs by testing only a small percentage of all such bulbs manufactured. (2) It is not economically feasible to secure all desired data from every citizen of the U. S. in the decennial census. Thus, many questions are answered by only a small percentage (5-10%) of all citizens.

When selecting a sample, an experimenter faces two immediate questions. One question involves the *number* (sample size of observations to be included in the sample. Much of this book answers that question by specifying n, the sample size, for each exercise or situation presented. Another question considers the *method of acquisition* (sample selection technique) to be used. The most frequent method of sampling is *simple random sampling*. However, in order to discuss that technique, we need to have some understanding of basic probability concepts. Thus, the purpose of this chapter is to present some definitions, notations, and rules of probability, which are then used to define and discuss simple random sampling.

2.2 PROBABILITY

To introduce some of the types of probability, we need to consider the following questions: (1) What is the probability of measurable precipitation occurring during the next 24 hours? (2) What is the probability that dog #2 will win the first race at the Juarez race track on August 7th of this year? (3) What is the probability that life exists on Mars? (4) What is the probability that "heads" will appear when a given coin is flipped? (5) What is the probability of 4 aces on 4 successive draws from a well-shuffled deck of 52 bridge cards? (6) What is the probability that "one" will appear when a die is rolled? Questions (1)-(3) refer to a *subjective* type of probability. These questions refer to situations for which the outcome is uncertain. Yet, we cannot study the outcomes by means of repeatable experiments. For example, the conditions of a dog race cannot be duplicated exactly, but odds on the race are posted and people place bets as they compare their own subjective probabilities with those of the house.

Since subjective probabilities vary from individual to individual and therefore cannot be discussed easily in general, we begin our study with *relative frequency* probabilities that can be used to answer questions similar to questions (4)-(6). In such cases, the uncertainty encountered can be studied by means of repeatable experiments. For example, suppose we conduct an experiment to study the probability that heads will appear on a given toss of a coin. Suppose that a set of students tosses the coin 10 times and records the number of heads, #(H), observed in the trials. (The students are encouraged to do this experiment and record their results.) Suppose the efforts of 20 students yield the following results, as shown in Table 2.1.

TABLE 2.1.
Experimental results show the number of heads observed on 10 tosses of a coin by 20 students.

Student	1	2	3	4	5	6	7	8	9	10	11	12	13	14	15	16	17	18	19	20
#(H)	7	6	6	7	6	7	8	4	5	4	3	4	5	3	4	4	6	7	4	4

Note that each student records an outcome for the experiment of flipping the coin 10 times. From that outcome, #(H), a *relative frequency* of heads can be obtained by

$$f(H) = \#(H)/10.$$

We consider each set of 10 trials as a sample of size 10 (i.e., n = 10). The result for each student or the pooling of some or all of the results yields an *empirical* (experimental) result to answer the question of the probability of heads appearing on the coin. Yet the answer we seek to the question is a *model* result. That is, the data obtained are only partial results because f(H) changes both as we shift from one student to another and as we change the sample size n. Thus, a *model* that can represent all of the *empirical* results is desired. For the model, let P(H) denote the probability of heads appearing when the coin is tossed. Then we define P(H) as the limit value of the frequency, f(H), as n becomes large. In symbols,

$$P(H) = \lim_{n\to\infty} \#(H)/n \qquad (2.1)$$

[the limiting value of #(H)/n as n increases is P(H)] whenever the indicating limit exists.

To gain some insight as to how the ratio, #(H)/n, fluctuates for increasing values of n, treat the data for the 20 students in Table 2.1 as a sequence of coin tosses for n = 10, 20, ..., 200. A graph of the results is shown in Figure 2.1.

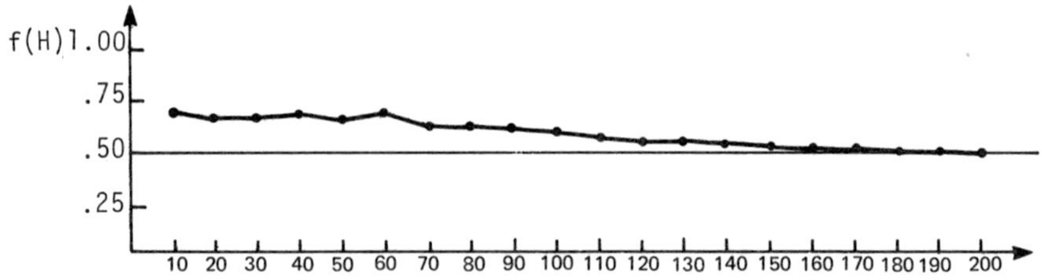

FIGURE 2.1. A sequential graph of #(H)/n for 200 flips of a coin.

After observing the changes that occur in f(H) as we increase n from 10 to 200 for the data in Table 2.1, it seems reasonable to have P(H) = .5. Thus, from a *relative frequency* viewpoint, P(H) = .5 is representative of all the data observed by repeatedly flipping the coin 10 times. Note that probabilities based on relative frequencies of different outcomes for an experiment are limit results that hold for large values of n. This property is responsible for the term, *long-run probabilities*, being associated with the situation we have considered.

Another method for arriving at P(H) = .5 is to use an argument of *symmetry*. Note that the coin has *two* sides (heads and tails). Since only one of the two sides can appear uppermost on any given flip, and since there is no obvious reason why one side should face upward more often than the other, it is assumed that each face should occur equally often in a series of n trials. We say that the two possibilities are *equally likely* outcomes for experiments. Thus, the model for flipping a coin is taken as P(H) = 1/2, P(T) = 1/2. We note that this discussion presupposes a standard coin rather than a coin that has been bent or defaced in an attempt to force one side to be more likely to occur on each flip.

Let us now present a comparison of similarities that exist between the empirical relative frequencies that we observe in the *experimental* (real) *world* and the probabilities (theoretical relative frequencies) that we assign as *models* for the observed frequencies. Table 2.2 presents the terms and symbols used for the two situations.

TABLE 2.2.
Terminology and symbols are given relating the observed experimental world to the idealistic model world.

Experimental world	Model world
1. Outcomes: $0_1, 0_2, \ldots$	1. Elementary events: $e_1, e_2, \ldots$
2. Sample space: S	2. Event space: E denotes the set of all elementary events, e_i
3. Relative frequencies of outcomes in S for n trials of the experiment: $f(0_1), f(0_2), \ldots$ where $f(0_i) = \frac{\#(0_i)}{n}$	3. Probabilities assigned to the elementary events in E $P(e_1), P(e_2), \ldots$
4. $0 \leq f(0_i) \leq 1$	4. $0 \leq P(e_i) \leq 1$
5. $\sum_{0_i \varepsilon S} f(0_i) = 1$	5. $\sum_{e_i \varepsilon E} P(e_i) = 1$

In Table 2.2, we find some properties that a good probability model should satisfy. First, if the model is to represent a real-world situation, it should have the same number of elementary events as the experiment has outcomes. That is, there should be a close term-to-term correspondence between the elements in E and S. Second, the assignment of probability to the elementary events, e_i, should be such that

$$P(e_i) = \lim_{n\to\infty} f(0i), \quad i = 1, 2, \ldots,$$

whenever we assume the experiment is performed in a random manner. Thus, we want the probability of an event, e_i, to be representative of the relative frequency with which the corresponding outcome, O_i, occurs in the random experiment. These two properties are necessary for a model to be a useful representation of a real-world experiment. However, the only properties that a model must satisfy to be valid mathematically are the following:

$$P(e_i) \geq 0 \quad \text{for all } e_i \varepsilon E, \tag{2.2}$$

$$\sum_{e_i \varepsilon E} P(e_i) = 1.$$

To further illustrate these concepts, suppose we consider the two probability models given in Table 2.3 as candidates to represent the experiment of tossing a coin. Let events e_1 and e_2 correspond to the outcomes of heads and tails respectively.

TABLE 2.3.
Two possible models are shown for an experiment of tossing a coin.

Model I		Model II	
e_i	$P(e_i)$	e_i	$P(e_i)$
e_1	1/4	e_1	1/2
e_2	3/4	e_2	1/2

Clearly both models satisfy the mathematical criteria [see Eq. (2.2)] for a probability model. Hence, we cannot choose one model over the other from a mathematical standpoint. However, when representation of the real world is a factor for consideration, it appears that

Model II agrees more closely to the observations given in Table 2.1 and illustrated in Figure 2.1 than does Model I. (Note that the coin used in these experiments has not been altered.)

Further development of probability models occurs in Chapters 4 and 5, where we present the definitions needed for a discussion of probability.

EXERCISES

1. Flip a coin 50 times, and prepare a graph of $f(H) = \#(H)/n$ and n (the number of flips). See Figure 2.1 for assistance. Does $f(H)$ approach 1/2 as n increases?
2. On the graph for Exercise 1, plot #(H) against n, and draw the line connecting the points (0,0) and (50, 25).
3. Interpret your graph for Exercise 2. Is there any relationship between the line and the function f(H)?
4. Does the graph in Exercise 2 tend to crisscross the line or *remain* on one side consistently? Repeat the experiment, or compare your results with the results of others.
5. An experiment may result in any one of three possible outcomes, say O_1, O_2, and O_3. Which of the following distributions would be appropriate to serve as a probability model for the experiment?
   (a) $P(E_1) = .50$, $P(E_2) = .31$, $P(E_3) = .20$.
   (b) $P(E_1) = P(E_2) = P(E_3) = 1/3$.
   (c) $P(E_1) = .17$, $P(E_2) = .04$, $P(E_3) = .79$.
   (d) $P(E_1) = .35$, $P(E_2) = .25$, $P(E_3) = .30$, $P(E_4) = .10$.
6. Consider the following spinner game.

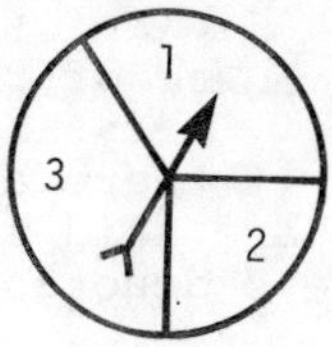

   Supply a probability model for repeated trials of the spinner under the assumption that all points on the circumference of the cicle are equally likely to occur on any given spin.

2.3 EVENTS

In the preceding section, probability was introduced as a means for modeling real-world experiments. There we listed *elementary events* as the elements of an event space (set), E. The elementary events for a model of an experiment are such that on any one trial of the experiment one and only one elementary event can occur. In the terminology of this section, we say that the elements of E are *mutually exclusive* and *exhaustive* for the experiment being modeled.

EXAMPLE 1: Consider the experiment of rolling a die. Let $E = \{1,2,3,4,5,6\}$, which is also the sample space, S. On each roll of the die, one and only one of the elements of E can occur. Thus, the elements of E are mutually exclusive and exhaustive.

DEFINITION 1: Any subset, say E, of E is called an *event*. Thus, an *event* is a collection of one or more elementary events. The designation of any event for the model implies an analogous collection of outcomes from the sample space of the experiment.

EXAMPLE 2: Consider the die experiment given in *Example 1*. Let $E = \{1,2,3,4,5,6\}$. The following sets are examples of events for the experiment: $E_1 = \{1, 2\}$, $E_2 = \{2,3,4\}$ and $E_3 = \{5,6\}$. There are, of course, many other events that can be defined for the experiment.

In example 2, we note that events E_1 and E_2 share the common point, 2, while E_2 and E_3 do not share any common points. Events E_2 and E_3 are said to be mutually exclusive events, while E_1 and E_2 are *not* mutually exclusive events. Furthermore, we observe that the combination of E_1, E_2, and E_3 includes every point in E at least once. Thus, we say that E_1, E_2, and E_3 are exhaustive events for the experiment. To give explicit definitions for mutually exclusive and exhaustive events, we need to define two set operations called *union* and *intersection*.

DEFINITION 2: The *intersection* of two events (sets), E_1 and E_2, is denoted by $E_1 \cap E_2$ and is the set of points common to both events. If there are no points common to both events, the events are said to be disjoint, and we write $E_1 \cap E_2 = \emptyset$, where $\emptyset$ denotes the empty event (null set).

DEFINITION 3: The *union* of two events (sets), E_1 and E_2, is denoted by $E_1 \cup E_2$ and is the set of point contained in either E_1 alone, *or* E_2 alone, *or both* E_1 *and* E_2.

The use of figures to graphically present events and the event space is often helpful for illustrating the foregoing set operations. Figure 2.2 contains what are called Venn diagrams, using a rectangle to represent the event space, E, and circles within the rectangle to denote events (subsets of E). The Venn diagrams show the union and intersection of two events, E_1 and E_2. *Example 3* demonstrates the use of union and intersection and illustrates the results by Venn diagrams.

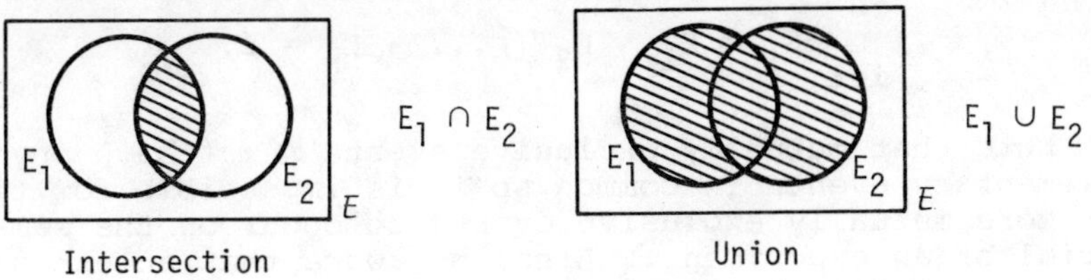

FIGURE 2.2. Venn diagrams illustrate the intersection and union for two events, E_1 and E_2.

EXAMPLE 3: From $E = \{1,2,3,4,5,6\}$ for the roll of a die, let $E_1 = \{1,2\}$, $E_2 = \{2,4,6\}$, $E_3 = \{2,5,6\}$, and $E_4 = \{1,5\}$. Then $E_1 \cap E_2 = \{2\}$, $E_2 \cap E_3 = \{2,6\}$, $E_1 \cup E_2 = \{1,2,4,6\}$, $(E_1 \cap E_2) \cap E_3 = \{2\}$, $(E_1 \cup E_2) \cup E_3 = \{1,2,4,5,6\}$, and $E_2 \cap E_4 = \emptyset$.

These relationships can be illustrated by a Venn diagram as follows.

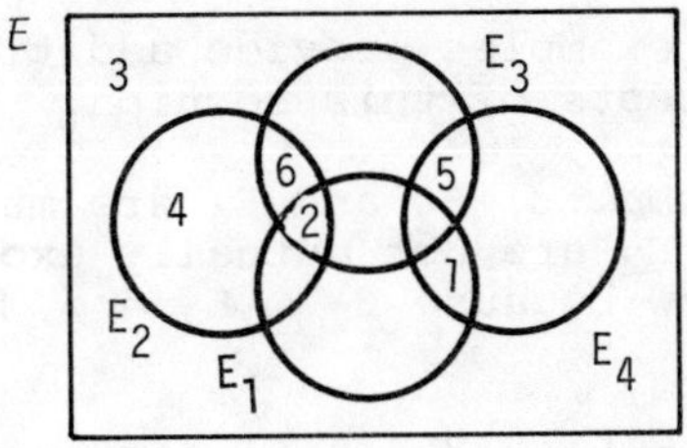

Note that overlapping circles exist where the two- and three-way intersections occur. However, the union includes those points that occur at least once in the sets involved. Also, since 3 is not an element of any of the events defined, it is written inside the rectangle but outside all circles. Furthermore, the fact that E_2 and E_4 are disjoint is represented by non-intersecting circles.

We now return to define the terms, *mutually* exclusive events and exhaustive events.

DEFINITION 4: Two events, E_1 and E_2, are said to be *mutually exclusive* if $E_1 \cap E_2 = \emptyset$.

DEFINITION 5: A set of events, E_1, E_2, ..., E_n, is *exhaustive* if

$$\bigcup_{i=1}^{n} E_i = E_1 \cup E_2 \cup \cdots \cup E_n = \mathcal{E}.$$

Note that mutually exclusive events *do not* have any elementary events in common so it is impossible for two or more mutually exclusive events to occur on the same trial of an experiment. Also, be aware that an exhaustive set of events must be such that every elementary event occurs in at least one of the events. In case the set of events is such that every elementary event is a member of one and only one event, we call the set of events a partition of $\mathcal{E}$. Thus, a partition of $\mathcal{E}$ is any set of events that are exhaustive and pairwise mutually exclusive for $\mathcal{E}$. The explicit definition of a partition is given in *Definition 6*.

DEFINITION 6: A set of events, E_1, E_2, ..., E_n, forms a *partition* of $\mathcal{E}$ if $E_i \cap E_j = \emptyset$ for every pair of events, E_i and E_j, and if

$$\bigcup_{i=1}^{n} E_i = \mathcal{E}.$$

The following examples provide additional experience in using the concepts of this section.

EXAMPLE 4: In *Example 3*, E_2 and E_4 are mutually exclusive, but E_1 and E_2 are not mutually exclusive. These conclusions follow because $E_2 \cap E_4 = \emptyset$, but $E_1 \cap E_2 = \{2\} \neq \emptyset$.

EXAMPLE 5: Let $\mathcal{E} = \{1,2,3,4,5,6\}$, and consider the events, $E_1 = \{1,2,3\}$, $E_2 = \{2,4,6\}$, $E_3 = \{1,3,5\}$, and $E_4 = \{3,4,5,6\}$. Sets E_1 and E_4 are exhaustive events. Sets E_2 and E_3 form a partition of $\mathcal{E}$ because $E_2 \cap E_3 = \emptyset$ and $E_2 \cup E_3 = \mathcal{E}$. Note that it is possible to indicate many different partitions for $\mathcal{E}$.

Now that we have defined events and indicated how they are derived from elementary events, the next step is to consider the assignment of probability to those events.

EXERCISES

1. Consider the elementary events associated with the experiment of rolling a die.
   (a) What is the event space?

(b) Indicate two events that are mutually exclusive but not exhaustive.
(c) Indicate two events that are exhaustive but not mutually exclusive.
(d) Indicate two events that partition the event space.

2. Consider the 16 elementary events associated with the flip of 4 coins: $\mathcal{E}$ = S = { (HHHH), (HHHT), (HHTH), (HTHH), (THHH), (HHTT), (HTHT), (HTTH), (THHT), (THTH), (TTHH), (HTTT), (THTT), (TTHT), (TTTH), (TTTT) }.
   (a) Let X = the number of heads in a sample point (elementary event). What are the possible values of the random variable, X? [i.e., What is the sample (event) space for X?]
   (b) Note that the variable, X, in part (a) partitions the elementary events into 5 events. List the elementary events associated with each possible value of X.
   (c) Suggest a probability model for the variable, X.

3. Repeat Exercise 2 using X = the number of girls in a family of 4 children. That is, let H correspond to a girl and T correspond to a boy.

2.4 PROBABILITY ASSIGNMENT FOR EQUALLY LIKELY ELEMENTARY EVENTS

Equation (2.1) assigns the probability for an outcome by determining the limit of the relative frequency of that outcome as the experiment is repeated an unlimited number of times. That technique is one possible method of probability assignment. Another approach is to begin our modeling at the level of *equally likely* events. For that method, we suppose that all the points in the sample space, S, are equally likely to occur on any trial of the experiment. Thus, it is reasonable to assume that all the elementary events in $\mathcal{E}$ (the model representation of S) have equal probabilities. Then any event, E, contained in $\mathcal{E}$ corresponds to a particular set of outcomes that are a subset of S. So it is often convenient to define the probability of E to be

$$P(E) = \#(E)/\#(\mathcal{E}), \tag{2.3}$$

where #(E) is the number of elements in the event, E, and #($\mathcal{E}$) is the number of elements in $\mathcal{E}$. In other words, #(E) corresponds to the number of outcomes in S associated with the event, E, while #($\mathcal{E}$) corresponds to the total number of outcomes in S. The correspondence between the sets, $\mathcal{E}$ and S, and subsets of those sets follows from our desire to assign probabilities to events in $\mathcal{E}$ associated with sets of outcomes for an experiment having a sample space, S.

In those cases where Eq. (2.3) is applicable for probability assignments, the task of selecting the appropriate probability for an event, E, reduces to a counting problem. That property and the use of Eq. (2.3) are illustrated by the following examples.

EXAMPLE 6: Let the experiment of interest be the roll of a die. Then $S = \mathcal{E} = \{1,2,3,4,5,6\}$. Let $E_1 = \{1,2,5\}$, $E_2 = \{2\}$, and $E_3 = \{7\}$. Then $P(E_1) = 3/6$, $P(E_2) = 1/6$, $P(E_3) = 0$, $P(S) = 1$.

EXAMPLE 7: Let the experiment be the selection of a card from a well-shuffled deck of 52 bridge cards. Then $S = \mathcal{E}$ = {ace of spades, king of spades, ..., deuce of spades, ace of hearts, ..., deuce of hearts, ace of diamonds, ..., deuce of diamonds, ace of clubs, ..., deuce of clubs}. Let E_1 = {ace}, E_2 = {4}, E_3 = {spade}, E_4 = {ace *or* king}, E_5 = {spade *or* ace}, and E_6 = {ace *and* spade}. Then $P(E_1) = 4/52$, $P(E_2) = 4/52$, $P(E_3) = 13/52$, $P(E_4) = 8/52$, $P(E_5) = 16/52$, and $P(E_6) = 1/52$.

Note that events E_4, E_5, and E_6 in *Example 7* involve combinations of events. For example, E_4 occurs if *either or both* of the two events, ace and king, occur, while E_6 occurs if *both* of the events, ace and spade, occur. These combinations of events are simply the union and intersection, respectively, designated by *Definitions 2* and *3*.

In the preceding examples, we could easily count the number of elementary events associated with the various events and use Eq. (2.3) to determine their probabilities. However, the examples given later in Section 2.7 demonstrate that this is not always the case. Thus, the next two sections present rules for calculating probabilities of functions of events and rules for counting the number of ways in which various events can occur.

2.5 GENERAL RULES OF PROBABILITY

In the preceding section, a technique is presented for assigning probability to any event, E, when the elementary events, e_i, are assumed equally likely. In this section we generalize that concept to suppose that each elementary event, e_i, is assigned a probability, $P(e_i)$. Since each event, E, is a set of elementary events, and since elementary events are mutually exclusive, it is natural that we define P(E) as follows:

$$P(E) = \sum_{e_i \varepsilon E} P(e_i) \tag{2.4}$$

for every event, E, that is a subset of $\mathcal{E}$.

Since ϕ is the empty set and contains none of the e_i, and $\mathcal{E}$ is the universal set and contains all of the e_i, a direct consequence of Eqs. (2.2) - (2.4) is that

$$P(\phi) = 0, \quad P(\mathcal{E}) = 1.$$

We refer to ϕ as an *impossible* event and $\mathcal{E}$ as a *certain* event. In fact, any event, E, for which P(E) = 0 is called an impossible event, while any event, E, for which P(E) = 1 is called a certain event.

Previously we assigned probability to an event, E, by considering it as a collection of elementary events. However, it is sometimes more convenient to determine the probability of an event of interest by expressing the event in terms of two or more other events (not necessarily elementary events). In that light, the two most common combinations of events are the *union* and the *intersection*, as explained in *Definitions 2* and *3*. Recall that the union of events, E_1 and E_2, say E_3, is written as $E_3 = E_1 \cup E_2$ and occurs if either E_1 *or* E_2 is observed, *or* if *both* E_1 and E_2 are observed. Thus, the union of n events, say the event, G, is written as

$$G = \bigcup_{i=1}^{n} E_i$$

and occurs if *at least one* of the E_i's is observed. Also, the intersection of two events, E_1 and E_2, say the event, E_4, is written as $E_4 = E_1 \cap E_2$ and represents the *simultaneous* occurrence of E_1 *and* E_2. We refer to E_4 as the *joint* event of E_1 and E_2. Similarly, the intersection of n events, say the event, F, is written as

$$F = \bigcap_{i=1}^{n} E_i$$

and designates the simultaneous occurrence of *all* n of the events, E_i. Using these ideas, we look again at Example 7 for illustrations.

EXAMPLE 8: Refer again to the experiment in Example 7, and let E_1 = {ace}, E_2 = {king}, and E_3 = {spade}. Then $E_4 = E_1 \cup E_2$ = {ace *or* king}; $E_5 = E_1 \cup E_3$ = {ace *or* spade}; $E_6 = E_1 \cap E_3$ = {ace *and* spade}; $E_7 = E_1 \cap E_2 = \phi$; and $E_8 = E_1 \cup E_2 \cup E_3$ = {ace *or* king *or* spade}.

Notice that E_1 and E_2 are mutually exclusive (disjoint) events as their intersection is the empty (null) set, while E_1 and E_3 are not mutually exclusive since their intersection includes one card, the ace of spades. The importance of whether or not two events are mutually exclusive or not is illustrated by the following rules and examples.

With this brief review of terminology, we now provide some rules for calculating the probability of an event, which is expressed as a combination of other events.

RULE 1a: If E and F are mutually exclusive, then

$$P(E \cup F) = P(E) + P(F). \tag{2.5}$$

RULE 1b: In general,

$$P(E \cup F) = P(E) + P(F) - P(E \cap F). \tag{2.6}$$

EXAMPLE 9: Using E_1, E_2, E_3, E_4, E_5, E_6, and E_7 as defined for Example 8 and the definition of probability as given in Eq. (2.4) with $P(e_i) = 1/52$ for all e_i, we find that $P(E_1) = 4/52$, $P(E_2) = 4/52$, $P(E_3) = 13/52$, $P(E_6) = 1/52$, and $P(E_7) = 0$.

Using *Rule 1a*,

$$P(E_4) = P(E_1) + P(E_2) = 8/52.$$

Using *Rule 1b*,

$$P(E_5) = P(E_1) + P(E_3) - P(E_1 \cap E_3)$$
$$= 4/52 + 13/52 - 1/52 = 16/52.$$

Note that the results for Example 9 agree with those for Example 7.

Two special events that form a partition of the outcomes for any experiment are E and $\overline{E}$, where $\overline{E}$ denotes the complement of E. That fact follows directly from the definition of a complement, namely that $E \cap \overline{E} = \phi$ and $E \cup \overline{E} = \mathcal{E}$. Using Eq. (2.4) and *Rule 1a*, we have

$$1 = P(\mathcal{E}) = P(E \cup \overline{E}) = P(E) + P(\overline{E}).$$

Therefore,

$$P(\overline{E}) = 1 - P(E) \tag{2.7}$$

for any event, E. Equation (2.7) is often useful for calculating probabilities of events and is discussed with more examples given in Section 2.7.

Further illustration of the results in Eqs. (2.5) - (2.7) is provided by Venn diagrams. To do this we assume that the rectangle depicts the sample space and is therefore assigned a probability value of 1. The circles inside the rectangle depict probabilities for the events of interest. The result is shown in Figure 2.3.

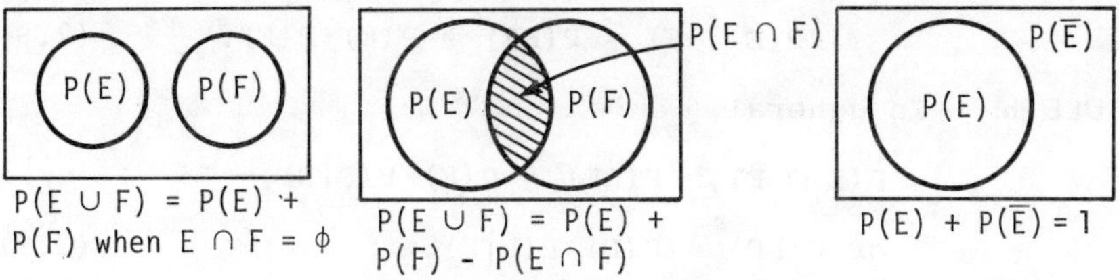

FIGURE 2.3. Venn diagrams are used to represent the probability results given in Eqs. (2.5) - (2.7).

Note that in *Rule 1b* the $P(E \cap F)$ is included in both $P(E)$ and $P(F)$. Thus, it must be subtracted from $P(E) + P(F)$ to provide the correct probability for $P(E \cup F)$.

In the examples and discussion presented so far, the probability of an event expressed as the intersection of two or more events has been determined by counting the number of elementary events in the designated intersection. To generalize our approach and to state rules for calculating the probability of intersections of events, we establish the following terminology and notation.

TYPES OF PROBABILITIES: (1) The probability of any event, E, is written as $P(E)$ and is frequently called the *marginal* probability of E. (2) The probability of any two events, E and F, occurring simultaneously is written as either $P(E \cap F)$ or $P(EF)$ and is called the *joint probability* of E and F. The definition extends directly to more than two events. (3) The probability that an event, E, occurs when an event, F, is known to have occurred is written as $P(E|F)$ and is called the *conditional probability* of E, given that F has occurred. [The notation $P(E|F)$ is often read as the probability of E, given F.]

The concept of *independent* events, which is basic to the statement of probability rules for the intersection of two or more events, is now defined in terms of the preceding notation.

DEFINITION 7: Two events, E and F, are *independent* events if $P(E|F) = P(E)$, or equivalently, if $P(F|E) = P(F)$. Events that are not independent are called *dependent* events.

In words, *Definition 7* says that two events, E and F, are independent if the probability of occurrence for either event is not affected by the occurrence of the other event. Expressing that idea in symbols gives two rules by which we can calculate the probability of joint events to be used alone or in equations such as Eq. (2.6).

RULE 2a: If two events, E and F, are independent, then

$$P(E \cap F) = P(EF) = P(E)\cdot P(F). \quad (2.8)$$

RULE 2b: In general,

$$P(E \cap F) = P(EF) = P(E)\cdot P(F|E),$$

$$\text{or } P(EF) = P(F)\cdot P(E|F). \quad (2.9)$$

Rule 2b provides the following definition for conditional probabilities.

DEFINITION 8: For any two events, E and F,

$$P(E|F) = P(EF)/P(F) \text{ if } P(F) > 0,$$

$$\text{and } P(F|E) = P(EF)/P(E) \text{ if } P(E) > 0. \quad (2.10)$$

Let us now consider a sampling situation that explores some implications of the preceding definitions and rules. Suppose that an urn contains 3 red and 4 blue marbles which, except for color, are identical. Suppose 2 marbles are selected in sequence from the urn. Let R_i, i=1, 2 denote the appearance of a red marble on the ith selection. Let B_i, i=1, 2 denote the appearance of a blue marble on the ith selection. Then determine $P(R_1)$, $P(B_1)$, $P(R_2)$, and $P(B_2)$ for the following two cases.

CASE 1. *Sampling with replacement:* Suppose the first marble is observed and returned to the urn and mixed with the other marbles before the second marble is selected. Under this sampling scheme, the contents of the urn are exactly the same for each selection. Therefore, Eq. (2.3) gives $P(R_1) = 3/7$, $P(B_1) = 4/7$, $P(R_2) = 3/7$, and $P(B_2) = 4/7$. Furthermore, $P(R_2|R_1) = P(R_2|B_1) = 3/7$ and $P(B_2|R_1) = P(B_2|B_1) = 4/7$ because under the replacement scheme, the urn contains 3 red and 4 blue marbles, regardless of what happened on the first selection. Thus, in terms of *Definition 7*, we find that R_1 and R_2, R_1 and B_2, B_1 and R_2, and B_1 and B_2 are pairs of independent events. Thus, by using Eq. (2.8), we find that $P(R_1\ R_2) = 3/7 \cdot 3/7 = 9/49$, $P(R_1\ B_2) = 3/7 \cdot 4/7 = 12/49$, $P(B_1\ R_2) = 4/7 \cdot 3/7 = 12/49$, $P(B_1\ B_2) = 4/7 \cdot 4/7 = 16/49$. Note that the 4 joint events, $(R_1 \cap R_2)$, $(R_1 \cap B_2)$, $(B_1 \cap R_2)$, and $(B_1 \cap B_2)$, are mutually exclusive and exhaustive events for the experiment outlined. Thus, the 4 events should and do have assigned probabilities whose total is 1.

CASE 2. *Sampling without replacement:* Suppose the first marble remains outside the urn, while the second marble is selected. (Notice that this scheme is equivalent to selecting the 2 marbles simultaneously.) Now we can still use Eq. (2.3) to obtain $P(R_1) = 3/7$ and $P(B_1) = 4/7$, but the determination of $P(R_2)$ and $P(B_2)$ offers quite a different problem. In particular, the event, R_2, can occur in either one of two ways. R_2 can occur in the joint event, $R_1 \cap R_2$, or the joint event, $B_1 \cap R_2$. Since these two events are mutually exclusive, Eq. (2.5) gives

$$P(R_2) = P(R_1\ R_2) + P(B_1\ R_2). \tag{2.11}$$

Each term on the right side of the equal sign in Eq. (2.11) can be expanded by Eq. (2.9) to give

$$P(R_2) = P(R_1)\ P(R_2|R_1) + P(B_1)\ P(R_2|B_1).$$

Under the scheme of no replacement,

$$P(R_2|R_1) = 2/6 \text{ and } P(R_2|B_1) = 3/6$$

because the selection of 1 red marble from the urn leaves 2 red marbles, with a total of 6, while the selection of 1 blue marble leaves 3 red marbles, with a total of 6. Therefore,

$$P(R_2) = (3/7 \cdot 2/6) + (4/7 \cdot 3/6) = 3/7.$$

Similarly,

$$P(B_2) = P(R_1)\ P(B_2|R_1) + P(B_1)\ P(B_2|B_1)$$

$$= (3/7 \cdot 4/6) + (4/7 \cdot 3/6) = 4/7.$$

So the marginal probabilities, $P(R_1)$, $P(B_1)$, $P(R_2)$, and $P(B_2)$, are the same for both cases. However, for *Case 2* we have joint probabilities, $P(R_1\ R_2) = 3/7 \cdot 2/6 = 6/42$, $P(R_1\ B_2) = 3/7 \cdot 4/6 = 12/42$, $P(B_1\ R_2) = 4/7 \cdot 3/6 = 12/42$, and $P(B_1\ B_2) = 4/7 \cdot 3/6 = 12/42$, which shows that the joint probabilities for the two cases differ. Furthermore, under the scheme of sampling without replacement, $P(B_2|R_1) \neq P(B_2)$, $P(B_2|B_1) \neq P(B_2)$, $P(R_2|R_1) \neq P(R_2)$, and $P(R_2|B_1) \neq P(R_2)$, so the members of each pair of events, $R_1 \cap B_2$, $B_1 \cap B_2$, $R_1 \cap R_2$, and $B_1 \cap R_2$, are dependent.

The probabilities associated with the urn example can best be summarized in a 2 × 2 table as shown in Table 2.4.

TABLE 2.4.
A 2 × 2 probability table shows the marginal and joint probabilities.

		Result of 2nd selection R_2	B_2	
Result of 1st selection	R_1	$P(R_1\ R_2)$	$P(R_1\ B_2)$	$P(R_1)$
	B_1	$P(B_1\ R_2)$	$P(B_1\ B_2)$	$P(B_1)$
		$P(R_2)$	$P(B_2)$	1

In Table 2.4 the joint probabilities occur in the body of the table at the intersection of the two events involved, while the marginal probabilities occur as border values of the table. Any desired conditional probability can be obtained by using Eq. (2.10) and by taking a ratio of the appropriate joint and marginal probabilities. For example, $P(R_2|R_1) = P(R_1\ R_2)/P(R_1)$ and $P(R_1|R_2) = P(R_1\ R_2)/P(R_2)$. Notice that the effect of the condition is to restrict interest to a particular row or column of the table.

By tabulating information such as that in Table 2.4 for *Cases 1* and *2*, we obtain the following for Table 2.5.

TABLE 2.5.
Probability tables are shown for *Cases 1* and *2* of the urn sampling problem.

Case 1 (independent events)

	R_2	B_2	
R_1	9/49	12/49	3/7
B_1	12/49	16/49	4/7
	3/7	4/7	1

Case 2 (dependent events)

	R_2	B_2	
R_1	6/42	12/42	3/7
B_1	12/42	12/42	4/7
	3/7	4/7	1

In Table 2.5, the joint probabilities are given by multiplying the marginal probabilities for *Case 1*, but the joint probabilities for *Case 2* cannot be obtained that way. This simply illustrates the difference between Eqs. (2.8) and (2.9). Furthermore, $P(R_2|R_1) = 6/42 \div 3/7 = 2/6$, just as it was presented by applying Eq. (2.3) to an urn containing 2 red and 4 blue marbles.

The preceding discussion demonstrates by example that sampling with replacement yields independent events, while sampling without replacement yields dependent events. The sampling concepts involved are discussed more completely in Section 2.8. Other examples of independent events are those events which represent the outcomes for experiments of tossing coins and rolling dice. For example, the probability of rolling 3 consecutive 6's on 3 rolls of a fair die is $P(666) = 1/6 \cdot (1/6) \cdot (1/6) = 1/216$. *Example 10* shows more independent events. Additional illustrations of the foregoing concepts and rules are provided by the exercises in this section and the examples in Section 2.7.

EXAMPLE 10: Referring to the card experiment in *Example 8*, let $E_1 = \{ace\}$ and $E_2 = \{spade\}$. Then $P(E_1) = 4/52$, $P(E_2) = 13/52$, and $P(E_1 \cap E_2) = 1/52 = 4/52 \cdot 13/52 = P(E_1) \cdot P(E_2)$. Thus, the events, ace and spade, are independent. That is, knowing that a randomly selected card is a spade does not alter the probability that it is an ace. Likewise, knowing that a randomly selected card is an ace does not change the probability that the card is a spade.

The preceding discussion defines the probability of an event, E, from a relative frequency viewpoint, from an equally likely events approach, and from a more general model, such as that used in Eq. (2.4). The discussion also provides rules for calculating *marginal* probabilities, say $P(E)$; *joint* probabilities, say $P(E \cap F)$; *either-or* probabilities, say $P(E \cup F)$; and *conditional* probabilities, say $P(F|E)$. Furthermore, all of the rules and techniques apply directly if the $P(e_i)$ values used in Eq. (2.4) have been assigned. However, if we resort to Eq. (2.3) and assume equally likely events to assign probabilities, $P(E)$, then we may encounter difficulties because it is not easy to determine either $\#(E)$, or $\#(\mathcal{E})$, or both for the events associated with the outcomes of many experiments. To alleviate that problem, the next section is devoted to the presentation of techniques that are useful in counting the number of different ways in which various events can occur.

EXERCISES

1. If 2 dice are rolled, what is the probability that the following occur?
   (a) Two 6's appear.
   (b) At least one 6 appears.
   (c) The sum is equal to 5.
   (d) The sum is equal to 1.
   (e) At least one even number appears.

2. If the probability that M. Clumsy will have an accident during the next year is .6, and if the probability that B. Graceful will have an accident during the next year is .4, what is the probability that *both* individuals will have an accident during the next year? What is the probability that *either* M. Clumsy *or* B. Graceful will have an accident next year? (Assume that the individuals have accidents independently.)

3. An automatic screw machine makes screws and fills boxes with them. If one box in 100 has at least one defective screw in it and the outcomes are independent, what is the probability that each of the next 3 boxes has 1 or more defective screws? What is the probability that all 3 boxes contain no defective screws?

4. If E and F are mutually exclusive events, what is the value of $P(E|F)$? Does the answer seem reasonable?

5. If $P(E_1) = .6$ and $P(E_2) = .2$, and if E_1 and E_2 are independent events, what is the probability that the following occur?
   (a) E_1 and E_2 occur.
   (b) Either E_1 or E_2 occurs.
   (c) Either E_1 or $\overline{E}_2$ occurs.
   (d) At least one of the events, E_1, E_2, occurs.
   (e) At most, one of the events, E_1, E_2, occurs.

6. If E and F are independent events, show that $P(E \cup F) = P(E) + P(F) \cdot P(\overline{E})$ whenever E and F are not mutually exclusive.

7. Table 2.4 presents an analysis of a sampling experiment by means of a 2 × 2 table. *Case 1* of that experiment can be analyzed by using a *tree diagram* as follows. Recall that the urn contains 3 red and 4 blue marbles.

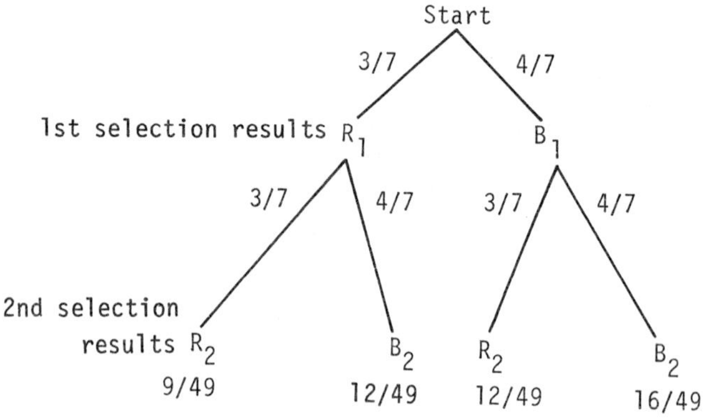

Case 1

Notice that each branch of the tree represents one of the four possible joint events for the experiment, and the numbers along each branch indicate the probability of following that branch. Next, the probability at the end of each branch (obtained by multiplication) represents the probability of that joint event. Find the tree diagram for *Case 2* of the urn sampling problem.

8. The following table provides the joint probabilities associated with the occurrence (p) [or nonoccurrence ($\overline{p}$)] of measurable precipitation either for today or tomorrow.

		tomorrow p	tomorrow $\overline{p}$
today	p	.06	.04
	$\overline{p}$	.04	.86

Are the events in the table independent? What is P (p today, p tomorrow)? Find the P (p today). What is P (p tomorrow$|p$ today)?

2.6 COUNTING TECHNIQUES

There are four counting techniques that are useful when applying Eq. (2.3) to obtain probabilities. Three of the techniques are specializations of the general counting rule.

GENERAL RULE OF COUNTING: If one event, A, can occur m different ways, and if for each of those different ways by which A can occur, an event B can occur n different ways, then the two events, A and B (i.e., $A \cap B$), can occur simultaneously mn different ways.

EXAMPLE 11: An experiment consists of rolling a die and flipping a coin. The coin can fall showing either H or T, while the die will show one element of the set 1,2, 3,4,5,6. Thus, the experiment can yield $2 \cdot 6 = 12$ different outcomes. It is sometimes convenient to present the outcomes in a table as follows:

		Die 1	Die 2	Die 3	Die 4	Die 5	Die 6
Coin	H	(H,1)	(H,2)	(H,3)	(H,4)	(H,5)	(H,6)
	T	(T,1)	(T,2)	(T,3)	(T,4)	(T,5)	(T,6)

PERMUTATION RULES: Frequently the different outcomes for an experiment are simply *rearrangements* (permutations) of a set of outcomes. For example, to determine the probability that a track star, Joe Fast, will draw lane #1 in a field of 8 runners, we need to know two things. First, the number of ways in which 8 runners can be assigned to 8 lanes is needed. Second, subject to the restriction that Joe Fast is assigned lane #1, the number of ways in which 8 runners can be assigned to 8 lanes is needed. Note that we can rephrase the preceding information to imply that we need to know how many ways there are for arranging 8 unique objects in a line with no restrictions, and how many ways the same 8 objects can be arranged in a line with the restriction that a given object occurs on a stated position. In this example, the given object is the individual, Joe Fast, and the stated position is lane #1. The latter formulation presents an abstract view of the original example, which places that example in a category of problems. That is, it does not matter whether we are assigning track stars to lanes of a track, aligning pictures on the wall of an art gallery, assigning subjects to psychological experiments, adding ingredients to a pet food ration, ordering a set of tests for presentation to prospective students, or any other number of situations in which different outcomes simply imply different arrangements. The counting problems associated with an arrangement type of problem can be handled conveniently by the technique called *permutations*. To discuss permutations we first need to define the term, *factorial*.

DEFINITION 9: The symbol n! is read as *n factorial* and denotes the product of the first n positive integers. That is,

$$n! = n(n - 1)(n - 2) \ldots 2 \cdot 1.$$

DEFINITION 10: The number of *permutations* (different arrangements) of n unique objects when all n objects are used is denoted by the symbol, ${}_nP_n$, and is given by

$${}_nP_n = n!.$$

Be aware that *Definition 10* is a direct application of the general counting rule in that we have n objects to be used in filling n positions. First, there are n objects available to occupy the first position. Second, after filling the first position, there are (n - 1) remaining objects available to fill the second position. Third, after using two objects to fill the first and second positions, there are (n - 2) objects available to fill the third position. This sequential procedure for filling the positions can be continued until finally there is only one object available for the nth position. Thus, by repeatedly applying the general counting rule, the number of ways in which n objects can be arranged (permuted) is $n(n - 1)(n - 2) \ldots 2 \cdot 1 = n!$. We can illustrate the idea as follows. Let N_i denote the number of objects available for placement in position i. Then we have the following illustration.

N_i	n	n-1	...	n-(i+1)	...	1
position #	1	2		i		n

EXAMPLE 12: Consider again the problem of determining the probability that Joe Fast is assigned to lane #1 when 8 runners are racing in 8 lanes. First, there are 8! ways in which the 8 runners can be assigned without restriction. Second, there are 7! ways in which the remaining 7 runners can be assigned to lanes 2-8 when Joe Fast is placed in lane #1. Thus, P(Joe Fast is assigned lane #1) = $7!/8! = (7\cdot6\cdot5\cdot4\cdot3\cdot2\cdot1)/(8\cdot7\cdot6\cdot5\cdot4\cdot3\cdot2\cdot1) = 1/8$.

It is worth noting that arguments similar to that used in Example 12 would indicate that Joe Fast has a probability of 1/8 for being assigned to any designated lane. Furthermore, each of the other runners also has an equal probability, namely 1/8, for being assigned to any given one of the 8 available lanes. Thus, we would say that the race is fair in that the runners are randomly assigned to the available lanes. That is, each runner has an equal chance of being the runner assigned to a given lane.

Definition 10 handles the counting problem when all n unique objects in a set are permuted. However, it is frequently necessary to count the number of permutations available when some number, say r, less than n of the objects is permuted.

DEFINITION 11: The number of *permutations* of n unique objects in groups of size r is denoted by the symbol, ${}_nP_r$, and is given by

$$ {}_nP_r = n!/(n - r)!. $$

Upon inspection, we find that the formula for ${}_nP_r$ is again a straightforward application of the general counting rule. That is, there are n choices for an object to occupy the first of r positions, (n - 1) choices for objects to occupy the second position after the first position has been filled, etc., until there are (n - r + 1) objects available to occupy the rth position. Thus, we have ${}_nP_r = n(n-1) \ldots (n - r + 1)$.

N_i	n	n-1	...	n-(r+1)
position #	1	2		r

When this result is multiplied by $(n - r)!/(n - r)!$, it becomes the more convenient form of ${}_nP_r = n!/(n - r)!$, as given in *Definition 11*.

EXAMPLE 13: Suppose there are 10 runners, of which 5 are to be selected and assigned lanes 1-5 for the first heat of a 440-yard dash. How many orderings of the 10 runners in groups of 5 exist?

SOLUTION:

$$ {}_{10}P_5 = \frac{10!}{5!} = (10\cdot 9\cdot 8\cdot 7\cdot 6\cdot 5!)/5! = 30,240. $$

Alternately, ${}_{10}P_5 = 10\cdot 9\cdot 8\cdot 7\cdot 6 = 30,240$. For convenience and consistency of notation, we define

$$ 0! = 1. \qquad (2.12) $$

With that definition of 0!, we find that *Definition 10* is a special case of *Definition 11* when r = n. Thus,

$$ {}_nP_r = n!/(n - r)!, \quad r = 0, 1, 2, \ldots, n. \qquad (2.13) $$

Also notice that by definition, 0! = 1! = 1.

The foregoing results, as summarized in Eq. (2.13), hold only for permutations of unique objects. So a further question to raise is the following one. What results are known for counting permutations of objects that are not unique? We provide an answer only for the case when *all* of the n objects are permuted, and that result is given in the next definition.

DEFINITION 12: Consider n objects, of which there are x_1 of a first kind, x_2 of a second kind, ..., x_k of a kth kind, and where

$$ \sum_{i=1}^{k} x_i = n. $$

The number of permutations of the n objects is then

$$ n!/(x_1!\ x_2!\ \ldots\ x_k!). $$

EXAMPLE 14: How many permutations of the letters, S-T-A-T-I-S-T-I-C-S, exist?

SOLUTION: For this example, $n = 10$, $x_1 = \#(S) = 3$, $x_2 = \#(T) = 3$, $x_3 = \#(A) = 1$, $x_4 = \#(I) = 2$, and $x_5 = \#(C) = 1$. Thus, there are $10!/(3!3!1!2!1!) = 50{,}400$ different arrangements of the letters that spell statistics.

EXAMPLE 15: Assuming that a child arranges letters at random in a line, what is the probability that she/he arranges the letters to spell statistics when given 3 S's, 3 T's, 1 A, 2 I's, and 1 C?

SOLUTION: $P(S\text{-}T\text{-}A\text{-}T\text{-}I\text{-}S\text{-}T\text{-}I\text{-}C\text{-}S) = 1/50{,}400$.

Additional uses of the results presented in *Definitions 10-12* are illustrated in Section 2.7, where various probability problems are presented.

COMBINATION RULES: There are sampling situations for which the ordering within the sample is not important, but the critical part is whether or not a given individual is included in the sample. That is, we are primarily concerned with the number of different *groups* (combinations or samples) of a specified size (sample size) that can be obtained from n distinct objects. For example, suppose the instructor of an elementary statistics class randomly selects three students from the class to be excused (with a grade of A) from each examination. The pertinent question to ask is how many different groups of three students exist, since ordering within the selected group is not of interest. Assume that a particular student does not care whether he is first, second, or third to be excused. His interest is only that he is among those excused. Thus, the question is that of how many different *groups* (combinations) of r objects can be selected from n objects? That question can be answered by *Definition 13*.

DEFINITION 13: The number of *combinations* of r objects selected from n distinct objects is denoted by $\binom{n}{r}$ and is given by

$$\binom{n}{r} = \frac{n!}{r!(n-r)!}; \quad r = 0,1,2, \ldots, n.$$

One way of rationalizing the equation in *Definition 13* is to recall that each combination of r distinct objects can be permuted r! ways, and notice that

$$_nP_r = \binom{n}{r} r!. \tag{2.14}$$

In other words, Eq. (2.14) states that the number of permutations of n objects taken r at a time is equal to the number of combinations of r objects from n objects multiplied by the number of permutations for each of those combinations of r objects. Then, by using Eqs. (2.13) and (2.14) to solve for $\binom{n}{r}$, we obtain *Definition 13.*

EXAMPLE 16: Use *Definition 13* to obtain solutions for the following combinatorials:

$$\binom{6}{2} = \frac{6!}{2!4!} = \frac{6 \cdot 5 \cdot 4!}{2 \cdot 1 \cdot 4!} = 15.$$

$$\binom{10}{1} = \frac{10!}{1!9!} = 10.$$

$$\binom{5}{3} = \frac{5!}{3!2!} = 10.$$

Before leaving the topic of combinations, a few properties should be considered. First,

$$\binom{n}{0} = \binom{n}{n} = 1. \tag{2.15}$$

There is only one combination that selects none of the objects, and there is only one combination that selects all of the objects. Second, the symmetry expressed by Eq. (2.15) is only a special case of a general symmetry given by

$$\binom{n}{r} = \binom{n}{n-r}. \tag{2.16}$$

It is easily verified by *Definition 13* that a group of n distinct objects yields the same number of combinations of size r as size (n - r). That property is intuitively plausible in that each time a group of size r is selected from the n objects, a group of size (n - r) is not selected (remains) and vice versa.

Third, the combinatorials. $\binom{n}{r}$, satisfy the Pascal triangle property,

$$\binom{n}{r-1} + \binom{n}{r} = \binom{n+1}{r}. \tag{2.17}$$

Equation (2.17) can be proved directly by *Definition 13* and illustrated by Pascal's triangle as shown in Table 2.6. Subsequent rows of Table 2.6 can be generated by increasing n by 1 and using Eq. (2.17). It is easy to

TABLE 2.6.
A table of $\binom{n}{r}$ illustrates Pascal's triangle for n = 0, 1, 2, 3, 4, 5.

n/r	0	1	2	3	4	5
0	1					
1	1	1				
2	1	2	1			
3	1	3	3	1		
4	1	4	6	4	1	
5	1	5	10	10	5	1

verify that the element in the rth column of the nth row of Table 2.6 is $\binom{n}{r}$.

A fourth property of $\binom{n}{r}$ is that

$$\sum_{r=0}^{n} \binom{n}{r} = 2^n. \tag{2.18}$$

This property is verified for n = 0, 1, 2, 3, 4, 5 by Table 2.6. In Chapter 4, Eq. (2.18) is verified in general by using the binomial theorem. Equation (2.18) denotes the total number of subsets of n objects. This is true because $\binom{n}{0} = 1$ indicates that there is one subset containing none of the elements; $\binom{n}{1} = n$ indicates that there are n different subsets containing one element each; and finally, $\binom{n}{n} = 1$ indicates that there is one subset containing all the elements. Thus, if we think of events as subsets of an event space E containing n elements, then there are 2^n different events that can be defined.

Finally, we define

$$\binom{n}{r} = 0, \tag{2.19}$$

whenever r > n, and n is an integer. Equation (2.19) simply states that there are no groups of a size greater than n that can be selected from n objects.

EXAMPLE 17: Let n = 6, and verify Eqs. (2.15)-(2.17) for r = 2.

SOLUTION:

$$\binom{6}{0} = \frac{6!}{0!6!} = 1, \quad \binom{6}{6} = \frac{6!}{6!0!} = 1.$$

Therefore, Eq. (2.15) holds for n = 6.

$$\binom{6}{2} = 15, \quad \binom{6}{4} = 15.$$

Therefore, Eq. (2.16) holds for n = 6 and r = 2.

$$\binom{6}{1} = 6, \quad \binom{6}{2} = 15, \text{ and } \binom{7}{2} = 21.$$

Thus, Eq. (2.17) holds for n = 6 and r = 2.

POWER RULES: When considering experiments that consist of independent repetitions of simple identical experiments, it is best to count the number of outcomes by what may be called a *power* formula. For example, consider an experiment in which 3 independent coins are tossed. Each coin has 2 distinct outcomes, namely H or T. Thus, two applications of the general counting rule show that the experiment has $2 \cdot 2 \cdot 2 = 2^3$ different outcomes. *Definition 14* gives a general result.

DEFINITION 14: An experiment involving k independent repetitions of an experiment having n outcomes yields a total number of outcomes given by $\#(\text{outcomes}) = n^k$.

EXAMPLE 18: How many outcomes are possible when 3 dice are rolled?

SOLUTION: Since each of the 3 dice has 6 possible outcomes, and since the 3 dice fall independently of one another, there are $6^3 = 216$ different outcomes possible. It is sometimes useful to let x denote the outcome on one die, let y denote the outcome on a second die, and let z denote the outcome on the third die. Then represent the set of 216 possible results as

$S = \{(x,y,z) \mid x,y,z$ are integers between 1 and 6, inclusively$\}$.

The power rule is most useful for calculating the number of outcomes possible for an experiment that involves tossing a number of coins or rolling a number of dice. Such experiments have a fixed number of outcomes for each die, and the coins and dice are assumed to be independent.

Another general category where the power rule is useful is that of *sampling with replacement.* Consider drawing 2 cards from a deck of 52 playing cards such that the first card is observed and replaced in the deck, which is then shuffled before the second card is selected. The replacement and shuffling make the 2 cards independent. That is, the outcome of the first draw has no influence on the second card selected. This characteristic is in contrast to the dependence that exists when the 2 cards are obtained by *sampling without replacement*. Further discussion of sampling with and without replacement is delayed until Section 2.8. Selected examples which illustrate the concepts of this section are presented in Section 2.7.

EXERCISES

1. Twenty subjects are available for a psychological experiment designated to test learning by 2 different teaching methods.
   (a) In how many different ways can the 20 individuals be divided into 2 groups having 10 subjects each?
   (b) If the groups include 10 men and 10 women, what is the probability that the random assignment of subjects to teaching methods yields a separation of the sexes?
2. Evaluate the following quantities:

   $\binom{10}{3}$, $\quad {}_{10}P_3$, $\quad 2^{10}$, $\quad 10!/3!2!4!$.

3. How are the evaluations of $\binom{10}{3}$ and ${}_{10}P_3$ related?
4. An experiment consists of flipping a fair coin, rolling a balanced die, and drawing a card from a well-shuffled deck of 52 playing cards. How many outcomes exist for the experiment? Also, P(head, 5, ace of spades) = ?
5. A fair coin is tossed 10 times. What is the probability that at least 1 head appears?
6. A review board for airing grievances within a certain company is to consist of 5 members. Of the 20 people eligible for board membership, 5 represent management, 10 represent sales, and 5 represent advertising.
   (a) How many different review boards can be formed?
   (b) How many review boards can be formed to include 2 representatives from management, 2 representatives from sales, and 1 representative from advertising?
   (c) What is the probability that the review board includes exactly 3 management representatives when random selection is used to select board members?
7. A police line-up is to present 3 women and 3 men.
   (a) How many arrangements of the 6 individuals are possible?
   (b) What is the probability that a random arrangement presents the 6 individuals so that the men and women alternate?

2.7 COUNTING AND PROBABILITY EXAMPLES

The beauty, intrigue, and excitement of the preceding concepts can best be realized through examples. Thus, we devote this section to the presentation and discussion of a set of examples that illustrate the foregoing results. The examples presented here exhibit varying degrees of difficulty and are representative of various types of probability calculations.

EXAMPLE 19: Consider the 5 digits, 1, 2, 3, 4, and 5.
(a) How many different 4-digit numbers can be formed when no digits are repeated?
(b) How many different 4-digit numbers can be formed when we allow repetition of digits within the numbers?
(c) How many different numbers can be formed when no digits are repeated?
(d) How many different numbers having 5 or fewer digits can be formed when repetition of digits is allowed?

SOLUTION: (a) In counting numbers, we look at different arrangements as being new elements in the event set of interest. Thus, the answer desired is the number of ways in which 5 digits can fill the 4 positions in a 4-digit number. That answer is ${}_5P_4 = 5!/1! = 120$.
(b) As in part (a), we look at different numbers, but now we allow the numbers of interest to include multiple digits. That is, a number may have 2, 3, or even 4 digits alike. Thus, there are 5 ways for filling each of the 4 positions of a number. Hence, we see that $5^4 = 625$.

REMARK: By not allowing digits to be repeated, the number of possible 4-digit numbers is greatly reduced.

(c) Since the only restriction is that the digits not be repeated within a number, the total quantity of numbers is the sum of the quantity of 1-digit numbers, the quantity of 2-digit numbers, ..., and the quantity of 5-digit numbers, such that all digits within a given number are unique. The summing operation follows because any number that we form under the conditions stated must consist of 1 digit, 2 digits, 3 digits, 4 digits, *or* 5 digits. Furthermore, these 5 events are mutually exclusive. Thus, the answer is

$${}_5P_1 + {}_5P_2 + {}_5P_3 + {}_5P_4 + {}_5P_5 = 5 + 20 + 60 + 120 + 120 = 325.$$

(d) As in part (c), we need to add the quantity of 1-, 2-, 3-, 4-, and 5-digit numbers when we allow the repetition of digits within each number. Hence, answer is

$$5 + 5^2 + 5^3 + 5^4 + 5^5 = 5 + 25 + 125 + 625 + 3125 = 3905.$$

REMARK: Again there is a tremendous increase in the quantity of numbers possible when we allow repeated digits. Also, by allowing the repetition of digits, we effectively have an endless supply of digits and can form numbers of any quantity of digits. Thus, an upper limit (5) was placed as the maximum quantity of digits to be used in each number.

EXAMPLE 20: What is the probability that a 4-digit number formed from the digits, 1, 2, 3, 4, and 5, is even? What is the probability that it is odd?
(a) Assume no repetition of digits.
(b) Consider the repetition of digits.

SOLUTION: (a) By Eq. (2.3), we see that P(even number) = #(even 4-digit numbers)/#(4-digit numbers), and P(odd number) = #(odd 4-digit numbers)/#(4-digit numbers).

From part (a) of Example 19, we have the denominator, 120. Now to determine the numerator we first note that whether a number is even or odd is determined by the last digit. Thus, we can specify that the number be even by first filling the last digit in one of the two (either 2 or 4) possible ways. After filling that position, any 3 of the remaining 4 digits can be used to fill the other 3 positions of the number. Thus, the number of even 4-digit numbers is given by $2 \cdot {}_4P_3 = 48$. Similarly, the number of odd 4-digit numbers is found to be $3 \cdot {}_4P_3 = 72$. Therefore, P(even number) = 48/120 = 2/5.

REMARK: The ratio of even to odd 4-digit numbers is 2:3, which is exactly the same ratio of even to odd digits in the 5 digits, 1, 2, 3, 4, and 5.

(b) Using arguments similar to those in part (a), we obtain #(even numbers) = $2 \cdot 5^3 = 250$, and #(odd numbers) = $3 \cdot 5^3 = 375$ when we allow the repetition of digits.

Therefore, P(even number) = 250/625 = 2/5, P(odd number) = 375/625 = 3/5.

REMARK: The remark for part (a) of Example 20 is also valid when digits are repeated.

EXAMPLE 21: What is the probability that a student will spell the word, education, by randomly arranging the letters, {a, c, d, e, i, n, o, t, u}?

SOLUTION: Since all letters are unique, they can be permuted 9! ways. Only one of these 9! permutations spells education. Thus, P(education) = 1/9! = 1/362,880.

EXAMPLE 22: A student is given the following set of 8 letters for random arrangement: {B, E, I, N, S, S, S, U}.

(a) What is the probability that one of the following permutations appears?

E = {SINBUSES, SUBSINES, SUNSIBES, SUBNESIS, BUSINESS}

(b) What is the probability that the 3 S's appear together in a random arrangement?

SOLUTION: (a) By *Definition 12*, the total number of permutations of the 8 letters is 8!/3! = 6720. Of those 6,720 permutations, only 5 appear in E. Therefore, P(E) = 5/6720 = 1/1344.
(b) The question here is how many of the 6,720 permutations will place the 3 S's in one block? To answer that question, we simply determine the number of permutations possible for the 6 unique symbols {B, E, I, N, [S S S], U}. That number is 6! or 720. Hence, P(3 S's appear together) = 720/6720 = 3/28.

REMARK: The technique just shown is often useful when the restrictions placed on the objects being permuted are those requiring various objects to appear together in a subgroup. In some cases, the total number of such permutations may be enlarged when the elements in the subgroup can themselves give rise to additional arrangements through permutations within the subgroup. That was not the case here because there is only one possible arrangement for the 3 S's.

EXAMPLE 23: There are 5 women and 4 men available, from whom 3 are to be selected to represent MYU at the winter conference of student government associations in Miami.
(a) What is the probability that the MYU delegation consists of all women? men?
(b) What is the probability that the MYU delegation includes at least 1 woman *and* at least 1 man?

SOLUTION: (a) First we must determine the number of different delegations that exist. Since order within the group is not important, that result is given by the number of combinations of 3, which can be obtained from the 9 available individuals. Thus, the denominator of the desired probabilities is

$$\binom{9}{3} = 84.$$

Second, we need to determine how many of these 84 possible delegations include 3 women and no men. That is just the number of combinations of 3 which can be selected from the 5 women available. Thus,

$$P(3 \text{ women in delegation}) = \binom{5}{3}\binom{4}{0}/\binom{9}{3} = 5/42.$$

Similarly,

$$P(\text{3 men in delegation}) = \binom{5}{0}\binom{4}{3}/\binom{9}{3} = 1/21.$$

(b) To satisfy the condition that at least 1 woman *and* at least 1 man be included in the delegation, there are 2 disjoint possibilities. The delegation must be composed of either 1 woman *and* 2 men *or* 2 women *and* 1 man. By using the general counting rule and *Rule 1a* for the union of two disjoint events, we find that

$$P(\text{at least 1 woman and at least 1 man})$$

$$= \binom{5}{1}\binom{4}{2}/\binom{9}{3} + \binom{5}{2}\binom{4}{1}/\binom{9}{3} = (30 + 40)/84 = 5/6.$$

EXAMPLE 24: Three dice are rolled, and the sum of the spots appearing on the uppermost faces is the variable, X.
(a) Find $P(X = 3)$.
(b) Find $P(X = 6)$.

SOLUTION: (a) First, there are $6^3 = 216$ different outcomes for the experiment of rolling 3 dice. Second, only one of those outcomes can produce the result, $X = 3$. (To get $X = 3$, we must observe one spot on each die.) Thus, $P(X = 3) = 1/216$.
(b) There are still 216 possible outcomes. So the problem is to determine the number of those outcomes that yield a sum of 6. One systematic approach is to first observe that there are certain combinations of the digits (1, 2, 3, 4, 5, 6) which, in groups of 3, add to 6. Second, we need to determine the number of possible permutations for each of those combinations. Each permutation represents a different assignment of the 3 elements in a given combination to the 3 available dice and is therefore a new outcome for counting purposes. Thus, we have the following:

Combinations with X = 6	# (permutations)
(1, 1, 4)	3
(1, 2, 3)	6
(2, 2, 2)	1
Total	10

Thus, $P(X = 6) = 10/216 = 5/108$.

REMARKS: It should be observed that the preceding example is posed as if 3 *different dice* were rolled independently. However, the analysis and results are identical if we consider the sum on 3 *independent rolls of the same die*. That is to say, rolling k dice once simultaneously and rolling 1 die k times are identical experiments. A similar statement can be made for coin-tossing experiments. The technique of, first, listing the combinations of events that satisfy a stated situation and, second, applying counting techniques to those combinations as just used is frequently useful in other exercises.

EXAMPLE 25: There were 200 students who took an elementary statistics course and a general history course. A review of the students' records showed that 12 failed statistics, 15 failed history, and 7 failed both statistics and history.
(a) What is the probability that a randomly selected student from the group of 200 passed both courses?
(b) What is the probability that a randomly selected student passed statistics?
(c) What is the probability that a student who failed history passed statistics?

SOLUTION: A most convenient form for summarizing the available data is a two-way table, such as Table 2.4. The information available is that 200 students are being classified with respect to their performance in two courses. In each course a student is classified as having passed or failed. The given data (the data appearing in circles) are sufficient to complete the following table:

		History		
		Passed	Failed	Total
Statistics	Passed	180	8	188
	Failed	5	(7)	(12)
	Total	185	(15)	(200)

(a) The table shows that 180 out of 200 students passed both courses. Therefore, P(passed history and passed statistics) = 180/200 = .9.
(b) Since 188 out of 200 students passed statistics, P(passed statistics) = 188/200 = .94
(c) Of the 15 students who failed history, 8 passed statistics. Thus, P(passed statistics | failed history) = 8/15.

REMARKS: A Venn diagram can be used to represent the data in the preceding table by defining the following two events. A denotes failed history. B denotes failed statistics. Then A ∩ B denotes those individuals who failed both subjects. The diagram is as follows:

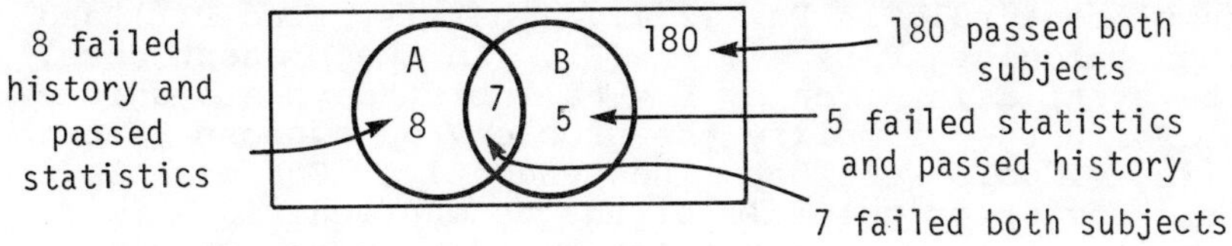

By using a table such as the preceding one, or such as Tables 2.4 and 2.5, the terms, joint probability, marginal probability, and conditional probability are quite descriptive. The joint event, passed history and passed statistics, involves the intersection of the row and the column of the table in which those students who passed *both* courses are tabulated. Yet when we ask a question concerning the result in statistics (regardless of the outcome in history), that marginal probability is determined by marginal values of the table. Any conditional probability question restricts interest to a particular column (or row) of the table. In part (c), the only information required appears in the column for failed history. Of those 15 individuals who failed history, 7 failed statistics and 8 passed statistics. Finally, we point out that failing statistics and failing history are *dependent* events. This follows because P(failed statistics) = 12/200, P(failed history) = 15/200, and P(failed history and failed statistics) = 7/200, but $7/200 \neq 12/200 \cdot 15/200$.

The next two examples demonstrate that intuition is not always to be trusted in the area of probability. The examples can be discussed and appreciated best by performing the experiments presented at the beginning of each example.

EXAMPLE 26: A classical birthday problem. What is the probability that at least 2 people in a group of r individuals share a common birthday (same month and day, but not necessarily same year)?

Two related experiments: (1) Find a group of 30 people, and check to see whether at least 2 people have the same birthday. (2) Check the birthday of the first name to appear on each of the first 30 pages of an edition of *Who's Who* to see if you find any repetitions of birthdays.

SOLUTION: In attempting to answer the question directly, we must determine the probability that the group of r people includes 2 people with the same birthday, *or* 3 people with the same birthday, *or* 4 people with the same birthday, *or* 4 people with 2 each having the same birthday, or, ..., or finally r people with the same birthday. Whenever we seek the probability for a set E that is the union of many sets, it is often best to use the complement rule in Eq. (2.7). That is, find $P(\bar{E})$ and then calculate $P(E) = 1 - P(\bar{E})$. In the present case, the event $\bar{E}$ is given by $\bar{E}$ = {all birthdays are unique}.

To further simplify the problem, we suppose that the r individuals were born independently. For example, we treat twins or multiple births as one individual. The problem of leap year is also avoided by considering a 365-day year. Under these assumptions, we seek the probability that no 2 individuals in a group of r share a common birthday. That is just the probability that all r individuals have distinct birthdays. Thus, we can take the individuals sequentially and reason as follows: The probability that the first individual has a birthday is 365/365. The probability that the second individual has a birthday different from the first individual is 364/365. The probability that the rth individual has a birthday different from the first (r-1) individuals is (365 - r + 1)/365. Putting these results together by multiplication (since we seek their simultaneous occurrence) gives

$$P(\bar{E}) = \frac{365(364)\ \dots\ (365 - r + 1)}{365^r}$$

$$= \frac{364(363)\ \dots\ (365 - r + 1)}{(365)^{r-1}}.$$

Therefore, by the complement rule in Eq. (2.7), we find that

$$P(E) = 1 - P(\bar{E}) = 1 - \frac{364(363)\dots(365 - r + 1)}{(365)^{r-1}}, \quad r = 2, 3, \dots .$$

REMARKS: Of interest here is the exact value of P(E) for various values of r. What is surprising to many people is that for r = 23, P(E) is slightly greater than 1/2. Thus, in a group of only 23 people, events E and $\bar{E}$ are almost equally likely. The following table gives some indication of how P(E) changes with r.

Kemeny, Snell, and Thompson [1] discuss the birthday problem and present a more comprehensive table of probabilities, as well as fair betting odds. It seems somewhat surprising that the odds in favor of at least 1 birthday match in a group of 100 people are approximately 3,300,000 to 1. The magnitude of $P(\bar{E})$ becomes small very rapidly because each successive term is the preceding term multiplied by a freaction less than 1. For

TABLE 2.7.
Shown here is the probability that at least 2 people in a group of r people have the same birthday.

r	P(E)
2	.0027
5	.027
10	.117
20	.411
23	.507
30	.706
50	.970

example, the probability of $\overline{E}$ for r = 100 is the product of P($\overline{E}$) for r = 99 and 266/365 (= .729). Repeated multiplication by factors of size .7 and smaller cause a rapid decrease in the magnitude of P($\overline{E}$) as r increases beyond 100. Another property indicated by Table 2.7 is that among groups of 30 people, we expect to find groups without repeated birthdays and groups with at least 1 birthday match to occur in the ratio of 3 to 7. This result can be applied to observations for the two experiments described in *Example 26*.

Further references on the birthday problem and its generalization include Naus [2], Weaver [3], and Gehan [4].

Another classic probability problem is sometimes posed as follows. Consider 3 identical small chests with 2 identical drawers each. One coin is placed in each drawer such that 1 chest has a gold coin in both drawers, 1 chest has a silver coin in both drawers, and the remaining chest has a gold coin in 1 drawer and a silver coin in the other drawer. A chest is picked at random, and inspection of 1 drawer reveals a gold coin. What is the probability that the other drawer contains a gold coin also? Since building 3 chests and securing the necessary coins to conduct the foregoing experiment may prove difficult, we propose the following analogous (but less glamorous) experiment that can be carried out with readily available resources.

EXAMPLE 27: Consider this problem. Cut out 3 small, identical cardboard cards. Write gold on both sides of the first card, silver on both sides of the second card, and silver on one side and gold on the other side of the third card. Mix the cards well by turning and shuffling. Then place a randomly selected card on a table. If the exposed word is gold, what is the probability that the other side also shows gold?

SOLUTION: To answer this question, define the following events: E_1 = (gold appears on the upper face of the selected card), E_2 = (gold appears on the lower face of the selected card). Then the desired probability is

$$P(E_2|E_1) = [P(E_1 \cap E_2)]/P(E_1).$$

Now $P(E_1 \cap E_2)$ = P(the selected card has gold on both sides) = 1/3 because the 3 cards should be selected with equal probability. Furthermore, $P(E_1) = 3/6$ because the upper face of the selected card should be a random selection of the 6 card faces, and 3 of those 6 faces carry the word, gold. Therefore,

$$P(E_2|E_1) = (1/3)/(3/6) = 2/3.$$

REMARKS: The student is encouraged to try the card experiment either with a group of people or repeating the previous experiment in order to verify that the result of that experiment is the relative frequency answer to the question. A common, but incorrect, answer to the question is 1/2. That answer is determined by saying that when gold shows, the selected card is either the gold-gold card or the gold-silver card. Thus, the chances of having gold on the other side are 1 in 2. That reasoning is not valid because it fails to recognize the randomness used in selecting the face that appears.

The examples in this section illustrate uses for the techniques of probability calculations discussed in earlier sections. The concluding section of this chapter presents further uses for probability techniques as they relate to sample selection.

EXERCISES

1. Of 20 boys who went on a picnic, 5 were sunburned, 8 were bitten by mosquitoes, and 10 returned home without either mishap.
   (a) What is the probability that a boy was both sunburned and bitten by mosquitoes upon arrival home?
   (b) What is the probability that a boy was bitten by mosquitoes, given that he was sunburned?
2. If 15 pool balls are lined along the side of a pool table, what is the probability that the number 7 and number 8 balls occur together?

3. What is the smallest number of tosses of a fair coin necessary to assure that the probability of at least 1 head exceeds .9?
4. List 3 situations in daily living where probability either is being used or is applicable. Explain how probability enters into each situation.
5. Browse through some publications (e.g., magazines, newspapers, etc.), and collect uses of probability in news stories, advertising, weather reports, and sports results.

2.8 SAMPLING TECHNIQUES

As has been alluded to previously, a sample is often the basis for statistical decision-making. This section presents a discussion of the most widely used forms of sampling. Before proceeding, let us define a sample.

DEFINITION 15: A sample is a subset of some specified population where the elements in the subset are determined by some rule (sampling rule).

The most frequently used sampling rule and the one that is basic to most statistical decision-making is called *simple random sampling* and is described in *Definition 16*.

DEFINITION 16: A sample of size n selected from a set of N objects so that every possible sample of size n has equal probability of being the selected sample is called a *simple random sample.*

NUMBER OF SAMPLES: To apply *Definition 16*, we need to determine the number of samples of size n that exist and then select the sample in such a way that P(any given sample) = 1/#(samples). The number of samples depends on how the sample is selected. If the sample is selected *without replacement* (no duplication of sample points), then the number of unordered samples is

$$\binom{N}{n} = \frac{N!}{n!(N-n)!}, \quad n = 0, 1, 2, \ldots, N. \qquad (2.20)$$

Note that n = 0 refers to a vacuous sample and has been included only for consistency. If the *order* within the sample is important and items are selected *without replacement*, then the number of ordered samples is

$${}_N P_n = N!/(N-n)!, \quad n = 0, 1, 2, \ldots, N. \qquad (2.21)$$

When the sample is selected *with replacement* (repetition of sample points allowed), the number of ordered samples is

$$N^n, \quad n = 0, 1, 2, \ldots. \tag{2.22}$$

After determining the number of samples that exist under a given sampling scheme, it remains to select a sample so that all possible samples are equally likely. For small values of N (population size), the different samples can be enumerated and the random selection accomplished by mixing poker chips or similar, easily mixed objects that correspond to the individual samples. This method is sometimes called the *fish bowl* technique.

EXAMPLE 28: Describe a method for selecting a simple random sample of 2 from A, B, C, D when sampling is accomplished without replacement. Describe such a method when sampling with replacement.

SOLUTION: (a) Without replacement, the possible samples are (AB), (AC), (AD), (BC), (BD), and (CD). These 6 samples are written on 6 poker chips that are placed in a convenient container for mixing. One chip is selected from the container, and the sample that it represents is the selected sample. This procedure gives a simple random sample because each of the 6 possible samples has a probability of 1/6 for being selected.
(b) With replacement, there are 16 possible samples, namely, (AA), (AB), (AC), (AD), (BA), (BB), (BC), (BD), (CA), (CB), (CC), (CD), (DA), (DB), (DC), and (DD). These samples are identified with 16 poker chips. After their being mixed in a container, 1 chip (sample) is selected. This procedure affords each sample a probability of 1/16 for being the selected sample and is therefore a simple random sample.

The fish bowl technique previously described is useful when the number of samples is small, but useless or at least impractical as the number of samples increases. The following example indicates some situations for which the preceding sample enumeration method would be tedious, if not impossible.

EXAMPLE 29: Consider a deck of 52 playing cards.
(a) How many different 13-card (bridge) hands can be dealt?
(b) How many different 5-card (poker) hands can be dealt?

SOLUTION: (a) The number of 13-card hands is

$$\binom{52}{13} = \frac{52!}{13!\ 39!} = 635,013,559,600. \tag{2.23}$$

(b) The number of 5-card hands is

$$\binom{52}{5} = \frac{52!}{5!\,47!} = 2{,}598{,}960. \qquad (2.24)$$

Clearly, the task of identifying and preparing poker chips for the 2,598,960 different poker hands would not appeal to even the most ardent poker player. If that task should appear possible, the task of securing a very large container and mixing the chips still looms ahead. The magnitude of that mixing problem is illustrated by observing that a stack of 2,598,960 chips stands 5.13 miles high (assuming that each chip is 1/8-inch thick).

The solution to our problem is to use the fish bowl technique in another way. Suppose that we identify each of the N elements to be sampled by a poker chip. Then select the sample of size n by mixing these chips in a fish bowl. The selection can be made one at a time with the replacement of the selected chip and the remixing of chips between each draw for *sampling with replacement*. For *sampling without replacement*, we may either select the n chips simultaneously or one at a time without replacement. However, even this procedure becomes laborious if N is large. Thus, we seek other methods for selecting random samples. One such technique, which is based on the identification method of one tag (chip) for each element, is to use a table of random digits as discussed in the next section.

RANDOM DIGITS

EXPERIMENT: Consider the following scheme for generating a table of digits. Write the integers, 0, 1, 2, ..., 9, on 10 poker chips, ping pong balls, or other easily mixed objects. Place these objects in a bowl or other container, and then select 1 object to observe the integer on it. Record that integer, replace the object in the container, mix the objects, randomly select another object from the container, and then record a second integer. Repeat this sequence until 50 integers are generated. The 50 integers so generated form a small table of random digits. On each selection, all 10 digits should have the same probability of being selected. That is, each digit occurs with a probability of 1/10 on each draw.

By independence of the draws and the use of *Rule 2a* in Section 2.5, we find the probability that any 1 of the 100 possible 2-digit numbers, 00, 01, 02, ..., 99, occurs with a probability of $1/10 \cdot 1/10 = 1/100$. Therefore, when taken 2 at a time, the digits yield a simple random sample of all 2-digit numbers. The preceding arguments extend to imply that the random digits are random when taken 1 at a time, 2 at a time, 3 at a time, etc.

Random digits are more effectively generated by computer through the use of so-called random number generators. Table 2.8 has been produced by such a generator and is a set of 2,500 random digits that can be used for sample selection. To use Table 2.8, we must tag (identify) each of the N objects in our population by assigning an identification number. The type of number used in the identification process depends on the magnitude of N. Table 2.9 provides guidelines for label assignments with various population sizes. A more extensive table of random digits is provided in Table A2 of Appendix A.

TABLE 2.8.
This is a table of 2,500 random digits.

Row	Column									
	01-05	06-10	11-15	16-20	21-25	26-30	31-35	36-40	41-45	46-50
01	72949	37133	37539	84676	47058	27980	21623	50126	67831	58481
02	59085	14310	97630	11734	66340	84797	73501	62007	80884	65150
03	00844	17813	08782	00817	86653	61668	15472	39078	04543	99101
04	44885	20344	87048	16680	05722	39202	22158	33580	87528	62947
05	00332	96400	85094	29091	30132	77208	50505	00735	18614	73009
06	05347	14815	56063	96088	05820	09502	30635	53968	43892	14720
07	64214	41260	70536	22981	41683	45986	24130	78388	57465	17555
08	08841	18875	57015	17656	13560	91899	99369	01209	32498	42531
09	94832	10841	01370	68453	04905	76962	86397	00136	63967	17820
10	53097	29243	55990	62532	07834	06510	52653	17143	11753	97078
11	12350	23900	89161	80801	20399	44375	51882	20867	47357	63456
12	85121	28500	66046	60303	22793	55662	93251	83710	68865	87350
13	94795	95820	45553	73880	60958	24548	16712	36732	25015	02270
14	35174	02603	85863	45186	45037	52594	10891	48134	86346	18348
15	79561	01487	66601	86277	71020	08617	90260	39931	52522	21725

Row	Column									
	01-05	06-10	11-15	16-20	21-25	26-30	31-35	36-40	41-45	46-50
16	64801	32144	25639	18407	85686	59663	72333	63024	68276	35527
17	86982	21861	60782	14514	16419	37490	11920	07004	38851	31679
18	95239	64274	78846	94221	57041	45047	85847	33864	68622	56266
19	00608	14270	01065	29378	68025	38125	81465	75755	47289	71001
20	11330	10311	02485	43022	39418	69208	22807	04194	93498	07202
21	85704	54267	77847	88808	09650	03507	89475	31155	45480	41866
22	79890	61935	97019	65431	84816	20903	13373	77441	20474	33379
23	77962	05016	57118	10418	30239	53765	94014	03662	53749	79483
24	15110	25979	75050	59443	29558	52163	45035	96055	79797	74107
25	17622	18614	64537	20846	20536	90691	37723	29798	74939	86902
26	02812	23488	38519	26990	26814	53343	06685	59183	88321	91184
27	38724	47530	72555	55063	16035	68975	39577	06401	90249	32412
28	50122	91318	38435	41339	97673	88496	59879	23034	75446	88624
29	50302	08174	12964	65313	01294	82451	07252	56423	96126	36187
30	54172	88552	85653	43839	11163	69455	83381	33019	94061	10755
31	19237	70799	02684	19113	84894	18804	71841	91464	92440	57012
32	25454	81551	53831	73593	49138	41108	73253	73337	51446	67062
33	57927	63888	24108	20474	45037	89545	45503	30196	23223	85310
34	92639	94377	94453	80533	83177	78906	02570	13348	78992	04557
35	70237	97754	81206	19548	85932	25636	90580	79916	16651	37469
36	68611	85684	44859	55077	61360	63427	35556	72951	71354	42875
37	01178	76336	31375	40191	00964	42664	76916	94004	46567	36765
38	31769	63342	39743	88599	88371	47097	56134	04039	53627	52969
39	94624	45484	78761	57496	16716	89066	97281	52109	94931	24241
40	23076	57543	57566	28613	08566	67568	31795	93710	29923	07031
41	75937	61955	72192	22537	91697	12632	31419	48347	60675	42069
42	06535	79209	11437	55192	51801	02033	07761	42666	66992	52317
43	52771	50608	17137	59170	63814	82628	91581	66522	61212	14711
44	12115	23123	35454	96239	64814	49620	92669	06348	66998	93258
45	06062	90462	57702	41080	42226	47351	98963	57799	94576	21960
46	29055	08529	27858	84698	57277	29536	50470	91816	98744	38316
47	64570	54286	17695	81775	82001	18499	29566	77399	63461	72810
48	88579	67411	06734	72504	72424	68244	69156	69781	39143	20909
49	25135	02087	01491	67454	58702	21303	79885	32591	09843	95934
50	96223	83071	23814	91273	73587	09301	72660	13547	11299	53340

TABLE 2.9.
These are identification numbers for populations of various sizes.

Population size	Identification numbers
$1 \leq N \leq 10$	0,1,2,...(N or 9)
$11 \leq N \leq 100$	00,01,02,...(N or 99)
$101 \leq N \leq 1000$	000, 001 002,..., (N or 999)
$1001 \leq N \leq 10000$	0000, 0001, 0002,..., (N or 9999)

The number of digits needed for identification is 1 for a population of 10 or fewer individuals, 2 for a population of 11 to 100 individuals, etc. Usually the first identification number is 0. However, that need not be the case unless the population size is exactly 10, 100, 1000, ... Any set of N numbers can be used as identification tags for the elements of the population being sampled, but such numbers as those given in Table 2.9 are quite convenient. If unassigned numbers are encountered while selecting a sample, they are ignored without influencing the randomness. Likewise, when sampling is done *without* replacement, then identification numbers that are repeated before the complete sample is selected are ignored.

The numbers in a table of random digits are random whether taken in groups of 1, 2, 3, etc., and whether read from left to right, top to bottom, or diagonally. Thus, there is great flexibility in using Tables 2.8 and A2. One question, however, always occurring is that of where do we start? Clearly, we should not start at the beginning of the table each time we seek a sample. Yet, we could begin with the upper left-hand corner and read left to right to select our first sample. Then we could begin the selection of a second sample at the point where we finished the first sample and continue the process for subsequent samples.

An alternative to the preceding method of starting is to start *at random*. For example, we could "randomly" place a pencil at some point in the table and use that number as a starting point. Or we could use the group of 5 digits selected by pencil point to determine the starting point. For example, suppose the first 2 digits indicate the *row* of a table, the second 2 digits indicate the *column* of the table, and the fifth digit indicates the *direction* for reading numbers from the table. (If the fifth digit is *even*, then read from left to right. Otherwise, read from top to bottom.)

To illustrate the preceding concepts, we have generated an hypothetical data set of 100 numbers that appear in Table 2.10. The labels, 00, 01, ..., 99, have been assigned to the elements of the population. Thus, 2-digit random numbers are needed from Table 2.8 to select a sample. The next example demonstrates two methods for using Table 2.8.

EXAMPLE 30: Select a random sample of 5 elements (without replacement) from Table 2.10.

SOLUTION: Method I. Using the upper left-hand corner of Table 2.8 as a starting point, we read the random digits, 72, 94, 93, 71, and 33. Thus, our sample consists of 0.186, -0.212, -1.098, 0.159, and 0.038.

Method II. Pick a 5-digit number from Table 2.8. Suppose the selected number is 45037, which appears in row 33 and column 21 - 25. The starting point is determined as follows:

45	03	7.
row	column	direction

TABLE 2.10.
An hypothetical data set is shown consisting of 100 observations on a continuous random variable, X.

Label	x	Label	x	Label	x	Label	x
00	0.415	25	0.351	50	-0.020	75	-2.269
01	-0.132	26	0.537	51	-0.229	76	0.115
02	0.061	27	-0.025	52	-1.930	77	-0.953
03	-0.412	28	0.230	53	-0.043	78	-1.274
04	-0.019	29	-0.080	54	0.387	79	-1.396
05	1.276	30	0.033	55	0.159	80	0.195
06	1.812	31	0.926	56	0.041	81	0.617
07	0.509	32	-0.462	57	-1.920	82	-0.858
08	-0.018	33	0.038	58	0.049	83	0.190
09	0.514	34	-0.833	59	1.107	84	0.533
10	0.667	35	0.178	60	0.053	85	-1.749
11	-0.785	36	0.287	61	-0.434	86	1.322
12	0.136	37	0.102	62	-0.537	87	-1.437
13	-0.280	38	0.586	63	0.800	88	-0.547
14	-0.396	39	-0.088	64	-1.401	89	-1.498
15	0.179	40	-0.556	65	-0.294	90	0.777
16	-0.463	41	0.121	66	0.192	91	-1.351
17	0.774	42	0.512	67	-2.599	92	-0.028
18	0.847	43	-0.695	68	0.992	93	-1.098
19	0.776	44	-0.885	69	2.777	94	-0.212
20	-0.074	45	1.564	70	-0.193	95	-0.604
21	0.786	46	1.731	71	0.159	96	1.239
22	-0.060	47	-0.580	72	0.186	97	1.334
23	0.803	48	-1.562	73	-0.327	98	-0.197
24	-0.105	49	0.984	74	-1.572	99	-0.389

Thus, we start in row 45, column 03, and read 2-digit numbers down (7 is an odd number) the page. The random numbers selected are 06, 05, 57, 13, and 22. Thus, our sample includes 1.812, 1.276, -1.920, -0.280, and -0.060.

REMARK: We ignored the second 57 in Method II because the sampling was without replacement. The observance of duplicate numbers when sampling with replacement provides repetition of sample elements.

OTHER TYPES OF SAMPLING: There are many situations for which the consideration of random samples is either impractical, impossible, or undesirable. For example, how do we select a random sample of the eligible voters for a U.S. presidential election? We must agree that such a task would be difficult and costly, even if desirable. Yet the information desired in such a case can be useful only if the sample represents the attitude of all voters. Thus, the emphasis is shifted to that of obtaining a sample that accurately represents the population being studied, rather than a strictly random sample of all elements in the population. The three specialized types of sampling to be discussed are widely used to secure representative samples.

STRATIFIED SAMPLING: Frequently the population being studied can be divided into homogeneous subunits (strata). When this is the case, only a few observations from each subunit are needed. The important idea is that *all* subunits are represented in the sample. A basic principle in stratified sampling is that elements within a subunit tend to vary less than elements from different subunits. It is that homogeneity within and heterogeneity among subunits that dictate a sample composed of a few elements from each of the subunits. National opinion polls are usually based on such stratified samples because opinions on many topics tend to vary more from locality to locality (county, state, section of the U.S.) than within localities.

CLUSTER SAMPLING: The idea of cluster sampling is almost opposite to that of stratified sampling. Here the population is assumed to be composed of heterogeneous subunits (clusters). In fact, each cluster is thought to be a "miniature" population. Thus, we can view the population better and more economically by observing *all* of a few clusters (possibly even one cluster) than by observing a part of many clusters. Two examples of populations for which cluster sampling would be applicable are the following. A study of nursing homes (100-200 beds) is being made to determine the training of staff members. It could be assumed that the staff of one nursing home for the indicated size is approximately a miniature of all such staffs. Thus, results obtained by studying the entire staff for a few nursing homes may be a good representation of the entire population.

The second example involves a biologist who is interested in determining the age composition of a prairie-dog population that resides within the limits of a particular national park. If we assume that each colony or prairie-dog town is a small replica of the complete population, then we need only study a few (perhaps only one) colonies to obtain the age composition information desired.

In cluster sampling, the clusters for the sample are selected at random. For those populations having clusters so large that it is impractical to obtain data from all elements of a selected cluster, it is common to randomly select elements from the cluster for the sample. The latter technique is called two-stage cluster sampling. In the first stage, a set of clusters is randomly selected. Then the second stage involves the random selection of elements from the selected clusters.

SYSTEMATIC SAMPLING: Another frequently used sampling technique is called systematic sampling. Here the sampling rule involves an initial random selection followed by some system. Common examples are as follows: (1) To sample the student body of MYU, one name is selected at random from the first 100 names listed in the current student directory. After that initial selection, every 100th name is taken for the sample. The resulting sample will include 1% of the student body. (2) Randomly select a number between 1 and 20, say 17. Then select taxpayers numbered 17, 37, 57, 77, ... for a sample to represent all taxpayers in a given taxing unit. (3) Select a random number between 1 and 100, say 39. Then select names from the local telephone directory that correspond to numbers 39, 139, 239, ... for a sample. (4) Randomly select a name from among the first 1,000 names on the list of subscribers for a popular magazine. Then select every 1,000th name thereafter for a sample. (5) Randomly select a first customer to interview when leaving a given business. Then interview every 15th customer who leaves the store after the initial interview.

Clearly the method of systematic sampling is convenient. Yet its effectiveness depends on how representative the selected sample is of the population being studied. For example, if the student body of MYU is the population being considered, then the selection of every 10th name from the directory could yield a useful sample. On the other hand, such a sample might be useless for studying attitudes of local businessmen.

The foregoing sampling techniques represent a large portion of all samples used for statistical decision-making. As we proceed in a study of basic statistical concepts, it is hoped that a strong appreciation will develop for the important role that sampling plays in all statistical inference procedures. Briefly, an analysis is only as good as the data put into it, and the quality of the data is dependent on the attention given to data collection through sampling.

EXERCISES

1. Use Table 2.8 to select a random sample of 10 items from Table 2.10 (sample without replacement).
   (a) How many samples of size 10 exist?
   (b) How many of the sample elements are negative?
   (c) Find the probability that a sample includes 5 negative numbers. (There are 50 negative and 50 positive numbers in Table 2.10.)
2. Select a random sample of 3 from (A, B, C, D, E, F, G, H, I, J) with replacement.
   (a) How many different samples of size 3 exist?
   (b) What is the probability that the sample consists of elements A, C, and E in some order?
3. Indicate an appropriate type of sampling for obtaining the following samples, giving reasons for the choices made.
   (a) A sample of the student body at MYU to study voting attitudes of students.
   (b) A sample of fish from Lake Retaw to study types of fish in the lake.
   (c) A sample of tomato plants to check tomato production.
   (d) A sample of beef carcasses for studying meat quality.
   (e) A sample of students from MYU to show attitudes toward the determination of parking permit eligibility based on the distance between a student's residence and the campus.

2.9 SUMMARY

The basic ideas for calculating probabilities are developed and discussed by means of numerous examples. Definitions for marginal, joint, and conditional probabilities are presented, as well as rules for calculating probabilities for mutually exclusive, independent, dependent, and complementary events. To assist with the necessary counting that is a part of probability calculations, various counting techniques such as permutations, combinations, and powers are included.

The following formulas summarize the important results of this chapter.

The probability of an event, E, is

$$P(E) = \#(E)/\#(\mathcal{E}) \qquad \text{from Eq. (2.3).}$$

If E and F are mutually exclusive, then $P(E \cup F) = P(E) + P(F)$ from Eq. (2.5). In general, $P(E \cup F) = P(E) + P(F) - P(E \cap F)$ from Eq. (2.6). If the two events, E and F, are independent, then

$$P(E \cap F) = P(EF) = P(E) \cdot P(F)$$

from Eq. (2.8). In general,

$$P(E \cap F) = P(EF) = P(E)\ P(F|E), \text{ or } P(EF) = P(F) \cdot P(E|F)$$

from Eq. (2.9). For conditional probabilities,

$$P(E|F) = P(EF)/P(F) \text{ if } P(F) > 0, \text{ and}$$

$$P(F|E) = P(EF)/P(E) \text{ if } P(E) > 0 \text{ from Eq. (2.10).}$$

Permutations:

$${}_nP_r = n!/(n-r)!,\ r = 0, 1, 2, \ldots, n \qquad \text{from Eq. (2.13).}$$

Combinations:

$$\binom{n}{r} = n!/r!(n-r)! \qquad \text{from } \textit{Definition 13} \text{ on page 47.}$$

The number of samples of size n from N unique items is $\binom{N}{n}$ when sampling without replacement and ignoring order, ${}_NP_n$ when sampling without replacement and considering order, and N^n when sampling with replacement and considering order.

Sampling methods, including the use of random digit tables, are presented and illustrated. These techniques are used throughout the remainder of the book.

2.10 REFERENCES

[1] Kemeny, J. G; Snell, J. L.; and Thompson, G. L. 1957. *Introduction to finite mathematics.* Englewood Cliffs, N.J.: Prentice-Hall, pp. 124-126.

[2] Naus, J. I. 1968. An extension of the birthday problem. *The American Statistician* 22: 27-28.

[3] Weaver, W. 1963. *Lady luck: the theory of probability.* Garden City, N.Y.: Doubleday, pp. 132-135.

[4] Gehan, E. A. 1968. Note on the birthday problem. *The American Statistician* 22: 28.

2.11 REVIEW EXERCISES

1. Suppose the serial number on an issue of $1 bills consists of a letter followed by 8 numbers and another letter. For example, G30262755B. What is the monetary value of one issue of $1 bills?

2. Suppose a license plate consists of 2 letters followed by 4 numbers (e.g., AB1234). How many different plates can be made using this system?

3. How many different 7-card hands can be dealt from a 52-card deck of playing cards?

4. How many different arrangements of M-A-V-E-R-I-C-K are possible?

5. How many different outcomes are possible for an experiment consisting of flipping 3 coins, rolling 2 dice, and drawing 2 cards (with replacement) from a 52-card deck?

6. How many different ways can 4 people be seated in a row? around a table?

7. A senate subcommittee consists of 4 Democrats, 3 Republicans, and 2 Independent members. Find the probability that a randomly selected visitation team of 4 includes the following:
   (a) 2 Democrats, 1 Republican, and 1 Independent;
   (b) 2 Republicans and 2 Independents;
   (c) 4 Democrats.

8. If 2 dice are thrown, what is the probability that the following occur?
   (a) Two 6's appear;
   (b) at least one 2 appears;
   (c) one 6 and one 5 appear;
   (d) one 6 or one 5 (but not both) appears;
   (e) at least one 6 appears.

9. If 3 dice are thrown, what is the probability the following occur?
   (a) Two 1's and one 2 appear;
   (b) the sum on the 3 dice is 4;
   (c) the sum on the 3 dice is either 11 or 12.

10. A box contains 7 red, 5 white, and 6 blue balls. What is the probability that the following will occur?
    (a) If 1 ball is drawn, it will be white.
    (b) If 1 ball is drawn, it will be nonwhite.
    (c) If 3 balls are drawn without replacement, all 3 will be red.

(d) If 3 balls are drawn with replacement, all 3 will be the same color.

(e) If 5 balls are drawn without replacement, 2 will be blue and 3 will be red.

11. Two urns contain, respectively, 3 black and 2 white balls, and 2 black and 3 white balls. One ball is transferred from the first urn into the second, after which a ball is drawn from the second urn. What is the probability that it is white?

12. One urn contains 8 red and 5 black balls. A second urn contains 3 red and 4 black balls. If an urn is selected at random and a ball is drawn from it, what is the probability that a red ball is selected?

13. Three urns identical in appearance contain, respectively, 3 white and 2 black balls, 2 white and 5 black balls, and 4 white and 2 black balls.
   (a) One urn is selected, and a ball is taken from it. What is the probability that this ball is white?
   (b) Two urns are selected, and a ball is taken from each of these urns. What is the probability that 2 black balls are drawn?

14. From 7 men and 5 women, a committee of 4 is chosen. Find the probability that the committee will contain the following: 4 women; 2 men and 2 women; at least 3 men; at least 2 women.

15. A first bag contains 6 green and 7 red balls, and a second bag contains 4 red and 5 black balls. If a ball is drawn from each bag, find the probability that the following will occur: 1 green and 1 red ball; a black ball and not a black ball; 2 red balls; at most, 1 red ball.

16. In dealing cards from a deck, what is the probability that the first spade occurs at the fifth card?

17. Independently, a coin is tossed, a card is drawn from a deck, and a die is thrown. What is the probability that we observe a head on the coin, an ace from the deck, and a 5 on the die?

18. In a shoe factory, uppers, soles, and heels are manufactured separately and randomly assembled into single shoes. Of the shoes made, 5% of the uppers, 4% of the soles, and 1% of the heels have flaws. What percent of the pairs of shoes are flawless in these 3 parts?

CHAPTER 3

Descriptive Statistics

3.1 A LOOK BACK, THEN AHEAD

The preceding development characterizes a statistical analysis as a study that begins with the formation of questions arising from the examination of some phenomenon, continues with the identification of measurable variables containing information about those questions, and then proceeds with an experiment for accumulating measurements on those variables. Such a study concludes with an analysis of the data. The procedures for securing samples, when necessary, are available in Chapter 2. Thus, we have the techniques available for identifying statistical questions that can be answered through the use of numerical data. Therefore, the next step is to begin the development of methods for analyzing data.

The present chapter is devoted to *descriptive statistics*. This means that our objective is to summarize sets of numerical data by calculating *key* statistics (numbers) that represent important characteristics of the data. The set of data may be either a sample or a population. But whatever the case may be, the statements pertain only to the set of numbers being studied. Hence, the aim of this chapter is to prepare a set of measurements for presentation. Our goal is to provide the best possible representation of the data in the most usable and concise form available.

Before proceeding, one comment pertaining to notation is needed. Since the emphasis is on the description of a data set by means of descriptive statistics, we use the notation usually adopted for sampling statistics. The analogous terms, properties and notation, that correspond to parameters of a population (theoretical) or model distribution are presented in Chapters 4 and 5.

Throughout the development of various descriptive statistics, we use hypothetical data to illustrate the necessary arithmetic, without becoming too involved and dependent on arithmetical calculations. It is important that the student understand the caculation steps and the interpretation of each statistic presented. The interpretation is more easily comprehended and appreciated when applied to topics and data familiar to the student. Therefore, a list of data-collection projects is provided in the next section.

It is recommended that each student select a project of personal interest and collect the data indicated. The greatest benefit will likely come from those projects where the students conduct their own data-collecting experiments. However, when that is not a feasible approach, much insight and appreciation can be gained by securing journal-type data in a field of interest. (See Chapter 1, page 22, *Exercises 1* and *2*, for sources of data.) The data obtained either by personal experimentation or from a published journal should be subjected to any applicable techniques of analysis as the study of this chapter progresses.

3.2 DATA-COLLECTING SITUATIONS (arranged alphabetically by discipline)

Agriculture: Record the birth weights, X, for 25 calves. (Preferably the calves are of the same breed and from dams having produced at least one other calf.)

Biology: Determine the blood types for 25 students. Classify them by 0-A group, 1-B group, 2-AB group, and 3-0 group.

Business: Count the number of customers leaving Joe's Grocery per 5-minute intervals between 10:00 and 11:00 a.m. Repeat the experiment until 20 observations have been recorded.

Chemistry: Measure the percent, X, of flouride in 20 tubes of Shiny Bright Toothpaste.

Economics: Record the closing price of any stock of interest for the last market day of each month for the past two years.

Education: (1) Record the scores, X, of a classroom examination in any class. (2) Record the scores, X, on the basic intelligence test for a class of elementary students.

Engineering: (1) Test the breaking strength, X, of a 1/4-inch steel cable. Repeat the experiment 20 times. (2) Observe the traffic flow through the intersection of Main and Side Streets. Count the number of passing cars per minute, X, on Main Street for 30 1-minute intervals.

Entomology: Apply the recommended dosage of GO FLY fly spray to 50 samples of 10 flies, and then record the number of flies killed by the spray.

Home Economics: Determine the price per pound for hamburger in 20 different stores.

Management: Determine the age, X, of 25 salesmen who work for the ABC Company. Then determine the number of years of service, X, for 25 salesmen who work for the ABC Company.

Milling: Record the weights of 25 50-pound bags of chick starter.

Political Science: Survey 100 students to determine whether they are 0-Democrat, 1-Republican, 2-Independent, or 3-Other.

Psychology: (1) Obtain the entrance exam scores for 25 students. (2) Administer a psychological test to your class, and record their scores.

General: (1) Draw a line on a blank sheet of paper, and have 25 people estimate its length, X, to the nearest tenth of an inch. (2) Flip 3 coins 32 times, and record the number of heads, X, that occur for each of the trials. (3) Shuffle a deck of 52 playing cards, and select a sample of 3 cards, with replacement from the deck. (Reshuffle before each draw.) Repeat the experiment 16 times, and record the number of spades, X, in each sample of 3 cards. (4) Measure the height of 25 male (or female) students. (5) Measure the weights of 25 male and 25 female students. (This could be done by having a bathroon scale at the entrance of a classroom building.) (6) Sample 100 names from the student directory at MYU, and classify students by 1-freshman, 2-sophomore, 3-junior, 4-senior, 5-graduate, and 6-other.

The situations and experiments described here are only suggestions. They can easily be changed to meet the needs of a particular class or individual. For example, the suggested sample sizes can be changed without destroying the value of the experiment. What is important is that each student gain experience in collecting data that have personal interest for her/him. Furthermore, data of sufficient types and quantities should be acquired so that most of the techniques of this chapter are applicable to one or more sets of data.

The variables involved in the preceding data-collection experiments are of two types. Those variables requiring *counts* are said to be *discrete*, while those variables requiring *measurements* of weight, height, time, etc., are called *continuous*. In general, a discrete variable is one that takes on distinct values such that two values are either equal or differ by at least some fixed unit. Thus, the sample space for a discrete variable, X, consists of a set of distinct points, say (x_1, x_2, ...). On the other hand, a variable that may conceivably take on any value in an interval (continuum) is called a continuous variable. The sample space for a continuous variable, X, is of the form, $a \leq X \leq b$, where a and b are any numbers such that $-\infty < a \leq b < \infty$. These concepts are developed further in this chapter and in Chapters 4 and 5.

EXERCISES

1. Identify the types of variables being observed in parts (1) - (3) under the general area in the list for Section 3.2, *Data-Collecting Situations*.

2. Select at least two of the situations in the preceding list, and collect a set of observations from each situation. Be sure that you have at least one set of continuous data (i.e., observations on a continuous variable) and at least one set of discrete data (i.e., observations on a discrete variable).

3.3 DATA PREPARATION AND PRESENTATION

A large set of data, say 25 or more observations, is difficult to assess without some organization and/or condensation. The purpose of this section is to develop methods for presenting data so they can be more readily interpreted.

Since techniques vary, depending on whether we have discrete data or continuous data, two distributions of 100 observations are presented for reference throughout the chapter. Tables 3.1 and 3.2 illustrate these two distributions.

ARRAYS AND ORDER STATISTICS: One convenient method for organizing data when n, the sample size, is not too large is that of an *ordered array*. The ordering may be either from largest to smallest or from smallest to largest. When the data, $(x_1, x_2, \ldots, x_n)$, are ordered to give $x_{(1)} \leq x_{(2)} \leq x_{(3)} \leq \cdots \leq x_{(n)}$, then the values, $x_{(1)}, x_{(2)}, \ldots, x_{(n)}$, are called the *order statistics* of the data. In particular,

$$x_{(1)} = \min\,(x_1, x_2, \ldots, x_n) = \text{the smallest observation},$$

$$x_{(n)} = \max\,(x_1, x_2, \ldots, x_n) = \text{the largest observation} \tag{3.1}$$

are special-order statistics that are used later. An (ordered) array of the data and the order statistics present the data so that it is easy to determine the largest observation, $x_{(n)}$; the smallest observation, $x_{(1)}$; the relative magnitude of the observations; and other properties that are discussed later.

TABLE 3.1.
Shown here are 100 measurements on a continuous variable, X.

Label	X	Label	X	Label	X	Label	X
00	3.9	25	3.9	50	3.5	75	1.2
01	3.4	26	4.0	51	3.3	76	3.6
02	4.1	27	3.5	52	1.6	77	2.5
03	3.1	28	3.7	53	3.5	78	2.2
04	3.5	29	2.4	54	3.9	79	2.1
05	4.8	30	3.5	55	3.7	80	3.7
06	5.3	31	2.5	56	3.5	81	4.1
07	4.0	32	3.0	57	1.6	82	2.6
08	3.5	33	3.5	58	3.5	83	3.7
09	4.0	34	2.7	59	4.6	84	4.0
10	4.2	35	3.7	60	3.4	85	1.8
11	2.7	36	3.8	61	3.1	86	4.8
12	3.6	37	3.6	62	3.0	87	2.1
13	3.2	38	4.1	63	4.3	88	3.0
14	3.1	39	3.4	64	2.1	89	2.0
15	3.7	40	2.9	65	3.2	90	4.3
16	3.0	41	3.6	66	3.7	91	2.1
17	4.2	42	4.0	67	0.9	92	3.5
18	4.3	43	2.8	68	4.5	93	2.4
19	4.3	44	2.6	69	6.3	94	3.3
20	3.4	45	1.9	70	3.3	95	2.9
21	4.3	46	1.8	71	3.7	96	4.7
22	3.4	47	2.9	72	3.7	97	4.8
23	4.3	48	1.9	73	3.2	98	3.3
24	3.4	49	4.5	74	1.9	99	3.1

EXAMPLE 1: Use Table 2.8 to select 10 observations from Table 3.1. Form an ordered array of the data. List the order statistics of the sample.

SOLUTION: Sample: $x_1 = 2.1$, $x_2 = 2.0$, $x_3 = 5.3$, $x_4 = 4.3$, $x_5 = 3.7$, $x_6 = 4.8$, $x_7 = 3.4$, $x_8 = 4.7$, $x_9 = 3.9$, and $x_{10} = 2.5$. Array: 2.0, 2.1, 2.5, 3.4, 3.7, 3.9, 4.3, 4.7, 4.8, and 5.3. Order statistics: $x_{(1)} = 2.0$, $x_{(2)} = 2.1$, $x_{(3)} = 2.5$, $x_{(4)} = 3.4$, $x_{(5)} = 3.7$, $x_{(6)} = 3.9$, $x_{(7)} = 4.3$, $x_{(8)} = 4.7$, $x_{(9)} = 4.8$, and $x_{(10)} = 5.3$.

When the number of data points becomes moderate to large, an array is quite awkward for purposes of interpretation. Therefore, a more compact method of organization is needed. The most widely used method is that of forming a distribution for the data. A *distribution* is simply a method by which we divide the available data

TABLE 3.2.
Shown here are 100 measurements on a discrete variable, X.

Label	X	Label	X	Label	X	Label	X
00	3	25	2	50	3	75	2
01	4	26	3	51	3	76	1
02	4	27	4	52	3	77	4
03	4	28	2	53	4	78	0
04	2	29	2	54	2	79	2
05	1	30	2	55	3	80	2
06	1	31	3	56	2	81	3
07	4	32	3	57	3	82	1
08	2	33	3	58	4	83	2
09	2	34	4	59	3	84	2
10	4	35	3	60	3	85	0
11	3	36	5	61	2	86	2
12	3	37	2	62	2	87	4
13	4	38	1	63	2	88	2
14	0	39	1	64	2	89	3
15	4	40	3	65	1	90	3
16	1	41	3	66	1	91	3
17	2	42	1	67	3	92	2
18	2	43	3	68	2	93	1
19	2	44	2	69	2	94	1
20	4	45	3	70	4	95	4
21	2	46	1	71	2	96	3
22	2	47	5	72	3	97	2
23	5	48	2	73	1	98	2
24	3	49	3	74	4	99	3

into *categories* (classes) and count the number of data points in each category. The problem for the discrete data is discussed next, followed by a discussion of the same problem for continuous data.

DISTRIBUTIONS OF DISCRETE DATA: By the definition of discrete data, only distinct values of the variables, X, are observed. Thus, a distribution of discrete data consists of listing the possible values of X, as well as the *frequency* (number of occurrences) of each X. The necessary counting to determine the frequency of each X is best completed by using a tally column in the distribution table. To illustrate the technique, we select a sample of 25 observations from Table 3.2 by using the random digits in Table 2.8.

Data: 3, 3, 4, 1, 3, 3, 1, 2, 2, 4, 4, 1, 1, 3, 2,

3, 2, 2, 2, 3, 5, 3, 1, 4, 2.

By inspection there are 5 different values of X, namely, 1, 2, 3, 4, 5. These values are the categories (classes) of the distribution. Presented in Table 3.3 are 3 different distributions: frequency, (f_i), percentage, (p_i), and cumulative percentage, (c_i).

TABLE 3.3.
Frequency, percentage, and cumulative percentage distributions are given for a random sample of 25 observations from Table 3.2.

x_i	Tally	Frequency, f_i	Percentage, p_i	Cumulative percentage, c_i
1	~~1111~~	5	20	20
2	~~1111~~ 11	7	28	48
3	~~1111~~ 111	8	32	80
4	1111	4	16	96
5	1	1	4	100
Total		25	100	

REMARKS: The *frequency distribution* of X consists of the two columns, x_i and f_i, as shown in Table 3.3. Similarly, columns x_i and p_i show the *percentage* distribution, and columns x_i and c_i show the *cumulative percentage distribution*. It is convenient to present all three tables in one. However, there are situations when only one or two of the tables would be given.

The frequency column gives the actual count for each value of x_i. For distributions having fewer than 50 observations, the actual count is small enough that we can readily compare the frequencies of the different categories. That is, the frequency distribution often provides a better representation of small numbers of observations than a percentage distribution.

The percentage distribution is obtained from the frequency distribution by $p_i = (f_i/n)\cdot 100$. Percentage distributions are useful either when the counts in each category become large or when we want to compare two distributions based on different total frequencies. A comparison of Table 3.4 (based on the full set of values in Table 3.2) and Table 3.3 illustrates the latter case.

The cumulative percentage distribution provides an easy way of finding what percent of the sample has a value equal to or less than a particular value of X. For example, 80% of the samples have a value of 3 or less, 48% have a value of 2 or less, etc.

A second example of a distribution for a discrete variable is given in Table 3.4. The data, as previously mentioned, are the values in Table 3.2.

TABLE 3.4.
Frequency, percentage, and cumulative percentage distributions are given for the discrete data in Table 3.2.

x_i	Tally	f_i	p_i	c_i
0	111	3	3	3
1	~~1111~~ ~~1111~~ 1111	14	14	17
2	~~1111~~ ~~1111~~ ~~1111~~ ~~1111~~ ~~1111~~ ~~1111~~ 1111	34	34	51
3	~~1111~~ ~~1111~~ ~~1111~~ ~~1111~~ ~~1111~~ 1111	29	29	80
4	~~1111~~ ~~1111~~ ~~1111~~ 11	17	17	97
5	111	3	3	100
Total		100	100	

REMARKS: Compare the percentage and cumulative percentage distributions for Tables 3.3 and 3.4. Notice that they exhibit similar patterns. The fact that the distribution of a sample from a population resembles the distribution of the population from which the sample has been obtained is basic to many statistical analyses. That property is demonstrated again later in the text.

The techniques for forming distributions used thus far work very well for discrete data. However, their usefulness is limited if the number of possible categories for X becomes large. In fact, a variable, X, that exhibits 20 or more values cannot be effectively represented by the foregoing methods. That is, as the number of categories increases, our ability to make comparative observations decreases. So in those cases where a discrete variable can assume a large number of values, we treat the variable as if it is a continuous variable and apply the techniques of the next section.

DISTRIBUTIONS OF CONTINUOUS DATA: Whenever we have measurements on a continuous variable, they are necessarily discrete because any measurement must be to the nearest inch, nearest tenth of a pound, nearest thousandth of a second, etc. Regardless of the finest unit of measurement specified, there is usually enough variation from individual to individual to produce a large set of diferent values for the continuous variable being measured. This is the property that leads us to summarize discrete data having a large number of possible values in the same way in which we summarize continuous data.

For continuous data, the problem of forming distributions begins with the selection of *class intervals* into which the data are grouped. The choice of class intervals was not necessary for the discrete data considered in the preceding section because the variable, X, assumed only a few possible values. Yet the choice of intervals corresponds somewhat to the set of X values for a discrete variable in that after grouping our continuous data into classes, we proceed as if the data within a class have a common value called the *midpoint* of the interval.

In the actual selection of classes for a distribution, we usually consider the following thoughts. First, the intervals must be selected so that they are *mutually exclusive* and *exhaustive* for the data set. That is, the intervals must not overlap, and they must span the data. In other words, every data point must fall into *one and only one* class interval. Second, the *number* of intervals should be neither too large nor too small. The number will be an increasing function of the sample size and is usually between 4 and 15. Third, intervals having integer midpoints are more convenient than those having fractional midpoints. Fourth, from a convenience viewpoint, intervals of equal lengths are desirable. Fifth, open-ended intervals, such as those over 40 and ones under 10, should be avoided if possible. However, in all of this discussion, *the important idea is that a reasonable number of mutually exclusive and exhaustive intervals presenting a realistic picture of the data be chosen for the distribution.*

A useful suggestion that may be of some assistance in choosing the number of intervals is given by Sturges' [1] rule. This rule states that when grouping n observations, the number of intervals needed is approximately

$$k = 3.3\ (\log n) + 1, \tag{3.2}$$

where log n is the logarithm (base 10) of n. For example, use 5-6 intervals for 25 observations, use 7-8 intervals for 100 observations, and use 10-11 intervals for 1,000 observations. It should be kept in mind that the values of k produced by Eq. (3.2) are suggestions rather than requirements. (Refer to Appendix A and Table A3 for use of logarithms.)

A technique for obtaining the approximate length of an interval is to determine the *range* of the data, R, by

$$\begin{aligned} R &= \text{maximum observation} - \text{minimum observation} \\ &= x_{(n)} - x_{(1)}, \end{aligned} \tag{3.3}$$

where $x_{(1)}$ and $x_{(n)}$ are defined by Eq. (3.1). Then the length of each interval must be at least L = R/k, where k is the number of intervals determined by Eq. (3.2).

A random sample of 25 observations is selected from Table 3.1 to illustrate the foregoing discussion.

Data: 3.9, 4.2, 2.1, 1.2, 4.3, 3.6, 2.5, 3.5,
3.1, 4.0, 4.0, 4.0, 3.0, 4.7, 3.3, 2.8, 3.7,
3.0, 3.5, 3.6, 2.6, 3.3, 3.7, 2.7, 3.7.

For convenience in writing the endpoints of the class intervals, we agree that the interval a-b is to include all points greater than or equal to a, but less than b. Thus, the lower endpoint is included in the interval, but the upper endpoint is excluded. We note that this method may not be quite as precise as the use of *class limits* (values that the measurements exhibit). However, the approximation involved is slight in most cases.

Since the preceding data vary between 1 to 5, we use 4 intervals: 1-2, 2-3, 3-4, and 4-5. Note that these intervals do not have integer midpoints, but they are of equal length. Table 3.5 presents frequency, percentage, and cumulative percentage distributions for the 25 observations.

REMARKS: The *frequency distribution* consists of the three columns: class intervals, x_i, and f_i. Similarly, the *percentage distribution* uses the columns for class intervals, x_i, and p_i, while the *cumulative percentage distribution* consists of class intervals, x_i, and c_i.

Following the grouping, we proceed as if we have 1 value = 1.5, 5 values = 2.5, 13 values = 3.5, and 6 values = 4.5.

TABLE 3.5.
Frequency, percentage, and cumulative percentage distributions are given for a random sample of 25 observations from Table 3.1.

Class intervals	Tally	Mid-points, x_i	Frequency, f_i	Percentage, p_i	Cumulative percentage, c_i
1-2	1	1.5	1	4	4
2-3	~~1111~~	2.5	5	20	24
3-4	~~1111~~ ~~1111~~ 111	3.5	13	52	76
4-5	~~1111~~ 1	4.5	6	24	100
Total			25	100	

The *Remarks* following Table 3.3 are also relevant for continuous data.

Table 3.6 gives the frequency, percentage, and cumulative percentage distributions for the 100 observations in Table 3.1 as a second example of how distributions are formed for continuous data. Seven classes are used to form the distributions.

TABLE 3.6.
Frequency, percentage, and cumulative percentage distributions are given for Table 3.1.

Class intervals	Tally	Mid-points, x_i	Fre-quency, f_i	Percen-tage p_i	Cumulative percentage, c_i
0-1	1	0.5	1	1	1
1-2	~~1111~~ 111	1.5	8	8	9
2-3	~~1111~~ ~~1111~~ ~~1111~~ 111	2.5	18	18	27
3-4	~~1111~~ ~~1111~~ ~~1111~~ ~~1111~~ ~~1111~~ ~~1111~~ ~~1111~~ ~~1111~~ ~~1111~~ 111	3.5	48	48	75
4-5	~~1111~~ ~~1111~~ ~~1111~~ 1111 111	4.5	23	23	98
5-6	1	5.5	1	1	99
6-7	1	6.5	1	1	100
Total			100	100	

As noted for Tables 3.3 and 3.4, there is some similarity between the cumulative percentage distributions presented in Tables 3.5 and 3.6.

Now that we have techniques available for summarizing data in distributions, the next step is to discuss some techniques for graphically presenting the distributions tabulated in the foregoing tables. The significant idea is that a large quantity of data presented in a graph is more easily interpreted than the same data presented in a table.

GRAPHIC PRESENTATION OF DISTRIBUTIONS: The purpose of this section is to show some graphic methods for presenting distributions of data. The methods depend on whether the data are discrete or continuous and whether the distribution is one of frequency or of cumulative percentage.

We begin by presenting a *line graph* for the frequency distribution of a discrete variable, as shown in Figure 3.1. Notice that the lines are drawn at each value of X having a positive frequency with height equal to the frequency of that value. Figure 3.1 gives the line graphs for the data in Tables 3.3 and 3.4.

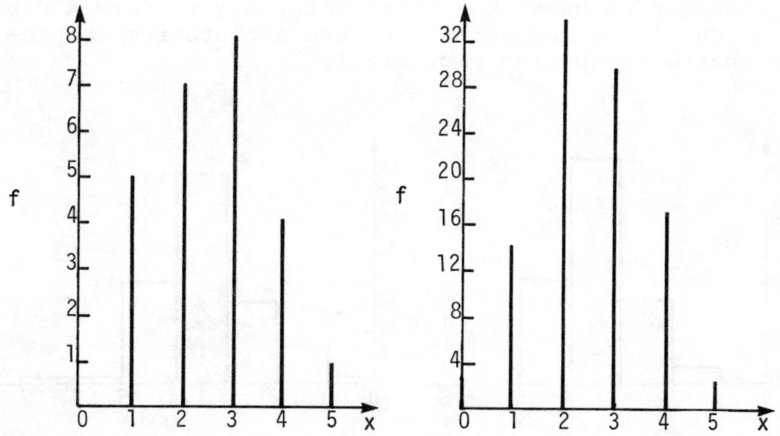

FIGURE 3.1. Line graphs are presented for the frequency distributions in Tables 3.3 and 3.4.

The main motivation for using lines to represent the frequencies is to emphasize the discreteness of the variable. That is, there are no values of the variables except at the integers 0, 1, 2, 3, 4, 5, so the graph indicates a 0 frequency for those values of X other than the observed values of X.

Next we graph the frequency distribution of a continuous variable by means of a *histogram*. A histogram is a *bar graph* constructed so that *equal frequencies* are given *equal areas*. Thus, if one class is twice as long as another class, then the bar height for that class is 1/2 the class frequency. The use of equal class intervals is desirable when constructing histograms. Figure 3.2 presents histograms for the frequency distributions in Tables 3.5 and 3.6.

Since we assume that our variable is continuous, it is conceivable that any value of X within the distribution limits is observable. That property of continuity is represented by drawing a bar over the interval. The implication is that all values within a certain interval are equally likely to occur. Thus, we can compare the relative likelihood of various values of X by comparing the height of the bar graph at the points of interest. In that light, we observe that bars having equal areas and equal class widths represent equally likely values of X.

We should note that the earlier remarks about treating discrete data with a large number of values as continuous data still applies in our graphing scheme. That is, whenever we have used class intervals to form a distribution, it is appropriate to use a histogram to present that distribution pictorially.

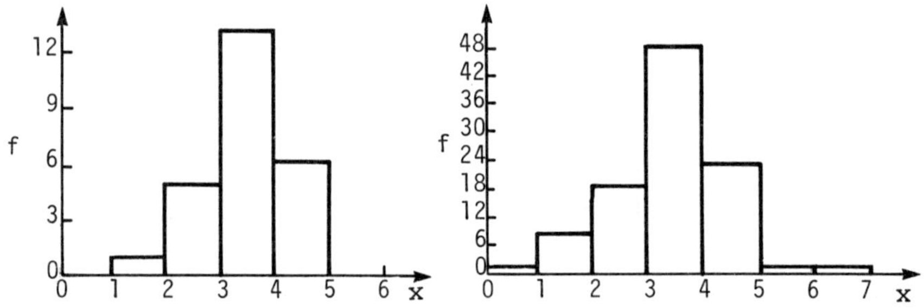

FIGURE 3.2. Histograms are presented for the frequency distributions in Tables 3.5 and 3.6.

A third method for graphic presentation of data is to use an *ogive* curve to portray a cumulative percentage distribution. Here again, we have the graph form changing for the two types (discrete and continuous) of data. First, the accumulation of discrete data occurs at distinct points of X. For example, all of the accumulation in Table 3.4 occurs at the integers 0, 1, 2, 3, 4, 5, and there is no accumulation otherwise. Second, the accumulation of continuous data is continuous throughout the range of the data. That is, as X increases, we accumulate a larger portion of the sample. To allow for graphic demonstration of these differences, we use a horizontal line to show regions of no accumulation and positively sloping lines to depict regions of accumulation. Thus, the cumulative percentage distribution graph for discrete data takes the appearance of a step function, while for continuous data it is a series of connected, inclined lines (a broken line segment). Figure 3.3 illustrates these ideas for the data in Tables 3.4 and 3.6.

REMARKS: The flat steps between the integers in the graph for Table 3.4 indicate no growth between the integers. All growth occurs at the integers 0, 1, 2, 3, 4, and 5.

The sloping lines between the integers in the graph for Table 3.6 indicate continuous growth throughout each interval. That is, we enter a given class with one cu-

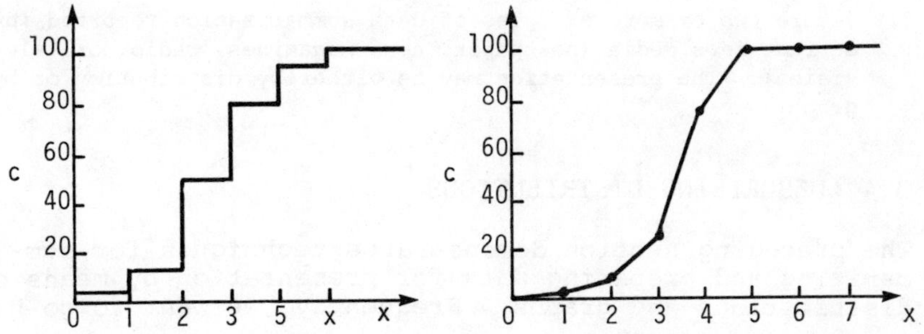

FIGURE 3.3. Presented are graphs of the cumulative percentage distributions in Tables 3.4 and 3.6

mulative frequency and exit that class with a higher cumulative frequency, and we recognize that the increase occurs throughout the entire interval.

In conclusion, we can say that graphic techniques are an effective vehicle for presenting summaries of large sets of data. Yet they are also easily distorted and can be effective tools for misrepresenting data. References for and a brief discussion of data distortion through misuses of graphs and statistics can be found in Section 3.6.

EXERCISES

1. Construct graphs for the cumulative percentage distributions in Tables 3.3 and 3.5.
2. Use Table 2.8 to select a random sample of 25 observations from Table 3.1.
   (a) Form frequency, percentage, and cumulative distributions for the sample.
   (b) Draw graphs for frequency and cumulative percentage distributions.
3. Use Table 2.8 to select a random sample of 25 observations from Table 3.2.
   (a) Find the range of the data.
   (b) Form frequency, percentage, and cumulative percentage distributions for the sample.
4. Form distributions and draw graphs to summarize and present any data collected for Exercise 2 on page 77.
5. Secure two or more examples of data summarization reported in current news media (newspaper, news magazines, radio, or television). The presentation may be either by distribution or by graph.

3.4 DESCRIBING DISTRIBUTIONS

The preceding section demonstrates techniques for summarizing and preparing data for presentation by means of distributions and graphs. Frequently, we want to go beyond presentation of distributions and graphs to describe the distribution of a set of data. The descriptions may be used for the purpose of *data reduction* (summarizing the data by presenting key statistics) or for the purpose of *comparing one distribution with another distribution*. To achieve these goals, we proceed to describe distributions relative to their form, location, and dispersion.

FORMS OF DISTRIBUTIONS: When forming distributions for sets of data, there are many different *shapes*, *patterns*, or *forms* that appear. It is sometimes useful to describe different distributions by means of these shapes. Thus, we say that a distribution is *symmetric* if there is a value of X, say x_o, such that the frequency of X is the same for all values of X equidistant on either side of x_o. Any distribution that is not symmetric is called *asymmetric* or *skewed*. Figure 3.4 shows sketches of some symmetric and some asymmetric distributions. (Note that the lines sketching the frequency curves can be thought of as a continuous line drawn through the midpoints of the bars of a histogram as the numbers of bars and the total frequency become large.)

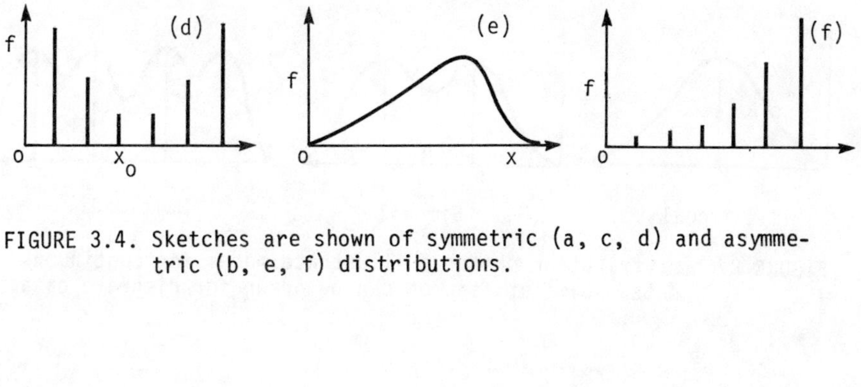

FIGURE 3.4. Sketches are shown of symmetric (a, c, d) and asymmetric (b, e, f) distributions.

Another way of describing the form of a distribution is to designate its shape. Figure 3.5 identifies some general shape terms used for distribution classification.

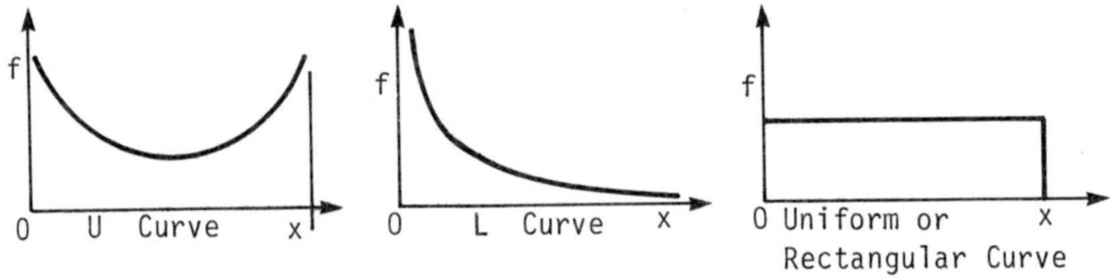

FIGURE 3.5. Sketches are shown of three types of distributional forms for continuous data. Similar figures can be drawn for discrete data.

A third method of distribution description is related to the number of *modes* in the distribution. A *mode* is a value of X, say x_m, such that those values of X adjacent to x_m on both sides are less frequent than x_m. Thus, a mode is a relative maximum frequency. A distribution having one mode is called a *unimodal* distribution. A *bimodal* distribution has two modes. Any distribution having more than two modes is said to be *multimodal*. Figure 3.6 presents sketches of distributions having various numbers of modes.

The terms introduced in this section are useful for describing the form of a distribution. However, they

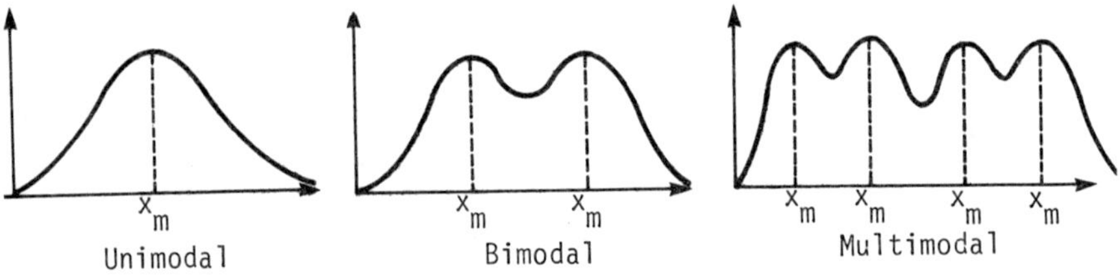

FIGURE 3.6. Distribution sketches illustrate modes for continuous data. Similar figures can be drawn for discrete data.

only describe the general shape of a data set. Therefore, we turn to more specific measures for describing distributions in the next section.

EXERCISE

1. Draw sketches to exhibit distributions having the following forms:
   (a) symmetric unimodal (discrete data),
   (b) symmetric unimodal (continuous data),
   (c) symmetric bimodal (continuous data), and
   (d) uniform (discrete data).

MEASURES OF LOCATIONS: In everyday conversation, the word, average, occurs frequently. For example, today's temperature is average for this time of year, last year's rainfall was slightly above average, Joe's score on the statistics examination was average, etc. Given a set of numerical data, there are many types of averages that can be calculated. Several so-called averages are presented in this section, as well as a discussion of their merits and their uses. However, before considering specific averages, we want to set forth the general ideas for calculating an average.

The following questions are frequently raised: What is the average family income for My Town, U.S.A.? What is the average cost per pound for sirloin steak? What is the average temperature for April 1st in Happy Corners? The implication in each question is that *one number* is desired for an answer. That number must in some sense represent all the numbers in a potential data set. That is, we seek *one number* to represent the size, magnitude, or value of the complete data set. Any number representing an average value for a data set is called a *measure of location*. The term, location, arises in that the type of representation we seek for the data is one which indicates where on the X scale (variable scale) the data points are located. In that light, the reasonable measures of location should in some sense yield middle values of the variable. With that brief introduction for background information, we now continue with our discussion of some specific measures of location.

ARITHMETIC MEAN: The most common and most widely used measure of location is the arithmetic mean. We use $\bar{x}$ to denote the arithmetic mean of a data set and n to denote the number of points in the set.

DEFINITION 1: The *arithmetic mean* for a set of observations, $x_1, x_2, \ldots, x_n$, is defined as

$$\bar{x} = \left(\sum_{i=1}^{n} x_i\right)/n. \qquad (3.4)$$

EXAMPLE 2: Find the arithmetic mean for the data selected for Table 3.3.

$$\bar{x} = \frac{3 + 3 + 4 + \ldots + 1 + 4 + 2}{25} = \frac{64}{25} = 2.56.$$

EXAMPLE 3: Find $\bar{x}$ for the data in Table 3.2.

$$\bar{x} = \frac{3 + 4 + 4 + 4 + 2 + \ldots + 3 + 2 + 2 + 3}{100} = \frac{252}{100} = 2.52.$$

The formula for the arithmetic mean as given in *Definition 1* refers to ungrouped data. Since most problems that require extensive calculations can be solved with the aid of a desk calculator or computer, Eq. (3.4) is the most frequently used method for obtaining $\bar{x}$. However, there are situations for which only a distribution of the data (rather that the individual observations) is available for analysis. When that is the case, Eq. (3.4) is unnecessarily clumsy to use. The following definition of $\bar{x}$ can be used for grouped data.

DEFINITION 2: If n observations are grouped into k classes such that the ith class has midpoint x_i and frequency f_i, then the arithmetic mean is

$$\bar{x} = \left(\sum_{i=1}^{k} f_i\, x_i\right)/n. \qquad (3.5)$$

EXAMPLE 4: Find $\bar{x}$ for the data in Table 3.5.

$$\overline{x} = \frac{(1)(1.5) + (5)(2.5) + (13)(3.5) + (6)(4.5)}{25} = \frac{86.5}{25} = 3.46.$$

EXAMPLE 5: Find $\overline{x}$ for the data in Table 3.3.

$$\overline{x} = \frac{5(1) + 7(2) + 8(3) + 4(4) + 1(5)}{25} = 2.56.$$

REMARKS: The arithmetic mean for discrete data is the same for grouped and ungrouped data (see *Examples 3* and *5*), but that is not true in general for continuous data. In grouping discrete data, we merely count the number of times each value of X occurs, but in continuous data all values in an interval are treated as being equal to the midpoint of the interval. (See *Exercise 1*, which follows.)

The relative frequency of x_i in a distribution is defined by

$$r_i = f_i/n, \tag{3.6}$$

so another formula for $\overline{x}$ when considering grouped data is

$$\overline{x} = \sum_{i=1}^{k} r_i \, x_i. \tag{3.7}$$

EXERCISES

1. Find $\overline{x}$ for the 25 observations on which Table 3.5 has been based. Compare the values of $\overline{x}$ from this exercise and from Example 4. Explain any differences observed in your comparison.
2. Find $\overline{x}$ for the observations in Table 3.6.
3. Select a random sample of 10 observations from Table 3.1. Find $\overline{x}$.
4. Select a random sample of 10 observations from Table 3.2. Find $\overline{x}$.
5. Find $\overline{x}$ for the data in Exercise 2 on page 89.
6. Find $\overline{x}$ for the data in Exercise 3 on page 89.
7. Find $\overline{x}$ for any data collected for Exercise 2 on page 77.

MEDIAN: The second most widely used measure of location is the *median*. In words, the median of a set of data is the middle value in the ordered data. Thus, for ungrouped data we define the median in terms of the order statistics as explained in *Definition 3*.

DEFINITION 3: If the order statistics of the data are $x_{(1)}$, $x_{(2)}$, ..., $x_{(n)}$, then the median is denoted by x_{md} and is defined to be

$$x_{md} = x_{[(n+1)/2]} \tag{3.8}$$

when n is an odd integer,

$$x_{md} = \left[x_{(n/2)} + x_{(n/2\ +\ 1)}\right]/2$$

when n is an even integer.

In *Definition 3*, we see that the median is the middle order statistic for the data sets having an odd number of elements and the mean of the *two* middle order statistics for data sets containing an even number of elements.

EXAMPLE 6: The number of automobiles sold per day by Reliable Used Cars for the week of October 23rd was 4, 2, 3, 5, 3. The week of October 30th showed sales of 1, 6, 4, 2, 5. Find the median daily sales for each week individually and for the combined two-week period.

SOLUTION: The ordered data for the week of October 23rd is; 2, 3, 3, 4, 5. Therefore, the median is

$$x_{[(5+1)/2]} = x_{(3)} = 3.$$

The ordered data for the week of October 30th is: 1, 2, 4, 5, 6. Therefore, the median is $x_{(3)} = 4$. The combined ordered data is: 1, 2, 2, 3, 3, 4, 4, 5, 5, 6. Therefore, $x_{(5)} = 3$ and $x_{(6)} = 4$. Thus, $x_{md} = (3+4)/2 = 3.5$.

EXAMPLE 7: Find the median for the sample, 3.5, 2.7, 0.9, 3.1, 4.3.

SOLUTION: The ordered sample is 0.9, 2.7, 3.1, 3.5, 4.3. Since n = 5, then (n+1)/2 = 3. That is, the median is the third term in the ordered sample. Thus, x_{md} = 3.1.

Definition 3 is not applicable in the determination of the median (middle value) of data that have been grouped into a distribution. Thus, we give the following definition to determine the median for grouped data.

DEFINITION 4: Any value, x_{md}, of a variable, X, such that at least 50% of the values of X are *less than or equal to* x_{md}, and at least 50% of the values of X are *greater than or equal to* x_{md}, is called a median for the distribution of X.

The use of *Definition 4* depends on the type of data being grouped. If the median is desired for a set of discrete data, then we simply look at the cumulative percentage distribution and take x_{md} to be the smallest value of x_i such that $c_i \geq 50$. Clearly the value selected in that manner satisfies *Definition 4*. However, if there exists an x_i such that $c_i = 50$, then there are two values of X, namely, x_i and x_{i+1}, which satisfy *Definition 4*. That case is analogous to the even sample size part of *Definition 3*, so that we use $x_{md} = (x_i + x_{i+1})/2$ as the unique value for the median of such a distribution.

EXAMPLE 8: Find the median for the data in Table 3.4.

SOLUTION: The cumulative distribution in Table 3.4 is as follows:

x_i	0	1	2	3	4	5
c_i	3	17	51	80	97	100

Thus, it follows that 51% of the values of X are less than or equal to 2. Furthermore, 83% of the values of X are greater than or equal to 2. So $x_{md} = 2$ by *Definition 4*. (Note that if c_i for 2 were 50 rather than 51, then both 2 and 3 would satisfy *Definition 4*. In that case, x_{md} would be $(2 + 3)/2 = 2.5$.)

The technique illustrated by *Example 8* for grouped discrete data is appropriate in that discrete data assume only elements from a set of distinct (and in many cases, integer) values. However, that property is not exhibited by grouped continuous data. In fact, we assume that the continuous variable, X, may assume any value in an interval. Now, with that in mind, we proceed to present an interpolation method for determining x_{md}, which is a good approximation to the median as defined in *Definition 4*. We begin by designating the first class for which $c_i \geq 50$ as the *median class*. That is, the class that includes the median value is the median class. After isolating the median class, we use interpolation within the median class to determine the median value, x_{md}. The required interpolation formula is

$$X_{md} = L_m + \{ [50 - c_{(m-1)}]/P_m\} W, \quad (3.9)$$

where m is the index for the median class, (m-1) is the index for the class preceding the median class, L_m is the lower limit of the median class, $c_{(m-1)}$ is the cumulative percentage of the class preceding the median class, $p_m = (100f_m/n)$ is the percentage for the median class, and W is the width of the median class. The following example illustrates the use of Eq. (3.9).

EXAMPLE 9: Find the median for the following data from Table 3.6.

Classes	x_i	p_i	c_i
0 - 1	.5	1	1
1 - 2	1.5	8	9
2 - 3	2.5	18	27
3 - 4	3.5	48	75
4 - 5	4.5	23	98
5 - 6	5.5	1	99
6 - 7	6.5	1	100

SOLUTION: For these data, the symbols in Eq. (3.9) are as follows. The median class is the interval, 3 - 4, where W = 1, $L_m = 3$, $p_m = 48$, $c_{(m-1)} = 27$. Therefore,

$$x_{md} = 3 + [(50-27)/48]\ (1) = 3 + (23/48) = 3.48.$$

REMARKS: By observing the c_i column, we see that 27% of the observations are less than 3 and 75% of the data are less than 4. Therefore, the desired 50% point must be somewhere between 3 and 4. To reach the desired 50% from the 27% accumulated prior to the median class, we must use 23% (50-27) of the 48% available in the median class.

Thus, the fraction $(50 - c_{m-1})/p_m$ represents that portion of the median class needed to reach the median (50%) value of X. That fraction, when multiplied by the width (W) of the median class, gives the proper adjustment to be added to the lower limit (L_m) of the median class in order to achieve the median, x_{md}.

The median for grouped data may also be determined graphically by using the cumulative percentage graph. To do so, we find that value of X which satisfies *Definition 4* by entering the c_i scale at 50 and reading the corresponding value, x_{md}, from the x scale. This method is illustrated by the following example.

EXAMPLE 10: Find the median for the two data sets presented in Tables 3.4 and 3.6.

SOLUTION: In Figure 3.3, we find the graphs for the cumulative percentage distributions, as repeated here.

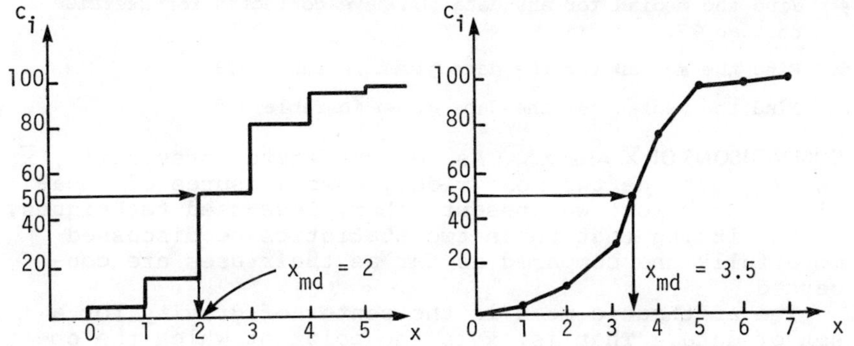

The precision with which we determine x_{md} by the graphic method described depends on the precision used in constructing the graphs for the cumulative percentage distribution. By using graph paper, we can obtain accuracy that is sufficient for most purposes. This method is particularly useful when a quick approximation for x_{md} is desired, and the necessary graph is already available from an earlier presentation of the data.

EXERCISES

1. Find the median of 1, 9, 6, 6, 3.
2. Find the median of 3, 1, 5, 8, 1, 8.
3. A check of 100 randomly selected income tax returns from My Town, U.S.A. revealed the following distributions for taxable incomes (yearly data).

Class intervals	x_i	p_i	c_i
Under $4,000	?	3	3
$4,000-$6,000	$5,000	12	15
$6,000-$8,000	$7,000	20	35
$8,000-$10,000	$9,000	36	71
$10,000-$12,000	$11,000	22	93
$12,000-$14,000	$13,000	5	98
Over $14,000	?	2	100

(a) Find the median for the income data.
(b) Can you find $\bar{x}$ for the data? Why or why not?

4. Find the median for any data you have collected for Exercise 2 on page 77.

5. Find the median for the data given in Table 3.3.

6. Find the median for the data given in Table 3.5.

COMPARISONS OF $\bar{X}$ AND X_{MD}: As we have stated previously, $\bar{x}$ and x_{md} are the two most widely used measures of location. So before we present other, less-used techniques, it is fitting that these two statistics be discussed more fully and compared as far as their uses are concerned.

The arithmetic mean is the center of gravity for a set of data. That is, $\bar{x}$ is the point at which the observations (of equal weight) will balance. Figure 3.7 depicts that property for 5 observations, 3, 1, 4, 7, 6. Another way of demonstrating that property is to observe that the sum of deviations between the observations and $\bar{x}$ is 0.

That is,

$$\sum_{i=1}^{n} (x_i - \bar{x}) = 0, \text{ when } \bar{x} = (\sum_{i=1}^{n} x_i)/n. \qquad (3.10)$$

Proof: $\sum_{i=1}^{n} (x_i - \bar{x})$

$$= \sum_{i=1}^{n} x_i - n\bar{x} \text{ by Eqs. (1.2) and (1.4)}$$

$$= \sum_{i=1}^{n} x_i - n\,[(\sum_{i-1}^{n} x_i)/n] \quad \text{by definition of } \bar{x}$$

$$= \sum_{i=1}^{n} x_i - \sum_{i=1}^{n} x_i = 0.$$

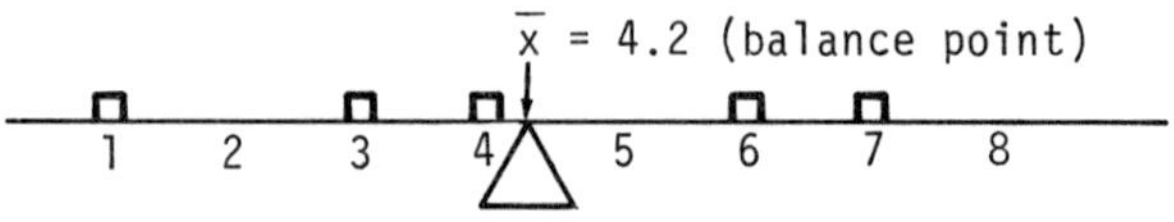

FIGURE 3.7. $\bar{x}$ is the balance point of the data.

For the preceding sample of 3, 1, 4, 7, 6, we have

$$\sum_{i=1}^{5} (x_i - \bar{x})$$

$$= (3 - 4.2) + (1 - 4.2) + (4 - 4.2) + (7 - 4.2) + (6 - 4.2)$$

$$= -1.2 - 3.2 - .2 + 2.8 + 1.8$$

$$= -4.6 + 4.6 = 0.$$

Corresponding to $\bar{x}$ splitting the data at the balance point, x_{md} divides the distribution of data into equal areas. That is, 50% of the data are smaller than x_{md}, and 50% are larger than x_{md}. Figure 3.8 gives an illustration of that property.

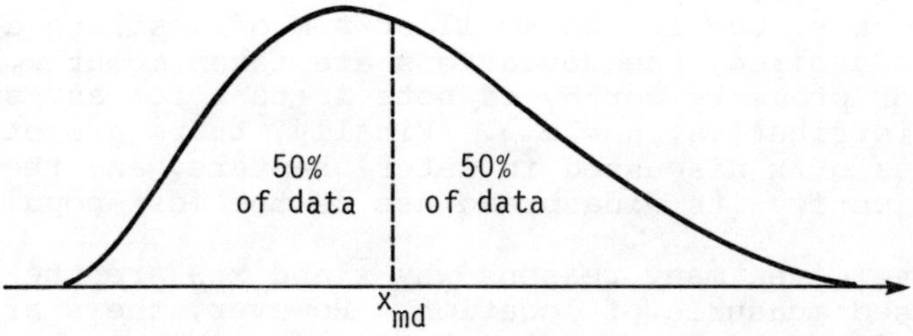

FIGURE 3.8. x_{md} is the midpoint of area in a distribution.

One consequence of the foregoing properties is that $\bar{x}$ is more sensitive to extreme values in the data than is x_{md}. For example, the largest observation in the data set, $x_{(n)}$, contributes the largest quantity, $x_{(n)}/n$, of any of the observations to $\bar{x}$, but it is equivalent to any other observation when calculating x_{md}. For illustration, suppose we consider the four numbers (1, 2, 4, 7) to determine $\bar{x} = 14/4 = 3.5$ and $x_{md} = (2+4)/2 = 3$. Now if we add one number to the data set to obtain (1, 2, 4, 7, 31), then the new values of $\bar{x}$ and x_{md} are $\bar{x} = 45/5 = 9$ and $x_{md} = 4$. We observe that $\bar{x}$ is no longer a middle value for the data set in that four of the five data points are less than $\bar{x}$, and only one point is greater than $\bar{x}$. Yet the effect of 31 on x_{md} is simply to add one more point to the data set. That lack of sensitivity to large (or small) extreme values by the median makes it a better central value (measure of location) than $\bar{x}$ for data distributions that are highly skewed. Huff [2] gives a good discussion of this property and suggests that skewed distributions of variables, such as age and income, are often more accurately described by medians than by arithmetic means.

Another comparison of $\bar{x}$ and x_{md} is provided by the following properties. First, the value of c (a constant) that minimizes

$$\sum_{i=1}^{n} (x_i - c)^2$$

is $c = \bar{x}$. Second, the value of K (a constant) that minimizes

$$\sum_{i=1}^{n} |x_i - K|$$

is $K = x_{md}$. Thus, if we want to get the smallest sum of squared deviations for a set of data, we take the deviations about $\bar{x}$, but if the smallest sum of absolute deviations is desired, the deviations are taken about x_{md}.

Another property worthy of note is that for any symmetric distribution, $\bar{x} = x_{md}$. Finally, there are other properties of $\bar{x}$ discussed in later chapters, and they tend to justify its widespread use as the most popular average.

We have given many reasons why $\bar{x}$ and x_{md} are the most widely used measures of location. However, there are types of data and types of averaging for which other statistics are more appropriate measures of location. Some of these methods are discussed in the subsequent development.

EXERCISES

1. Consider the data, 1, 3, 7, 9, 10.
   (a) Find $\bar{x}$, x_{md}.
   (b) Find

$$\sum_{i=1}^{5} (x_i - \bar{x})^2, \quad \sum_{i=1}^{5} (x_i - x_{md})^2.$$

   Is $\Sigma(x_i - \bar{x})^2 < \Sigma(x_i - x_{md})^2$?

   (c) Find

$$\sum_{i=1}^{5} |x_i - \bar{x}|, \quad \sum_{i=1}^{5} |x_i - x_{md}|.$$

   Is $\Sigma |x_i - \bar{x}| > \Sigma |x_i - x_{md}|$?

2. Suppose Bill, Jim, Ray, Carl, and John live at the indicated points along the scale of Interstate Highway 70.

Bill	Jim		Ray	Carl		John
0	50	100	200	300		500

(a) Where should a meeting be scheduled so that the total distance traveled is minimized?
(b) Where should a meeting be scheduled so that the total squared distance traveled is minimized?
(c) Which meeting place is more reasonable?

GEOMETRIC MEAN: Suppose the price of a commodity has been as follows during the past 3 years:

Year	2 years ago	Last year	This year
Price	\$2	\$4	\$2
% change from preceding year	—	200	50

It is clear that the current price is exactly the same as the price 2 years ago. However, when we find the average percent change in price by the arithmetic mean, we find that $\bar{x} = (50 + 200)/2 = 125\%$. Thus, we see that the arithmetic mean gives more weight to increases (doubling) than it does to decreases (halving). The appropriate statistic for averaging the preceding percentages is the *geometric mean*. For the positive observations, x_1 and x_2, the geometric mean is defined by

$$\bar{x}_g = \sqrt{x_1 \cdot x_2}\,. \tag{3.11}$$

Therefore, the geometric mean of 50 and 200 is $\bar{x}_g = \sqrt{(50)(200)} = \sqrt{10,000} = 100$. This example illustrates the fact that the geometric mean gives equal weight to increases and decreases when they are expressed as percentages. The general definition for the geometric mean follows.

DEFINITION 5: If $x_1, x_2, \ldots, x_n$ are positive observations, then their geometric mean is

$$\bar{x}_g = \sqrt[n]{x_1 \cdot x_2 \cdot \ldots, x_n}, \tag{3.12}$$

where $\sqrt[n]{\ }$ denotes the nth root of the quantity included.

In words, the geometric mean is the nth root of the product of the observations. In general, finding nth roots can best be accomplished by using logarithms. (Refer to Appendix A to review logarithms if necessary.) By taking logarithms,

$$\log \bar{x}_g = (1/n) \log (x_1 \cdot x_2 \cdot \ldots, x_n)$$
$$= (1/n) \sum_{i=1}^{n} \log x_i. \qquad (3.13)$$

Thus, the logarithm of the geometric mean is the arithmetic mean of the logarithms of the data. Thus, given a set of data, we first find the arithmetic mean of the logarithms of the data. Then we find $\bar{x}_g$ by taking the antilog of that arithmetic mean. The necessary steps are illustrated by the following example.

EXAMPLE 11: Find the geometric mean of the following percentages: 100, 90, 120, 130.

Calculation table

x	log x	
100	2.000	$\log \bar{x}_g = \frac{8.1473}{4} = 2.0368$
90	1.9542	
120	2.0792	
130	2.1139	$\therefore \bar{x}_g = 108.8$

Note that $\bar{x} = 440/4 = 110$ for the data shown in the preceding calculation table. Thus, $\bar{x}_g \leq \bar{x}$.

It can be shown in general that $\bar{x}_g \leq \bar{x}$ for all samples of positive observations. The equality occurs if and only if all the observations are identical. The difference between $\bar{x}$ and $\bar{x}_g$ increases as the observations vary more widely and is only slight for observations that exhibit a narrow range of values. Thus, it is questionable whether the added precision gained by using the geometric mean is offset by its additional arithmetic complexity over the arithmetic mean. Yet, $\bar{x}_g$ is the appropriate mean for averaging the change type percentages indicated at the beginning of this section.

EXERCISES

1. The changes in beef prices as a percentage of the preceding year are as follows for the last 5 years.

Year	1	2	3	4	5 (this year)
Percentage change	120	150	125	110	75

   Find $\bar{x}_g$. Find $\bar{x}$. Is $\bar{x}_g < \bar{x}$?

2. Compare $\bar{x}$ and $\bar{x}_g$ for the two numbers, $x_1 = 100$ and $x_2 = 10{,}000$.

HARMONIC MEAN: Suppose we drive 10 miles at 30 mph and another 10 miles at 60 mph. What is the average rate of speed for the 20-mile trip? Here the arithmetic and geometric means are

$$\bar{x} = 90/2 = 45 \text{ mph}, \quad \bar{x}_g = \sqrt{(30)(60)} = \sqrt{1800} = 42.4 \text{ mph.}$$

The question then is whether or not either $\bar{x}$ or $\bar{x}_g$ is appropriate for averaging rates. The answer is no in that neither the arithmetic mean nor the geometric mean allows for units of measures such as miles per hour, cents per dozen, or other rate values. In miles per hour, we must account for the time required to travel the 20 miles because we know that distance (d) = rate (r) × time (t). Therefore, r = d/t. Now traveling at 30 mph for 10 miles requires 20 minutes, and traveling 10 miles at 60 mph requires 10 minutes. So 30 minutes (.5 hours) are required for the trip. Thus, the average rate of travel is r = 20/.5 = 40 mph.

The appropriate statistic for averaging rates of travel for equal distances is the *harmonic mean.* For two positive observations, x_1 and x_2, the harmonic mean is

$$\bar{x}_h = 2/(1/x_1 + 1/x_2). \tag{3.14}$$

For the preceding hypothetical trip, we have

$$\bar{x}_h = 2/(1/30 + 1/60) = 2/(3/60) = 2(60/3) = 40 \text{ mph.}$$

The general definition for the harmonic mean of a data set is presented next.

DEFINITION 6: If $x_1, x_2, \ldots, x_n$ are positive observations, the harmonic mean is

$$\bar{x}_h = n/\left(\sum_{i=1}^{n} \frac{1}{x_i}\right). \tag{3.15}$$

In words, the harmonic mean is the reciprocal of the arithmetic mean of the data reciprocals. That is,

$$1/\bar{x}_h = \left(\sum_{i=1}^{n} 1/x_i\right)/n,$$

where the material on the right side of the equal sign is simply the arithmetic mean of the reciprocals of the data.

EXAMPLE 12: Find the harmonic mean of 3, 4, and 5.

$$\bar{x}_h = 3/(1/3 + 1/4 + 1/5) = 3/(47/60) = 180/47 = 3.83.$$

The use of the harmonic mean is appropriate for averaging rates when equal distances are traveled at different rates, when equal amounts of money are spent for purchasing items in bulk at different rates, etc. The important requirement is that equal amounts of money, equal distances, etc., occur at each of the different rates.

In general, it can be shown that for positive observations, $\bar{x}_h \le \bar{x}_g \le \bar{x}$, where the equality holds if and only if all the observations are identical.

EXERCISES

1. Suppose BNRL (Build Now, Repair Later) Company purchases items in \$1,000 orders. If the prices per item for the last 4 orders are \$1.50, \$2.00, \$1.75, \$2.25, what is the average cost per item for the last 4 orders?

2. Consider a hill which is one mile up and one mile down. Assume that instant acceleration to any desired speed is possible at the top of the hill. If the average speed going up the hill is 30 mph, what average speed is needed for going down hill to produce an average speed of 60 mph while traveling over the hill?

MIDRANGE: Sometimes we want to obtain a quick measure of location. That is to say, we do not want to perform the necessary arithmetic or ordering to find $\bar{x}$, $\bar{x}_g$, $\bar{x}_h$, or x_{md}, but we still need an indication of an average value for the variable being studied. One such quick statistic is called the *midrange*.

DEFINITION 7: If $x_{(1)}$ = minimum $(x_1, x_2, \ldots, x_n)$ and $x_{(n)}$ = maximum $(x_1, x_2, \ldots, x_n)$, then the midrange of the sample is

$$x_{mr} = \left[x_{(n)} + x_{(1)}\right]/2. \qquad (3.16)$$

Thus, we can say that the midrange of a sample is the arithmetic mean of the largest and smallest observations. One disadvantage of the midrange is that it is entirely dependent on extreme values, so it is likely to fluctuate a great deal from sample to sample. One positive aspect of the midrange is its ease of computation. That property allows us to quickly assess the level or location of a data set without needless delays for arithmetic operations.

EXAMPLE 13: Find the midrange for the data in Table 3.1 on page 78. Upon inspection, we see that $x_{(1)} = 0.9$ (item #67) and $x_{(n)} = 6.3$ (item #69). Therefore, $x_{mr} = (6.3 + .9)/2 = 3.6$.

EXERCISES

1. Find the midrange for the data in Table 3.4 on page 81. How does x_{mr} compare to $\bar{x}$ and x_{md}?
2. Find the midrange for the numbers 7, 2, 6, 5, 9, 10.

CONCLUSION: In the preceding development, five different statistics are presented as measures of location for sets of observations. None of these techniques is appropriate for all types of data. Yet the most frequently used measures are $\bar{x}$ and x_{md}. The measures presented are sometimes called measures of central tendency in that all of them produce a center value of the variable.

At this point, we can review our objectives for this chapter. A set of statistics is desired by which we can represent (describe) a set of data. The first statistic included in the set of descriptive statistics is a measure of location, usually $\bar{x}$ or x_{md}. Our philosophy is that we can summarize our data as far as location (size) is concerned by using *one key statistic*. Thus, we have reduced our data to one number that represents the average size of the variable. All of the statistics presented in the preceding sections produce values that are within the range of data points. Hence, the one number computed to represent all numbers relative to size is a middle value for the data.

The representation of a distribution or set of data by one measure of location can distinguish distributions *only* with respect to location. Although there are many distributions that have the same value of $\bar{x}$, they are quite different distributions. Figure 3.9 illustrates six different distributions that have $\bar{x} = 4$.

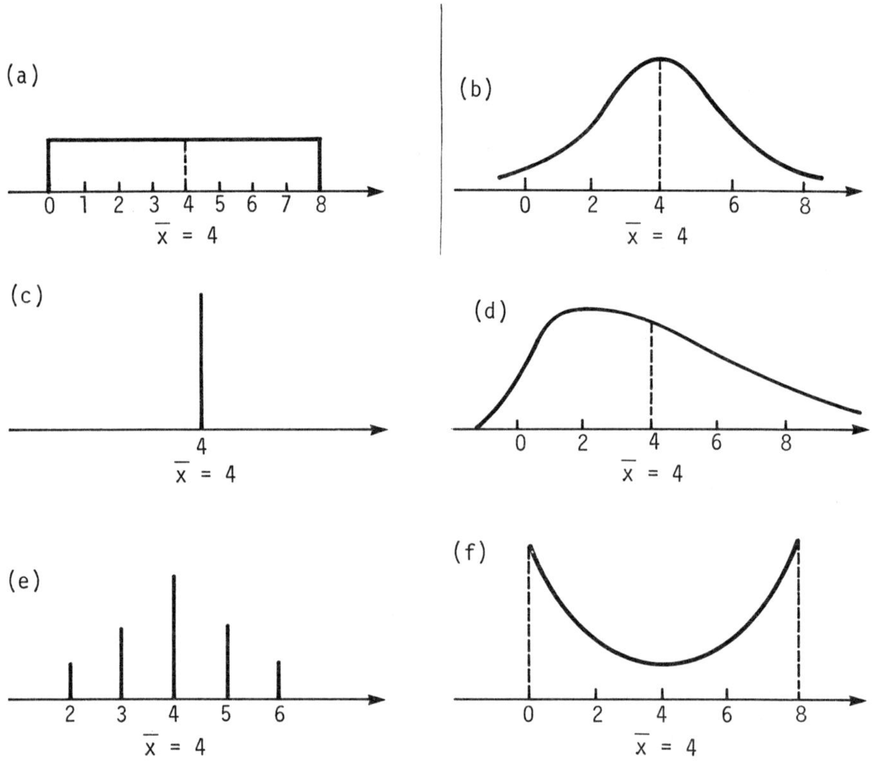

FIGURE 3.9. Six different distributions with $\bar{x} = 4$ are shown here.

To distinguish distributions from one another, such as those in Figure 3.9, we need more than a descriptive statistic for location. Therefore, we ask for a second descriptive statistic to represent the amount of variation present in the data. For example, there is no variation among sample points selected from (c) in Figure 3.9 (all values are 4), but there is much variation among samples from (f) since the values away from 4 occur with high frequencies. Thus, the next step in our study is to develop measures of variation that can be used to distinguish distributions from one another when they have the same mean.

MEASURES OF DISPERSION: Our attention is now focused on the determination of statistics that represent the dispersion or the spread of the elements in a sample or population. In this section the range, variance, and standard deviation are discussed. There are other measures, to be sure, but these are the most common ones.

RANGE: As defined in Eq. (3.3), the range of a data set is the difference between the largest and smallest observations. In terms of $x_{(1)}$ and $x_{(n)}$ as defined by Eq. (3.1), we have

$$R = x_{(n)} - x_{(1)}.$$

Equation (3.3) applies directly for ungrouped data but must be given some interpretation for grouped data. The usual interpretation for grouped data is that

$$R = U_k - L_1, \tag{3.17}$$

where U_k is the upper limit of the kth (last) class, and L_1 is the lower limit of the first class. One consequence of that interpretation is that the range for a grouped set of data is always greater than the range of the same data ungrouped. For example, the data in Table 3.1 show $x_{(n)} = 6.3$ and $x_{(1)} = 0.9$, so $R = 6.3 - .9 = 5.4$ for the ungrouped data, but $R = 7 - 0 = 7$ for the grouped data as presented in Table 3.6.

The use of the range is a quick method for comparing the variability in two or more samples or populations. If the range of one data set is greater than the range of another data set, we say that the first set has greater dispersion (variation) than the second set.

VARIANCE AND STANDARD DEVIATION: Since we have available measures of location for any data set we wish to describe, it is reasonable to discuss the dispersion of the data relative to a location statistic. The most common measure of dispersion is called the *variance*. The variance for ungrouped data is defined as follows.

DEFINITION 8: If $x_1, \ldots, x_n$ represents a set of n observations on a variable X, the variance of X is denoted by s^2 or V(X) and is defined by

$$s^2 = \sum_{i=1}^{n} [(x_i - \bar{x})^2]/(n-1), \text{ where } \bar{x} = (\sum_{i=1}^{n} x_i)/n. \tag{3.18}$$

REMARKS: Recall that from Eq. (3.10),

$$\sum_{i=1}^{n} (x_i - \bar{x}) = 0,$$

so that quantity is the same for all distributions and is of no value in describing distributions. Hence, the deviations $(x_i - \bar{x})$ are squared before being summed. That way Eq. (3.18) gives a different measure for different sets of data.

The division by (n - 1), rather than n, in Eq. (3.18) is for subtle theoretical reasons which are in part illustrated in Chapters 6 and 7.

If the observations, x_i and $\bar{x}$, are integers, then the use of Eq. (3.18) is convenient for calculating a variance. However, when fractions and decimals are involved, the following derivation provides a more useful formula given in Eq. (3.19).

Derivation: By the rules of Section 1.6, we find that

$$\Sigma(x_i - \bar{x})^2 = \Sigma x_i^2 - 2\bar{x}\Sigma x_i + n\bar{x}^2 .$$

Using $\bar{x} = \Sigma x_i/n$, we can write

$$\Sigma(x_i - \bar{x})^2 = \Sigma x_i^2 - [(\Sigma x_i)^2/n].$$

Therefore,

$$s^2 = \{\Sigma x_i^2 - [(\Sigma x_i)^2/n]\}/(n - 1). \qquad (3.19)$$

EXAMPLE 14: Use Eqs. (3.18) and (3.19) to find the variance of 2, 4, 6, 8, 10.

SOLUTION: Using Eq. (3.18),

$$s^2 = \Sigma(x_i - 6)^2/4 = [(-4)^2 + (-2)^2 + 0^2 + 2^2 + 4^2]/4$$

$$= 40/4 = 10.$$

Using Eq. (3.19),

$$\Sigma x_i = 2 + 4 + 6 + 8 + 10 = 30,$$

$$\Sigma x_i^2 = 4 + 16 + 36 + 64 + 100 = 220,$$

$$s^2 = \{220 - [(30)^2/5]\}/4 = (220 - 180)/4 = 40/4 = 10.$$

When defining the variance for grouped data, we use the frequency and the relative frequency symbols in the following definition.

DEFINITION 9: If n observations are grouped into k classes such that the ith class has midpoint x_i, frequency f_i, and relative frequency $r_i = f_i/n$, then the variance of the data set is

$$s^2 = [\sum_{i=1}^{k} f_i(x_i - \overline{x})^2]/(n - 1)$$

$$= [n/(n-1)] \sum_{i=1}^{k} r_i(x_i - \overline{x})^2, \qquad (3.20)$$

where

$$\overline{x} = \sum_{i=1}^{k} f_i x_i/n = \sum_{i=1}^{k} r_i x_i.$$

REMARKS: Since $f_i = nr_i$ by definition, the two forms of s^2 in Eq. (3.20) are identical.

For computational convenience, it is useful to prove the following result:

$$s^2 = \{\Sigma f_i x_i^2 - [(\Sigma f_i x_i)^2/n]\}/(n - 1)$$

$$= [n/(n - 1)](\Sigma r_i x_i^2 - \overline{x}^2). \qquad (3.21)$$

Proof:

Using the summation rules in Section 1.6, we find that

$$\Sigma f_i(x_i - \overline{x})^2 = \Sigma f_i x_i^2 - 2\overline{x}\Sigma f_i x_i + \overline{x}^2 \Sigma f_i.$$

Using $\Sigma f_i = n$ and $\overline{x} = \Sigma f_i x_i/n$, we have

$$\Sigma f_i(x_i - \overline{x})^2 = \Sigma f x_i^2 - [(\Sigma f_i x_i)^2/n].$$

That proves the first part of Eq. (3.21). The second part follows by using $\Sigma r_i = 1$ in a similar set of steps.

The computations necessary to apply the variance formula for grouped data can be completed in a computation table. The computational procedure is illustrated by the following example.

EXAMPLE 15: Find the variance for the data in the following table.

Computation table						
Classes	x_i	f_i	$f_i x_i$	fx_i^2	$(x_i - \overline{x})^2$	$f_i(x_i - \overline{x})^2$
0 - 2	1	3	3	3	16	48
2 - 4	3	7	21	63	4	28
4 - 6	5	10	50	250	0	0
6 - 8	7	7	49	343	4	28
8 - 10	9	3	27	243	16	48
Totals		30	150	902		152

Equation (3.5) gives $\overline{x} = 150/30 = 5$. By Eq. (3.20), $s^2 = 152/29 = 5.24$. By Eq. (3.21),

$$s^2 = \{902 - [(150)^2/30]\}/29 = 152/29 = 5.24.$$

REMARK: The preceding example demonstrates the use of Eqs. (3.20) and (3.21). By inspection, it seems that Eq. (3.20) requires less arithmetic than Eq. (3.21). However, if $\overline{x}$ were a fraction rather than an integer, the reverse would be true.

The variance is a descriptive statistic that yields a measure of how the elements in a data set are dispersed around their mean. If the items vary over a wide range of values, the variance is large because some of the deviations, $(x_i - \overline{x})$, are large in magnitude. These large differences are then squared so their magnitude is amplified to produce a large s^2. On the other hand, squaring small differences (less than 1) makes them even smaller, so the variance is quite small for homogeneous points which are all close to or equal to their mean. Thus, the variance discriminates among distributions by the way in which the elements of the data vary around their mean (a location descriptive statistic).

One awkward aspect about the variance is that it is a squared quantity. That is, whenever X is measured in pounds, the variance is expressed in squared pounds, which is an unfamiliar unit of measure. So it is useful to obtain a descriptive statistic that is expressed in units of the variable for describing the dispersion in our data. The most common statistic for that purpose is the *standard deviation*, which is defined as follows.

DEFINITION 10: The standard deviation of a data set is the positive square root of its variance. Thus,

$$s = \sqrt{s^2} \tag{3.22}$$

is the standard deviation of a data set having variance s^2.

EXAMPLE 16: Find the standard deviation for the data in Example 15.

SOLUTION: $s = \sqrt{5.24} = 2.29$.

REMARKS: If the variance of a variable is greater than one, then the standard deviation of the variable is less than its variance. However, the opposite result is true when the variance is less than one.

In summarizing data, it is common to use two descriptive statistics. These statistics are usually the arithmetic mean and the standard deviation (or the variance). For most data, the arithmetic mean as a measure of location and the standard deviation as a measure of dispersion summarize in two numbers much of the information about the variable being studied.

A part of the usefulness of $\bar{x}$ and s as descriptive statistics can be argued as follows. First, $\bar{x}$ indicates the average size of the variable X being studied. Second, s is a measure of how the observations vary around $\bar{x}$. To better interpret the meaning of dispersion around $\bar{x}$, we use the Chebychev inequality statement that follows.

STATEMENT OF THE CHEBYCHEV INEQUALITY: If $\bar{x}$ and s are the arithmetic mean and standard deviation, respectively, for a set of data, then the relative frequency of data, which are within k standard deviations of $\bar{x}$, is greater than or equal to $1 - (1/k^2)$, where $k > 1$. In symbols, we have

$$\#\text{ (observations in the interval, } I_k)/n \geq 1 - (1/k^2), \tag{3.23}$$

where I_k is the interval from $\bar{x} - ks$ to $\bar{x} + ks$.

REMARKS: For most data sets, the bound given by (3.23) is a loose bound, but it can be shown that it is impossible to obtain a better bound that holds for *all* data sets.

Since the left-side quantity in (3.23) is always greater than or equal to 0, the Chebychev inequality is useful only for $k > 1$. In particular, (3.23) states that at least 3/4 (75%) of all data elements are in the interval $(\bar{x} - 2s, \bar{x} + 2s)$. Also, at least 8/9 (≈89%) of all data elements are in the interval $(\bar{x} - 3s, \bar{x} + 3s)$.

By putting the two preceding remarks together, we can give some rule-of-thumb guidelines useful in interpreting the magnitude and range of a data set when the descriptive statistics $\bar{x}$ and s are available. That is to say, for most of the data sets we consider in our studies and for data sets collected in most practical situations, we can expect to find the following:

$$
\begin{aligned}
&50 - 75\% \text{ of the data between } \bar{x} - s \text{ and } \bar{x} + s,\\
&75 - 95\% \text{ of the data between } \bar{x} - 2s \text{ and } \bar{x} + 2s,\\
&90 - 100\% \text{ of the data between } \bar{x} - 3s \text{ and } \bar{x} + 3s.
\end{aligned}
\tag{3.24}
$$

More examples to illustrate the use of $\bar{x}$ and s for distinguishing one distribution from another are provided in subsequent chapters, so we conclude this section with an example to illustrate (3.24).

EXAMPLE 17: Refer to the data in Example 15 and the results in Examples 15 and 16 to calculate and interpret the intervals $(\bar{x} - s, \bar{x} + s)$, $(\bar{x} - 2s, \bar{x} + 2s)$, and $(\bar{x} - 3s, \bar{x} + 3s)$.

SOLUTION: From Examples 15 and 16, we obtain $\bar{x} = 5$ and $s = 2.29$. Therefore, the desired intervals are $(\bar{x} - s, \bar{x} + s) = (2.71, 7.29)$, $(\bar{x} - 2s, \bar{x} + 2s) = (.42, 9.58)$, and $(\bar{x} - 3s, \bar{x} + 3s) = (-1.87, 11.87)$. A rough interpolation within the classes of Example 15 shows that out of 30 observations there are approximately 18 observations (60%) between 2.71 and 7.29, 28 observations (93%) between .42 and 9.58, and 30 observations (100%) between -1.87 and 11.87. Thus, the percentages given in (3.24) are satisfied by the data for Example 15.

EXERCISES

1. Find the range, variance, and standard deviation of the following sample: 2, 6, 10, 14, 18.

2. Select a random sample of size 10 from Table 3.1, and find the range, variance, and standard deviation for the sample.
3. Repeat *Exercise 2* for 5 different samples, or compare the results for *Exercise 2* with similar results for 5-10 classmates.
4. Find the arithmetic mean, range, variance, and standard deviation for each of the following samples.

(a)

Classes	f_i
0 - 2	10
2 - 4	4
4 - 6	4
6 - 8	10

(b)

Classes	f_i
0 - 2	4
2 - 4	10
4 - 6	10
6 - 8	4

(c)

Classes	f_i
0 - 2	7
2 - 4	7
4 - 6	7
6 - 8	7

(d)

Classes	f_i
0 - 8	28

5. Rank the 4 distributions in *Exercise 4* relative to the amount of variation (low to high) with respect to range, variance, and standard deviation.
6. Evaluate the intervals $(\bar{x} - s, \bar{x} + s)$ and $(\bar{x} - 2s, \bar{x} + 2s)$ for the data in parts (a), (b), and (c) in *Exercise 4*. Do the suggested percentages in (3.24) hold for these distributions?
7. If a set of exam scores shows $\bar{x} = 75$ and $s = 10$, suggest limits between which we might expect to find the following percentages of scores: 60-75%, 85-95%, and 100%.
8. Find the range, variance, and standard deviation for any data available for *Exercise 2* on page 77. Combine those statistics with the earlier mean values obtained in *Exercise 7* on page 93 and check to see if the results in (3.24) hold for the data sets.

COEFFICIENT OF VARIATION: Are the weights of fleas more variable than the weights of elephants? On the surface, we observe that fleas in general cannot vary a great deal in weight. That is, the range of possible weights (in pounds) for fleas is very narrow. However, the range of possible weights (in pounds) for adult elephants is quite wide. Thus, we expect the variance for a sample of flea weights to be smaller than the variance for a sample of elephant weights. In answering questions of this type, we desire a measure of variability relative to the size of the observations. The measure should be a pure number in that the result should not depend on any particular unit of measure. The *coefficient of variation* is such a statistic and is defined as follows.

DEFINITION 11: If $\overline{x}$ and s are the mean and standard deviation, respectively, for a set of positive observations, then the *coefficient of variation* is denoted by C and defined to be

$$C = [(s/\overline{x})\ 100]\%. \tag{3.25}$$

REMARKS: Since s and $\overline{x}$ have the same units of measure, the ratio $s/\overline{x}$ is a pure number. Thus, we express the coefficient of variation as a percent. In words, C is a measure of the variability in a variable relative to the arithmetic mean of the variable. C can be used to compare the variabliity relative to the arithmetic mean for two unlike groups of data, e.g., weights of fleas and elephants.

EXAMPLE 18: Suppose a sample of 10 fleas gave $\overline{x}$ = .001 pounds and s = .0001 pounds, while a sample of 10 cow elephants produced $\overline{x}$ = 4000 pounds and s = 500 pounds. Find C_f, the coefficient of variation for fleas, and C_e, the coefficient of variation for elephants. Interpret the results.

SOLUTION: $C_f = [(.0001/.001)\ 100]\ \% = 10\%;$

$$C_e = [(500/4000)\ 100]\ \% = 12.5\%.$$

Thus, the variability of elephant weights is 12.5% of the mean weight, while the variability of the flea weights is only 10% of their mean weight. Thus, when the variability for weights of fleas and elephants is adjusted for different mean values and expressed as a percentage of their mean values, we observe comparable coefficients of variations.

EXERCISES

1. Find the coefficient of variation for 2, 6, 10, 14, 18.
2. Find the coefficients of variation for the distributions in parts (a), (b), and (c) of *Exercise 4* on page 113.

3. Find the coefficient of variation for sample data secured for *Exercise 2* on page 77.

CONCLUSION: The purpose of this section has been to present some descriptive statistics that illustrate the dispersion among elements of a data set. Thus, by combining the techniques in the section, *Measures of location* (page 91), and in the section, *Measures of dispersion* (page 107), we find several choices for pairs of descriptive statistics such that one element of the pair describes location, while the other element describes variability. However, the most widely used pair of statistics is $(\bar{x}, s)$. Sometimes s is replaced by s^2, but that change is simply a matter of personal preference since both values offer the same type of information about the data set.

In Figure 3.9, we present graphs of several different distributions that have the same value of $\bar{x}$. Figure 3.9 provides our motivation for including a measure of dispersion in our descriptive statistics. In a similar fashion, it is possible (but more difficult than the case in Figure 3.9) to exhibit two or more distributions that have the same mean and variance, so that our pair of descriptive statistics $(\bar{x}, s)$ would not distinguish any one of the distributions from the other distributions. In such cases, we can define higher order statistics based on $\Sigma(x_i - \bar{x})^3$ and $\Sigma(x_i - \bar{x})^4$ to use as additional descriptive statistics for comparing the distributions. Those techniques are not as widely used as the statistics already discussed and are not discussed here, but further information concerning them is available in Perles and Sullivan [3].

*3.5 CODING DATA BY LINEAR TRANSFORMATION

Whenever computers and calculators are not available for doing the arithmetic required for statistical analyses, it is sometimes desirable to transform the data so that the arithmetic for calculating descriptive statistics can be performed on smaller numbers. The usual coding is a linear transformation of the form,

$$y_i = A + Bx_i, \tag{3.26}$$

where A and B are constants chosen for convenience of calculations. If we use $\bar{y}$ and s_y^2 to denote the mean and variance of y, and $\bar{x}$ and s_x^2 to denote the mean and variance of x, respectively, then

$$\bar{x} = (\bar{y} - A)/B, \tag{3.27}$$

$$s_x^2 = s_y^2/B^2, \quad s_x = s_y/B. \tag{3.28}$$

Proof: Summing both sides of Eq. (3.26) and dividing the result by n give

$$\Sigma y_i/n = nA/n + B(\Sigma x_i/n), \quad \bar{y} = A + B\bar{x}. \tag{3.29}$$

Solving Eq. (3.29) for $\bar{x}$ gives Eq. (3.27). By Eq. (3.18),

$$(n - 1)\, s_y^2 = \Sigma\,(y_i - y)^2 = \Sigma\,[(A + Bx_i) - (A + B\bar{x})]^2$$

$$= \Sigma\,[B(x_i - x)]^2 = B^2\, \Sigma(x_i - x)^2.$$

Therefore,

$$s_y^2 = B^2\, s_x^2. \tag{3.30}$$

Solving Eq. (3.30) for s_x^2 gives Eq. (3.28).

Equations (3.27) and (3.28) provide the necessary formulas for decoding the descriptive statistics calculated for the coded data. That is, we can apply Eq. (3.26) to transform the data on X to a set of data on Y, then use *Definitions 1* and *8* to calculate $\bar{y}$ and s_y^2. These values can then be decoded by Eqs. (3.27) and (3.28) to give the desired $\bar{x}$, s_x^2, and s_x for the original data. Notice that the addition of a constant, A, does not change the variance of the data. It simply means that changing the location of a variable by an additive constant does not change the dispersion of the elements in the sample.

EXAMPLE 19: Find $\bar{x}$ and s_x^2 for 500, 750, 1000, 1250, 1500.

SOLUTION: First, find $y_i = -4 + (1/250)x_i$ to obtain (-2, -1, 0, 1, 2). Second, $\bar{y} = (-2-1+0+1+2)/5 = 0$ and $s_y^2 = (4+1+0+1+4)/4 = 10/4 = 2.5$. Third, use Eq. (3.27) to find $\bar{x} = 250(0 + 4) = 1000$, and use Eq. (3.28) to find $s_x^2 = (250)^2\,(2.5) = 156{,}250$.

EXERCISES

1. Suggest values for A and B to code the following sets of data. Find $\bar{x}$ and s_x^2 for the sets directly and by Eqs. (3.27) and (3.28).
   (a) X: {3, 7, 11, 14}.
   (b) X: {-1/2, -1/4, 0, 1/4, 1/8}.
   (c) X: {2.7, 2.4, 2.1, 3.0}.
2. Is coding useful in calculating descriptive statistics for any data collected for *Exercise 2* on page 77? If so, find the mean and variance by coding the data.

3.6 MISUSES OF STATISTICS

Much of the popular criticism of statistics is due to the misuse or abuse of legitimate statistical techniques. In fact, statistics has gained a rather infamous reputation over the years, and that reputation is mostly due to misuse of statistics. Furthermore, statistical misuses have probably been the basis for some of the following catchy phrases coined by various people. For example, Benjamin Disraeli is reported to have said, "There are three kinds of lies - lies, damned lies, and statistics." During the late 1960s, a Blue Cross Blue Shield commercial began with, "Statistics are dull ... ." Other comments about statistics include the following: "Statistics can prove anything," "Figures don't lie, but liars figure," and even, "Statistics are like a bikini - what they reveal is enticing, but what they conceal is vital." These ideas and comments make interesting reading. They also point out the fact that statistics can be, and are, misused. The primary objective of this text is to present correct uses of statistics. However, we take a brief aside here to give a few examples of statistical misuses and abuses.

The standard procedure for misuses of statistics is to misrepresent the data through distorted graphs, misleading numbers, and deceptive reporting. Two of the most frequently used methods for graph distortions are called no scale and no zero. For the no-scale technique, we simply present a graph where one or both scales are not labeled. The no-zero presentation requires the omission of zero on one or both scales and frequently uses an expanded scale to imply rapid and large increases. Figure 3.10 presents an example of each type of graph distortion.

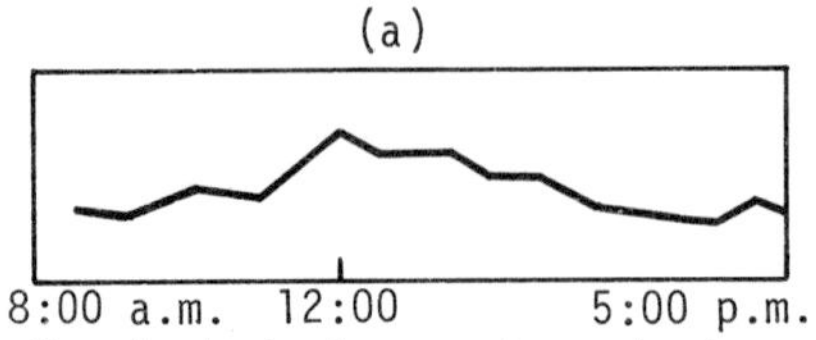

Hourly body temperature during the day for a hospital patient.

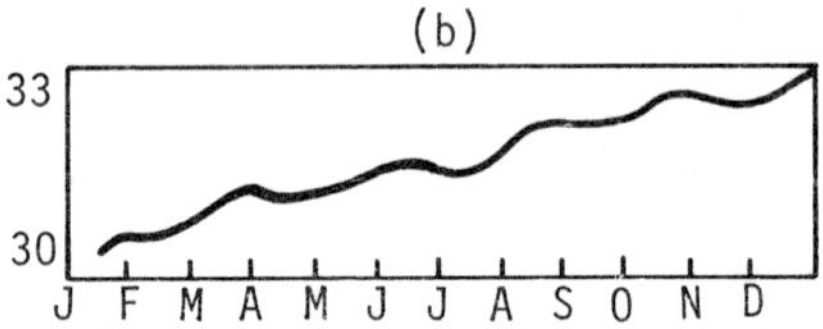

Company sales for last year in thousands of dollars.

FIGURE 3.10. Two examples of graph distortions show the no-scale and no-zero techniques.

In part (a) of Figure 3.10, we cannot determine what the temperature is without a vertical scale. From the graph, it appears that the temperature increases until noon and then tapers off, but it is impossible to determine the magnitude of the increase without a scale. For part (b) of Figure 3.10, a quick glance might indicate a tremendous increase in sales when, in reality, the increase is less than 10%. Notice that a zero for the vertical scale of part (b) would require a graph at least 11 times as high as the presented one.

Another frequently employed misuse of statistics is to present percentages for very small sample sizes. For example, it is correct (but misleading) to say that a Ph.D. degree has been earned by 50% of the individuals who entered the first grade with the author. Certainly, stating the 50% figure is more impressive than is the fact that one of the two people who entered the first grade with the author has earned a Ph.D. Huff [2] quotes a similar statistic by saying that during the first year that The Johns Hopkins University was coeducational, 33 1/3% (1 out of 3) of the coeds married faculty members. Those 2 examples show that percentages can be misleading for small samples.

Choosing the sample that tells the desired story is also a method for misrepresenting data. A common phrase used in advertising is, "9 out of 10 doctors prefer ...." The advertiser hopes the reader will interpret the statement to mean that 9 out of *every* 10 doctors prefer .... Of course, by selecting the responses received, we can get 9 out of 10, 6 out of 10, or even 10 out of 10, so the statement really means nothing. Other types of number games in advertising include the use of irrelevant figures, such as 32% fewer cavities or 20% more energy. These figures are used without any reference to a comparative base. For example, it may not be difficult to find one child who has 32% fewer cavities than another child, but to find a toothpaste that shows 32%

fewer cavities than any other toothpaste when tested on two independent and similar groups of children is quite another matter.

Using the wrong average is another standard distortion procedure. For example, including two or three retired senior citizens who go to college to begin degree work could result in showing an average age of students in a freshman English class to be some unreasonable age such as 24 years. The preceding data could result from the use of an arithmetic mean, while the median for the same class could give a more representative value of 18 years.

Thus, there are many ways in which data can be misrepresented through the use of graphs, percentages, and descriptive statistics. Some of these techniques can be humorous, and many provide new insights into the basic ideas of statistics. We encourage the reader to refer to one or more of the following excellent references for more detailed discussions of statistical misuses: Huff [2], Moroney [4], Wallis and Roberts [5], and Wheeler [6].

3.7 SUMMARY

The purpose of this chapter has been to present techniques for use in distinguishing among data sets. Descriptive statistics for measures of location and dispersion are given. Finally, some common misuses of descriptive statistics are provided. The ideas of descriptive statistics in this chapter and the concepts of probability from Chapter 2 are combined in Chapter 4 to present and describe some basic discrete probability models.

3.8 REFERENCES

[1] Sturges, H. A. 1926. The choice of a class interval. *Journal of the American Statistical Assoc.* 21: 55-56.

[2] Huff, D. 1954. *How to lie with statistics.* New York: W. W. Norton & Co., Inc.

[3] Perles, B., and Sullivan, C. 1969. *Freund and Williams' modern business statistics.* Englewood Cliffs, N. J.: Prentice-Hall, Inc., Chapter 5.

[4] Moroney, M. J. 1956. *Facts from figures.* Baltimore: Penguin Books, Inc.

[5] Wallis, W., and Roberts, V. 1965. *The nature of statistics.* New York: The Free Press, Chapter 4.

[6] Wheeler, M. 1976. *Lies, damn lies, and statistics: the manipulation of public opinion in America.* New York: Liveright.

3.9 REVIEW EXERCISES

1. Table 3.7 presents a data set of 100 hypothetical observations on a discrete variable, X.
   (a) Form frequency and cumulative percentage distributions for the data in Table 3.7.
   (b) Graph the frequency and cumulative percentage distributions for part (a).
   (c) Find the arithmetic mean, median, and midrange for X.
   (d) Find the variance, standard deviation, and range for X.
   (e) Describe the form of the frequency distribution of X.
   (f) Select a random sample of 25 observations from Table 3.7, and repeat parts (a)-(e) for the sample.
   (g) Compare the answers to parts (c)-(e) for the full data set and the sample.

TABLE 3.7
A hypothetical population of 100 observations is shown on a discrete variable, X.

Label	x	Label	x	Label	x	Label	x
00	6	25	3	50	0	75	2
01	1	26	1	51	5	76	0
02	2	27	2	52	4	77	1
03	3	28	2	53	2	78	4
04	3	29	1	54	1	79	2
05	3	30	3	55	3	80	2
06	0	31	5	56	1	81	5
07	3	32	2	57	4	82	1
08	1	33	2	58	3	83	2
09	1	34	2	59	6	84	3
10	4	35	1	60	2	85	4
11	4	36	4	61	3	86	2
12	1	37	1	62	1	87	3
13	1	38	3	63	3	88	5
14	0	39	0	64	4	89	7
15	5	40	2	65	4	90	2
16	1	41	5	66	2	91	4
17	2	42	1	67	3	92	4
18	3	43	2	68	2	93	1
19	1	44	3	69	1	94	2
20	2	45	0	70	5	95	3
21	3	46	3	71	4	96	0
22	4	47	2	72	3	97	1
23	0	48	3	73	2	98	6
24	2	49	2	74	2	99	1

2. Table 3.8 presents a data set of 100 hypothetical observations on a continuous variable, X. Answer parts (a)-(g) of *Exercise 1* for the data in Table 3.8.

TABLE 3.8.
A hypothetical population of 100 observations is shown on a continuous random variable, X.

Label	x	Label	x	Label	x	Label	x
00	69.15	25	68.51	50	64.80	75	42.31
01	63.68	26	70.37	51	62.71	76	66.15
02	71.01	27	64.75	52	45.70	77	55.47
03	60.88	28	67.30	53	64.57	78	52.26
04	64.81	29	54.20	54	68.87	79	51.04
05	77.76	30	65.33	55	66.59	80	66.95
06	83.12	31	74.26	56	65.41	81	71.17
07	70.09	32	60.38	57	45.80	82	56.42
08	64.82	33	65.38	58	65.49	83	66.90
09	70.14	34	56.67	59	76.07	84	70.33
10	71.67	35	66.78	60	65.53	85	47.51
11	57.15	36	67.87	61	60.66	86	78.22
12	66.36	37	66.02	62	59.63	87	50.63
13	62.20	38	70.86	63	73.00	88	59.53
14	61.04	39	64.12	64	50.99	89	50.02
15	66.79	40	59.44	65	62.06	90	72.77
16	60.37	41	66.21	66	66.92	91	51.49
17	72.74	42	70.12	67	39.01	92	64.72
18	73.47	43	58.05	68	74.92	93	54.02
19	72.76	44	56.15	99	92.77	94	62.88
20	64.26	45	80.64	70	63.07	95	58.96
21	72.86	46	82.31	71	66.59	96	77.39
22	64.40	47	59.20	72	66.86	97	78.34
23	73.03	48	49.38	73	61.73	98	63.03
24	63.95	49	74.84	74	49.28	99	61.11

CHAPTER 4

Discrete Probability Models For Empirical Data

4.1 INTRODUCTION

In the preceding chapters, we have discussed the use of *experiments* to collect data for statistical analyses. The concepts of probability as they relate to the selection of samples from populations are given in Chapter 2, while the summarization, description, and presentation of data are developed in Chapter 3. Also, Section 2.2 includes a brief description of probability models. The objective of this chapter and of Chapter 5 is to present a more complete motivation for using statistical models, as well as some specific families of models useful in modeling observations on discrete variables (in Chapter 4) and continuous variables (in Chapter 5).

The basic idea to understand about probability (statistical) models is that when studying some phenomenon by means of an experiment, we need to know that the random nature of the experiment is such that we do not expect two or more independent studies to reveal identical results. That is, since we cannot predict the outcome of any trial of an experiment with certainty, we do not expect the independent, experimental results collected by different investigators to agree exactly. These concepts are illustrated by the coin-tossing experiment in Section 2.2. In particular, Table 2.1 reveals that only

2 (#9 and #13) of the 20 trials agree with the model that one-half of the results are heads. However, as we pool the samples together, we see that our results tend to group around the model value. In fact, 9 samples show values of #(H) less than 5, 2 samples have #(H) equal to 5, and 9 samples exhibit #(H) greater than 5. It is important that we first note the variation from sample to sample and then note the tendency for our samples to group around the model value of 5. That is, the purpose of a model is to describe an underlying law for all possible results of an experiment rather than provide an exact expression for one particular study.

To further illustrate the variety of data sets that a number of investigators may observe while independently conducting identical experiments, let us consider the following experiment.

EXPERIMENT: Place 3 coins in a cup and, after shaking the cup several times, toss the coins onto a desk or other hard, flat surface. Let X be the variable that denotes the number of heads showing after each toss. Record X for 16 independent trials of the experiment, and present the results in distribution form. (Clearly the sample space for X is 0, 1, 2, and 3.)

Table 4.1 presents 5 independent samples observed for the preceding experiment. Notice that no two sample distributions are alike. Yet there seems to be a common pattern among the frequencies. Furthermore, all 5 samples exhibit some departure from the frequencies given in the model column. However, these model frequencies are somewhat representative of all 5 samples.

TABLE 4.1.
Empirical frequency distributions are shown, as well as the model frequency for the number of heads, X, observed on 16 independent tosses of 3 coins.

x	Sample 1	Sample 2	Sample 3	Sample 4	Sample 5	Model
0	2	3	1	4	1	2
1	7	5	6	7	8	6
2	6	6	7	4	7	6
3	1	2	2	1	0	2
Total	16	16	16	16	16	16

One more observation can be made of the samples in Table 4.1. Since all the samples are independent, we can pool the observations from any 2 or more samples and obtain empirical results for increasing sample sizes. These pooled frequencies can then be compared to multiples of the model frequencies. For example, pooling all 5 samples together gives empirical frequencies of eleven 0's, thirty-three 1's, thirty 2's, and six 3's, which can be compared to model frequencies of ten 0's, thirty 1's, thirty 2's, and ten 3's. In considering models for empirical data, we desire a better agreement between model and empirical frequencies as the number of trials, n, increases.

The preceding experiment and discussion illustrate some general properties for a probability (statistical) model. However, the remaining question is that of how the model frequencies are determined. Or more precisely, how is a model for empirical observations determined? The answer to such questions is usually provided in two parts. First, the assumptions under which the experiment is conducted usually specify a general type of probability models that describe the observations produced by repeated experimental trials. Second, the data may be used to estimate (select values for) any parameters (model characteristics) present in the indicated family of models to specify a particular model. Thus, it is the purpose of this chapter and of Chapter 5 to present several families of models that are useful in describing empirical observations. The problem of selecting a specific model from a given family of models is presented in Chapter 7.

It is necessary to establish some definitions and notation before beginning a formal development of specific models. A part of that symbolism is given in Table 2.2, but, for the sake of easy reference, we present a complete set of definitions and notation in the next section.

4.2 RANDOM VARIABLES AND EXPECTED VALUES

As we enter the world of models, it is good to emphasize the fact that a model is an abstraction of the real (experimental) world. Thus, to avoid confusion, we define terms and symbols for models that have counterparts in the experiments. These analogies are noted as the necessary definitions are presented.

DEFINITION 1: A *random variable* is a rule (function) that assigns a real number to each outcome of an experiment.

The customary notation for random variables is to

use a capital letter for the variable and the corresponding lower-case letter for the values of the variable. For example, let X be the random variable that denotes the number of heads observed when 3 coins are tossed. Then the possible values of X are $x_1 = 0$, $x_2 = 1$, $x_3 = 2$, and $x_4 = 3$. Notice that one value of the random variable corresponds to each outcome of the experiment. Namely, x_1 corresponds to the outcome showing 0 heads and 3 tails, x_2 corresponds to the 3 outcomes showing 1 head and 2 tails, x_3 corresponds to the 3 outcomes showing 2 heads and 1 tail, and x_4 corresponds to the outcome showing 3 heads and 0 tails. Thus, the set of possible values for a random variable corresponds to the sample space of an experiment. For reference, let $\mathcal{X}$ denote the set of values for a random variable. Thus, when specifying a model for empirical data, we first define a random variable, X, such that $\mathcal{X}$ corresponds to the sample space, S, of the experiment used to collect data. Second, probabilities are assigned to the elements of $\mathcal{X}$ in such a way that they model (represent) the relative frequencies of the respective experimental outcomes under repeated trials.

In this chapter, we suppose that the variable under investigation is discrete and that

$$\mathcal{X} = \{x_1, x_2, \ldots\}.$$

Note that $\mathcal{X}$ consists of a set of distinct points and may be either a finite or an infinite set. We can now define a probability model (function).

DEFINITION 2: The function, P, is called a *probability model (function)*, and we write $P(x) = P(X = x)$ to denote the probability that the variable, X, assumes the value, x, if

$$0 \leq P(x_i) \text{ for } x_i \in \mathcal{X}, \tag{4.1}$$

and

$$\sum_{x_i \in \mathcal{X}} P(x_i) = 1.$$

The conditions stated in Eq. (4.1) are the only conditions that a function, P, must satisfy to be a potential model for a set of discrete experimental data. That is, any function satisfying Eq. (4.1) is a valid model in a mathematical sense. However, the usefulness of a function for modeling data is judged by whether or not it effectively represents observations from the experimental world. Questions of representation of the real world are called goodness of fit questions and are discussed in detail in Chapter 9. At that time we will have gained some experience with and knowledge of several families of models.

We refer to a table containing the elements of $\mathcal{X}$ and the probabilities assigned to these elements as a *probability distribution*. Table 4.2 presents four different probability distributions having $\mathcal{X}$ such that they could serve to model the coin-flipping experiment in Table 4.1.

TABLE 4.2
Shown here are 4 probability distributions.

(a)		(b)		(c)		(d)	
w	P(w)	x	P(x)	y	P(y)	z	P(z)
0	.25	0	.125	0	.3	0	.1
1	.25	1	.375	1	.2	1	.4
2	.25	2	.375	2	.2	2	.4
3	.25	3	.125	3	.3	3	.1
Total	1.00		1.000		1.0		1.0

All four distributions in Table 4.2 assign probabilities to the set of points (0, 1, 2, 3), which is $\mathcal{X}$ for the variable, X, representing the number of heads observed on the toss of 3 coins. These examples illustrate 4 of a possible infinity of ways in which a total probability of 1.00 can be assigned to the points, 0, 1, 2, 3. The model given in Table 4.1 is based on distribution (b) in Table 4.2. That is, P(0) = 1/8, P(1) = 3/8, P(2) = 3/8, and P(3) = 1/8. Thus, in 16 trials of the experiment, we expect two 0's, six 1's, six 2's, and two 3's. The motivation for selecting (b) as the modeling distribution is explained in Section 4.4.

Since there are many models (probability distributions), it is important that we have a method by which we can compare such distributions. The logic used in Chapter 3 to motivate the use of descriptive statistics as a means for comparing empirical distributions suggests that we seek measures of location and dispersion for use in comparing probability distributions. The usual measure of location is the distribution *mean* or *expected value*, while the distribution *variance* is a common measure of dispersion.

DEFINITION 3: The expected value (or mean) of a discrete random variable, X, is denoted by either E(X) or μ

and defined by

$$\mu = E(X) = \sum_{x_i \varepsilon X} x_i \ P(x_i). \tag{4.2}$$

DEFINITION 4: The variance of a discrete random variable, X, is denoted by either σ^2 or V(X) and defined by

$$V(X) = \sigma^2 = E(X-\mu)^2 = \sum_{x_i \varepsilon X} (x_i-\mu)^2 \ P(x_i). \tag{4.3}$$

EXAMPLE 1: Find the mean and variance for (c) and (d) in Table 4.2.

SOLUTION: Distribution (c):

$$\mu = 0(.3) + 1(.2) + 2(.2) + 3(.3) = 1.5.$$

$$\begin{aligned}\sigma^2 &= (0 - 1.5)^2 \ (.3) + (1 - 1.5)^2 \ (.2) \\ &\quad + (2 - 1.5)^2 \ (.2) + (3 - 1.5)^2 \ (.3) \\ &= 2.25(.3) + .25(.2) + .25(.2) + 2.25(.3) \\ &= 1.45.\end{aligned}$$

Distribution (d):

$$\mu = 0(.1) + 1(.4) + 2(.4) + 3(.1) = 1.5.$$

$$\begin{aligned}\sigma^2 &= (0 - 1.5)^2 \ (.1) + (1 - 1.5)^2 \ (.4) \\ &\quad + (2 - 1.5)^2 \ (.4) + (3 - 1.5)^2 \ (.1) \\ &= .65.\end{aligned}$$

REMARK: Notice that μ and σ^2 help to distinguish between distributions (c) and (d). First, both distributions are located at $\mu = 1.5$. Second, the fact that (d) is more concentrated about 1.5 than (c) is indicated by the smaller variance.

Since the expectation definition for discrete random variables is a summation operation, we can state rules for manipulating expectations just as we gave rules for manipulating summations. Two such rules are presented here.

RULE 1: For any constant, a, we have E(a) = a.

RULE 2: For any constant, a, and any variable, X, we have E(aX) = a E(X).

A most useful result follows by combining *Rules 1* and *2.* That is, if a and b are constants, then

$$E(aX + b) = a\,E(X) + b. \tag{4.4}$$

The usefulness of Eq. (4.4) can best be summarized as follows. Let X be a random variable with mean, μ, and variance, σ^2. Let a and b be constants, and define the random variable, $Y = a X + b$. Then

$$\mu_y = E(Y) = a\mu + b,$$

$$\sigma_y^2 = a^2\,\sigma^2, \tag{4.5}$$

$$\sigma_y = |a|\sigma.$$

Proof: $\mu_y = a\mu + b$ is a direct consequence of Eq. (4.4).

By (4.3), $\sigma^2_y = E(Y - \mu_y)^2$.

Now $Y - \mu_y = (a X + b) - (a\mu + b)$

$= a\,(X - \mu)$.

Therefore,

$\sigma^2_y = E[a^2\,(X - \mu)^2] = a^2\,E(X - \mu)^2$

$= a^2\,\sigma^2$.

By taking the positive square root of both sides of $\sigma_y^2 = a^2\sigma^2$, we find that

$\sigma_y = |a|\sigma$.

It is often convenient to calculate σ^2 by rewriting Eq. (4.3) to read

$$V(X) = \sigma^2 = E(X^2) - \mu^2, \tag{4.6}$$

where $E(X^2)$ is defined by

$$E(X^2) = \sum_{x_i \in X} x_i^2\,P(x_i). \tag{4.7}$$

Proof: By Eq. (4.3), $\sigma^2 = E(X - \mu)^2$.

Expanding the square, we find that

$$\sigma^2 = E(X^2 - 2\mu X + \mu^2).$$

Using *Rules 1* and *2* gives

$$\sigma^2 = E(X^2) - 2\mu E(X) + \mu^2.$$

Since $\mu = E(X)$, we have

$$\sigma^2 = E(X^2) - \mu^2.$$

EXAMPLE 2: Find the mean and variance for the following probability distribution:

u	5	10	15	20
P(u)	.3	.2	.2	.3

SOLUTION:

$$\mu = 5(.3) + 10(.2) + 15(.2) + 20(.3)$$
$$= 12.5.$$
$$\sigma^2 = E(X^2) - (12.5)^2$$
$$= 25(.3) + 100(.2) + 225(.2) + 400(.3) - 156.25$$
$$= 36.25.$$

REMARK: The value of σ^2 can be calculated by using Eq. (4.3) rather than Eq. (4.6).

EXAMPLE 3: Let $V = U/5 - 2$, where U is the random variable defined in *Example 2*. Find the mean and variance of V.

SOLUTION: By *Example 2*, E(U) = 12.5 and the variance of U = 36.25. Thus, Eq. (4.4) gives E(V) = 12.5/5 - 2 = .5, and Eq. (4.5) provides $\sigma_V^2 = 36.25/25 = 1.45$.

In addition to the mean and variance, we can use the expectation of higher powers of the variable, e.g., $E(X^r)$, r = 3, 4, ..., to distinguish one distribution from another. However, those techniques are not discussed in this text since E(X) and $E(X^2)$ are adequate for the problems of interest in this introductory course.

Now that the necessary terms and notation are defined, we can proceed to a discussion of specific families of models.

EXERCISES

1. Which of the following are probability distributions?

(a)

t	P(t)
1	.6
5	.3
7	.2

(b)

u	P(u)
1	.4
2	−.1
3	.3
4	.2
5	.2

(c)

w	P(w)
−2	.25
−1	.40
0	.25
1	.10

(d)

x	P(x)
0	.3
2	.3
4	.3

(e)

y	P(y)
1	.50
2	.25
3	.125
4	.0625
5	.0625

2. Criticize the following probability assignments:
   (a) Mr. Van Meter drives a taxi in New York and assigns the following probabilities to the variable, N, which represents the number of accidents he will experience during the next 12 months: $P(N = 0) = .80$, $P(N = 1) = .05$, $P(N = 2) = .05$.
   (b) Mr. Figures teaches an elementary statistics course and gives the following probabilities to the variable, G, which denotes the course grade received: $P(G = A) = .25$, $P(G = B) = .30$, $P(G = C) = .40$, $P(G = D) = .05$.
   (c) The Oidar Radio Shop sells a transistor radio and assigns the following probability distribution to N, the number of failures during the first two years of service: $P(N = 0) = .50$, $P(N = 1) = .25$, $P(N = 2) = .20$, $P(N = 3) = .10$.
3. (a) Find the mean and variance of the probability distribution (b) in Table 4.2.
   (b) Interpret your results for part (a) in view of the distribution serving as a model for the coin-tossing experiment in Table 4.1.

4.3 DISCRETE UNIFORM MODELS

The first family of models we develop can be illustrated by conducting one or both of the following experiments. Although a few possible samples for the experiments are presented, each student should conduct her/his own trials to compare with the results of other students. Such an experience demonstrates how widely experimental results obtained by independent experimenters can vary.

EXPERIMENT 1: Select a cube (die) such that each of the 6 faces exhibits 1, 2, 3, 4, 5, or 6 dots. (The cube should not show any obvious bias.) Let X denote the number of dots showing on the uppermost face after each roll of the cube. Conduct 30 trials of the experiment, and record the results in a frequency distribution.

EXPERIMENT 2: Select 50 random digits (1-digit numbers) from Table 2.8, and record your results in a frequency distribution.

Tables 4.3 and 4.4 present the results for 4 samples generated by repeating each of the above experiments 4 times. Also indicated in each table is a suggested model frequency distribution.

TABLE 4.3
Empirical frequencies and model frequencies are given for 4 samples of 30 rolls of a die (cube).

x	Sample 1	Sample 2	Sample 3	Sample 4	Model
1	4	6	9	1	5
2	3	2	3	6	5
3	8	6	4	7	5
4	5	6	2	7	5
5	3	6	4	6	5
6	7	4	8	3	5
Total	30	30	30	30	30

The question now to be faced is the following: What probability model represents the data generated by the two experiments previously described? One approach is to reason from symmetry and assign equal probability to

TABLE 4.4.
Empirical frequencies and model frequencies are given for 4 samples of 50 random digits.

x	Sample 1	Sample 2	Sample 3	Sample 4	Model
0	4	6	8	5	5
1	5	6	4	5	5
2	4	3	5	1	5
3	7	4	5	3	5
4	4	5	4	4	5
5	6	4	6	7	5
6	5	5	6	11	5
7	5	5	2	4	5
8	7	9	6	5	5
9	3	3	4	5	5
Total	50	50	50	50	50

each possible outcome in the two sample spaces. That is, there is no obvious reason why any one side of the die should occur more or less often than any other side. Neither is there any obvious reason why any one digit should be more or less frequent than any other digit. Thus, the probability models we assign to the two experiments are as follows.

Model for rolling a die:

x	1	2	3	4	5	6
P(x)	1/6	1/6	1/6	1/6	1/6	1/6

Model for selecting a random digit:

x	0	1	2	3	4	5	6	7	8	9
P(x)	.1	.1	.1	.1	.1	.1	.1	.1	.1	.1

The preceding models are two special cases of a more general model called the *uniform probability model*, which assigns equal probability to all points in $\mathcal{X}$. Thus, if X is a uniform random variable, and $\mathcal{X}$ consists of K distinct points, then the probability model for X is

$$P(x_i) = 1/K, \ i = 1, 2, \ldots, K. \tag{4.8}$$

Clearly the probability function defined by Eq. (4.8) satisfies the requirements of Eq. (4.1). Furthermore, the die-rolling and random-digit experiments are examples of uniform probability models with K = 6 and 10, respectively. Another example of a uniform model is the assignment of equal probabilities (.5) to each of the two possible outcomes resulting from the flip of a coin.

Notice that the only quantity needed to determine a particular uniform model is K, the number of different points in $\mathcal{X}$. Thus, to select a uniform model for an experiment, we choose the uniform model with K equal to the number of points in S, the sample space of the experiment being modeled.

We should further note that the models we proposed are probability models (distributions). Thus, to obtain the model frequencies for n trials of an experiment as presented in Tables 4.1, 4.3, and 4.4, we simply multiply the probability of each point in $\mathcal{X}$ by the number of experimental trials conducted. That is, the model probabilities represent the proper allocation of frequency to each value of the variable so that the total number of the model frequencies is equal to the number of trials of the experiment.

Frequently, the sample space of the experiment of interest consists of the first K positive integers, i.e., S = {1, 2, ..., K}. When this is the case, the appropriate uniform model becomes

$$P(i) = 1/K, \ i = 1, 2, \ldots, K. \tag{4.9}$$

To study the properties of the distribution defined by Eq. (4.9), we need two algebraic identities, which are found in Cameron [1] and are defined in the next two equations. For any integer k,

$$\sum_{x=1}^{k} x = \frac{k(k+1)}{2}, \tag{4.10}$$

$$\sum_{x=1}^{k} x^2 = \frac{k(k+1)(2k+1)}{6}. \tag{4.11}$$

Thus, if X has the uniform distribution as defined in

Eq. (4.9), it follows directly from Eq. (4.10) that

$$E(X) = \sum_{x=1}^{K} x\left(\frac{1}{K}\right) = \frac{K+1}{2}. \qquad (4.12)$$

Furthermore, the use of Eqs. (4.11) and (4.12) in Eq. (4.6) gives

$$\sigma_x^2 = [(K+1)(2K+1)]/6 - [(K+1)^2/4]$$

$$= [(K+1)(K-1)]/12 = (K^2-1)/12. \qquad (4.13)$$

EXAMPLE 4: Find the mean and variance for the uniform model associated with the roll of a die.

SOLUTION: The distribution for modeling the die-rolling experiment is $P(x) = 1/6$, $x = 1, 2, \ldots, 6$. Therefore, Eq. (4.12) gives $E(X) = (6+1)/2 = 3.5$, and Eq. (4.13) gives $\sigma_x^2 = (36-1)/12 = 35/12$.

REMARKS: To visualize the implication of the mean of a distribution, suppose a friend will pay you $1 per dot on each roll of the die. The E(X) represents the average amount of money the friend can expect to lose and you can expect to gain on each roll of the die. That is, you expect to gain an average (mean) of $3.50 per roll of the die.

To interpret the variance of the model, recall the interpretations of s in Chapter 3. Namely, recall that frequently between 1/2 and 3/4 of a distribution is within one standard deviation of its mean. In this example, $\sigma = 1.708$. Thus, you can expect to win between $1.79 and $5.21 for 1/2 to 3/4 of the time. That is what the usual interpretation of σ tells us, but in reality we know that a gain of either $2, $3, $4, or $5 can be expected 2/3 of the time when a fair die is rolled.

EXAMPLE 5: What are the mean and variance of X, the value of a random digit?

SOLUTION: The model involved is given by Eq. (4.8) with K = 10 and defined by $x = 0, 1, 2, \ldots, 9$. Thus, the expected value of a random digit is

$$E(X) = \sum_{x=0}^{9} x\frac{1}{10} = \frac{1}{10}\ \frac{(9)(10)}{2} = 4.5.$$

Next, Eq. (4.11) gives

$$E(X^2) = \sum_{x=0}^{9} x^2 \left(\frac{1}{10}\right) = \frac{1}{10}\,\frac{9(10)(19)}{6} = \frac{57}{2}$$

and

$$V(X) = (57/2) - (81/4) = 33/4.$$

REMARK: The average value of a random digit is 4.5, while the standard deviation is about 2.87. From the Chebychev inequality, we expect between 1/2 and 3/4 of our trials to give digits between 1.63 and 7.37, while the uniform model leads us to expect that 60% of our trials will yield either 2, 3, 4, 5, 6, or 7.

The results given in Eqs. (4.12) and (4.13) apply only to the special uniform distribution in Eq. (4.9). For the general uniform random variable, X, having the distribution in Eq. (4.8), we return to Eqs. (4.2) and (4.6) to write

$$E(X) = \frac{1}{K}\sum_{i=1}^{K} x_i,$$

$$V(X) = \frac{1}{K}\sum_{i=1}^{K} x_i^{\,2} - \left(\frac{1}{K}\sum_{i=1}^{K} x_i\right)^2. \tag{4.14}$$

The uniform models discussed in this section are appropriate for representing those experimental situations in which we believe all possible outcomes are equally likely to occur. Since there are many situations lacking the characteristics of symmetry and uniformity required for the uniform models, we discuss other types of discrete probability models in the remaining sections of this chapter.

EXERCISES

1. A child's game contains a spinner on a circle which is divided into 5 equal sectors. The sectors have values as shown in the accompanying figure.

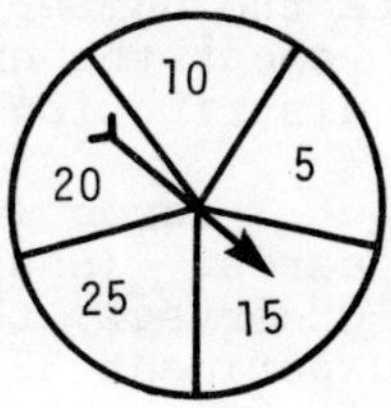

(a) What probability model would be appropriate for modeling repeated spins?
(b) Find the expected value and variance of the model suggested in part (a).

2. Provide answers to parts (a) and (b) of *Exercise 1* when the numbers 5, 10, 15, 20, and 25 are replaced by -5, 10, 15, 20, and -25, respectively.

3. Roll a die 48 times, and record the data in a distribution.
   (a) Find the mean and variance of the empirical data.
   (b) Specify a model for the experiment, and determine the model frequencies for the 6 possible outcomes.
   (c) Compare the mean, variance, and observed frequencies of the empirical distribution with the mean, variance, and model frequencies of the model.

4. Select 100 random digits from Table 2.8, and form a distribution of the data.
   (a) Find the mean and variance of the empirical data.
   (b) Specify a model for the experiment, and determine the model frequencies for the 10 possible outcomes.
   (c) Compare the mean, variance, and observed frequencies of the empirical distribution with the mean, variance, and model frequencies of the model.

4.4 BINOMIAL EXPERIMENTS AND MODELS

The second family of models we are going to develop is that of the *binomial* models. This family of distributions can be useful for modeling a wide range of experiments. Each student should collect at least one set of data for one or more of the following situations. These data can be compared with corresponding data which are either collected by other students or presented in the following experiments.

EXPERIMENT 1: Flip 4 coins and record X, the number of heads appearing. Repeat the experiment 32 times, and form a frequency distribution of the results. The coins should be placed in a cup or other container for easy mixing before being tossed onto a hard, flat surface.

EXPERIMENT 2: Roll a die 3 times, or roll 3 dice 1 time, and let X denote the number of 1's or 2's that appear on the 3 dice. Repeat the experiment 27 times, and record the empirical distribution of X.

EXPERIMENT 3: Select 3 cards (with replacement) from a deck of 52 playing cards. (Shuffle the 52 cards prior to each selection.) Observe X, the number of spades selected. Repeat the experiment 32 times, and form a frequency distribution of the results.

EXPERIMENT 4: (Class project) State a question regarding a topic of general interest (e.g., student government candidate; political candidate at local, state, or national level; environmental conditions; federal support of higher education; etc.), and let each student record the answers of 10 randomly selected individuals. Let X denote the number of yes answers in each group of 10. Record the frequency distribution of X for all persons in the class. Notice that the answers are to be classified as yes or not yes.

TABLE 4.5.
Empirical frequencies and model frequencies are given for 4 samples of 32 tosses of 4 coins, as in Experiment 1.

x	Sample 1	Sample 2	Sample 3	Sample 4	Model
0	2	3	4	1	2
1	10	8	7	8	8
2	10	13	13	8	12
3	8	5	5	10	8
4	2	3	3	5	2
Total	32	32	32	32	32

TABLE 4.6.
Empirical frequencies and model frequencies are given for 4 samples of 27 rolls of a die, as in Experiment 2.

x	Sample 1	Sample 2	Sample 3	Sample 4	Model
0	8	8	13	4	8
1	13	11	9	15	12
2	6	6	5	6	6
3	0	2	0	2	1
Total	27	27	27	27	27

Be aware that each trial (repetition) of the preceding experiments involves trials (repetitions) of a simpler experiment. For example, one trial of the first experiment (the flip of 4 coins) represents 4 trials of an experiment that involves the flipping of 1 coin. Thus, the word trial has two meanings. First, it refers to the *separate trials of a basic experiment*, such as the flip of 1 coin. Second, it refers to a *single trial of a composite experiment*, such as the flip of 4 coins. This dual usage of trial occurs frequently in the subsequent discussion.

Tables 4.5 and 4.6 present 4 samples for *Experiments 1* and *2*, respectively.

MOTIVATION AND DEFINITION: The results in Tables 4.5 and 4.6 indicate that the observed frequencies in Table 4.5 tend to be symmetric about x = 2, while those in Table 4.6 tend to cluster toward x = 0 and x = 1. Both of these situations and others can be modeled by binomial models. Thus, we now use the common properties exhibited by the preceding experiments to indicate the assumptions that experimental data must satisfy to be modeled by binomial distributions.

First, observe that each experiment is composed of trials of a basic experiment that produces *1 of 2* possible outcomes for each trial. In particular, each coin falls either with heads or tails appearing on each flip, each die falls either with 1 or 2 appearing, or not 1 or 2 appearing on each roll, each randomly selected card is either a spade or not a spade, and each question is answered by either yes or not yes. Thus, binomial models can represent experimental data from a composite experiment for which the number of possible outcomes for each basic experiment is 2. Note that the two acceptable categories of response may be natural outcomes for the basic experiment (e.g., coins), or they may be artificial categories formed by designating a portion of the sample space as one category and using the remainder of the sample space as the other category.

Second, observe that each experiment consists of a fixed number of trials of the basic experiment. In particular, *Experiment 1* involves 4 coins, *Experiment 2* calls for 3 dice (or 3 rolls of 1 die), *Experiment 3* uses 3 cards, and *Experiment 4* uses 10 responses. Traditionally, the number of trials of the basic experiment used to form the composite experiment is denoted by n.

Third, observe that the method for conducting trials of the basic experiment is such that all trials are independent. That is, the 4 coins are independent of each other, the 3 dice (or 3 rolls of the same die) are independent, and the 10 individuals surveyed should be responding independently.

Fourth, the probability of the designated category is the same for all n trials of the basic experiment. This common probability is usually denoted by p. Thus, we have p = 1/2 for the probability of heads appearing on the toss of a fair coin, p = 1/3 for the probability of 1 or 2 appearing on the toss of a fair die, and p = 1/4 for the probability that a randomly selected card is a spade. In *Experiment 4*, the value of p (i.e., the probability of a yes response) is probably not known. In fact, the estimation of p might be the primary reason for conducting *Experiment 4*. Estimation problems of this type are a basic part of statistical inference and are discussed in Chapter 7.

By summarizing the preceding remarks, we find that an experiment should exhibit the following properties in order to be modeled by a binomial distribution:

(1) The sample space of the basic experiment is partitioned into two sets. These sets are denoted by S (success) and F (failure), where F is the complement of S.

(2) n independent trials of the basic experiment are conducted to complete a single trial of the binomial experiment.

(3) P(S) = p for each trial of the basic experiment. [For notational convenience, we use P(F) = q = 1 - p.]

The binomial model is defined for a random variable, X, which denotes the number of successes observed in n independent trials of the basic experiment (single trial of the binomial experiment). The probability function (model) for X is called the binomial distribution and is given by

$$P(x) = \binom{n}{x} p^x q^{n-x}, \; x = 0, 1, 2, \ldots, n. \tag{4.15}$$

Equation 4.15 is motivated in the following manner. First, the number of successes must be an integer between 0 and n, inclusive. That is, it is clearly impossible to either have fewer that 0 S's or more than n S's result from n trials. Thus, x = 0, 1, 2, ..., n.

Second, any set of n independent trials provides a sequence of S and F. For example, 4 flips of a coin may yield any one of the following $2^4 = 16$ sequences, where S denotes heads and F denotes tails: SSSS, SSSF, SSFS, SFSS, FSSS, SSFF, SFSF, SFFS, FSFS, FSSF, FFSS, SFFF, FSFF, FFSF, FFFS, FFFF.

Third, the independence of trials implies that the probability of any sequence of outcomes is a product of p's and q's. For example, P(SSSS) = pppp = p^4, and P(SFSF) = pqpq = p^2q^2.

Fourth, we note that more than one sequence of outcomes may give the same value of X. In particular, SSSF, SSFS, SFSS, and FSSS all produce X = 3.

Even though different sequences give the same value of X, it is important to observe that all sequences giving a common value of x consist of exactly x S's and (n-x) F's. That is, the probability of each such sequence is p^xq^{n-x}. Hence, to determine the probability of x successes [i.e., P(x) = P(X=x)], we need to count the number of sequences that include exactly x S's. Since each sequence of x S's has the same probability, p^xq^{n-x}, and since the sequences are mutually exclusive,

$$P(X=x) = (\text{number of sequences of x S's}) \cdot p^x q^{n-x}.$$

However, the number of sequences having x S's is just the number of ways in which x of the available n positions can be selected for placing S. From Definition 13 on page 47, that number is $\binom{n}{x}$. Combining all of the preceding implications results in Eq. (4.15).

Another way of arriving at the coefficient $\binom{n}{x}$ in Eq. (4.15) is to view the problem as one of counting the number of possible permutations of n items consisting of x S's and (n-x) F's. By *Definition 12* on page 46, the result is n!/x!(n-x)!, which is exactly $\binom{n}{x}$. The advantage of this reasoning over the previous slot-selection reasoning becomes apparent when we discuss models for experiments composed of basic experiments having more than two responses.

Let us now use Eq. (4.15) to model the samples given in Tables 4.5 and 4.6. First, in *Experiment 1*, we have n = 4 (1 toss of *4* coins or *4* tosses of 1 coin), and we assume that heads and tails are equally likely to occur, i.e., P(S) = P (heads) = p = 1/2. Second, in *Experiment 2*, we have n = 3 (1 die rolled *3* times or *3* dice rolled 1 time), and by assuming symmetric dice, we arrive at P(S) = P(1 or 2) = p = 1/3. Thus, the desired probability models are as given in Table 4.7.

The model frequencies given in Tables 4.5 and 4.6 follow directly by multiplying the respective values of P(x) in Table 4.7 by the number of trials (repetitions) of each experiment, namely, 32 trials for Table 4.5 and 27 trials for Table 4.6.

As a matter of convenience from a notational viewpoint, we write

$$X \sim b(n, p) \tag{4.16}$$

TABLE 4.7.
Shown here are two binomial probability models.

n=4	p=1/2	n=3	p=1/3
x	P(x)	x	P(x)
0	1/16	0	8/27
1	4/16	1	12/27
2	6/16	2	6/27
3	4/16	3	1/27
4	1/16		
Total	1.00	Total	1.00

to indicate that the random variable, X, is a binomial variable based on n trials of a basic experiment with probability of success equal to p. Thus, we write $X \sim b(4, 1/2)$ for Table 4.5 and $X \sim b(3, 1/3)$ for Table 4.6.

Formula (4.16) emphasizes the fact that there is a different binomial model for each pair of values for n and p. Now n is the number of trials conducted so that quantity is a known integer. The quantity, p, is a probability and thus must exist between 0 and 1, inclusively. For any integer value of n, there is an infinity of binomial models determined by a choice of p. That is, *p is the parameter of a binomial model.* Later (Chapter 7) in our study, we are concerned with the estimation of p for model selection, but for the remainder of this discussion, we assume that p is either known or can be calculated by the elementary probability techniques of Chapter 2.

Equation (4.15) can be used to calculate the binomial probability distribution for any specified values of n and p. However, the necessary arithmetic increases rapidly with n. Thus, to facilitate easy reference, the binomial probability distributions for selected values of p and n are presented in Appendix B. Also included in Appendix B are tables for the cumulative probability distributions, F(x), for the same values of p and n, where

$$F(x) = P(X \le x) = \sum_{t=0}^{x} \binom{n}{t} p^t q^{n-t}. \qquad (4.17)$$

The following examples illustrate the use of Eqs. (4.15) and (4.17) and Table B1.

EXAMPLE 6: Suppose that boys and girls are born with equal probability. (Note that this is not quite true biologically.)
(a) What is the probability that a family of 5 children includes exactly 2 girls?
(b) What is the probability that a family of 5 children includes at most 2 girls?

(Note: To use the foregoing terminology and discussion, we must associate boys and girls with successes and failures. So, for the sake of solving our problem, we will arbitrarily associate a girl birth with success and a boy birth with failure. These associations can be reversed without difficulty by interchanging the roles of boys and girls in the questions to be answered.)

SOLUTION: The conditions of the problem suggest using $X \sim b(5, 1/2)$ as our model.
(a) Equation (4.15) gives

$$P(X=2) = \binom{5}{2}\left(\frac{1}{2}\right)^2\left(\frac{1}{2}\right)^3 = \frac{5!}{3!2!}\left(\frac{1}{4}\right)\left(\frac{1}{8}\right)$$

$$= 10/32 = .3125.$$

(b) Equation (4.17) gives

$$F(2) = P(X \leq 2) = \sum_{t=0}^{2} \binom{5}{t}\left(\frac{1}{2}\right)^t \left(\frac{1}{2}\right)^{5-t}$$

$$= 1/32 + 5/32 + 10/32$$

$$= 1/2 = .5.$$

REMARKS: The results given for parts (a) and (b) could also have been obtained directly from the binomial table (Table B1) for n=5 and p=.5 in Appendix B. After making the simplifying assumption that boy births and girl births are equally likely, we might conclude that the sex composition of families of n children can be modeled by the same distribution as the results of n coin flips. Theoretically the argument sounds good, but in practice it does not work too well. For example, an empirical distribution of 2-child families tends to show an abundance of boy-girl or girl-boy families and a deficiency of boy-boy and girl-girl families in relation to the binomial ratios of 1/4 boy-boy, 1/2 with 1 of each, and 1/4 girl-girl. An explanation of this discrepancy is that society values as ideal a family of 1 boy and 1 girl. Hence, such families remain families of 2 children, while parents of boy-boy or girl-girl families seek a bit more variety and become families of 3 or more children. It should be pointed out that even though a b(2, 1/2) model does not describe the frequencies we observe in families of 2 children, it may well describe the frequencies of the different possibilities for the first 2 children born in a random sample of families of 2 or more children. Thus, we might conclude that raising families and flipping coins are equivalent in a particular sense.

EXAMPLE 7: A die is rolled 25 times, and the event, S, is designated to be either 1 or 2.
(a) What is the probability that 10 S's are observed?
(b) What is the probability that 10 or fewer S's are observed?

SOLUTION: (a) Equation (4.15) gives

$$P(10) = P(X=10) = [25!/(10!\ 15!)]\,(1/3)^{10}\ (2/3)^{15}.$$

Obviously, the arithmetic necessary to reduce the preceding expression to a decimal is beyond the interest of the most interested student, so we refer to Appendix B and the binomial table with n = 25 and p = 1/3. The result is P(X=10) = .1264.
(b) Similarly, Eq. (4.17) gives us an expression as an answer to our question, but we are not interested in doing the arithmetic necessary to evaluate

$$F(10) = P(X \le 10) = \sum_{x=0}^{10} \binom{25}{x}\left(\frac{1}{3}\right)^{x}\left(\frac{2}{3}\right)^{25-x}.$$

Therefore, we read the answer directly from Appendix B and the binomial table for n = 25 and p = 1/3. The result is F(10) = .8220.

Further application of Appendix B follows the discussion of some properties of the binomial model.

PROPERTIES OF A BINOMIAL DISTRIBUTION: Since p is the only parameter in a binomial model, it is natural that the properties we discuss depend on p and the number of trials, n.

The first property we give is that the function defined by Eq. (4.15) is a valid probability function. Clearly, $P(x) \ge 0$ for x = 0, 1, 2, ..., n, so it remains only to show that

$$\sum_{x=0}^{n} P(x) = 1. \qquad (4.18)$$

To establish Eq. (4.18) we first recall the binomial theorem from algebra. Cameron [1] states that for any quantities, a and b, and positive integer, n,

$$(a + b)^n = \sum_{x=0}^{n} \binom{n}{x} b^x a^{n-x}. \qquad (4.19)$$

[Note that Eq. (4.19) is usually proved by mathematical induction.] Using Eq. (4.19), we have

$$\sum_{x=0}^{n} P(x) = \sum_{x=0}^{n} \binom{n}{x} p^x q^{n-x} = (q + p)^n.$$

However, by definition of q and p, we have $q + p = 1$. Therefore, Eq. (4.18) is satisfied, and Eq. (4.15) defines a valid probability function.

A second property follows from the result for combinatorials given by Eq. (2.18). We note that each trial of the basic experiment has 2 possible outcomes, so n trials produce a binomial experiment with 2^n outcomes. Furthermore, we have said that exactly $\binom{n}{x}$ of these outcomes exhibit the value, x, for the binomial variable, X. Thus, Eq. (2.18) states that

$$\sum_{x=0}^{n} \binom{n}{x} = 2^n$$

and proves that our method of assigning probabilities has accounted for all the points in the sample space for the n trials of the basic experiment.

A third property of a binomial model is that

$$E(X) = np. \qquad (4.20)$$

Proof: By definition,

$$E(X) = \sum_{x=0}^{n} x\,P(x) = \sum_{x=1}^{n} x\,P(x)$$

$$= \sum_{x=1}^{n} x \frac{n!}{x!\,(n-x)!} p^{x} q^{n-x}$$

$$= np \sum_{x=1}^{n} \frac{(n-1)!}{(x-1)!\,(n-x)!} p^{x-1} q^{n-x}$$

$$= np \sum_{x=1}^{n} \binom{n-1}{x-1} p^{x-1} q^{(n-1)-(x-1)}$$

$$= np \sum_{y=0}^{n-1} \binom{n-1}{y} p^{y} q^{(n-1)-y}$$

$$= np\,(q+p)^{n-1} \quad \text{by Eq. (4.19)}$$

$$= np.$$

A fourth property of a binomial model is that

$$V(X) = npq. \tag{4.21}$$

Proof: By definition, we need $E(X^2) - n^2p^2$.

To determine $E(X^2)$, it is easier to find

$$E\,[X(X-1)] = \sum_{x=0}^{n} x(x-1)\,P(x) = \sum_{x=2}^{n} x(x-1)\,P(x)$$

$$= \sum_{x=2}^{n} x(x-1) \frac{n!}{x!\,(n-x)!} p^{x} q^{n-x}$$

$$= n(n-1)p^2 \sum_{x=2}^{n} \binom{n-2}{x-2} p^{x-2} q^{(n-2)-(x-2)}$$

$$= n(n-1)p^2\,(q+p)^{n-2} \quad \text{by Eq. (4.19).}$$

Then

$$E(X^2) = E[X(X-1)] + E(X)$$
$$= n(n-1)p^2 + np.$$

Thus,

$$V(X) = n(n-1)p^2 + np - n^2p^2$$
$$= np - np^2$$
$$= np(1-p)$$
$$= npq.$$

Before proceeding to the next property, we observe that the mean of X increases from near 0 to near n as p varies from 0 to 1. However, the variance of X increases from near 0 to n/4 as p varies from 0 to 1/2, then decreases to near 0 as p increases to 1. Thus,

$$V(X) \leq n/4 \tag{4.22}$$

for all values of p.

A final property of a binomial model is that of symmetry. The binomial model is symmetric about n/2 for p = 1/2 and for all values of n. This symmetry is evident in Table 4.7 for n = 4 and p = 1/2. If $p \neq 1/2$, then the distribution is asymmetric with higher probabilities assigned to small values of X for $p \leq 1/2$ and large values of X for $p \geq 1/2$. Examples of the symmetry and asymmetry are exhibited by the many distributions tabled in Appendix B. When studying these tabled distributions, notice that even though the binomial distribution is asymmetric for $p \neq 1/2$, it becomes more symmetric in appearance as n increases.

EXAMPLE 8: About 1/10 of all new cars are returned to the dealer within the first week for adjustments and/or repairs.

(a) If the We Trade Fairly dealership delivers 36 cars in one day, how many of these cars should they expect to be returned during the next week?
(b) What is the variance of the number of cars to be returned?

SOLUTION: (a) By assuming that the number of returned cars is a variable, $X \sim b\ (36, .1)$, we have $E(X) = 36(.1) = 3.6$ from Eq. (4.20).
(b) Similarly, Eq. (4.21) gives $V(X) = 36(.1)(.9) = 3.24$.

REMARK: The preceding *Solution* answers the questions posed, but we should go a bit beyond the answers and ask of what value the results are to the We Trade Fairly dealership. To interpret these results we go back to the reasoning of Chapter 3. First, note that the standard deviation of X is $\sqrt{3.24} = 1.8$. Second, we know that a large percentage (at least 75%) of the observed values of X lie within two standard deviations of the mean. Thus, it is most likely that the number of returned cars will be between 0 and 7 (or 7.2). Stating this another way, it is unlikely that the dealer will have 8 or more cars returned during the next week.

EXAMPLE 9: A rubber band machine produces rubber bands with a defective rate of 10%.

(a) In a sample of 20, what is the expected number of defective rubber bands?

(b) What is the variance of the number of defective rubber bands?

(c) What is the probability that the number of defective rubber bands is greater than or equal to 4?

SOLUTION: (a) Let X = #(defective rubber bands). By Eq. (4.20), $E(X) = 20(.1) = 2$.

(b) By Eq. (4.21), $V(X) = 20(.1)(.9) = 1.8$.

(c) By definition,

$$P(X \geq 4) = \sum_{x=4}^{20} \binom{20}{x} (.1)^x (.9)^{20-x}.$$

However, by complementation, $P(X \geq 4) = 1 - P(X \leq 3)$, where $P(X \leq 3)$ can be obtained from the binomial table in Appendix B with $p = .1$ and $n = 20$. In fact, $F(3) = P(X \leq 3) = .8670$. Thus, $P(X \geq 4) = .1330$.

Before presenting some exercises which further illustrate the usefulness of binomial models, we need to make a few comments. First, the family of binomial models is widely used in experiments of a survey type because frequently a survey is conducted with a desired answer in mind. That is, the responses are often classified as either favorable or unfavorable and thus satisfy the constraint of two possible outcomes. Furthermore, the responses of different individuals are assumed to be independent, which meets that assumption for the binomial model. Second, binomial models are useful for political polls, product testing, product purchasing, and other types of situations where each response is categorized in one of two classes.

A third general area where binomial models are used and are useful is for *sampling with replacement* experiments. The replacement retains a constant probability for the two outcomes from one draw to the next. A final comment is simply to indicate that Appendix B gives a small selection of binomial distributions, but more comprehensive tables with many values of both n and p are given by the National Bureau of Standards [2] and Romig [3].

EXERCISES

1. Let X denote the number of 1's that appear on a roll of 3 dice.
   (a) What are the possible values of X?
   (b) Select a binomial model that might describe the relative frequency of each possible value of X.
   (c) Table the probability distribution and cumulative probability distribution selected in part (b).
   (d) What is the E(X)?
   (e) What is V(X)?

2. Use $n = 3$, and give an example of a symmetric binomial model and an example of an asymmetric binomial model.

3. A butcher believes that 50% of all meat shoppers buy the hamburger that is sale-priced.
   (a) What is the probability that 8 of the next 20 shoppers buy the sale-priced hamburger?
   (b) What assumptions did you make to arrive at your answer?

4. A fair coin is tossed 4 times, and X denotes the number of heads that occur.
   (a) Find the probability distribution for X.
   (b) What are E(X) and V(X)?
   (c) If coin flipping and family compositions are assumed to be (abstractly) equivalent experiments, determine model frequencies for 160 families of 4 children.

5. Use the binomial models in Appendix B to answer the following questions when $W \sim b(n, p)$.
   (a) If $n = 10$ and $p = .5$, then $P(W=6)$; $P(W\leq 7)$; $P(W\geq 2)$; $P(5\leq W\leq 8)$; most likely value(s) of W.
   (b) If $n = 20$ and $p = 1/3$, then $P(W=6)$; $P(W\leq 10)$; $P(W>6)$; $P(3<W<9)$; most likely value(s) of W.
   (c) If $n = 5$ and $p = .5$, what is the most likely value(s) of W?
   (d) If $n = 25$ and $p = .75$, then $P(W=20)$; $P(W\leq 15)$; $P(W\geq 18)$; $P(11\leq W\leq 18)$; most likely value(s) of W.
   (e) If $n = 16$ and $p = 2/3$, then $P(W=7)$; $P(W=16)$; $P(W<20)$; $P(W>7)$; $P(5<W<12)$; most likely value(s) of W.

6. Indicate at least one study (possibly in your field of interest) that satisfies the assumptions for a binomial model.

*4.5 MULTINOMIAL EXPERIMENTS AND MODELS

Forcing the response for a question to fall into one of only two available categories is sometimes unnatural.

For example, a survey question might be best answered by yes, no, or undecided, rather than simply yes or no. Classification of achievement in a class could perhaps be better represented by the 5 letters, A, B, C, D, and F, rather than simply by P(pass) or F(fail). By relaxing the assumption that only two categories of response are allowed, while retaining the other assumptions for a binomial distribution, we can generate a new set of statistical models. Those models that describe experiments having three or more responses are called *multinomial* models. The discussion here is divided into two parts. First, we consider only three categories to present the necessary extensions of notation, as well as a study of trinomial models. Second, we present a summary of results for general multinomial models.

TRINOMIAL EXPERIMENTS AND MODELS: As an introduction to trinomial experiments and a motivation for trinomial models, the following experiment is suggested.

EXPERIMENT: Roll a die 3 times, and at the end of 3 rolls, record the following data:

x_1 = number of times the event $A_1 = \{1\}$ occurs,
x_2 = number of times the event $A_2 = \{2, 3\}$ occurs,
$3-x_1 - x_2$ = number of times the event $A_3 = \{4, 5, 6\}$ occurs.

Repeat the experiment 27 times, and record the observed frequencies in a distribution.

As we perform an experiment such as the one just outlined, some of the differences between a binomial model and a trinomial model become apparent. One difference is that additional notation is needed for discussion. In the binomial case, we used X and (n-X) to denote the number of successes and the number of failures, respectively, in n trials of the experiment, so only one variable, X, was needed. However, in the trinomial case, we must use two variables. X_1 denotes the number of times category 1 (event A_1) occurs; X_2 denotes the number of times category 2 (event A_2) occurs; and $n-X_1-X_2$ denotes the number of times category 3 (event A_3) occurs when the experiment consists of n independent trials that can result in any one of three possible outcomes. Since there are three possible outcomes for each trial of the experiment, we need to assign probabilities to three events. Thus, we use

$$p_1 = P(A_1) \text{ on every trial,}$$

$$p_2 = P(A_2) \text{ on every trial,}$$

$$q = 1-p_1-p_2 = P(A_3) \text{ on every trial.} \quad (4.23)$$

The use of $q = 1-p_1-p_2$ is simply a notational convenience.

Turning now to a second difference between a binomial distribution and a trinomial distribution, let us consider the sample spaces for the variables involved. First, a binomial variable, X, is an integer which takes on values x such that $0 \le x \le n$. Second, a trinomial distribution involves two variables, X_1 and X_2, which are integers such that their values, x_1 and x_2, satisfy

$$0 \le x_1 + x_2 \le n. \tag{4.24}$$

The preceding comments can be summarized in the following assumptions for a trinomial model:

(1) The sample space of the experiment is partitioned into three sets. These sets are denoted by A_1, A_2, and A_3.

(2) n independent trials of the experiment are conducted.

(3) For each trial of the experiment, we assume that $P(A_1) = p_1$, $P(A_2) = p_2$, and $P(A_3) = 1-p_1-p_2 (=q)$.

We use variables

$X_1 = \#(A_1)$ occurring in n trials,
$X_2 = \#(A_2)$ occurring in n trials,
$n-X_1-X_2 = \#(A_3)$ occurring in n trials.

Then the trinomial probability model is defined as

$$P(x_1, x_2) = P(X_1 = x_1, X_2 = x_2)$$

$$= \frac{n!}{x_1!x_2!(n-x_1-x_2)!} p_1^{x_1} p_2^{x_2} q^{n-x_1-x_2} \tag{4.25}$$

for all pairs of points, (x_1, x_2), such that $0 \le x_1 + x_2 \le n$.

Equation (4.25) can be rationalized in the following way. Any sequence of n terms such that $\#(A_1) = x_1$, $\#(A_2) = x_2$, and $\#(A_3) = n-x_1-x_2$ has (by independence) probability $p_1^{x_1} p_2^{x_2} q^{n-x_1-x_2}$. Furthermore, *Definition 12* on page 46 indicates that the number of permutations of such sequence is

$$n!/[x_1!x_2!(n-x_1-x_2)!].$$

Thus, the coefficient in Eq. (4.25) gives the number of outcomes that show exactly $\#(A_1) = x_1$ and $\#(A_2) = x_2$ [necessarily $\#(A_3) = n-x_1-x_2$], while the remaining terms involving p_1, p_2, and q provide the probability of any one of these outcomes.

Let us return now to the experiment given at the beginning of this section. For the experiment being considered, we have $n = 3$, $p_1 = 1/6$, $p_2 = 1/3$, and $q = 1/2$. Thus, the sample space consists of all pairs of integer points (x_1, x_2) such that $0 \leq x_1 + x_2 \leq 3$. These points and their corresponding probabilities are given in Table 4.8.

TABLE 4.8.
A trinomial probability distribution is given with $n = 3$, $p_1 = 1/6$, $p_2 = 1/3$, and $q = 1/2$.

(x_1, x_2)	$P(x_1, x_2)$
(0,0)	$[3!/(0!0!3!)](1/6)^0 (1/3)^0 (1/2)^3 = 1/8$
(0,1)	$[3!/(0!1!2!)](1/6)^0 (1/3)^1 (1/2)^2 = 1/4$
(0,2)	$[3!/(0!2!1!)](1/6)^0 (1/3)^2 (1/2)^1 = 1/6$
(0,3)	$[3!/(0!3!0!)](1/6)^0 (1/3)^3 (1/2)^0 = 1/27$
(1,0)	$[3!/(1!0!2!)](1/6)^1 (1/3)^0 (1/2)^2 = 1/8$
(1,1)	$[3!/(1!1!1!)](1/6)^1 (1/3)^1 (1/2)^1 = 1/6$
(1,2)	$[3!/(1!2!0!)](1/6)^1 (1/3)^2 (1/2)^0 = 1/18$
(2,0)	$[3!/(2!0!1!)](1/6)^2 (1/3)^0 (1/2)^1 = 1/24$
(2,1)	$[3!/(2!1!0!)](1/6)^2 (1/3)^1 (1/2)^0 = 1/36$
(3,0)	$[3!/(3!0!0!)](1/6)^3 (1/3)^0 (1/2)^0 = 1/216$
Total	1

To further illustrate a trinomial model, Table 4.9 presents empirical distributions for four repetitions of the experiment outlined at the beginning of this section. The model frequencies are produced by multiplying each probability in Table 4.8 by 27.

TABLE 4.9.
Empirical frequencies and model frequencies are given for 4 samples of 27 rolls of 3 dice with $X_1 = \#(1)$ and $X_2 = \#(2,3)$.

(x_1,x_2)	Sample 1	Sample 2	Sample 3	Sample 4	Model
(0,0)	4	4	4	3	3.375
(0,1)	6	4	9	6	6.750
(0,2)	4	4	2	4	4.500
(0,3)	0	3	3	1	1.000
(1,0)	6	4	2	6	3.375
(1,1)	3	5	5	5	4.500
(1,2)	2	1	0	2	1.500
(2,0)	1	0	1	0	1.125
(2,1)	1	2	1	0	.750
(3,0)	0	0	0	0	.125
Total	27	27	27	27	27.000

The empirical results in Table 4.9 seem to agree reasonably well with the model frequencies. Thus, it seems that the proposed trinomial model is appropriate for describing the results for the preceding dice experiment.

To conclude this section, we present some properties of a trinomial model. First, the probabilities in Eq. (4.25) total to 1 because a generalization of Eq. (4.19), which could be called the trinomial theorem, states that

$$\sum_{\{(x_1,\, x_2)\,|\,0 \le x_1 + x_2 \le n\}} \sum P(x_1, x_2) = (q + p_1 + p_2)^n = 1^n = 1. \tag{4.26}$$

Table 4.8 verifies the result for $n = 3$, $p_1 = 1/6$, $p_2 = 1/3$, and $q = 1/2$. Second, we look at means and variances of the two variables, X_1 and X_2. Here it can be shown that

$$E(X_1) = np_1, \; E(X_2) = np_2,$$
$$V(X_1) = np_1(1-p_1), \; V(X_2) = np_2(1-p_2). \tag{4.27}$$

The results in Eq. (4.27) are exactly as if $X_1 \sim b(n,p_1)$ and $X_2 \sim b(n,p_2)$, and this *is* the case. That is, X_1 and X_2 are jointly distributed as a trinomial distribution, but marginally they are binomially distributed. The following examples help to clarify that statement.

EXAMPLE 10: For the joint distribution of X_1 and X_2 given in Table 4.8, show that the marginal distributions of X_1 and X_2 are b(3, 1/6) and b(3, 1/3), respectively.

SOLUTION: For convenience, the joint distribution is presented here in a two-way table rather than a column, as in Table 4.8.

$x_1 \backslash x_2$	0	1	2	3	Total
0	1/8	1/4	1/6	1/27	125/216
1	1/8	1/6	1/18	0	75/216
2	1/24	1/36	0	0	15/216
3	1/216	0	0	0	1/216
Total	8/27	4/9	2/9	1/27	1

When the joint distribution is presented as shown, the marginal distributions of X_1 and X_2 are found by adding across the columns and rows, respectively. Thus, the marginal distribution of X_1 is the set of totals on the right side of the table, and the marginal distribution of X_2 is given by the totals at the bottom of the table. Next, observe that these numbers are exactly the ones we obtain by evaluating the following binomial probabilities:

$$P(X_1) = \binom{3}{x_1}(1/6)^{x_1}(5/6)^{3-x_1}, \quad x_1 = 0, 1, 2, 3,$$

$$P(X_2) = \binom{3}{x_2}(1/3)^{x_2}(2/3)^{3-x_2}, \quad x_2 = 0, 1, 2, 3.$$

Thus, $X_1 \sim b(3, 1/6)$ and $X_2 \sim b(3, 1/3)$.

EXAMPLE 11: Find $E(X_1)$, $E(X_2)$, $V(X_1)$, and $V(X_2)$ when X_1 and X_2 have the trinomial distribution given in Table 4.8 (*Example 10*).

SOLUTION: By Eq. (4.27),

$$E(X_1) = 3(1/6) = .5, \quad E(X_2) = 3(1/3) = 1,$$

$$V(X_1) = 3(1/6)(5/6) = 5/12,$$

$$V(X_2) = 3(1/3)(2/3) = 2/3.$$

Note that these results also follow from Eqs. (4.20) and (4.21) when the results of *Example 10* are used.

EXERCISES

1. Suppose that X_1 and X_2 are trinomial variables with $p_1 = 1/4$ and $p_2 = 1/2$. If $n = 5$, find the following:

   (a) $P(X_1 = 2, X_2 = 2)$ (e) $E(X_1)$

   (b) $P(X_1 = 5, X_2 = 0)$ (f) $E(X_2)$

   (c) $P(X_1 = 0, X_2 = 5)$ (g) $V(X_1)$

   (d) $P(X_1 = 3, X_2 = 3)$ (h) $V(X_2)$.

2. Assume that the students of MYU are composed of 40% Republicans, 40% Democrats, and 20% Independents. In a random sample of 10 students, what is the probability that we find 5 Democrats and 5 Republicans? If 12 repeated samples of 10 each are observed, and the foregoing assumptions are correct, what are the expected numbers of Republicans, Democrats, and Independents in the combined samples?

3. Three balanced cubes (dice) are painted so that each has 1 red face, 2 blue faces, and 3 green faces. What trinomial distribution would be appropriate for modeling repeated rolls of the 3 cubes? (In specifying a model, indicate X_1, X_2, p_1, p_2, and n.) Does the selected model differ from the trinomial distribution in Table 4.8?

GENERAL MULTINOMIAL EXPERIMENTS AND MODELS: We now generalize the preceding trinomial model to include an arbitrary number of categories, k. The assumptions and notations are as follows.

(1) Each trial of an experiment results in an outcome that is classified into one of k categories. These categories are denoted by A_1, A_2, ..., A_k.

(2) n independent trials of the experiment produce values for the following (k-1) random variables:

$X_1 = \#(A_1),\ X_2 = \#(A_2),\ \ldots,\ X_{k-1} = \#(A_{k-1}),$

and $n - X_1 - X_2 \ldots - X_{k-1} = \#(A_k)$.

(3) On each and every trial of an experiment, we assume that

$p_1 = P(A_1),\ p_2 = P(A_2),\ \ldots,\ p_{k-1} = P(A_{k-1}),$

and $q = 1 - p_1 - p_2 - \ldots - p_{k-1} = P(A_k)$.

REMARK: It is frequently convenient to think of the classes (categories) in multinomial models as consisting of (k-1) classes of special interest and *one* class (the kth class) which includes all of the other possibilities. Thus, the kth class (pick-up class) gets the outcomes not assigned to the (k-1) classes of interest and is assigned probability available after the (k-1) classes have received their allotment. In that light, the notation adopted seems to be quite descriptive.

When the previously cited assumptions (1), (2), and (3) are met, we can use the logic of Eq. (4.25) to write the probability function for $X_1, X_2, \ldots, X_{k-1}$ as follows:

$$P(x_1, x_2, \ldots, x_{k-1}) = \frac{n!}{x_1!x_2! \ldots x_{k-1}!(n-x_1-x_2 \ldots -x_{k-1})!}$$

$$p_1^{x_1} p_2^{x_2} \ldots p_{k-1}^{x_{k-1}} \cdot q^{n-x_1-x_2- \ldots -x_{k-1}}, \qquad (4.28)$$

for all integer values of $x_1, x_2, \ldots, x_{k-1}$ satisfying $0 \le x_1 + x_2 + \ldots + x_{k-1} \le n$.

The properties for multinomial variables, $X_1, X_2, \ldots, X_{k-1}$, are analogous to those for the trinomial values in the section on trinomial models, page 149. In particular, the function defined in Eq. (4.28) adds to 1 when summed over all the points, $(x_1, x_2, \ldots, x_{k-1})$ satisfying the constraint

$$0 \le \sum_{i=1}^{k=1} x_i \le n. \qquad (4.29)$$

Furthermore,

$$E(X_i) = np_i,\ V(X_i) = np_i(1-p_i),\ i = 1, 2, \ldots, k-1. \qquad (4.30)$$

In fact, the marginal distribution of each X_i, i = 1, 2, ..., k-1 is binomial with the parameter, p_i, and the number of trials, n.

It is perhaps obvious that the amount of arithmetic required to exhibit the probabilities associated with a multinomial model increases rapidly as either k or n or both increase. That fact becomes painfully clear as an example is worked. Hopefully, *Example 12* can illustrate some of the problems involved. In particular, observe the increased number of points to which probabilities are assigned when changing from 3 (Table 4.8) to 4 categories while n is held at 3. This simply emphasizes the fact that the number of points $(x_1, x_2, \ldots, x_{k-1})$ satisfying Eq. (4.29) increases rapidly with k. We notice that it also increases rapidly with n.

EXAMPLE 12: Suppose 3 fair dice are rolled (or equivalently, 1 fair die is rolled 3 times), and the results are recorded as follows:

$$X_1 = \#(A_1), \text{ where } A_1 = \{1\},$$
$$X_2 = \#(A_2), \text{ where } A_2 = \{2, 3\},$$
$$X_3 = \#(A_3), \text{ where } A_3 = \{4, 5\},$$
$$3-X_1-X_2-X_3 = \#(A_4), \text{ where } A_4 = \{6\}.$$

Then the sample space becomes all ordered triplets (x_1, x_2, x_3) such that $0 \leq x_1 + x_2 + x_3 \leq 3$. Furthermore, in the notation just given, n = 3, $p_1 = 1/6$, $p_2 = 1/3$, $p_3 = 1/3$, and q = 1/6. Thus, we can substitute these values into Eq. (4.28) and generate the desired probability distribution given in Table 4.10.

TABLE 4.10.
A multinomial probability distribution is given with $n = 3$, $p_1 = 1/6$, $p_2 = 1/3$, $p_3 = 1/3$, and $q = 1/6$.

(x_1, x_2, x_3)	$P(x_1, x_2, x_3)$
(0,0,0)	$[3!/(0!0!0!3!)](1/6)^0\ (1/3)^0\ (1/3)^0\ (1/6)^3 = 1/216$
(0,0,1)	$[3!/(0!0!1!2!)](1/6)^0\ (1/3)^0\ (1/3)^1\ (1/6)^2 = 1/36$
(0,0,2)	$[3!/(0!0!2!1!)](1/6)^0\ (1/3)^0\ (1/3)^2\ (1/6)^1 = 1/18$
(0,0,3)	$[3!/(0!0!3!0!)](1/6)^0\ (1/3)^0\ (1/3)^3\ (1/6)^0 = 1/27$
(0,1,0)	$[3!/(0!1!0!2!)](1/6)^0\ (1/3)^1\ (1/3)^0\ (1/6)^2 = 1/36$

Table 4.10. (continued)

(x_1, x_2, x_3)	$P(x_1, x_2, x_3)$
(0,1,1)	$[3!/(0!1!1!1!)](1/6)^0\ (1/3)^1\ (1/3)^1\ (1/6)^1 = 1/9$
(0,1,2)	$[3!/(0!1!2!0!)](1/6)^0\ (1/3)^1\ (1/3)^2\ (1/6)^0 = 1/9$
(0,2,0)	$[3!/(0!2!0!1!)](1/6)^0\ (1/3)^2\ (1/3)^0\ (1/6)^1 = 1/18$
(0,2,1)	$[3!/(0!2!1!0!)](1/6)^0\ (1/3)^2\ (1/3)^1\ (1/6)^0 = 1/9$
(0,3,0)	$[3!/(0!3!0!0!)](1/6)^0\ (1/3)^3\ (1/3)^0\ (1/6)^0 = 1/27$
(1,0,0)	$[3!/(1!0!0!2!)](1/6)^1\ (1/3)^0\ (1/3)^0\ (1/6)^2 = 1/72$
(1,0,1)	$[3!/(1!0!1!1!)](1/6)^1\ (1/3)^0\ (1/3)^1\ (1/6)^1 = 1/18$
(1,0,2)	$[3!/(1!0!2!0!)](1/6)^1\ (1/3)^0\ (1/3)^2\ (1/6)^0 = 1/18$
(1,1,0)	$[3!/(1!1!0!1!)](1/6)^1\ (1/3)^1\ (1/3)^0\ (1/6)^1 = 1/18$
(1,1,1)	$[3!/(1!1!1!0!)](1/6)^1\ (1/3)^1\ (1/3)^1\ (1/6)^0 = 1/9$
(1,2,0)	$[3!/(1!2!0!0!)](1/6)^1\ (1/3)^2\ (1/3)^0\ (1/6)^0 = 1/18$
(2,0,0)	$[3!/(2!0!0!1!)](1/6)^2\ (1/3)^0\ (1/3)^0\ (1/6)^1 = 1/72$
(2,0,1)	$[3!/(2!0!1!0!)](1/6)^2\ (1/3)^0\ (1/3)^1\ (1/6)^0 = 1/36$
(2,1,0)	$[3!/(2!1!0!0!)](1/6)^2\ (1/3)^1\ (1/3)^0\ (1/6)^0 = 1/36$
(3,0,0)	$[3!/(3!0!0!0!)](1/6)^3\ (1/3)^0\ (1/3)^0\ (1/6)^0 = 1/216$
Total	1

By summing the appropriate probabilities in Table 4.10, we find that marginally $X_1 \sim b(3, 1/6)$, $X_2 \sim b(3, 1/3)$, and $X_3 \sim b(3, 1/3)$. Thus, we have

$$E(X_1) = 1/2, \; E(X_2) = E(X_3) = 1,$$

$$V(X_1) = 5/12, \; V(X_2) = V(X_3) = 2/3.$$

These are also the values indicated by Eq. (4.30).

EXERCISES

1. Indicate one or more studies that satisfy the assumptions of a multinomial model.

2. A political science class conducted a census of the student body at MYU and found that the students could be classified as:

   A_1: registered Republican

   A_2: registered Democrat

   A_3: registered Independent

   A_4: unregistered.

   They also determined that the percentages for each category were as follows: A_1, 20%; A_2, 30%; A_3, 10%; and A_4, 40%. What is the expected number of individuals for each category in samples of size 10? What is the probability that a random sample of 10 students produces 2 registered Republicans, 3 registered Democrats, 1 registered Independent, and 4 unregistered voters?

3. If a fair die is rolled 6 times, what is the probability that all faces are observed?

*4.6 HYPERGEOMETRIC EXPERIMENTS AND MODELS

In the preceding section, a family of models called multinomials was introduced. Those models were motivated by modifying one of the basic requirements needed for a binomial model; namely, that each outcome of an experiment be classified in one of only two categories. In this section, we relax the independence requirement for the binomial model. That is, the probability of various events is dependent from one trial of our experiment to another. These models are called *hypergeometric* distributions and are widely used in experiments where sampling is conducted without replacement. In fact, we sometimes see these models called *sampling without replacement* models in contrast to binomial and multinomial models, which may be called *sampling with replacement* models.

It is recommended that one or both of the following experiments be completed, either as an individual exercise or as a class exercise, to collect some data for which hypergeometric distributions are the appropriate models.

EXPERIMENT 1: Mark 4 ping-pong balls (or other easily mixed objects) as R, and mark 6 ping-pong balls as A. Mix the 10 balls thoroughly, and select 3 balls simultaneously (or equivalently one at a time *without* replacement). Let X = #(R) observed. Repeat the experiment 30 times, and form a distribution of your results. (Note: There are many substitutes for the ping-pong balls just mentioned. For example, we could select 4 hearts (R) and 6 spades (A) from a deck of playing cards and select 3 cards without replacement after shuffling the cards.)

EXPERIMENT 2: Mark 12 ping-pong balls as follows: 4 as C, 5 as A, and 3 as T. Mix the 12 marked balls, and select 3 without replacement. Let X_1 = #(C) and X_2 = #(A) [necessarily, $3-X_1-X_2$ = #(T)]. Repeat the experiment 55 times, and form a distribution for points (x_1, x_2) such that $0 \leq x_1 + x2 \leq 3$.

Notice that *Experiment 1* has only two classes of elements being sampled, while *Experiment 2* has three classes of elements in the population. Our classification of these two hypergeometric models corresponds to the binomial and multinomial (trinomial) models, respectively. Thus, we discuss hypergeometric models for populations consisting of two kinds of elements, then proceed to models for populations having k ($k \geq 3$) kinds of elements.

HYPERGEOMETRIC MODELS (TWO CLASSES): In experiments of the type just described, we consider a finite population of elements of two or more types being sampled without replacement. This section is concerned with those situations where the population elements are of two types, while the next section generalizes the results for populations having three or more types of elements.

In inspection sampling, where a few items from a bulk of items are inspected to determine whether or not the lot is accepted or rejected, the two classes of elements in the population are called *defective* and *nondefective*. Of course, these terms do not restrict the usefulness of hypergeometric models. They simply provide reference names for the two types of elements present in the population. In that light, we establish the necessary notation for a hypergeometric model as follows:

N = total number of elements in the population
D = number of defective elements in the population
n = sample size
X = number of defective elements in the sample.

(4.31)

Before stating some motivation for the hypergeometric probability model, we need to discuss the limits for the sample space of X. First, the upper bound on X, the number of defectives in the sample size, n, is either D or n because the number of defectives in the sample is bounded by either the sample size, n, or the total number of defectives in the population, D. In particular, X is bounded above by D whenever $D < n$ and n otherwise. Similarly, the lower bound on X is 0 whenever $N \geq n + D$ and $n + D - N$ otherwise. Thus, using the notation in (4.31) and the preceding limits for X, the hypergeometric probability function is as follows:

$$P(x) = P(X=x) = \left[\binom{D}{x}\binom{N-D}{n-x}\right] / \binom{N}{n},$$

$$x = \max(0, n+D-N), \ldots, \min(D,n), \qquad (4.32)$$

where max(a,b) and min(a,b) denote the larger and smaller, respectively, of the two numbers, a and b.

The functional form of Eq. (4.32) follows directly from the combinatorial results in Sections 2.6 and 2.8. First, there are $\binom{N}{n}$ ways in which a sample size n can be selected from the N elements in the population. Second, when the population is partitioned into D defective and (N-D) nondefective items, there are $\binom{D}{x}$ ways in which x defectives can be selected and $\binom{N-D}{n-x}$ ways for selecting (n-x) nondefectives. Thus, the probability that a sample of n contains x defectives [and (n-x) nondefectives] is simply the ratio of the number of ways for selecting x defectives *and* (n-x) nondefectives from D defectives and (N-D) nondefectives, respectively, to the total number of possible samples of size n. This reasoning gives Eq. (4.32) immediately.

For *Experiment 1*, we have N=10, D=4, and n=3. The corresponding probability model is found by using those values of N, D, and n in Eq. (4.32) for x=0, 1, 2, 3. Table 4.11 presents that hypergeometric model.

To compare data from *Experiment 1* with the model in Table 4.11, we determine the model frequencies by multiplying the probabilities in Table 4.11 by 30, the total frequency. The empirical frequencies for two sets of 30 samples from *Experiment 1* and the corresponding model frequencies are given in Table 4.12.

TABLE 4.11.
The hypergeometric probability model is given for N = 10, D = 4, and n = 3.

x	P(x)
0	$[\binom{4}{0}\binom{6}{3}]/\binom{10}{3} = 1/6$
1	$[\binom{4}{1}\binom{6}{2}]/\binom{10}{3} = 1/2$
2	$[\binom{4}{2}\binom{6}{1}]/\binom{10}{3} = 3/10$
3	$[\binom{4}{3}\binom{6}{0}]/\binom{10}{3} = 1/30$
Total	1

TABLE 4.12.
Empirical and model frequencies are given for results from *Experiment 1*.

x	sample 1	Sample 2	Model
0	6	2	5
1	12	16	15
2	12	10	9
3	0	2	1
Total	30	30	30

A quick glance at Table 4.12 seems to indicate that our model is a reasonable representation of the empirical results. To investigate the effects of increasing the sample size, we may pool together 2 or more independent sets of 30 observations. These pooled frequencies would then be compared with multiples of the model frequencies in Table 4.12.

Direct evaluation of E(X) and V(X) by means of Eqs. (4.2) and (4.3) for the distribution in Table 4.11 gives

$$E(X) = 0(1/6) + 1(1/2) + 2(3/10) + 3(1/30)$$
$$= 1.2,$$
$$V(X) = (-1.2)^2(1/6) + (-.2)^2(1/2) + (.8)^2(3/10) + (1.8)^2(1/30)$$
$$= .56.$$

These values can also be verified by using the following general formulas for hypergeometric distributions having two classes of elements:

$$E(X) = n(D/N). \tag{4.33}$$

$$V(X) = n(D/N)\ [1 - (D/N)][(N-n)/(N-1)]. \tag{4.34}$$

If we use D/N (the probability of defectives in the first selection) as p, then Eq. (4.33) corresponds exactly to Eq. (4.20). Also, the first three factors of Eq. (4.34) agree with Eq. (4.21) for p = D/N. The remaining factor of Eq. (4.34) is the *finite population correction factor.* Since we are sampling a finite population without replacement, we use the total population whenever n = N and there is no variation in X for that case. The finite correction factor is discussed in more detail in Chapter 6.

The similarities between the mean and variance of hypergeometric and binomial models are only a part of the full relationship. If we allow N and D to become large (approach ∞) so that p = D/N is a fixed value, then the hypergeometric probability model in Eq. (4.32) becomes the binomial probability model in Eq. (4.15). That is,

$$\lim_{\substack{p = D/N \\ D,\ N\to\infty}} \left[\binom{D}{x}\binom{N-D}{n-x}\right] \Big/ \binom{N}{n} = \binom{n}{x} p^x q^{n-x},$$

$$x = 0, 1, 2, \ldots, n.$$

The importance of this relationship is that binomial probabilities (see Table B1 in Appendix B) provide excellent approximations for hypergeometric probabilities when N is large. For example, when selecting 20 light bulbs from a population of 10,000 bulbs, of which 1,000 are defective, the probability of 6 defectives is

$$P(6) = \left[\binom{1000}{6}\binom{9000}{14}\right] \Big/ \binom{10000}{20} \approx \binom{20}{6}(.1)^6(.9)^{14}$$

$$= .0089 \quad \text{(see Table B1 in Appendix B).}$$

This completes our discussion of hypergeometric models for populations containing two types of elements. The next section extends the foregoing arguments to develop hypergeometric models for populations having more than two types of elements.

HYPERGEOMETRIC MODELS (K CLASSES): Just as it was not always natural to restrict observations to one of only two classes as required by the binomial model, it is also unnecessarily restrictive to require that the elements of a population be classified into one of only two categories. Thus, we extend the notation and results of the preceding section to produce models for sampling (without replacement) populations having an arbitrary k type of elements. Data generated by *Experiment 2* provide an example of sampling from such a population when k = 3. Following a few general definitions, we return to present two sets of empirical data, as well as model frequencies for *Experiment 2*.

To accommodate populations having k types of elements, the notation in Eq. (4.31) is extended as follows:

$$\begin{aligned}
N &= \text{total number of elements in the population}\\
D_1 &= \text{number of elements of a first kind}\\
D_2 &= \text{number of elements of a second kind}\\
&\vdots\\
D_{k-1} &= \text{number of elements of a (k-1)th kind}\\
n &= \text{sample size}\\
X_1 &= \text{number of first type elements in sample}\\
X_2 &= \text{number of second type elements in sample}\\
&\vdots\\
X_{k-1} &= \text{number of (k-1)th type elements in sample.}
\end{aligned} \tag{4.35}$$

Before stating the multidimensional, hypergeometric model using the notations of (4.35), the many restrictions on $x_1, x_2, \ldots, x_{k-1}$ (the values which the random variables $X_1, X_2, \ldots, X_{k-1}$ take on) are exhibited. First,

$$0 \le x_i \le \min(D_i, n), \quad i = 1, 2, \ldots, k - 1. \tag{4.36}$$

Second,

$$0 \leq x_1 + x_2 + \ldots + x_{k-1} \leq n. \qquad (4.37)$$

The conditions in (4.36) and (4.37) must be satisfied simultaneously. Now, using the notation in (4.35), the desired probability is

$$P(x_1, x_2, \ldots, x_{k-1})$$

$$= \frac{\binom{D_1}{x_1}\binom{D_2}{x_2}\cdots\binom{D_{k-1}}{x_{k-1}}\binom{N-D_1-D_2-\ldots-D_{k-1}}{n-x_1-x_2-\ldots-x_{k-1}}}{\binom{N}{n}} \qquad (4.38)$$

for all points $(x_1, x_2, \ldots, x_{k-1})$ satisfying the restrictions in (4.36) and (4.37). Using Eqs. (4.38), (4.2), and (4.3), it can be shown that

$$E(X_i) = n(D_i/N), \; i=1, 2, \ldots, k-1, \qquad (4.39)$$

$$V(X_i) = n(D_i/N)[1-(D_i/N)][(N-n)/(N-1)],$$

$$i=1, 2, \ldots, k-1. \qquad (4.40)$$

We note that the limiting form of the hypergeometric model in Eq. (4.38) is the multinomial model in Eq. (4.28). That is, Eq. (4.38) approaches Eq. (4.28) when $N, D_1, D_2, \ldots D_{k-1}$ approach ∞ in such a way that $p_i = D_i/N$, $i=1, 2, \ldots, k-1$ are fixed values. Thus, multinomial probabilities provide good approximations for hypergeometric probabilities.

For an example of a hypergeometric model with $k = 3$, return to *Experiment 2* on page 159. Here $N = 12$, $D_1 = 4$, $D_2 = 5$, $(N-D_1 - D_2 = 3)$, and $n = 3$. The probability model for those values of N, D_1, D_2, and n is given in Table 4.13. The empirical frequencies from 2 independent samples of 55 trials each and model frequencies for *Experiment 2* are also presented in Table 4.13.

Applying Eq. (4.39), we find $E(X_1) = 1$ and $E(X_2) = 1.25$. Also, Eq. (4.40) gives $V(X_1) = 6/11$ and $V(X_2) = 105/176$. These results could have been obtained from Table 4.13 by first finding the marginal distributions for X_1 and X_2, but using Eqs. (4.39) and (4.40) is considerably easier.

TABLE 4.13.
A hypergeometric distribution for $N = 12$, $D_1 = 4$, $D_2 = 5$, and $n = 3$, as well as empirical and model frequencies for 55 trials of *Experiment 2*.

(x_1,x_2)	$P(x_1,x_2)$	Frequencies		
		Sample 1	Sample 2	Model
(0,0)	$\binom{4}{0}\binom{5}{0}\binom{3}{3}/\binom{12}{3} = 1/220$	0	0	.25
(0,1)	$\binom{4}{0}\binom{5}{1}\binom{3}{2}/\binom{12}{3} = 15/220$	1	4	3.75
(0,2)	$\binom{4}{0}\binom{5}{2}\binom{3}{1}/\binom{12}{3} = 30/220$	7	12	7.50
(0,3)	$\binom{4}{0}\binom{5}{3}\binom{3}{0}/\binom{12}{3} = 10/220$	2	2	2.50
(1,0)	$\binom{4}{1}\binom{5}{0}\binom{3}{2}/\binom{12}{3} = 12/220$	4	2	3.00
(1,1)	$\binom{4}{1}\binom{5}{1}\binom{3}{1}/\binom{12}{3} = 60/220$	14	16	15.00
(1,2)	$\binom{4}{1}\binom{5}{2}\binom{3}{0}/\binom{12}{3} = 40/220$	12	12	10.00
(2,0)	$\binom{4}{2}\binom{5}{0}\binom{3}{1}/\binom{12}{3} = 18/220$	4	2	4.50
(2,1)	$\binom{4}{2}\binom{5}{1}\binom{3}{0}/\binom{12}{3} = 30/220$	9	5	7.50
(3,0)	$\binom{4}{3}\binom{5}{0}\binom{3}{0}/\binom{12}{3} = 4/220$	2	0	1.00
Total	1	55	55	55

The preceding example indicates that listing the full distribution for a multidimensional, hypergeometric model may be a difficult task. Yet the evaluation of any particular term of interest is quite straightforward. To simplify the evaluation of the combinatorials involved in any particular probability calculation, Table B2 in Appendix B may be used. The following example provides an illustration of the procedure involved.

EXAMPLE 13: The student governing board consists of 4 freshmen, 6 sophomores, 6 juniors, and 5 seniors. If 4 individuals are selected at random to attend a convention for developing better student governments, what is the probability that each class is represented on the convention delegation?

SOLUTION: Equation (4.38) gives

$$P(1,1,1) = \left[\binom{4}{1}\binom{6}{1}\binom{6}{1}\binom{5}{1}\right]\Big/\binom{21}{4}.$$

Either by direct calculation or by using Table B2 in Appendix B, we find

$$\binom{21}{4} = 5985,\ \binom{4}{1} = 4,\ \binom{6}{1} = 6,\ \text{and}\ \binom{5}{1} = 5.$$

Therefore, the probability that each class is represented on a randomly selected delegation is

$$(4 \cdot 6 \cdot 6 \cdot 5)/5985 = 16/133 = .12.$$

EXERCISES

1. In *Example 13*, what is the probability that the selected delegation contains only freshmen? sophomores? juniors? seniors?

2. The School of Business Administration at MYU has purchased 10 new typewriters for classroom use. The 10 machines have been selected from a local dealer's stock of 25 machines, of which 5 are defective. Find the following probabilities: that all 10 typewriters are nondefective; that 5 of the typewriters are defective; that 2 of the typewriters are defective.

3. From *Exercise 2*, find the mean and variance for the number of defective typewriters in the order of 10.

4. The members of a senate committee on foreign trade are classified as follows: 6 Democrats, 4 Republicans, and 2 Independents. A delegation of 4 is to be selected to attend a trade conference in Europe. What is the probability that the delegation consists of 2 Democrats and 2 Republicans? 2 Republicans and 2 Independents? 1 Democrat, 1 Republican, and 2 Independents? What is the probability that both Independents are members of the delegation? What is the expected number of Democrats to be selected for the delegation?

5. In *Exercise 4*, if the committee chairperson (a Democrat) is automatically a member of the delegation, what is the probability that the delegation consists of 2 Democrats and 2 Republicans?

6. In a bag of 20 apples, 2 are spoiled. Find the probability distribution for the variable, X, which denotes the number of spoiled apples in a sample of 3 apples selected for a weekend hike. Find E(X) and V(X) by two methods: first, by using Eqs. (4.2) and (4.3), and, second, by using Eqs. (4.33) and (4.34).

*4.7 POISSON EXPERIMENTS AND MODELS

With respect to all previous models, they can be derived from basic probability concepts quite easily. The Poisson probability distribution can also be derived, but its derivation is not as easily seen since it involves limits. Thus, we conclude this chapter with a discrete probability distribution that is named after Simeon Denis Poisson (1781-1840), a French mathematician and mathematical physicist. The distribution may be described a number of different ways, but one of those ways permits us to relate Poisson models to the binomial models discussed in Section 4.4. To do that, suppose the limit of the binomial probability function, as given in Eq. (4.15), is taken as $n \to \infty$ and $p \to 0$ in such a way that $np = \lambda$. That is, the number of trials becomes infinite, while the probability of success on any one trial becomes infinitesimal. Under the specified conditions, it can be shown that

$$\lim_{\substack{n \to \infty \\ p \to 0 \\ np = \lambda}} \binom{n}{x} p^x (1-p)^{n-x} = \frac{e^{-\lambda}\lambda^x}{x!}.$$

Thus, we define the Poisson probability model to be

$$P(x) = (e^{-\lambda}\lambda^x/x!), \quad x=0,1,2,\ldots, \qquad (4.41)$$

for any $\lambda > 0$, and where $e = 2.71828\ldots$, is the base for natural logarithms.

Notice that the sample space of a Poisson random variable is the infinite set of all nonnegative integers. From a practical point of view, however, the values of P(x) differ significantly from 0 for only a small subset of those integers. In particular, the probability is spread over more points as the one parameter, λ, increases in value. The only restriction on λ is that it be positive.

The following properties can be established for the Poisson probabilities in Eq. (4.41). First,

$$\sum_{x=0}^{\infty} P(x) = 1.$$

Second,

$$E(X) = \sum_{x=0}^{\infty} x\, P(x) = \lambda. \tag{4.42}$$

Third,

$$V(X) = \sum_{x=0}^{\infty} (x - \lambda)^2\, P(x) = \lambda. \tag{4.43}$$

Thus, we see that the mean and variance of a Poisson distribution are equal to λ. Hence, it is impossible to change the location or average value of a Poisson variable without changing its variability. In particular, Poisson variables with large means also have large variances. This property is the basis for our earlier claim that the total probability of 1 is assigned to more points as λ increases.

Another motivation for the Poisson model is to suppose that the n trials of a binomial experiment are conducted at times $t_1 = 1/n$, $t_2 = 2/n$, ..., $t_n = n/n = 1$. Then the number of successes, X, occurring in the n trials denotes the *number of successes per unit of time*. The variable, X, is a count variable, and, as n becomes large, it has the probability distribution given in Eq. (4.41). Thus, Poisson models are widely used to describe variables that count the number of events occurring per unit of time, space, area, etc. Some situations which involve a variable, X, that could be or have been modeled by Poisson models are as follows:

(1) X denotes the number of students entering the student union per minute.

(2) X denotes the number of chocolate chips per cookie.

(3) X denotes the number of incoming calls per minute on the main switchboard.

(4) X denotes the number of orders for type A transistors per day.

(5) X denotes the number of real estate loan applications per week received by the $ Savings and Loan Company.

(6) X denotes the number of sod webworms per square foot of sod (3 inches deep) on the Clean Greens Golf Course.

(7) X denotes the number of automobiles per minute passing the intersection of First and College Streets.

(8) X denotes the number of defective light bulbs per shipping carton (100 bulbs).

(9) X denotes the number of deer per square mile in the Evergreen National Forest.

(10) X denotes the number of left-handed individuals in random groups of 30 individuals.

Any of the preceding situations could be used as a data-gathering exercise for either an individual student or a class. Here the distribution would consist of the number of times each value of the variable, X, was observed in repeated observations (counts).

The parameter, λ, is the expected value of X and thus represents the average count per unit of time or space. Our problem then in selecting a Poisson model for a set of empirical data is the specification of a value for λ. In general, that choice is a problem of statistical estimation and is dealt with in Chapter 7, but for the rest of this discussion, λ is usually given. In the remaining cases, where a Poisson distribution is used to approximate a binomial distribution with n large and p small, $\lambda = np$.

To illustrate some of the foregoing points, let us consider a famous set of data collected by Bortkiewicz and reported by Fisher [4] and Fisz [5]. Over a period of 20 years, a group of 10 army corps was observed. The variable, X, was the number of army men in each corps per year who died as a result of being kicked by a horse. Thus, there are $10 \cdot 20 = 200$ data points, as recorded in Table 4.14. The model frequencies and probability distributions for two proposed models are also included in Table 4.14.

Model I is the model presented by both Fisher [4] and Fisz [5]. Their selection is based on the fact that $\bar{x} = .61$ for the data in Table 4.14. Hence, the mean of the model, λ, is chosen to equal the mean of the sample, $\bar{x}$. Model II is presented here for two reasons. First, the probabilities can be calculated by using Table C2 in Appendix C. Second, the effect of a slight increase in λ is to decrease the probability for $X = 0$ and to increase the probability for $X = 1, 2, 3$. However, both models exhibit a very good agreement between sample and model frequencies. Certainly, we would not question the usefulness of a Poisson variable for representing the random variable, X, as defined for these data.

TABLE 4.14.
Bortkiewicz's horse-kick data and two Poisson models are given.

Empirical data		Model I		Model II	
x	Sample frequencies	Poisson dist. $\lambda = .61$	Model frequencies	Poisson dist. $\lambda = .6$	Model frequencies
0	109	.544	108.8	.549	109.8
1	65	.331	66.2	.329	65.8
2	22	.101	20.2	.099	19.8
3	3	.021	4.2	.020	4.0
4	1	.003	0.6	.003	.6
Total	200	1.000	200.0	1.000	200.0

Appendix C presents both the probability [P(x)] and cumulative probability distributions [F(x)] for a selected set of λ values. To assist in the evaluation of Poisson probabilities not included in Table C1, Table C2 presents values for $e^{-\lambda}$ for a larger set of λ values. These values can be used in Eq. (4.41) to calculate additional probabilities. The following examples illustrate the use of Appendix C.

EXAMPLE 14: The Easyflow Money House expects to serve an average of 2 customers per 5-minute intervals. If a Poisson model is assumed to be appropriate for the variable, X, which represents the number of customers served in the next 5 minutes, find the following: $P(X=2)$; $P(X\leq 3)$; $P(X\geq 1)$.

SOLUTION: By the conditions of the problem, $\lambda = 2$. Using Table C1 with $\lambda = 2$, we find

$$P(X=2) = P(2) = .271;$$

$$P(X\leq 3) = F(3) = .857;$$

$$P(X\geq 1) = 1 - P(X=0) = 1 - P(0) = .865.$$

EXAMPLE 15: For the situation described in *Example 14*, find E(X), V(X), and the most likely value(s) of X.

SOLUTION: From Eqs. (4.42) and (4.43), we find that E(X) = 2 and V(X) = 2.

By inspecting the Poisson distribution for $\lambda = 2$ in Table C1, we find that X=1 and X=2 are equally likely and that their common probability exceeds the probability of any other value of X. Thus, 1 and 2 are the most likely values of X.

EXAMPLE 16: If X, the number of defective light bulbs per carton (100 bulbs), is assumed to be a Poisson variable with $\lambda = .65$, what is the probability that 3 defective bulbs are found in one randomly selected carton?

SOLUTION: Since $\lambda = .65$ is not included in Appendix C, we use Eq. (4.41) to write

$$P(X=3) = [e^{-.65}(.65)^3]/3!.$$

From Table C2, $e^{-.65} = .522$. Therefore,

$$P(X=3) = [(.552)(.274625)]/6 = .0239.$$

The tables in Appendix C provide P(x), F(x), and $e^{-\lambda}$ for only a selected set of λ values. Poisson distributions for a wider set of λ values are available in Molina [6] or Pearson and Hartley [7], while values of $e^{-\lambda}$ for more values of λ can be obtained from Selby [8].

To close our discussion of Poisson models, we present the techniques for using a Poisson distribution to approximate a binomial distribution. We first observe that direct evaluation of a binomial probability by Eq. (4.15) can result in some very heavy arithmetic. This is particularly true when n is large, so there are times when we seek a good approximation for the terms of a binomial distribution. Recall that at the beginning of this section we described the Poisson model as the limit of a binomial model as $n \to \infty$ and $p \to 0$, so that $np = \lambda$ remains constant. Thus, it is not surprising that the conditions under which a Poisson model gives a good approximation to binomial probabilities are large values of n and small values of p, with $\lambda = np$. To illustrate just how the convergence takes place, we have listed the Poisson probability distribution for $\lambda = 1$ and 5 binomial probability distributions with np = 1 in Table 4.15.

TABLE 4.15.
Illustrated are a Poisson distribution with $\lambda = 1$ and 5 binomial distributions with $np = 1$.

	Poisson	Binomial				
x	λ=1	n=10 p=.1	n=20 p=.05	n=50 p=.02	n=100 p=.01	n=1000 p=.001
0	.368	.349	.358	.364	.366	.368
1	.368	.387	.377	.372	.370	.368
2	.184	.194	.189	.186	.185	.184
3	.061	.057	.060	.061	.061	.061
4	.015	.011	.013	.015	.015	.015
5	.003	.002	.002	.003	.003	.003
6	.001	.000	.000	.000	.000	.001
Total	1.000	1.000	.999	1.001	1.000	1.000

In Table 4.15, we see that the approximation is exact to 3 decimal places for n=1000, quite good for n as small as 50, and does not differ by more than .02 for n as small as 10. That is to say, the smallness of p is more critical to the approximation than is the largeness of n. This property is joined in Chapter 5 by another approximation which is good for moderate values of n when p is neither near 0 nor 1. These two techniques provide approximations for binomial probabilities for all values of p because the Poisson approximation just presented may be used for large p by reversing the roles of p and q. That is, when p is large, q is small. So we find the Poisson probability for (n - x) with probability of success q rather than the Poisson probability for x with probability of success p.

EXAMPLE 17: The probability of a defective transistor is .01. What is the probability that a sample of 500 transistors contains 6 defectives? What is the probability that a sample of 500 contains 10 or more defectives?

SOLUTION: By Eqs. (4.15) and (4.17), the answers are as follows:

$$P(X=6) = \binom{500}{6}(.99)^{494}\ (.01)^{6},$$

$$P(X\geq 10) = \sum_{x=10}^{500} \binom{500}{x}(.01)^{x}\ (.99)^{500-x}$$

$$= 1 - \sum_{x=0}^{9} \binom{500}{x}(.01)^{x}\ (.99)^{500-x}$$

$$= 1 - F(9).$$

Clearly this approach to the questions requires a considerable amount of arithmetic. So let us consider Poisson approximations for answers. First, np = 5. So we use a Poisson model with $\lambda = 5$. From Table C1, we obtain

$$P(X=6) = .146,$$

$$p(X\geq 10) = 1 - .968 = .032.$$

EXAMPLE 18: Of all patients given CURE-ALL, 99% reported that they were cured completely. What is the probability that 98% of 100 treated patients are cured completely?

SOLUTION: Going first to the Poisson approximation, we have n = 100 and p = .99. Since p is near 1, we work with q = .01 and ask for the probability that 2 of the 100 patients were not cured. Then, using $\lambda = 100(.01) = 1$, we find

$$P(X=98) = P(100-X=2) = .184$$

from Table C1 (λ=1, X=2). We can check the accuracy of this approximation by noting that the true binomial probability as given in Table 4.15 is .185. Clearly the approximate values are accurate enough for all practical purposes.

EXERCISES

1. The average number of personal ads per issue of the *College News* is 4. Let X represent the number of personal ads per issue, and find the following probabilities: $P(X=10)$; $P(X=0)$; $P(X\leq 8)$; $P(X\geq 6)$. What are the most likely values of x?

2. (a) If X is a Poisson variable with $\lambda = 3$, show that $P(X=3) = P(X=2)$.
   (b) If X is a Poisson variable with parameter λ equal to an integer, show that $P(X=\lambda) = P(X=\lambda-1)$.

3. If the average number of chocolate chips per cookie is 3, what is the probability that the next cookie selected contains 8 chips? What is the variance of the number of chips per cookie?
4. A switchboard operator expects 3 calls per minute. How likely is it that she is swamped with 10 calls during the next minute? What is the probability that the next minute will produce 5 or fewer calls?
5. During rush hour, the expected number of cars passing the corner of First and College Streets per minute is 10. Would you consider it a rare minute when no cars passed?
6. The $ Savings and Loan Company expects 7 real estate loan applications per week. If next week shows 1 application, would it be reasonable to assume that the average number has decreased?
7. If the probability that a driver on the interstate highway is involved in an accident is .001, what is the probability that 2 of the next 1,000 drivers on that road are involved in an accident?
8. Find the mean and variance for the variables indicated in *Exercises 6* and 7.

4.8 REFERENCES

[1] Cameron, E. A. 1965. *Algebra and trigonometry*. New York: Holt, Rinehart and Winston, pp. 100-102.

[2] Nat'l. Bureau of Standards. 1949. *Tables of the binomial probability distribution*. Applied Math. Series No. 6. Washington, D.C.: U.S. Govt. Printing Office.

[3] Romig, H. G. 1953. *50-100 binomial tables*. New York: Wiley.

[4] Fisher, R. A. 1963. *Statistical methods for research workers*. 13th ed. New York: Hafner, p. 55.

[5] Fisz, M. 1963. *Probability theory and mathematical statistics*. 3rd ed. New York: Wiley, pp. 141, 444.

[6] Molina, E. C. 1942. *Poisson's exponential binomial limit*. New York: D. Van Nostrand.

[7] Pearson, E. S., and Hartley, H. O. 1970. *Biometrika tables for statisticians*. 3rd ed. London: Cambridge Univ. Press.

[8] Selby, S. M. 1973. *C.R.C. standard mathematical tables*. 23rd ed. Cleveland: Chemical Rubber Pub. Co., p. 218.

[9] Pedoe, D. 1963. *The gentle art of mathematics*. New York: Macmillan, pp. 36, 43-48.

4.9 REVIEW EXERCISES

1. Find E(X) and V(X) for the following distribution:

x	1	2	3	4	5
P(x)	.1	.1	.4	.2	.2

2. Suggest one or more experimental situations for which one of the following discrete distributions could serve as a model. Specify (approximate) values for the following model parameters: uniform; binomial; multinomial; hypergeometric; Poisson.

3. A spinner is constructed so that the 3 sections, R(red), G(green), and B(blue), are expected to occur with probabilities 1/3, 1/2, and 1/6, respectively.
   (a) What is the probability that 7 independent spins show 3R, 3G, and 1B?
   (b) What is the expected number of R on 15 spins?
   (c) What is the variance of the number of R on 15 spins?

4. In *Exercise* 3, let X = #(R) on 20 spins. What is the distribution of X? Find $P(X=3)$; $P(X\geq 7)$; $P(2<X\leq 8)$; E(X); V(X).

5. Three members of the ABC House Committee are to be picked for television appearances. If the committee consists of 10 liberals (L) and 5 conservatives (C), and if the selection is random, probabilities should be assigned to the following events: that 3 L's are selected; that 3 C's are selected; that 2 L's are selected.

   What is the most likely classification of the 3 individuals selected? Find the mean and variance for the #(C) selected.

6. Approximately 10% of the population write left-handed. In a classroom of 100 seats, how many seats should be left-handed? How do the binomial and Poisson approximations to the binomial answers agree for the probability that a class of 100 students includes 15 who write left-handed?

7. The Continental Biscuit Association expects an average of 1 customer per 5 minutes. How many customers should they expect during the next 30 minutes? What is the probability that 3 customers arrive during the next 5 minutes? What is the probability that 3 customers arrive in the next 30 minutes?

8. Suppose X is a binomial variable with n = 100 and p = 1/2. Let Y = 2X-3. Find E(X), σ_x^2, E(Y), σ_y^2, and σ_y.

9. The number (X) of customers served in 10 minutes by one teller at the Easy Flow Bank is assumed to be a Poisson variable with $\lambda = 3$. Let Y = 3X be the number of customers served in 10 minutes by 3 tellers. Find E(X), σ_x^2, E(Y), σ_y^2, and σ_y.

10. Pedoe [9] presents the following problem as one that Samuel Pepys posed to Isaac Newton: Individuals A, B, and C are going to roll dice according to the following plan. A rolls 6 dice and wants to observe at least one 6. B rolls 12 dice and wants

to observe at least two 6's. C rolls 18 dice and wants to observe at least three 6's. If all dice are fair (balanced), which individual has the best chance of realizing the desired outcome? (Pedoe presents some of the original correspondence between Pepys and Newton on this problem.)

CHAPTER 5

Continuous Probability Models For Empirical Data Results

5.1 INTRODUCTION AND DEFINITIONS

When variables such as time, weight, height, and length are measured, the data collected are necessarily discrete because any two unequal numbers must differ by at least the smallest unit of measure being used. For the variable, time, the smallest unit of measure may be minutes, seconds, tenths of seconds, or even ten-thousandths of seconds. This simply means that to write down an observed time for an event, we must cease writing decimal values at some point. Yet the variable under study, e.g., time, is assumed to be continuous, and it is only our inability to record measurements continuously that leads to the discrete data. Thus, this chapter is concerned with the presentation of definitions and properties associated with some selected continuous models. Throughout the development, we should think of the models being discussed as a subset of the total set of potential models for continuous data, just as the discrete models presented in Chapter 4 are a subset of all potential models for discrete data.

One way of expressing our objectives for this chapter is to recall the histograms used in Chapter 3 to pictorially represent a set of continuous data. Recall that the number of classes, k, used in forming a distribution was dependent on the number of observations, n, being grouped. Thus, an increase in n results in an increase in k. So imagine now a set of measurements on a variable, X, which can assume any value between 0 and 5. Suppose that samples of size n = 100, 500, 1000 grouped in k = 5, 10, 20 intervals yield histograms as presented in Figure 5.1.

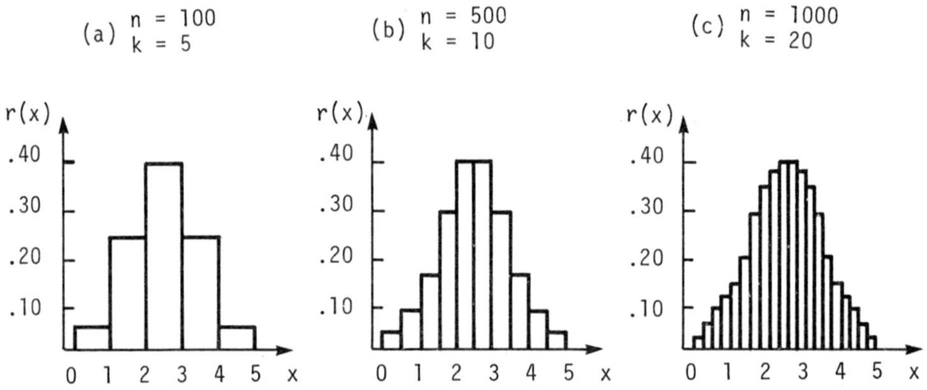

FIGURE 5.1. Three hypothetical histograms illustrate the effect of increasing k, the number of classes, and n, the number of observations.

In Figure 5.1, r(x) denotes the relative frequency of the x values in each interval [see Eq. (3.6) on page 93]. Furthermore, each histogram is constructed so that the sum of the areas of the bars for each graph is 1. For interpretation purposes, the area of each bar of the histogram denotes the probability that the variable, X, falls in the interval of x values determined by the base of the bar. For example, in Figure 5.1,

$P(2 \leq X \leq 3) = (1)(.39) = .39) = .39$ in part (a),

$P(2.5 \leq X \leq 3.0) = (.5)(.39) = .195$ in part (b), and

$P(2.75 \leq X \leq 3.00) = (.25)(.38) = .095$ in part (c).

The preceding area interpretation of probability makes it natural to assign $P(X=x) = 0$ for any point, x, in the sample space of a continuous random variable, X. That result follows because a point has no dimensions (e.g., width), and thus the line over the point, x, has area (and hence, probability) equal to 0. Thus, $P(2 < X < 3) = P(2 \leq X \leq 3) = P(2 < X \leq 3) = P(2 \leq X < 3)$

for the preceding continuous random variable, X. This property is in contrast to the property that $P(X=x) > 0$ for any point, x, in the sample space of a discrete variable, X.

Clearly, the procedure illustrated in Figure 5.1 could be extended so that we continually increase k as n increases. This leads us to ask the question, "What is the appearance of the histogram as n approaches ∞ and as k approaches ∞?" It does not seem unreasonable that the limit of the histogram for the assumed hypothetical variable, X, would appear as pictured in Figure 5.2.

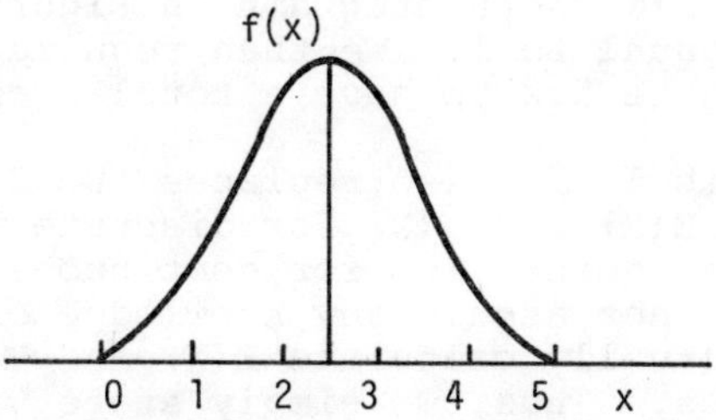

FIGURE 5.2. A continuous model is given for the histograms in Figure 5.1.

Additional mathematical requirements are needed to fully discuss the properties of general continuous models. One of these mathematical needs is the integral calculus which replaces the summation operation we use in discussing discrete models. Since we do not assume any knowledge of the calculus for our study, it is necessary to accept certain properties without proof. The omission of proofs, however, will not distract from the development of an appreciation for continuous type models needed to describe continuous data. Rather, it means that we depend on available tables for calculating probabilities of events of interest.

Before proceeding to the discussion of specific continuous models, we present some of the similarities and differences that exist between discrete and continuous models. For this reason, we compare the properties of each type model, using the notation given in the following definition to better facilitate these comparisons.

DEFINITION 1: Let X be a continuous random variable with a sample space $\mathcal{X}$, which is an interval of the real line. Then f(x) is a *probability density function* for the continuous random variable X, if

$$f(x) \geq 0, \qquad \text{for all } x, \tag{5.1}$$

$$\int_X f(x)dx = 1.$$

The results in Eq. (5.1) correspond exactly to those in Eq. (4.1) for the probability function, P, except that Σ has been replaced by the integral symbol, ∫. The sigma, Σ, simply implies that when all the probabilities are added together, we get a total probability of 1. The integral symbol, ∫, indicates that the total *area* under curve f(x) and above the x axis is 1. This is the reason for drawing all the histograms in Figure 5.1 to have a total area equal to 1. We then require the model f(x) graphed in Figure 5.2 to have a total area of 1 between 0 and 5.

The integral symbol, ∫, also replaces the Σ used in the definitions of E(X) and V(X) for discrete variables when we define these quantities for continuous variables. Since we do not assume any knowledge of the calculus, we cannot actually calculate E(X) and V(X) for continuous variables. Thus, we simply state values for E(X) and V(X) for the various continuous models discussed in this chapter. Also notice that all the properties for functions of random variables as stated in Eq. (4.5) hold for continuous variables, as well as discrete random variables.

We can now provide a further contrast between the concept of probability functions for discrete random variables and the concept of probability density functions for continuous random variables. As stated earlier, it is sensible to speak of quantities such as $P(X=x_i) = P(x_i)$ for a discrete random variable because $P(x_i) > 0$ for all points x_i in the sample space of X. However, when we assign probability over a continuous interval of points to model a continuous random variable X, we assign $P(X=x) = 0$ for every point x in the sample space of X. This is true since the rectangles of a histogram approach 0 width as the number of classes is increased and the grouped data approach a continuous distribution. Thus, it only makes sense to talk of the *density of a particular point and the probability of intervals* for any continuous variable. That is, we use f(x) to denote the density [height (ordinate) of the density function] of the variable X at the point x, and we calculate $P(a \leq X \leq b)$ to be the area between the density function f(x) and the X-axis from a to b. Figure 5.3 uses the continuous model pictured in Figure 5.2 to illustrate the foregoing points.

In Figure 5.3, $f(2.5) = .4$ is the density (not the probability) of the point $x = 2.5$. Similarly, the value of f(x) at any point $0 < x < 5$ gives the density of x.

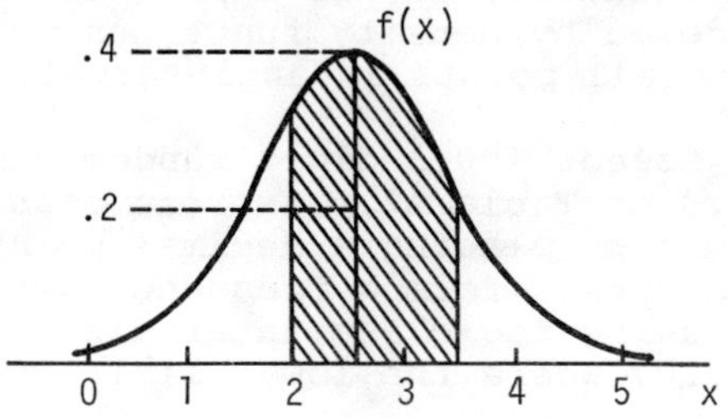

FIGURE 5.3. $P(2 \leq X \leq 3.5)$ is illustrated for the probability density function given in Figure 5.2.

For example, $f(1) \approx .1$. The shaded area in Figure 5.3 represents the probability that X falls between 2 and 3.5. Since a part of the boundary for the desired area is a curve (rather than a straight line), we cannot find the desired area exactly without using the calculus. However, a good approximation can be obtained from part (c) in Figure 5.1 by adding the areas for those rectangles between 2 and 3.5. Rough measurements give

$$P(2 \leq X \leq 3.5) = (1/4)(.38 + .40 + .40 + .38 + .34 + .28)$$
$$= .545.$$

Notice that similar approximations based on parts (a) and (b) in Figure 5.1 are .505 and .545, respectively.

Another difference between a probability function $P(x)$ and a density function $f(x)$ is that $f(x)$ may exceed 1 for some sample points x as long as the total area does not exceed 1, while $P(x)$ is always less than or equal to 1 for all the values of x in the sample space. That property and the other ideas presented in this brief introduction are illustrated further for specific continuous models in the following sections.

5.2 CONTINUOUS UNIFORM EXPERIMENTS AND MODELS

The first model we consider is the continuous model, which is analogous to the discrete uniform model discussed in Section 4.3. To assist in the discussion, we recommend that data for the following experiment be collected. These data should then be compared to both the results obtained by other students and to the model and empirical data presented in the development. The empirical data thus collected should appear to be such that they can be modeled by density functions which assign equal density to all points in an interval.

EXPERIMENT: Select 100 3-digit random numbers from either Table 2.8 or Table A2. Express each of the numbers as a decimal by placing a decimal point in front of each set of 3 digits. Form a frequency distribution for the numbers by using the class intervals 0 - .1, .1 - .2, ···, .9 - 1.0, where the lower limit of each class is included.

Table 5.1 presents 3 independent trials of the experiment, as well as the model frequencies for a uniform model.

TABLE 5.1.
Empirical and model frequencies are given for 100 random numbers between 0 and 1.

Intervals	Sample 1	Sample 2	Sample 3	model
0 - .1	7	5	12	10
.1 - .2	12	12	8	10
.2 - .3	8	16	7	10
.3 - .4	10	5	15	10
.4 - .5	13	12	11	10
.5 - .6	4	14	12	10
.6 - .7	10	9	7	10
.7 - .8	11	7	10	10
.8 - .9	10	8	9	10
.9 - 1.0	15	12	9	10
Total	100	100	100	100

To calculate the model frequencies given in Table 5.1, we assume that all numbers between 0 and 1 are equally likely to occur. For the preceding experiment, we have used 3-digit numbers so that each of the possible 1,000 different 3-digit numbers is assumed to have probability .001 of occurring on any given selection. Clearly, the experiment could have been performed with 4-digit numbers so that each of the 10,000 possible numbers would occur with probability .0001. Of course, we could use more and more digits so that more and more points between 0 and 1 would be possible. This simply indicates the fact that we are simulating the sampling of variables between 0 and 1. Thus, the model we assume is the continuous *standard uniform* model with f(x) defined to be

$$f(x) = 1, \ 0 < x < 1 \qquad (5.2)$$
$$= 0, \text{ elsewhere.}$$

Equation (5.2) defines the probability density function of X, a random number between 0 and 1. The *cumulative distribution* function of the standard uniform variable X for a point x is the probability that X < x. We denote that function by F(x) and write

$$F(x) = P(X < x) = \begin{cases} 0, & x \leq 0 \\ x, & 0 < x \leq 1 \\ 1, & 1 < x. \end{cases} \qquad (5.3)$$

Figure 5.4 presents graphs of both the probability density function, f(x), and the cumulative distribution function, F(x), for the standard continuous uniform variable, X.

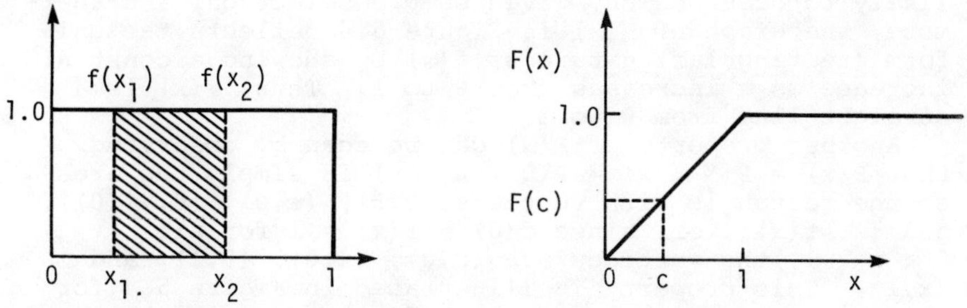

FIGURE 5.4. Graphs of f(x) and F(x) are given for a random variable, X, which is uniformly distributed between 0 and 1.

Now we can use the graphs in Figure 5.4 to calculate the model frequencies in Table 5.1 and to discuss some of the properties of f(x) and F(x). First, we note that

$P(x_1 \leq X < x_2)$ for any two points, x_1 and x_2, such that $0 \leq x_1 < x_2 \leq 1$ is the distance (x_2-x_1) multiplied by the height of $f(x)$, which is 1.0 for all $0 < x < 1$. That is, $P(x_1 \leq X < x_2)$ is the area of a rectangle determined by the four points $(x_1,0)$, $(x_2,0)$, $[x_1,f(x_1)]$, and $[x_2,f(x_2)]$. The desired area is shaded in Figure 5.4. Using such an argument, we see that the probability of X falling in the ith interval for Table 5.1 is

$$P[(i - 1)/10 \leq X \leq i/10] = (.1)(1) = .1,$$

$$i = 1, 2, 3, \ldots, 10.$$

Thus, when 100 independent observations of X are obtained, we would expect to find (.1)(100) = 10 of the selections in each interval. In general, for a continuous model which has been subdivided into intervals such that the ith interval has probability p_i of containing a randomly selected value of X, the expected (model) frequency for the ith class is

$$E(\text{frequency}) = np_i, \tag{5.4}$$

where n denotes the number of independent trials. Equation (5.4) applies to the other continuous models developed in this chapter, as well as the uniform models discussed in this section.

Another name for a uniform model is a *rectangular* model since the graph of $f(x)$ forms a rectangle. Note again that the consequence of the rectangular form is to assign equal likelihood to every point in an interval. Thus, we say that all points between 0 and 1 are equally likely to occur on any given sample selection. Furthermore, the graph of $F(x)$ in Figure 5.4 reflects the uniform (rectangular) nature of $f(x)$ by showing a constant increase as x increases from 0 to 1. Thus, $F(x)$ is a straight line from 0 to 1.

Another property of $F(x)$ can be seen by observing that $F(x) = P(X < x) = P(0 \leq X < x)$ is simply the area of the rectangle with vertices $(0,0)$, $(x,0)$, $[0, f(0)]$, and $[x, f(x)]$, or, since $f(0) = f(x) = 1$ for all $0 < x < 1$, the vertices are $(0,0)$, $(x,0)$, $(0,1)$, and $(x,1)$. This property is illustrated in Figure 5.4 for the point c. Further note that for any two points, x and y, such that $0 \leq x < y \leq 1$, we have

$$P(x \leq X < y) = F(y) - F(x).$$

We can use an argument of symmetry to show that the standard uniform variable X, having the density function given by Eq. (5.2) and graphed in Figure 5.4. has an expected value of

$$E(X) = 1/2. \qquad (5.5a)$$

It can also be shown that

$$V(X) = 1/12. \qquad (5.5b)$$

The preceding definitions and properties pertain only to the standard uniform density model, but they can easily be extended to a general uniform model for a variable X which assumes values between two limits, say, a and b. The probability density function, f(x), and cumulative probability distribution function, F(x), are defined by Eqs. (5.6) and (5.7), respectively, and graphed in Figure 5.5.

$$f(x) = \begin{cases} 1/(b-a), & a<x<b, \\ 0, & \text{elsewhere.} \end{cases} \qquad (5.6)$$

$$F(x) = \begin{cases} 0, & x<a, \\ (x-a)/(b-a), & a<x\leq b, \\ 1, & b<x. \end{cases} \qquad (5.7)$$

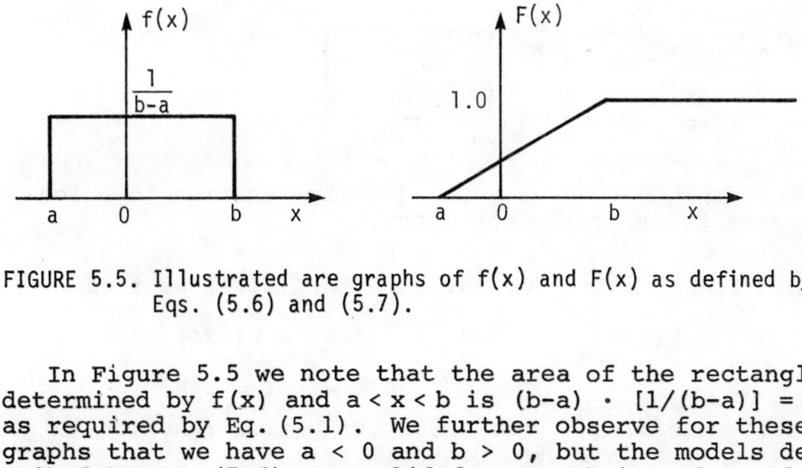

FIGURE 5.5. Illustrated are graphs of f(x) and F(x) as defined by Eqs. (5.6) and (5.7).

In Figure 5.5 we note that the area of the rectangles determined by f(x) and $a < x < b$ is $(b-a) \cdot [1/(b-a)] = 1$, as required by Eq. (5.1). We further observe for these graphs that we have $a < 0$ and $b > 0$, but the models described by Eq. (5.6) are valid for any choice of a and b such that $a < b$. That is, both a and b may be greater than 0, both a and b may be less than 0, or a may be less than 0 while b is greater than 0.

For any two values of X, say x_1, x_2, such that $a \leq x_1 < x_2 \leq b$, we have

$$P(x \leq X < x_2) = F(x_2) - F(x_1). \tag{5.8}$$

Again an appeal to symmetry gives

$$E(X) = (b + a)/2. \tag{5.9}$$

The calculus can be used to show that

$$V(X) = (b - a)^2/12. \tag{5.10}$$

Equations (5.9) and (5.10) for a continuous uniform model correspond to Eqs. (4.12) and (4.13) for a discrete uniform model. Furthermore, we observe that Eqs. (5.9) and (5.10) reduce to Eqs. (5.5a) and (5.5b) when $a = 0$ and $b = 1$.

The following examples and exercises help to demonstrate the concepts presented in this section.

EXAMPLE 1: Find $P(1/4 \leq X < 5/8)$ when X has the standard uniform density defined by Eq. (5.2).

SOLUTION: Method 1. By sketching the desired area, we see that

$$P(1/4 \leq X < 5/8) = (5/8 - 1/4)(1) = 3/8.$$

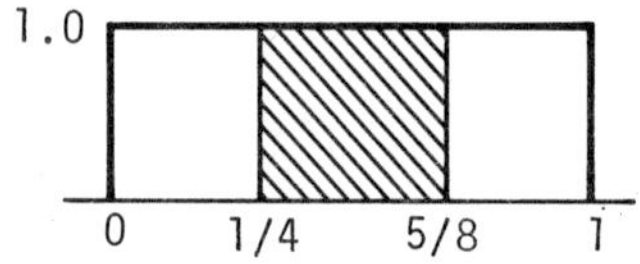

(Note: This method uses the density function to calculate the indicated probability.)

Method 2. By using Eq. (5.3) in Eq. (5.8), we find

$$P(1/4 \leq X < 5/8) = F(5/8) - F(1/4)$$
$$= 5/8 - 1/4 = 3/8.$$

(Note: This method uses the distribution function to calculate the indicated probability.)

EXAMPLE 2: Let X be a random variable having a uniform distribution with $a = -2$ and $b = 3$. (a) Define $f(x)$ and $F(x)$. (b) Find $E(X)$ and $V(X)$. (c) Find $P(-1 \leq X < 1)$.

SOLUTION: (a) From Eqs. (5.6) and (5.7), it follows that

$$f(x) = \begin{cases} 1/5, & -2<x<3 \\ 0, & \text{elsewhere.} \end{cases}$$

$$F(x) = \begin{cases} 0, & x\leq -2 \\ (x+2)/5, & -2<x\leq 3, \\ 1, & 3<x. \end{cases}$$

(b) By Eqs. (5.9) and (5.10),

$$E(X) = (3-2)/2 = .5,$$

$$V(X) = (3+2)^2/12 = 25/12.$$

(c) Method 1: $P(-1 \leq X < 1) = [1-(-1)](1/5) = 2/5.$

Method 2: $P(-1 \leq X < 1) = F(1) - F(-1) = 3/5 - 1/5$

$= 2/5.$

EXERCISES

1. Graph a continuous uniform density function and cumulative distribution function to model a variable that is uniformly distributed between 1 and 5.

2. Suggest a uniform density model for the illustrated spinner.
   (a) Find $f(x)$ and $F(x)$. (Assume that the circumference of the circle is 4 inches.)
   (b) What is the $P(.5<X\leq 2.5)$?
   (c) Find $E(X)$ and $V(X)$.

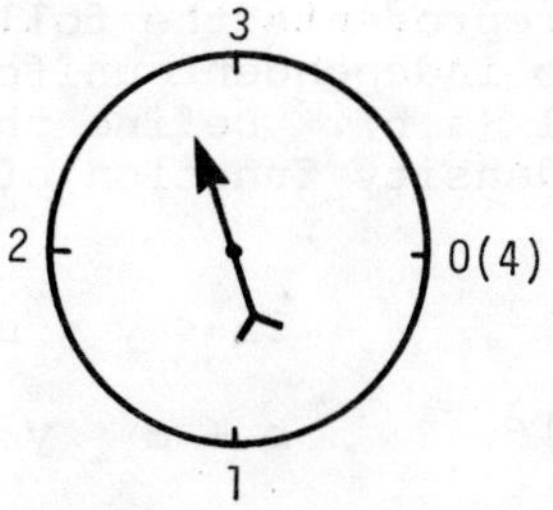

3. If X is a uniform random variable, what is the probability that X is within one standard deviation of the mean? two standard deviations of the mean?

4. Let X have the following density function:

$$f(x) = \begin{cases} 2, & .5<x<1, \\ 0, & \text{elsewhere.} \end{cases}$$

(a) Find F(x), E(X), and V(X).
(b) Find $P(.625 \leq X < .825)$.

5. Determine at least one set of values for a, b, and c so that the graph shows a valid uniform density model.

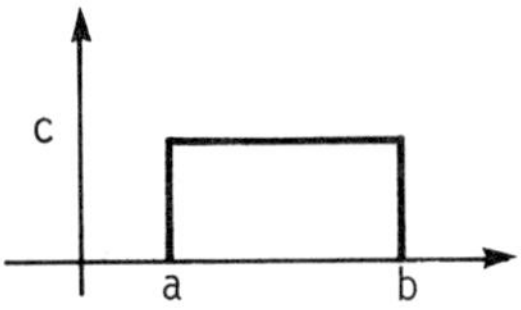

*5.3 TRIANGULAR MODELS

In the preceding section, we introduced uniform probability models and calculated the probability that a uniform variable X falls within specified limits. We could calculate those probabilities by finding the area of rectangles. Similarly, by using the formula for the area of a triangle, we can define triangular probability density models and answer probability questions about them. That is the purpose of this section.

SYMMETRIC (ISOSCELES) TRIANGLE MODELS: A symmetric (isosceles) triangle model presents a graph of f(y), which is a triangle with two sides of equal length. This type of model represents the following situation. Let X_1 and X_2 be two independent uniform random variables on the interval (a,b). Define the variable, $Y = X_1 + X_2$. Then the density function of Y is given by the triangular function

$$f(y) = \begin{cases} (y-2a)/(b-a)^2, & 2a < y < b + a, \\ (2b-y)/(b-a)^2, & b + a \leq y < 2b, \\ 0, & \text{elsewhere.} \end{cases} \qquad (5.11)$$

By using the fact that a triangle with base (w) and height (h) has area (A) determined by

$$A = wh/2, \qquad (5.12)$$

we can see that f(y) satisfies Eq. (5.1). Thus, f(y) is a valid density function.

The graph of f(y) is presented in Figure 5.6 and shows that f(y) is symmetric about y = b + a. Thus, E(Y) = b + a. Also, the maximum value of f(y) is f(b + a) = 1/(b - a).

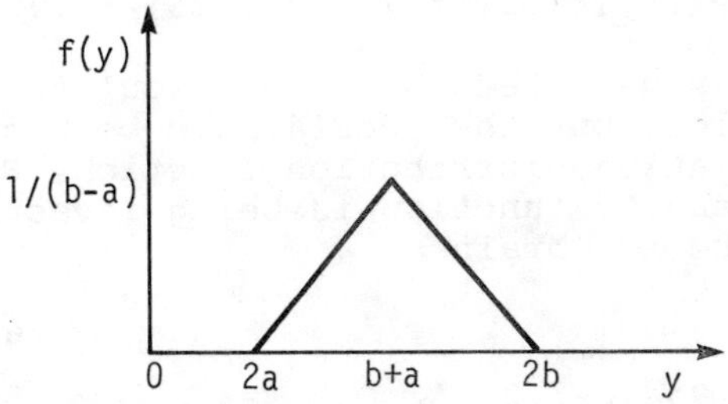

FIGURE 5.6. A graph of f(y), as in Eq. (5.11), is presented here.

EXAMPLE 3: Let X_1 and X_2 be two independent standard uniform variables. Let $Y = X_1 + X_2$. (a) Find f(y); (b) find E(Y); (c) find $P(.5 \leq Y < 1.5)$.

SOLUTION: (a) Using a = 0 and b = 1 in Eq. (5.11) gives

$$f(y) = \begin{cases} y, & 0<y<1, \\ 2-y, & 1<y<2, \\ 0, & \text{otherwise.} \end{cases} \quad (5.13)$$

(b) E(Y) = 1 + 0 = 1. (c) The desired probability is the shaded area in the figure shown. Using the symmetry in f(y), we find that

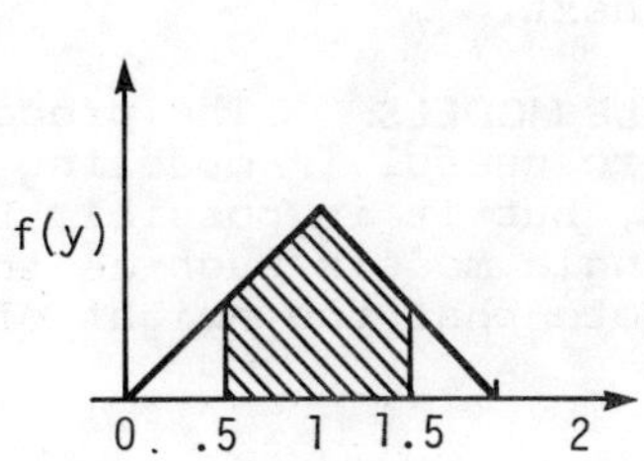

$$P(.5 \le Y < 1.5) = 1 - 2P(0 \le Y < .5).$$

From Eq. (5.12), it follows that

$$P(0 \le Y < .5) = (.5)(.5)/2 = .125.$$

Thus,

$$P(.5 \le Y < 1.5) = 1 - 2(.125) = .75.$$

Area formulas have been used to calculate the preceding probabilities, but they could have been obtained by using the cumulative distribution function, F(y). Since a continuous density function is being investigated, we use the calculus to obtain

$$F(y) = \begin{cases} 0, & y \le 2a, \\ (y-2a)^2/2(b-a)^2, & 2a < y \le b+a, \\ 1 - (2b-y)^2/2(b-a)^2, & b+a < y \le 2b, \\ 1, & 2b < y, \end{cases} \tag{5.14}$$

for the density f(y) as defined in Eq. (5.11). Also, the cumulative distribution function for f(y) as defined in Eq. (5.13) is

$$F(y) = \begin{cases} 0, & y \le 0, \\ y^2/2, & 0 < y \le 1, \\ 1 - (2-y)^2/2, & 1 < y < 2, \\ 1, & 2 < y. \end{cases} \tag{5.15}$$

By using Eq. (5.15), we can answer part (c) of Example 3 as follows:

$$p(.5 \le y < 1.5) = F(1.5) - F(.5) = (1 - 1/8) - (1/8) = 3/4.$$

The foregoing results for symmetric triangle models are simply a special case of the asymmetric triangle models presented next.

ASYMMETRIC TRIANGLE MODELS: The preceding symmetric triangle models are useful in modeling the sum of two uniform variables, but it is possible to study general (asymmetric) triangle models such as the one illustrated in Figure 5.7. Note that the height of the triangle

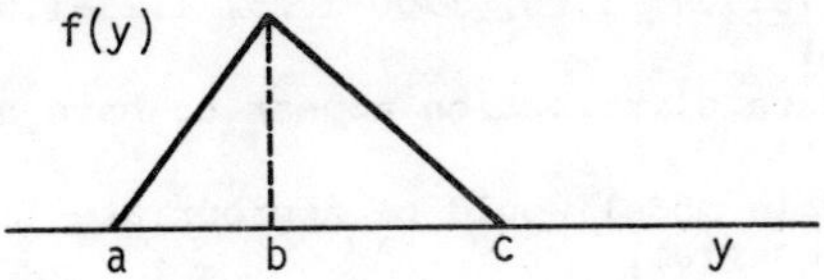

FIGURE 5.7. A graph of an asymmetric triangle model is shown.

is f(b). Thus, the area is given by

$$A = f(b)(c-a)/2.$$

So any values for a, c, and f(b) such that A = 1 [i.e., f(b) = 2/c-a] provide a valid (mathematically speaking) probability model for the variable, Y. Of course, the usefulness of that model depends on whether or not it is representative of a variable of interest. We leave that practical question and proceed by presenting the actual functional forms of the density, f(y), and distribution function, F(y), for a random variable, Y, having the density graphed in Figure 5.7. First,

$$f(y) = \begin{cases} 2(y-a)/(c-a)(b-a), & a < y < b, \\ 2(c-y)/(c-a)(c-b), & b \leq y < c, \\ 0, & \text{elsewhere}. \end{cases} \qquad (5.16)$$

Second, the calculus is used to find

$$F(y) = \begin{cases} 0, & y \leq a, \\ (y-a)^2/(c-a)(b-a), & a < y \leq b, \\ 1 - (c-y)^2/(c-a)(c-b), & b < y \leq c, \\ 1, & c < y. \end{cases} \qquad (5.17)$$

Before presenting the exercises that further illustrate the ideas of this section, we note that Eqs. (5.16) and (5.17) reduce to Eqs. (5.11) and (5.14) when b = (c + a)/2, i.e., when the distribution is symmetric.

EXERCISES

1. Select 50 pairs (x_1, x_2) of 2-digit numbers from either Table 2.8 or Table A2. Express each number as a decimal between 0

and 1. Find $y = x_1 + x_2$ for each pair of numbers. Form a distribution of the 50 values of y. (Use class intervals 0-.25, .25-.50, .50-.75, .75-1.00, 1.00-1.25, 1.25-1.50, 1.50-1.75, and 1.75-2.00.)

(a) Does the data distribution appear to have a triangular shape?
(b) What triangle model would be appropriate for the empirical data? [Find f(y)].
(c) Graph f(y).
(d) Find E(Y).
(e) Find $P(.25 \leq Y < 1.5)$.

2. Find a so that the triangle illustrated is a probability density function.

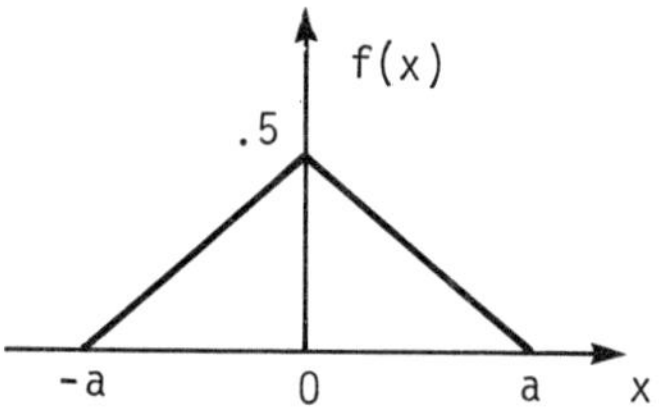

3. (a) For the triangle model illustrated, find $P(1 \leq X < 2)$.
(b) Find f(x).
(c) Find F(x).

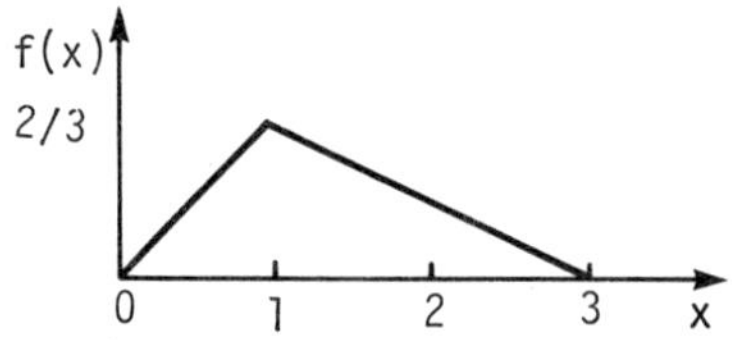

4. (a) For the triangle model illustrated, find f(2).
(b) What is f(1)?
(c) $P(1 \leq Y \leq 5)$.

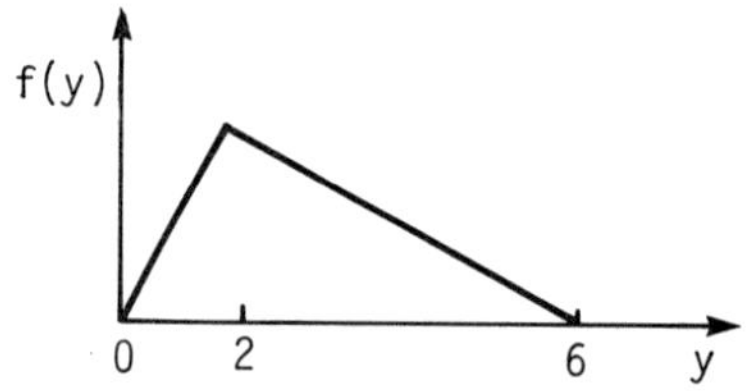

5.4 NORMAL EXPERIMENTS AND MODELS

The next model we present is probably the most widely used model in statistics. It was studied extensively as the limiting distribution for errors of measurements, for sums of independent random variables, and for binomial probabilities when p is neither close to 0 nor close to 1. Those early studies were conducted by the previously discussed scientists, Laplace, Gauss, and De Moivre. The distribution is termed as the law of Laplace and the Gaussian curve, in honor of the contributions made by those men, but the most common name used is the *normal* probability density function.

The purposes of this section are to discuss normal distributions for modeling empirical data, to develop a proficiency in reading normal tables for calculating probability for normal models, and to present an approximation for binomial probabilities using normal tables. Thus, we begin by indicating some experiments to be used in collecting data that could be modeled by normal distributions.

EXPERIMENT 1: Obtain the weight (in pounds) for 50 men (or women). Form a distribution of the weights using 5 or 6 classes.

EXPERIMENT 2: Draw a straight line on a blank piece of paper, and have 50 people guess its length (to the nearest tenth of an inch). Form a distribution of the data using 7 classes with the line's true length as the midpoint of the fourth interval.

EXPERIMENT 3: Select 25 sets of 12 random numbers between 0 and 1 from either Table 2.8 or Table A2. (Use 2-digit numbers preceded by a decimal point.) Find the sum for each set of 12 numbers. Form a distribution of the 25 sums using the following classes: 2.5-3.5, 3.5-4.5, 4.5-5.5, 5.5-6.5, 6.5-7.5, 7.5-8.5, 8.5,-9.5.

EXPERIMENT 4: Use Table A2 to select a sample of 100 random normal deviates from Table D3. Form a distribution with classes -4 to -3, -3 to -2, -2 to -1, -1 to 0, 0 to 1, 1 to 2, 2 to 3, 3 to 4.

Data for some of the preceding experiments should be collected for reference throughout this section. For the purpose of comparison and for subsequent reference, we use Tables 5.2 and 5.3 to present 4 samples each for Experiments 3 and 4.

TABLE 5.2.
Empirical frequencies and model frequencies are given for 25 sums of 12 random numbers between 0 and 1 (see Experiment 3).

Classes	Sample 1	Sample 2	Sample 3	Sample 4	Model
2.5-3.5	0	0	0	1	.25
3.5-4.5	1	2	3	1	1.5
4.5-5.5	4	6	7	9	6.0
5.5-6.5	8	10	8	9	9.5
6.5-7.5	10	6	7	5	6.0
7.5-8.5	1	1	0	0	1.5
8.5-9.5	1	0	0	0	.25
Total	25	25	25	25	25

TABLE 5.3.
Empirical frequencies and model frequencies are given for 100 random normal deviates from Table D3 (see Experiment 4).

Classes	Sample 1	Sample 2	Sample 3	Sample 4	Model
-4 to -3	0	0	0	0	.14
-3 to -2	4	4	4	1	2.14
-2 to -1	13	9	13	13	13.59
-1 to 0	31	29	38	38	34.13
0 to 1	34	39	35	27	34.13
1 to 2	15	16	6	16	13.59
2 to 3	3	3	3	5	2.14
3 to 4	0	0	1	0	.14
Total	100	100	100	100	100.00

As the data in Tables 5.2 and 5.3 are reviewed, we observe a tendency for the middle classes to have higher frequencies than the other classes. In fact, the middle class for Table 5.2 contains about 40% of the data points. In Table 5.3, we find that about 68% of the data occur in the two middle classes. Furthermore, Table 5.3 shows that about 50% of the data is negative and about 50% of the data is positive. These properties are discussed more explicitly after the normal probability density function is defined in the next subsection.

DEFINITIONS AND PROPERTIES OF NORMAL MODELS: The preceding discussion implies that the model being considered is symmetric in relation to some central value, with the density decreasing as it moves away from the point of symmetry. This means that most of the data are located close to the middle of the distribution. That is, values of the variable which are either much larger or much smaller than the center value are not likely to be observed.

The normal probability density function is proposed as a possible model for data that exhibit symmetry and are heavily concentrated around a central value. The normal probability density function is defined by

$$f(x) = [1/\sqrt{2\pi}\sigma] \quad e^{-(x-\mu)^2/2\sigma^2}, \quad -\infty < x < \infty, \qquad (5.18)$$

where $\pi = 3.14159 \cdots$ and $e = 2.71828 \cdots$. That is, π is the ratio of the circumference of a circle to its diameter, and e is the base of the natural (Naperian) logarithms. The normal density function defined by Eq. (5.18) depends on two parameters, μ and σ^2. The value of μ may be any real number, and σ^2 may be any positive real number. Thus, the family of potential normal models defined by Eq. (5.18) contains an infinite number of members. Therefore, our interest in normal models must eventually be directed at determining the best choices of μ and σ^2 for modeling a given set of data. As with the previous models discussed, that important topic is delayed until Chapter 7 because we must first learn to use the model and become familiar with its properties.

The particular normal model that has $\mu = 0$ and $\sigma^2 = 1$ is called the *standard normal density* and is defined by

$$f(z) = [1/\sqrt{2\pi}] \; e^{-z^2/2}, \quad -\infty < z < \infty, \qquad (5.19)$$

where π and e are as defined for Eq. (5.18).

Table D1 supplies values for f(z) as z varies from 0 to 3.50. Notice that f(-z) = f(z), so Table D1 also gives the values of f(z) as z ranges from -3.50 to 0. Some of those values are used to graph f(z) in Figure 5.8. The function f(x) in Eq. (5.18) is also graphed in Figure 5.8.

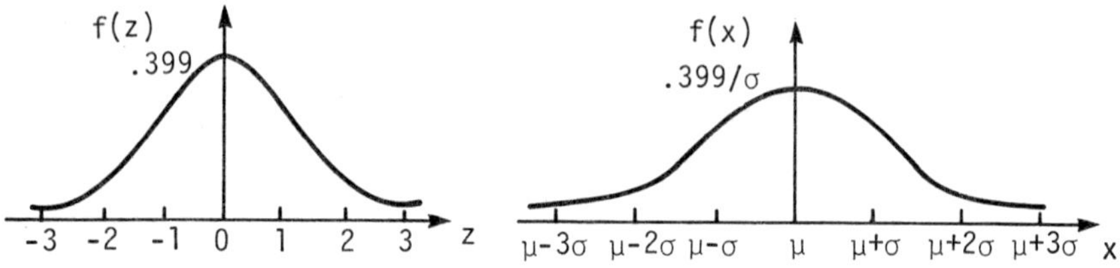

FIGURE 5.8. Graphs of f(z) and f(x) are shown as defined in Eqs. (5.19) and (5.18), respectively.

In Figure 5.8, note that f(z) is symmetric about $z = 0$ and that f(x) is symmetric about $x = \mu$. Thus, we can use symmetry to show that the parameter, μ, is the median of a normal model. It is also true that μ is the mode of a normal distribution, and it can be shown that

$$E(X) = \mu, \qquad (5.20)$$

$$V(X) = \sigma^2. \qquad (5.21)$$

Thus, the usual interpretation of the parameters of a normal distribution is that μ denotes the mean and σ^2 denotes the variance. Hence, μ is an expression of location, and σ^2 is an expression of dispersion. Furthermore, σ is the standard deviation of a normal model. Clearly, the standard normal model has $E(Z) = 0$ and $V(Z) = 1$. Also, the standard deviation of Z is 1.

The effect of changing μ by an amount, c, in Eq. (5.18) is simply a shift of the normal curve either to the right if c is positive or to the left if c is negative. In particular, we may subtract μ from each X value and obtain a distribution centered at 0. Thus, μ is truly the location parameter. On the other hand, any change in σ (or σ^2) changes the shape (scale) of f(x). Notice that the height of f(x) at $x = \mu$ depends on σ. If $\sigma > 1$, then $f(\mu) < .399$; but when $\sigma < 1$, then $f(\mu) > .399$. Thus, values of $\sigma > 1$ provide a model that has a lower peak and is more widely dispersed than the standard normal variable z, while $\sigma < 1$ yields a model that has a higher peak and is less dispersed than f(z). So σ^2 is the shape and scale parameter for the normal model.

For the convenience of easy reference, we introduce the notation, $X \sim N(\mu,\sigma^2)$, which indicates that the variable X is normally distributed with mean μ and variance σ^2. Thus, $X \sim N(2,4)$ implies that X is a normal random variable with mean 2 and variance 4.

Using the foregoing notation, we now indicate some relationships between $X \sim N(\mu,\sigma^2)$ and $Z \sim N(0,0)$. First, if $X \sim N(\mu,\sigma^2)$, then Z defined by

$$Z = (X-\mu)/\sigma \tag{5.22}$$

is such that $Z \sim N(0,1)$. Second, if $Z \sim N(0,1)$, then X defined by

$$X = \mu+\sigma Z \tag{5.23}$$

is such that $X \sim N(\mu,\sigma^2)$. Thus, Eq. (5.22) can be used to transform any value of a normal variable X into a value for a standard normal variable Z. Such values of Z are sometimes called *normal scores* or *Z-scores*. Likewise, Eq. (5.23) can be used to transform any given value of a standard normal variable into the corresponding value of a normal variable with mean μ and variable σ^2. Hence, we can generate a random sample from any given normal distribution by, first, selecting a random sample $z_1, z_2, \ldots, z_n$ from Table D3 and, second, calculating $x_i = \mu+\sigma z_i$, $i = 1, 2, \ldots, n$ for the desired values of μ and σ^2. The resulting values $x_1, x_2, \ldots, x_n$ represent a random sample from a $N(\mu,\sigma^2)$ distribution.

EXAMPLE 4: If $Z \sim N(0,1)$ with probability density function $f(z)$ as given in Eq. (5.19), find $f(0)$, $f(1.5)$, and $f(-2.05)$.

SOLUTION: From Table D1, we find directly that $f(0) = .3989$ and $f(1.5) = .1295$. Then by symmetry, $f(-2.05) = f(2.05)$. Therefore, $f(-2.05) = .0488$.

EXAMPLE 5: Sketch the normal curve for $f(x)$ when $X \sim N(1,4)$.

SOLUTION: Since $\sigma^2 = 4$, we have $\sigma = 2$. Now $\mu = 1$, so $x = 1$ is the mode of the distribution. Furthermore, $f(1) = .399/\sigma = .399/2 \approx .20$. Also, the bell-shaped curve must taper off from the mode to be close to the x-axis for $x = \mu-3\sigma = 1-3(2) = -5$ and $x = \mu + 3\sigma = 7$. Thus, we get the following illustration.

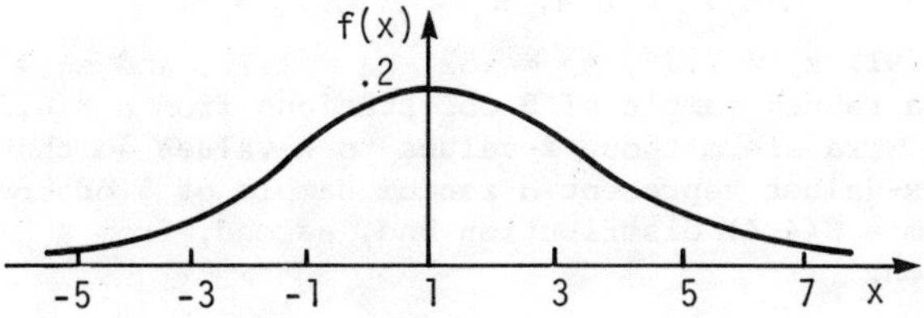

EXAMPLE 6: Find Z-score for $x = 10$ when $X \sim N(15,25)$.

SOLUTION: If $X \sim N(15,25)$, then $\mu = 15$ and $\sigma = 5$. Using these values of μ and σ, as well as $x = 10$ in Eq. (5.22), gives

$$Z = (10-15)/5 = -1.$$

EXAMPLE 7: If $Z \sim N(0,1)$ and $z = .5$, find the corresponding value of x such that $X \sim N(15,25)$.

SOLUTION: Using Eq. (5.23), we have

$$x = 15 + 5(.5) = 17.5.$$

EXERCISES

1. Sketch the following normal distributions:
   (a) N(0,1)
   (b) N(0,.25)
   (c) N(1,1)
   (d) N(-1,1)
   (e) N(0,4)
   (f) N(5, 1/16).

2. Discuss the similarities and differences for each of the following pairs of distributions sketched in Exercise 1:
   (a) N(0,1) vs. N(0,.25)
   (b) N(0,1) vs. N(0,4)
   (c) N(0,1) vs. N(-1,1)
   (d) N(0,4) vs. N(5, 1/16)
   (e) N(1,1) vs. N(-1,1).

3. Consider the following density function for X:

$$f(x) = [1/\sqrt{2\pi}3]\ e^{-(x-6)^2/18}, \quad -\infty < x < \infty.$$

   (a) What is E(X)?
   (b) What is V(X)?
   (c) What is f(3)?
   (d) If a random sample of 50 is selected from the f(x), classify the following values of x as either reasonable or unreasonable: 5, 6.8, -10, 10, 15, 20.

4. Find the Z-scores for the following values of x when $X \sim N(5,16)$: $x_1 = 5.0$, $x_2 = 6.4$, $x_3 = 2.3$, $x_4 = -8$.

5. Let $z_1 = -1.91$, $z_2 = 2.75$, $z_3 = .52$, $z_4 = 1.26$, and $z_5 = -2.02$ represent a random sample of 5 observations from a N(0,1) distribution. Transform those z-values to x-values so that the resulting x-values represent a random sample of 5 observations, first, from a N(4,4) distribution and, second, from a N(-3,9) distribution.

READING TABLES AND CALCULATING PROBABILITIES FOR NORMAL MODELS: One of the primary reasons for using a probability model is to make probability statements about the random variable being modeled. That is, we are interested in determining $P(c \le X \le d)$ for given values of c and d. We may also want to determine limits which, for all practical purposes, contain the values for X that are likely to be observed when studying X. These are some of the issues considered next.

We should first point out that the direct determination of $P(c \le X \le d)$ requires use of the calculus, so we must rely on techniques other than direct computation. Our methods require only the study of Table D2, a knowledge of the symmetry in normal densities, and the use of Eq. (5.22). In fact, the relationship expressed by Eq. (5.22) permits us to make probability statements about any normal variable X by using only the probability table (Table D2) for the standard normal variable Z. Thus, if $X \sim N(\mu,\sigma^2)$ and $Z \sim N(0,1)$, then

$$P(c \le X \le d) = P[(c-\mu)/\sigma \le Z \le (d-\mu)/\sigma]. \quad (5.24)$$

Proof: From algebra, we have the following results for inequalities: If a, b, and c are real numbers, then

$$a \le b \rightarrow a \pm c \le b \pm c, \quad (5.25)$$

$$a \le b \rightarrow a/c < b/c \text{ or } ac < bc, \text{ if } c > 0, \quad (5.26)$$

$$a \le b \rightarrow a/c \ge b/c \text{ or } ac \ge bc, \text{ if } c < 0. \quad (5.27)$$

Thus,

$$P(c \le X \le d) = P(c-\mu \le X-\mu \le d-\mu) \quad \text{by Eq. (5.25)}$$

$$= P[(c-\mu)/\sigma \le (X-\mu)/\sigma \le (d-\mu)/\sigma]. \quad \text{by Eq. (5.26)}$$

However, $Z = (X-\mu)/\sigma$ by Eq. (5.22). Therefore,

$$P(a \le X \le b) = P[(c-\mu)/\sigma \le Z \le (d-\mu)/\sigma].$$

Expanding upon Eq. (5.24), we now present some useful probabilities for events involving $X \sim N(\mu,\sigma^2)$ and $Z \sim N(0,1)$. First, the symmetry present in Eqs. (5.18) and (5.19) implies that

$$P(X \leq \mu) = P(Z \leq 0) = .5000,$$

$$P(X \geq \mu) = P(Z \geq 0) = .5000. \qquad (5.28)$$

Recall that the points $X = \mu$ and $Z = 0$ have 0 probabilities. Thus, the probabilities in Eq. (5.28) could be $P(X < \mu)$, $P(X > \mu)$, $P(Z < 0)$, and $P(Z > 0)$ and give the same answer of 1/2. Second, the functions, f(x) and f(z), are probability densities and thus satisfy Eq. (5.1). Therefore,

$$P(-\infty < X < \infty) = P(-\infty < Z < \infty) = 1.00000. \qquad (5.29)$$

Third, we present the following special probabilities:

$$P(\mu - \sigma \leq X \leq \mu + \sigma) = P(-1 \leq Z \leq 1) = .68268; \qquad (5.30)$$

$$P(\mu - 2\sigma \leq X \leq \mu + 2\sigma) = P(-2 \leq Z \leq 2) = .95450; \qquad (5.31)$$

$$P(\mu - 3\sigma \leq X \leq \mu + 3\sigma) = P(-3 \leq Z \leq 3) = .99730; \qquad (5.32)$$

$$P(\mu - 4\sigma \leq X \leq \mu + 4\sigma) = P(-4 \leq Z \leq 4) = .99994. \qquad (5.33)$$

Thus, any normal variable X is such that approximately 68% of its values are within 1 standard deviation of its mean, approximately 95% are within 2 standard deviations of its mean, and rarely (less than .3%) does a value of X differ from its mean by more than 3 standard deviations. Note also that even though we must extend the limits of either X or Z to $-\infty$ and $+\infty$ to obtain an interval with probability of exactly 1 [as given in Eq. (5.29)], for all practical purposes, the occurrence of values outside 4 or 5 standard deviations has 0 probability. That is, from the standpoint of practical statistics, $P(\mu - 4\sigma \leq X \leq \mu + 4\sigma)$ is essentially equal to 1.

Consider the next example to illustrate the foregoing ideas. Also, observe that in the example a sketch of the distribution is used to indicate the probability of interest. Such simple sketches help to avoid many mistakes when calculating probabilities.

EXAMPLE 8: Mr. Scitsitats has taught elementary statistics courses for several years and has found that weights (X) of male students in his class are approximately normally distributed with a mean of 170 pounds and a standard deviation of 15 pounds. Find the following probabilities for X: (a) $P(155 \leq X \leq 185)$; (b) $P(170 \leq X \leq 200)$; (c) $P(95 \leq X \leq 245)$.

SOLUTION: (a) Using Eqs. (5.24) and (5.30), we find that $P(155 \leq X \leq 185) = P(-1 \leq Z \leq 1) = .68268$.

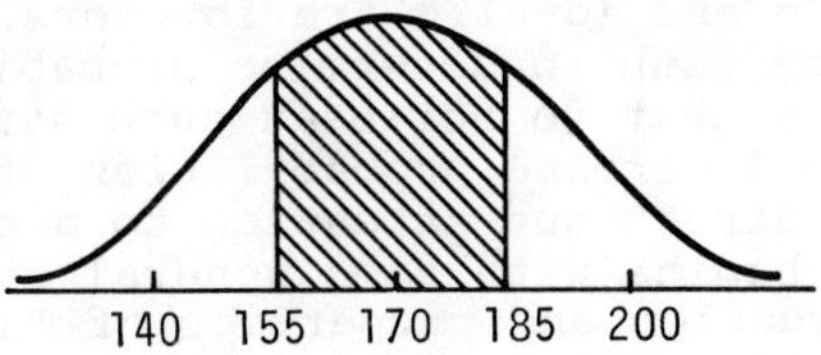

(b) Similarly, using Eq. (5.24), we find that

$$P(170 \leq X \leq 200) = P(0 \leq Z \leq 2).$$

Then by symmetry and Eq. (5.31), we have

$$P(170 \leq X \leq 200) = P(-2 \leq Z \leq 2)/2 = .95450/2 = .47725.$$

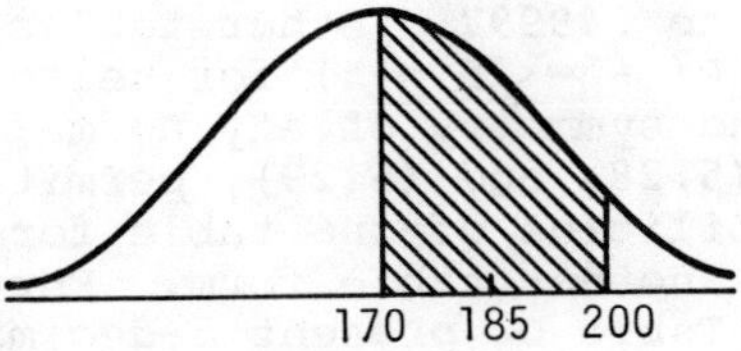

(c) By Eq. (5.24),

$$P(95 \leq X \leq 245) = P(-5 \leq Z \leq 5) \approx 1.$$

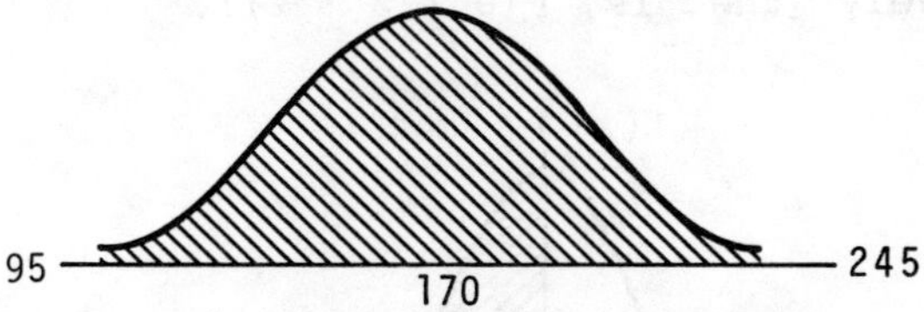

REMARK: By sketching the area associated with an indicated probability, it becomes apparent whether or not the desired answer is near 1, near 0, near .5, less than .5, more than .5, etc. Thus, we can quickly judge if our answer is reasonable or if an error may have been committed during the calculation. The value of such sketches becomes more obvious as we begin using Table D2.

The preceding discussion has provided methods by which the $P(c \le X \le d)$ can be calculated if c and d are such that $(c-\mu)/\sigma$ and $(d-\mu)/\sigma$ are integers. Clearly, experimenters may wish to determine probabilities for values of c and d that do not meet such strict requirements. Thus, to determine probabilities of a more general nature, we direct our attention to a study of Table D2. That study begins with some general statements about the construction and properties of Table D2 and continues with the presentation of numerous examples illustrating the various uses for Table D2.

As stated previously, the use of Eq. (5.22) allows us to express probabilities for any normal variable X in terms of probabilities for the standard normal variable Z. Hence, we need only obtain a table of probabilities for Z which can be used for any normal variable X. Table D2 is one such table. There are many so-called normal tables available in various books. They may differ in presentation and precision. For example, Table D2 gives $P(0 \le Z \le z)$ for a selected set of values, z(z = .00, .01, .02, ···, 3.99), and thus the probabilities vary from .00000 to .49997. Other tables may present $P(-z \le Z \le z)$ or $P(-\infty \le Z \le z)$ for selected values of z. Of course, the symmetry of any normal distribution, as well as Eqs. (5.28) and (5.29), permits us to change the normal probabilities of one table form into the probabilities of another table form. Furthermore, the probabilities in Table D2 present 5-decimal-place accuracy for changes in z as small as .01, while other sources (see National Bureau of Standards [1]) may provide up to 15-decimal-place accuracy for changes in z as small as .0001.

Figure 5.9 provides as illustration of the association between the area under the standard normal density curve f(z) and the probability that Z is between 0 and z, inclusively [that is, $P(0 \le Z \le z)$].

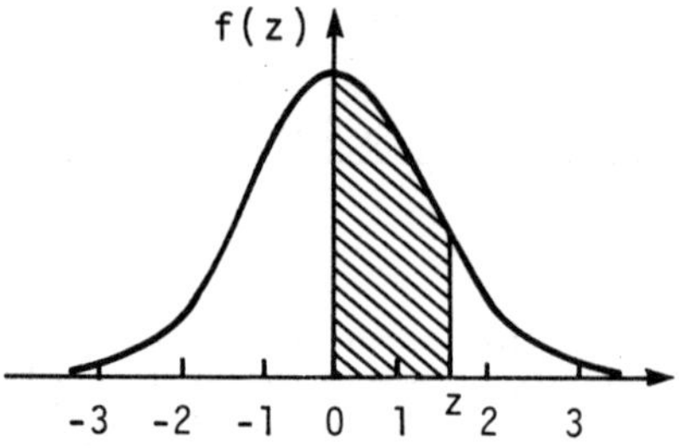

FIGURE 5.9. As given by Table D2, $P(0 \le Z \le z)$ is shown here as the area in the shaded region.

It is important that we keep firmly in mind the type of region for which probabilities are provided by Table D2 and illustrated in Figure 5.9 because any region for which the probability is desired must be expressed in terms of intervals such as $(0 \le Z \le z)$. Thus, we present the following examples for using Table D2 when calculating the probabilities for different events involving values of $Z \sim N(0,1)$. Then we can extend these techniques to determine probabilities for events involving $X \sim N(\mu,\sigma^2)$. Observe that the integer part and the first decimal place of z occur in the left column of Table D2 and determine a particular row of the table. Then the second decimal place occurs along the top of the table and selects a particular column. The point in the table at which the selected row and column intersect gives $P(0 \le Z \le z)$.

EXAMPLE 9: (a) Find $P(0 \le Z \le 1.00)$. (b) Find $P(-2.00 \le Z \le 0)$.

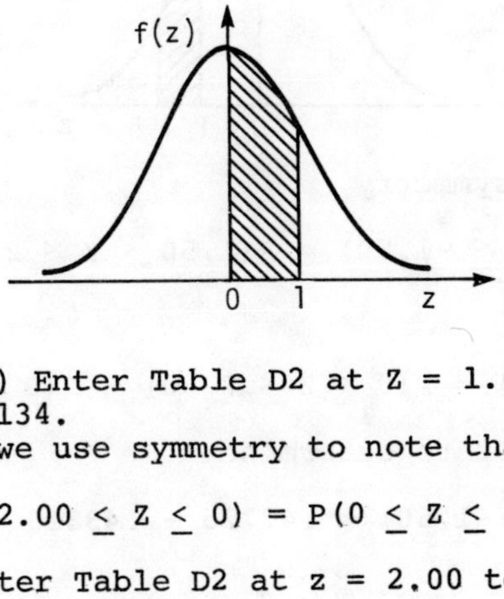

SOLUTION: (a) Enter Table D2 at Z = 1.00. Then $P(0 \le Z \le 1.00) = .34134$.

(b) First we use symmetry to note that

$$P(-2.00 \le Z \le 0) = P(0 \le Z \le 2.00).$$

Second, we enter Table D2 at z = 2.00 to determine

$$P(-2.00 \le z \le 0) = .47725.$$

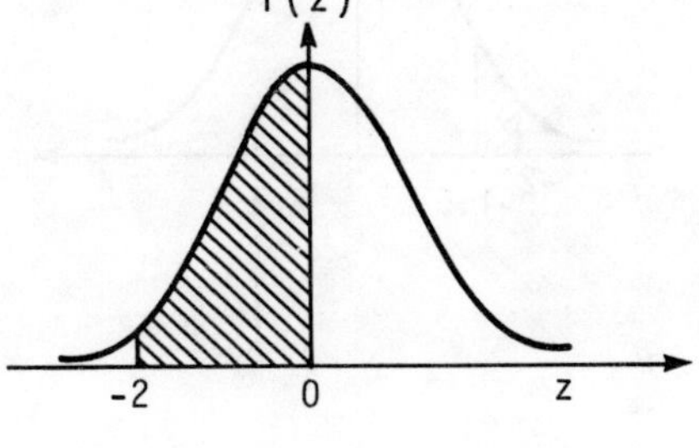

REMARK: Example 9 illustrates the techniques for finding the probability for an event of either the form $(0 \leq Z \leq d)$ where $d > 0$, or the form $(d \leq Z \leq 0)$ where $d < 0$. Note that the result is read *directly* from Table D2.

EXAMPLE 10: (a) Find $P(1.00 \leq Z \leq 1.50)$. (b) Find $P(-2.00 \leq Z \leq -1.50)$.

SOLUTION: (a) First, observe that

$$P(1.00 \leq Z \leq 1.50) = P(0 \leq Z \leq 1.50) - P(0 \leq Z \leq 1.00).$$

Second, enter Table D2 at $Z = 1.50$ and $Z = 1.00$ to find

$$P(1.00 \leq Z \leq 1.50) = .43319 - .34134 = .09185.$$

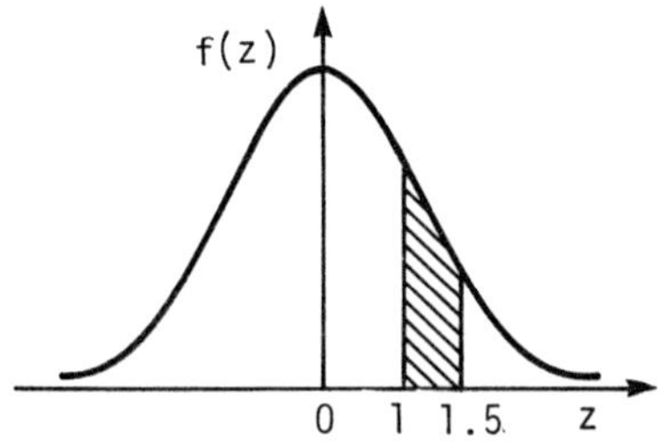

(b) First, by symmetry,

$$P(-2.00 \leq Z \leq -1.50) = P(1.50 \leq Z \leq 2.00).$$

Next, observe that

$$P(1.50 \leq Z \leq 2.00) = P(0 \leq Z \leq 2.00) - P(0 \leq Z \leq 1.50).$$

Then enter Table D2 to determine

$$P(-2.00 \leq Z \leq -1.50) = .47725 - .43319 = .04406.$$

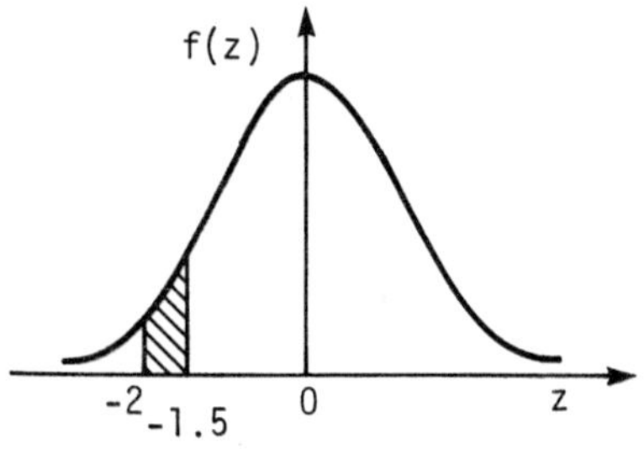

REMARK: Example 10 illustrates the technique for using Table D2 when working with events of the type $(c \le Z \le d)$, where c and d are either both positive or both negative. Be aware that the result is the *difference* between two table values.

EXAMPLE 11: Find $P(-1.25 \le Z \le 1.43)$.

SOLUTION: First, write

$$P(-1.25 \le Z \le 1.43) = P(-1.25 \le Z \le 0) + P(0 \le Z \le 1.43).$$

Second, by symmetry,

$$P(-1.25 \le Z \le 0) = P(0 \le Z \le 1.25).$$

Third, use Table D2 to write

$$P(-1.25 \le Z \le 1.43) = .39435 + .42364 = .81799.$$

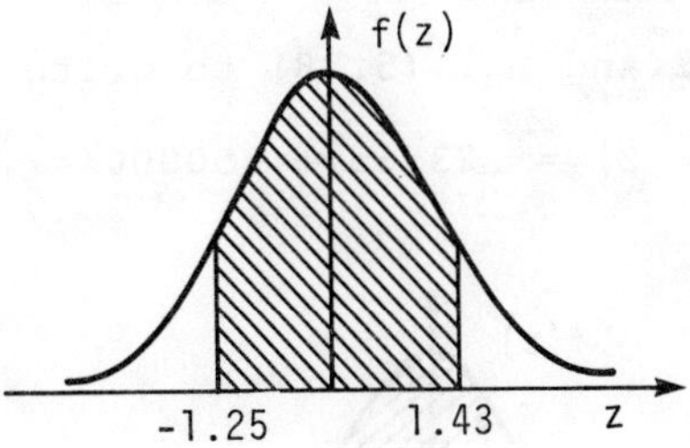

REMARK: Example 11 presents the technique for using Table D2 for events of the form $(c \le Z \le d)$ when $c < 0$ and $d > 0$. Be aware that the result is the *sum* of two table values.

EXAMPLE 12: (a) Find $P(1.45 \le Z)$. (b) Find $P(-.63 \le Z)$.

SOLUTION: (a) First note that

$$P(1.45 \le Z) = P(0 \le Z < \infty) - P(0 \le Z \le 1.45).$$

Second, use Eq. (5.28) and Table D2 to write

$$P(1.45 \le Z) = .50000 - .42647 = .07353.$$

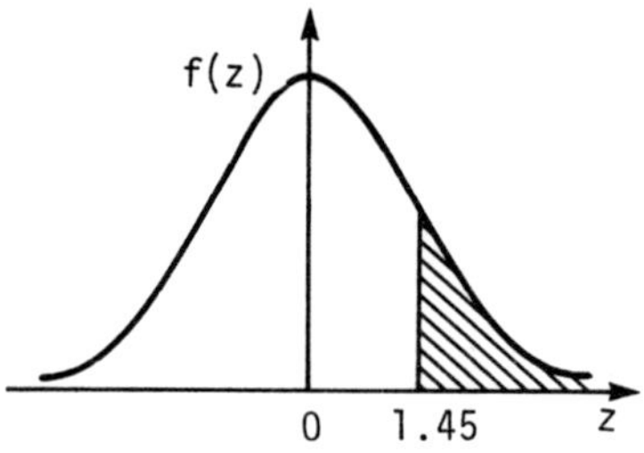

(b) First note that

$$P(-.63 \leq Z) = P(-.63 \leq Z < 0) + P(0 \leq Z < \infty).$$

Then, by symmetry,

$$P(-.63 \leq Z \leq 0) = P(0 \leq Z \leq .63).$$

Next use Table D2 and Eq. (5.28) to write

$$P(-.63 \leq Z) = .23565 + .50000 = .73565.$$

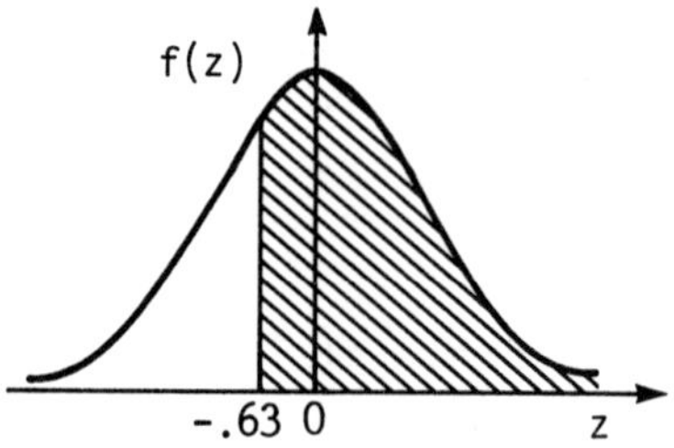

REMARK: Example 12 provides the method for using Table D2 for events of the type $c \leq Z$, where either $c > 0$ or $c < 0$. Notice that cases $c \geq 0$ and $c < 0$ require the calculation of a *difference* and a *sum*, respectively. Furthermore, for $c > 0$, the result is less than .50000, while $c < 0$ yields a probability greater than .50000.

EXAMPLE 13: (a) Find $P(Z \leq -2.05)$. (b) Find $P(Z \leq 1.17)$.

SOLUTION: (a) First, symmetry inplies that

$$P(Z < -2.05) = P(Z > 2.05).$$

Second,

$$P(Z > 2.05) = P(0 \leq Z < \infty) - P(0 \leq Z \leq 2.05).$$

Third, Eq. (5.28) and Table D2 give

$$P(Z < -2.05) = .50000 - .47982 = .02018.$$

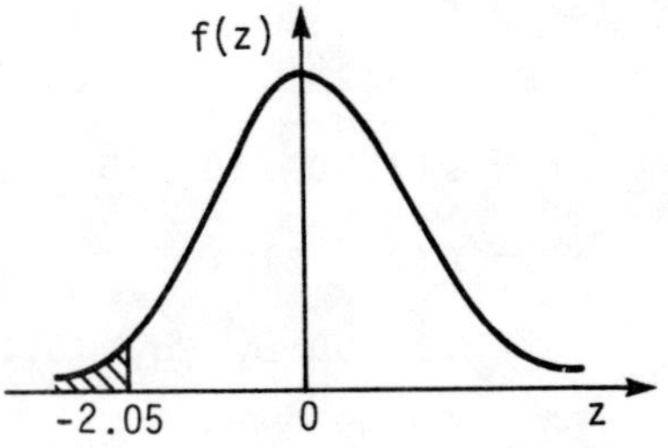

(b) First,

$$P(Z \leq 1.17) = P(-\infty < Z \leq 0) + P(0 \leq Z \leq 1.17).$$

Second, Eq. (5.28) and Table D2 give

$$P(Z \leq 1.17) = .50000 + .37900 = .87900.$$

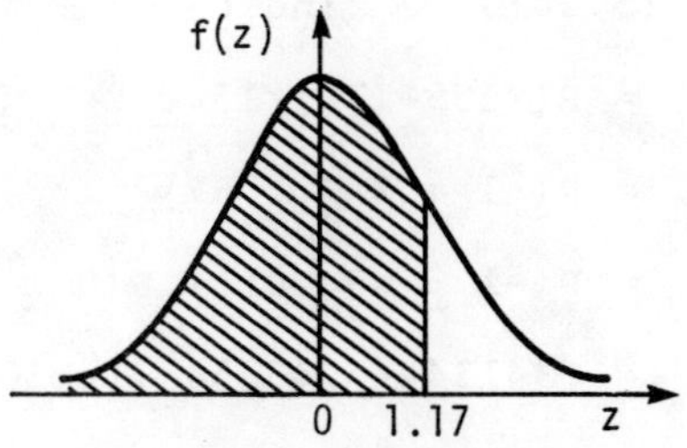

REMARK: Example 13 provides the methods for using Table D2 for events of the type $Z \leq d$, where either $d < 0$ or $d > 0$. Notice that the cases $d < 0$ and $d > 0$ require the calculation of a *difference* and a *sum*, respectively. Furthermore, $d < 0$ yields a probability less than .50000, while $d > 0$ gives a probability greater than .50000.

The preceding examples illustrate some methods for using Table D2 to obtain the probability for various types of events involving a standard normal variable Z.

Any problem with one or more probabilities for events involving a normal variable X is reduced directly to one of the types of problems illustrated by Examples 9-13 by one application of Eq. (5.24). The necessary steps are illustrated in Example 14.

EXAMPLE 14: Consider again the normal distributions of weights (X) described in Example 8, where $\mu = 170$ and $\sigma = 15$. Find the following: (a) $P(170 \leq X)$; (b) $P(155 \leq X \leq 195)$; (c) $P(X \leq 200)$.

SOLUTION: (a) Using Eq. (5.24), we obtain

$$P(170 \leq X) = P[(170-170)/15 \leq Z]$$
$$= P(0 \leq Z)$$
$$= .50000 \text{ by Eq. (5.28).}$$

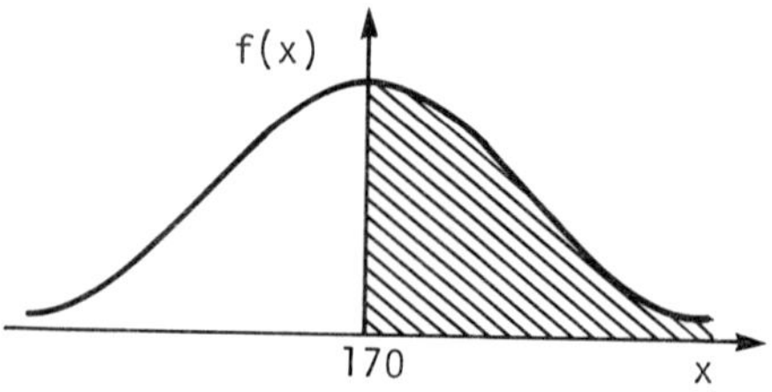

(b) Using Eq. (5.24), we find

$$P(155 \leq X \leq 195) = P[(155-170)/15 \leq Z \leq (195-170)/15]$$
$$= P(-1 \leq Z \leq 1.67)$$
$$= P(-1 \leq z \leq 0) + P(0 \leq Z \leq 1.67)$$
$$= .34134 + .45254 \text{ (Table D2)}$$
$$= .79388.$$

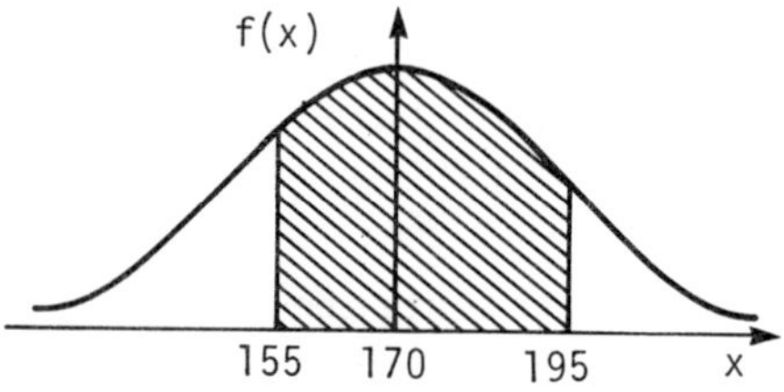

(c) Using Eq. (5.24) provides

$$P(X \leq 200) = P[Z \leq (200-170)/15]$$
$$= P(Z \leq 2)$$
$$= P(-\infty < Z \leq 0) + P(0 \leq Z \leq 2)$$
$$= .50000 + .47725$$
$$= .97725.$$

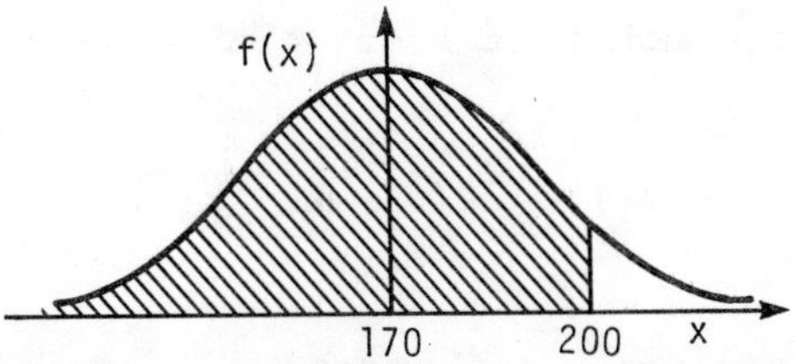

Thus far we have used Table D2 to determine probabilities for given values of Z, but the table may be used to determine the value of Z which corresponds to a given probability. That is, a problem of interest is as follows: Given that $P(0 \leq Z \leq z_1) = .42220$, find z_1. By searching through the body of Table D2, we find that .42220 corresponds to $z_1 = 1.42$. Problems of this type require the use of Table D2 in reverse to obtain answers. Furthermore, given the value of Z which satisfies a given probability statement, we can use Eq. (5.23) to transform that Z-value into a meaningful X-value. As might be expected, the steps used in Examples 9-14 are reversed in the next examples.

EXAMPLE 15: Find z_1 such that $P(Z \leq z_1) = .88100$.

SOLUTION: $P(Z \leq z_1) = .88100$ implies that $z_1 > 0$. Thus, we write

$$P(Z \leq z_1) = P(-\infty < Z \leq 0) + P(0 \leq Z \leq z_1) = .88100$$
$$= .50000 + P(0 \leq Z \leq z_1).$$

Therefore,

$$P(0 \leq Z \leq z_1) = .38100.$$

Searching the body of Table D2 shows that $z_1 = 1.18$.

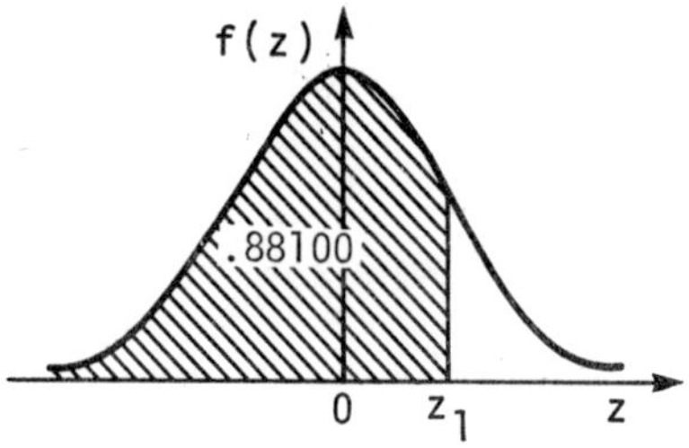

EXAMPLE 16: Find z_1 such that $P(-z_1 \leq Z \leq z_1) = .85014$.

SOLUTION: We note that

$$P(-z_1 \leq Z \leq z_1) = 2P(0 \leq Z \leq z_1).$$

Therefore,

$$P(0 \leq Z \leq z_1) = .85014/2$$

$$= .42507.$$

Searching Table D2, we find $z_1 = 1.44$.

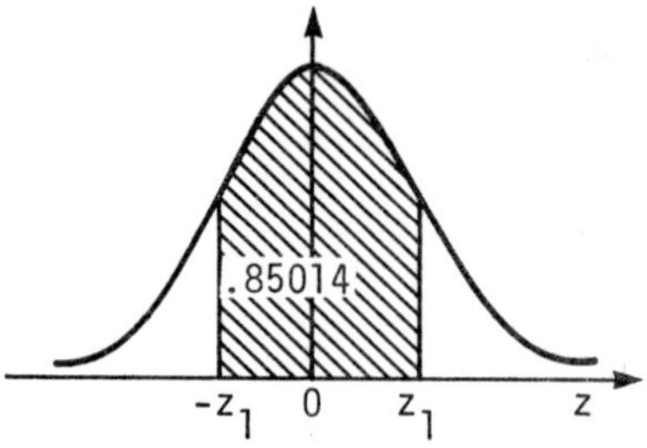

REMARK: As in the case of Examples 9-13, it is recommended that sketches be drawn to help determine the sign and size of an unknown Z-value.

EXAMPLE 17: Suppose the test scores (X) for an introductory statistics course are assumed to be normally distributed with mean 75 and variance 16. Find the value of X, say x_0, such that $P(X < x_0) = .94520$.

SOLUTION: First we transform to a probability statement in terms of Z. Thus,

$$P(X \le x_0) = P[Z \le (x_0 - 75)/4] = .94520.$$

Now let $z_1 = (x_0 - 75)/4$, and follow the procedures in Examples 15 and 16 to find z_1. That is,

$$P(Z \le z_1) = P(-\infty < Z \le 0) + P(0 \le Z \le z_1)$$

gives $.94520 = .50000 + P(0 \le Z \le z_1)$. Thus,

$$P(0 \le Z \le z_1) = .44520.$$

Searching Table D2 provides $z_1 = 1.60$.

Next we transform back to x_0 by solving $(x_0 - 75)/4 = 1.60$. Thus,

$$x_0 = 75 + 4(1.60) = 81.40.$$

(This implies that 94.5% of all statistics test scores are below 81.4%.)

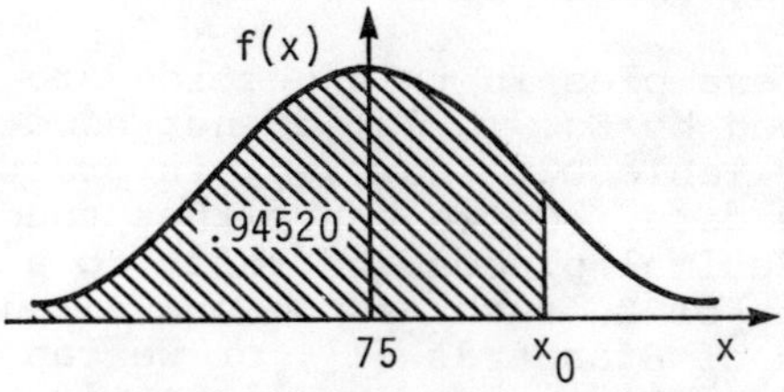

The final example in this section illustrates how the knowledge of two probabilities for a normal variable X can be used to determine the values of μ and σ.

EXAMPLE 18: Suppose $X \sim N(\mu, \sigma^2)$, and it is known that $P(X \le 12.8) = .91924$ and $P(X \le 8.2) = .18406$. Find μ and σ^2.

SOLUTION: Transforming the two probabilities to functions of Z gives

$$P[Z \le (12.8-\mu)/\sigma] = .91924,$$

$$P[Z \le (8.2-\mu)/\sigma] = .18406.$$

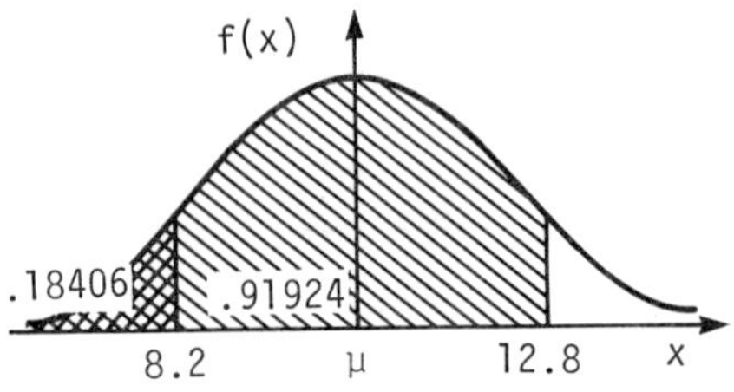

Let $z_1 = (12.8-\mu)/\sigma$ and $z_2 = (8.2-\mu)/\sigma$. Then

$$P(0 \leq Z \leq z_1) = .41924,$$

$$P(0 \leq Z \leq -z_2) = .31594.$$

Searching Table D2 provides $z_1 = 1.40$ and $-z_2 = .90$. Therefore, μ and σ must satisfy $(12.8-\mu)/\sigma = 1.4$ and $(8.2-\mu)/\sigma = -.9$.

Rewriting this, we have $1.4\sigma + \mu = 12.8$ and $.9\sigma - \mu = -8.2$. Adding the two equations gives $2.3\sigma = 4.6$. Therefore, $\sigma = 2$, and by substitution $\mu = 10$. Hence, $X \sim N(10, 4)$.

Most of the examples in this section are such that Z-values produced by Eq. (5.22) do not need to be rounded off to enter Table D2. However, in practice this need not be the case. So we can either round off the result to two decimal places and refer to a table having finer increments of Z, such as the one cited for the National Bureau of Standards [1], or we can interpolate within Table D2. The tables in [1] provide probability for Z in steps of .0001 between 0 and 1 and steps of .001 for Z between 1 and 7 or 8.

A similar situation may arise when we work from known probabilities to determine values for Z and X. Again either rough interpolation or simple round-off techniques are usually sufficient for most problems. However, if the problem merits greater precision, the tables in [1] can be used.

EXERCISES

1. Find the following probabilities for $Z \sim N(0,1)$:
   (a) $P(0 \leq Z \leq 1.75)$;
   (b) $P(1.35 \leq Z \leq 2.54)$;
   (c) $P(Z \leq 1.07)$;
   (d) $P(-.75 \leq Z \leq .50)$;
   (e) $P(-.67 \leq Z \leq .67)$;
   (f) $P(-.82 \leq Z)$;

(g) $P(2.50 \leq Z)$;
(h) $P(Z \leq -.73)$.

2. Find the values for z that satisfy the following probability statements when $Z \sim N(0,1)$:
   (a) $P(0 \leq Z \leq z) = .15910$;
   (b) $P(-z \leq Z \leq z) = .95$;
   (c) $P(Z \leq z) = .15866$;
   (d) $P(z \leq Z) = .05$.

3. Indicate one or more variables in your field of interest that could be modeled by normal distributions. For each variable, specify approximate values for its mean and variance.

4. Suppose the diameters (X) of 1/2-inch bolts are normally distributed with mean 1/2 inch and standard deviation .001 inches.
   (a) What percent of the bolts have diameters between .498 and .500 inches? What percent have diameters between .4995 and .5005 inches?
   (b) Find $P(.501 \leq X)$.
   (c) If a randomly selected bolt has a diameter of .51 inches, is there any reason to doubt that the machine is still in control?

5. In Exercise 4, find the diameter x such that $P(x \leq X) = .025$.

6. Suppose Mr. Scitsitats wants to grade his students on a normal curve so that the scale includes 10% A's, 20% B's, 40% C's, 20% D's, and 10% F's. Assume that $\mu = 75$ and $\sigma = 8$. Then determine the range of scores for each letter grade.

7. Each box of Wake-Up breakfast cereal lists the net weight (X) as 15 ounces. If 5% of all boxes of Wake-Up show net weights of 15.3645 ounces or more, and if $P(X \leq 15.004) = .025$, what are μ and σ when X is assumed to be a normal variable? What percent of the boxes contain 15 ounces or less of Wake-Up cereal?

USING NORMAL MODELS TO APPROXIMATE BINOMIAL MODELS: In Section 4.7, a Poisson model is proposed as an approximation for a binomial model with n large and p small (or q small). In that case, we approximate a discrete distribution by another discrete distribution where the sample space of each distribution consists of nonnegative integers. In this section, we propose the normal model (a continuous distribution) as an approximation for a binomial model (a discrete distribution) when n is large and p is neither close to 0 nor close to 1. Thus, we combine the normal approximation to the binomial discussed here with the Poisson approximation discussed in Section 4.7 to get a method for approximating binomial probabilities for all values of p when n is large.

The full justification for using normal models to approximate binomial probabilities is furnished by the central limit theorem discussed in Chapter 6. However, the tendency for a variable $X \sim b(n, p)$ to exhibit symmetry makes it reasonable to consider normal approximations to the binomial. Furthermore, we recall that binomial distributions are symmetric for all values of n when p = .5 and become symmetric as n increases when p $\neq$ .5. Thus, we may expect a normal approximation to the binomial to provide the best results when p is either equal to .5 or near .5.

Since the normal model requires two parameters for specification, and since those parameters are the mean, μ, and the variance, σ^2, it follows that to approximate a binomial model with sample size n and parameter p, we equate the mean and variance of the binomial model to μ and σ^2, respectively. Thus, we use

$$\mu = np,$$

$$\sigma^2 = npq. \quad (5.34)$$

Then if X is a binomial variable (n, p),

$$Z = (X - np)/\sqrt{npq} \quad (5.35)$$

is approximately a N(0,1) variable for large values of n and values of p neither close to 0 nor close to 1.

Now we consider a basic difference between normal and binomial models. First, the binomial assigns positive probabilities to a finite set of integers, while the normal model assigns density to an interval of infinite length. Yet the bulk of the normal density is assigned to an interval of length 6σ [i.e., $P(\mu - 3\sigma \leq X \leq \mu + 3\sigma) = .9974$]. Second, we recall that the probability of any point has 0 probability for a normal model. Thus, to obtain a positive probability for a binomial integer value, say x, from a normal density, we must associate the *point* x with an *interval* under the normal model. Of course, there are many ways in which the interval between two consecutive integers can be allocated to those integers, and naturally the different allocations yield approximations with different properties. However, a simple and useful rule is to divide each interval evenly between its integer end points. Thus, any integer value, x, for a binomial variable is assigned the area under the normal curve between x - .5 and x + .5 as its probability. Hence, if $X \sim b(n,p)$, then

$$P(X=x) \approx P[(x-.5-np)/\sqrt{npq} \leq Z \leq (x+.5-np)/\sqrt{npq}], \quad (5.36)$$

where $Z \sim N(0,1)$. Furthermore, for any two integers, x_1 and x_2, and $X \sim b(n,p)$

$$P(x_1 \leq X \leq x_2) \approx$$

$$p[(x_1-.5-np)/\sqrt{npq} \leq Z \leq (x_2+.5-np)/\sqrt{npq}], \quad (5.37)$$

where $Z \sim N(0,1)$.

One final caution must be stated concerning the normal approximation to the binomial. When working with *continuous* variables, the events $(x_1 \leq X \leq x_2)$ and $(x_1 < X < x_2)$ have the same probability because $P(X=x_1) = P(X=x_2) = 0$. However, for *discrete* variables the inclusion or exclusion of the equality sign changes the probability of an event. For example, consider x_1 and x_2 to be values of a variable X on the integers. Then the event $(x_1 < X < x_2)$ has the same probability as the event $(x_1 + 1 \leq X \leq x_2 - 1)$. That is, $P(X=x_1)$ and $P(X=x_2)$ must be excluded from the probability, and for a discrete variable, such probabilities are, in general, positive numbers. Thus, the improper calculation for this event produces a probability which is too large.

The foregoing properties, as well as the proper use of Eqs. (5.36) and (5.37), are illustrated by the following examples.

EXAMPLE 19: If it is known that 64% of the students at MYU favor the construction of special bicycle parking on campus, what is the probability that the number (X) of students in a random sample of 100 who favor construction of bicycle parking is equal to 64?

SOLUTION: Note that $\mu = np = 100(.64) = 64$ and that $\sigma^2 = 100(.64)(.36) = 23.04$. By Eq. (5.36),

$$
\begin{aligned}
P(X=64) &\approx P[(63.5-64)/\sqrt{100(.64)(.36)} \leq Z \\
&\qquad \leq (64.5-64)/\sqrt{100(.64)(.36)}] \\
&= P(-.5/4.8 \leq Z \leq .5/4.8) - P(-.104 \leq Z \leq .104) \\
&= 2(.04142)\text{ (by interpolation)} \\
&= .08284.
\end{aligned}
$$

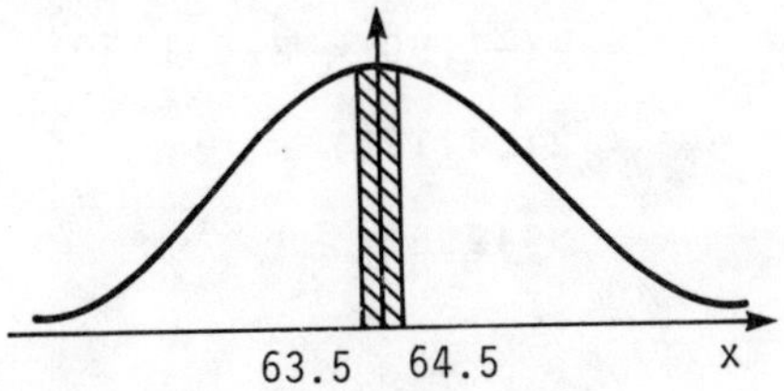

REMARK: Romig [2] gives the exact probability as $P(X=64) = .082882$.

EXAMPLE 20: Find $P(50 < X)$ for the sampling situation described in Example 19.

$$\begin{aligned} P(50 < X) &= P(51 \leq X) \\ &\approx P[(50.5-64)/4.8 \leq Z] \\ &= P(-13.5/4.8 \leq Z) \\ &= P(-2.81 \leq Z) \\ &= .50000 + .49752 = .99752. \end{aligned}$$

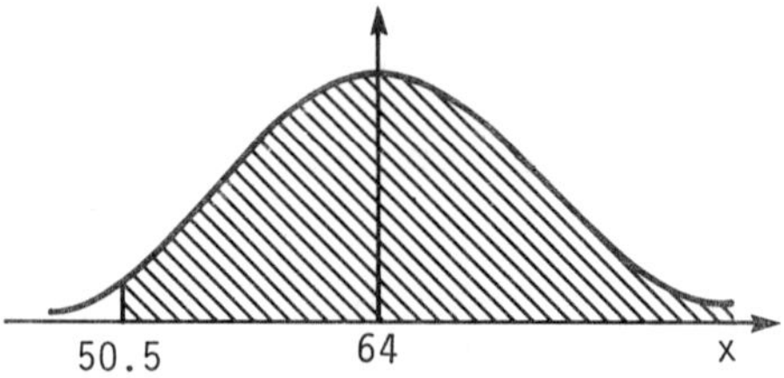

REMARK: Romig [2] gives the exact probability to be $P(51 \leq X) = .997142$.

EXAMPLE 21: Find the probability that 100 flips of a fair coin ($p = .5$) result in a number of heads, x, between 40 and 60, exclusively.

SOLUTION: Note that $\mu = 100(.5)=50$, $\sigma^2 = 100(.5)(.5) = 25$, and $\sigma = 5$. By Eq. (5.37),

$$\begin{aligned} P(40 < X < 60) &= P(41 \leq X \leq 59) \\ &\approx P[(40.5-50)/5 \leq Z \leq (59.5-50)/5] \\ &= P(-1.90 \leq Z \leq 1.90) \\ &= 2(.47128) \\ &= .94256. \end{aligned}$$

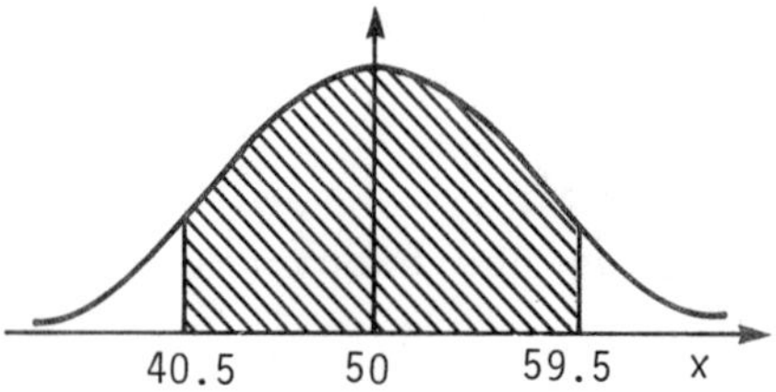

REMARK: Romig [2] gives the exact probability as $P(40 < X < 60) = .943112$.

A primary concern for any approximation is its accuracy. For Eqs. (5.36) and (5.37), Raff [3] has shown that the error involved is never greater than $.14/\sqrt{np(1-p)}$. However, this bound is much too large when p is near .5. To see that result we conclude this section with a table showing exact values and normal approximations for the binomial probabilities when $n = 4$ and $p = .5$. The results in Table 5.4 indicate a very good approximation, even though n is very small. In fact, the largest error is about .008. Similar accuracy is available for any intervals with integer end points.

TABLE 5.4.
A comparison is shown of exact probabilities and normal approximations to those probabilities for a variable $X \sim b(4,.5)$.

x	Z-interval	P(X=x)	P(Z-interval)
0	-2.5 to -1.5	.0625	.06060
1	-1.5 to -0.5	.2500	.24173
2	-0.5 to 0.5	.3750	.38292
3	0.5 to 1.5	.2500	.24173
4	1.5 to 2.5	.0625	.06060
Total		1.0000	.98758

EXERCISES

1. Use Table B1 and Eqs. (5.36) and (5.37) to calculate and compare the exact probabilities and their normal approximations for the indicated events.
   (a) $X \sim b(50, .5)$; $P(X = 25)$; $P(20 \leq X \leq 30)$.
   (b) $X \sim b(50, .2)$, $P(X = 12)$; $P(X \leq 15)$.
   (c) $X \sim b(10, .5)$; $P(X = 5)$; $P(3 \leq X \leq 7)$.
   (d) $X \sim b(10, .25)$; $P(X = 2)$; $P(X \leq 4)$.

2. Of all students, 50% favor Candidate A over his opponents. If a random sample of 100 students is questioned, what is the probability that the number (X) favoring Candidate A is equal to 50? $P(51 \leq X)$?

3. If 40% of all voters favor an increase in federal taxes to support education, what is the probability that the Pullag Poll

based on a random sample of 1,200 voters shows a majority in favor of this tax increase?

4. According to the genetic theory, 25% of Dr. F. Fly's fruit flies should have blue eyes. A random sample of 300 flies produces 90 blue-eyed flies. Find the probability that 90 or more blue-eyed flies occur in samples of 300, and explain whether or not the observed sample casts doubt on the assumed 25% blue-eyed flies hypothesis.

5. Of all customers at the Food Corner, 64% buy the special Bread of the Day. If X is the number of customers buying the special and the store has 900 customers today, what is $P(X \leq 600)$?

5.5 THE CHI-SQUARE DISTRIBUTION

Another continuous distribution used in statistical inference is called the chi-square distribution. The pioneering efforts of F. R. Helmert in 1875 and K. Pearson in 1900 in developing that distribution are referenced by Johnson and Kotz [4]. Pearson is responsible for the chi-square distribution's wide use in statistical applications.

One way of describing a chi-square variable is to consider the square of a standard normal variable. That is, if $Z \sim N(0,1)$, then Z^2 is distributed as a chi-square variable with a parameter equal to 1. Furthermore, if $Z_1, Z_2, \cdots, Z_n$ are independent $N(0,1)$ variables, then

$$Y = \sum_{i=1}^{n} Z_i^2 \tag{5.38}$$

has a chi-square distribution with parameter n. In symbols, we write

$$Y \sim \chi^2(n) \tag{5.39}$$

to denote that Y has a chi-square distribution with parameter n. Historically, the parameter n for a chi-square variable is referred to as degrees of freedom. Practically speaking, it represents the number of free variables being squared and summed to give the chi-square variable. That property has more relevance as we use the chi-square distribution for statistical inference in Chapters 7, 9, and 10.

In view of the foregoing description of a chi-square variable, the following experiments generate random samples of size 50 from selected chi-square distributions.

EXPERIMENT 1: Select a random sample of size 50 from Table D3, say, $z_1, z_2, \ldots, z_{50}$. Then calculate the square of each z_i, say, $y_1 = z_1^2$, $y_2 = z_2^2$, ... $y_{50} = z_{50}^2$. Form a distribution of the 50 observations using as class intervals 0-1, 1-2, 2-3, 3-4, 4-5, and 5-6.

EXPERIMENT 2: Select 50 sets of 5 N(0,1) variables, say, $(z_{1,1}, z_{1,2}, z_{1,3}, z_{1,4}, z_{1,5})$, $(z_{2,1}, z_{2,2}, z_{2,3}, z_{2,4}, z_{2,5})$ ..., $(z_{50,1}, z_{50,2}, z_{50,3}, z_{50,4}, z_{50,5})$. Compute

$$y_1 = \sum_{i=1}^{5} z_{1,i}^2, \; y_2 = \sum_{i=1}^{5} z_{2,i}^2, \; \ldots, \; y_{50} = \sum_{i=1}^{5} z_{50,i}^2.$$

Form a distribution of the 50 values of y_i using as class intervals 0-2, 2-4, 4-6, 6-8, 8-10, 10-12, 12-14, and 14-16.

The preceding experiments may be conducted as either individual or class exercises. The results can then be compared with the results presented in Tables 5.5 and 5.6.

TABLE 5.5
Empirical and model frequencies for the squares of 50 standard normal deviates from Experiment 1.

Classes	Sample 1	Sample 2	Sample 3	Sample 4	Model
0-1	35	33	35	38	34.2
1-2	9	10	6	6	8.0
2-3	3	3	4	4	3.7
3-4	2	4	3	1	1.9
4-5	1	0	2	0	1.0
5-6	0	0	0	1	.6
Total	50	50	50	50	49.4

TABLE 5.6.
Empirical and model frequencies for 50 sums of squares of 5 standard normal deviates from Experiment 2.

Classes	Sample 1	Sample 2	Sample 3	Sample 4	Model
0-2	11	8	8	10	7.5
2-4	17	14	14	13	15.0
4-6	9	14	11	13	12.0
6-8	8	3	6	7	7.5
8-10	2	5	8	2	4.0
10-12	1	5	2	4	2.0
12-14	2	0	0	1	1.0
14-16	0	1	1	0	.5
Total	50	50	50	50	49.5

If Y is a χ^2 variable with d degrees of freedom [i.e., $Y \sim \chi^2(d)$], then the χ^2 probability density function of Y is defined by

$$f(y) = [(y/2)^{(d/2)-1} e^{-(y/2)}]/[2\Gamma(d/2)], \quad y>0, \tag{5.40}$$

where $\Gamma(d/2)$ is a gamma function such that

$$\Gamma(d/2) = \begin{cases} [(d/2)-1)][(d/2)-2)] \ldots (3/2)\cdot(1/2)\sqrt{\pi}, & \text{odd values of } d, \\ [(d/2)-1)]! & \text{, even values of } d. \end{cases} \tag{5.41}$$

Notice that the gamma function is defined for non-integer values of d, but the integer value results presented in Eq. (5.41) are adequate for our purposes. When $Y \sim \chi^2(d)$ with probability density given by Eq. (5.40), it can be shown that

$$\begin{aligned} E(Y) &= d, \\ V(Y) &= 2d. \end{aligned} \tag{5.42}$$

Thus, it is obvious that both the average value and the variance of a χ^2 variable increase as the parameter d increases. That is, we can expect a χ^2 density to exhibit more dispersion and larger values as d increases. In the preceding examples, d is equal to the sample size n. However, since that is not always the case for subsequent discussions and uses of χ^2 models, we use the letter d to denote the parameter (degrees of freedom) of an arbitrary χ^2 variable.

The data for the preceding experiments, as presented in Tables 5.5 and 5.6, are modeled by $\chi^2(1)$ and $\chi^2(5)$, respectively. These two density functions are sketched in Figure 5.10 to illustrate the following properties. The density for d = 1 is very heavily concentrated near 0, while the density for d = 5 is more widely dispersed around its mean, E(Y) = 5.

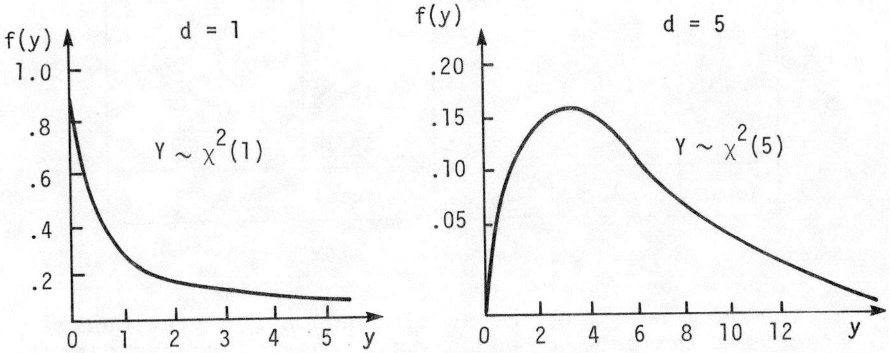

FIGURE 5.10. Shown are graphs of two χ^2 density functions with d = 1 and d = 5.

To determine the model frequencies for Tables 5.5 and 5.6, we need the probability values for the class intervals given. These probabilities can be obtained from the tables available in Abramowitz and Stegun [5] and are as given in Table 5.7. The model frequencies are then obtained by multiplying the probability of each interval by the total frequency, which is 50.

TABLE 5.7.
Shown are probabilities for the class intervals given in Tables 5.5 and 5.6 when $X \sim \chi^2(1)$ and $\chi^2(5)$, respectively.

d = 1		d = 5	
Class	Probability	Class	Probability
0-1	.683	0-2	.15
1-2	.160	2-4	.30
2-3	.074	4-6	.24
3-4	.038	6-8	.15
4-5	.020	8-10	.08
5-6	.011	10-12	.04
		12-14	.02
		14-16	.01
Total	.986	Total	.99

The data sets and the model frequencies previously discussed indicate a tendency for the chi-square distribution to become symmetric as d increases. That property makes it reasonable to consider approximating the distribution of $Y \sim \chi^2(d)$ by a normal variable when d gets large. The actual justification of that approximation is provided by the central limit theorem presented in Chapter 6. The exact form of the approximation is given as a footnote to Table E2.

The chi-square models are basic for making inference about the value of σ^2 for a normal model (as in Chapters 7 and 10). It is also used in Chapter 9 to determine whether or not a particular model offers a good description (fit) of an empirical data set. For those uses, we do not need a table of probabilities for all values of the chi-square variable, Y [i.e., $Y \sim \chi^2(d)$], such as provided in Table D2 for the standard normal model. Rather, we need to know the *α-level values* of Y for degrees of freedom, d = 1, 2, ..., say, $\chi^2(d, \alpha)$, for selected values of α, where $\chi^2(d,\alpha)$ is defined by

$$P[Y \geq \chi^2(d, \alpha)] = \alpha. \tag{5.43}$$

Table E2 provides those *α-level values* of $Y \sim \chi^2(d)$ for a selected set of α values. The use of Table E2 is demonstrated by the next examples and exercises.

EXAMPLE 22: Let $Y \sim \chi^2(5)$. Find a, b, c, and e such that $P(Y \geq a) = .95$, $P(a \leq Y \leq b) = .90$, $P(Y \leq c) = .10$, and $P(Y \leq e) = .99$.

SOLUTION: First, each event of interest must be expressed in terms of the type given in Eq. (5.43). Second, since d = 5, the values of interest occur in line 5 of Table E2. Thus, $P(Y \geq a) = .95$ implies that $a = \chi^2(5, .95)$. From the line for d = 5 and the column for α = .95 in Table E2, we find that

$$\chi^2(5, .95) = a = 1.1455.$$

Now, for a = 1.1455, we have $P(Y \leq a) = .05$. Therefore, $P(a \leq Y \leq b) = .90$ implies that $P(Y \geq b) = .05$. Thus, $b = \chi^2(5, .05)$. From the line for d = 5 and the column for α = .05, we find that

$$\chi^2(5, .05) = b = 11.0705.$$

Next, $P(Y \leq c) = .10$ implies that $P(Y \geq c) = .90$. So

$$c = \chi^2(5, .90) = 1.6103.$$

Similarly, $P(Y \leq e) = .99$ implies that $P(Y \geq e) = .01$. Then

$$e = \chi^2(5, .01) = 15.0863.$$

REMARKS: The interpretations for the preceding results are that when sampling a χ^2 variable with 5 degrees of freedom, we can expect about 5% of our observations to be less than 1.1455, while about 95% are greater than that value. Furthermore, roughly 90% of our observations are expected to fall between 1.1455 and 11.0705. Also, about 90% are expected to exceed 1.6103, while only 1% are likely to exceed 15.0863. These results can be applied to the data in Table 5.6 and any other data collected for Experiment 2.

EXAMPLE 23: Table E2 can also be read in reverse to find d or α when certain probability statements are given. (a) Find d when $P(8.5468 \leq Y \leq 22.3071) = .80$. (b) Find α such that $P(Y \geq 28.2412) = \alpha$ when d = 24.

SOLUTION: (a) Searching the columns of Table E2 for α = .90 and α = .10, we find $\chi^2(15, .90) = 8.5468$ and $\chi^2(15, .10) = 22.3071$. Thus, a χ^2 variable with d = 15 satisfies the given probability statement.

(b) In the line of Table E2 for d = 24, we find $\chi^2(24, .25) = 28.2412$. Thus, $\alpha = .25$ satisfies the given probability statement.

EXERCISES

1. Let $Z_1, Z_2, \cdots, Z_{10}$ be 10 standard normal deviates. If

$$Y = \sum_1^{10} Z_i^2,$$

what is E(Y)? What is V(Y)? Sketch the distribution of Y.

2. If $Y \sim \chi^2(15)$, find the following:
(a) $P(Y \geq 11.0365)$;
(b) $P(7.2609 \leq Y \leq 24.9958)$;
(c) $P(4.6009 \leq Y \leq 27.4884)$;
(d) $P(Y \leq 22.3071)$.

3. If $Y \sim \chi^2(d)$, find d to satisfy the following:
(a) $P(Y \leq 24.3366) = .50$;
(b) $P(Y \geq 18.4927) = .95$;
(c) $P(20.8434 \leq Y \leq 30.4346) = .50$;
(d) $P(Y \leq 14.4494) = .975$.

5.6 SUMMARY

This chapter present the basic ideas of continuous probability models. In particular, discussions of uniform, triangular, normal, and chi-square models are included. Now that we have introduced models for both discrete and continuous random variables, we turn our attention to sampling distributions in Chapter 6. The study of sampling variations in the next chapter is basic to our study of statistical inference beginning in Chapter 7.

We have used specific tables for the normal and chi-square distributions in this chapter. Two other excellent sources of statistical tables for use with various statistical distributions are Fisher and Yates [6] and Pearson and Hartley[7].

5.7 REFERENCES

[1] Natl. Bureau of Standards. 1952. *A guide to tables of the normal probability integral.* U.S. Dept. of Commerce, Applied Math. Series No. 21.

[2] Romig, H. G. 1953. *50-100 binomial tables.* New York: John Wiley & Sons.

[3] Raff, M. S. 1956. On approximating the point binomial. *The Journal of the American Statistical Assoc.* 51: 293-303.

[4] Johnson, N. L., and Kotz, S. 1970. *Continuous univariate distributions-1.* New York: Houghton Mifflin, Chapter 17.

[5] Abramowitz, M., and Stegun, I. A. 1964. *Handbook of mathematical functions.* U.S. Dept. of Commerce, Natl. Bureau of Standards, Applied Math. Series No. 55, pp. 978-984.

[6] Fisher, R. A., and Yates, F. 1963. *Statistical tables for biological, agricultural, and medical research.* 6th ed. New York: Hafner.

[7] Pearson, E. S., and Hartley, H. O. 1966. *Biometrika tables for statisticians.* 3rd ed. Vol. I. Cambridge: The University Press.

5.8 REVIEW EXERCISES

1. (a) Find c so that the figure below is a valid probability density function.
   (b) Find $P(-c \leq X \leq 0)$.

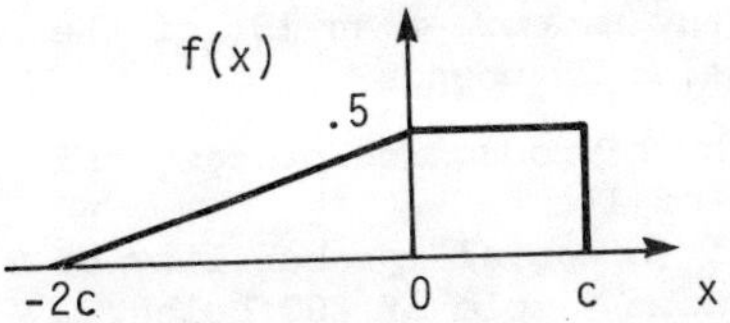

2. (a) Sketch F(x) for the distribution of X in the accompanying figure.
   (b) Find E(X).
   (c) Find V(X).
   (d) Find $P(1.5 \leq X \leq 4.0)$.

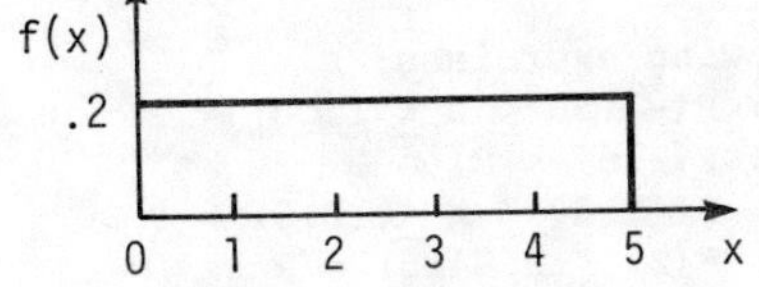

3. What is the effect on the graph of a normal curve when μ is doubled? when σ is doubled? when both μ and σ are doubled?

4. If $X \sim N(10,16)$, find the mean, variance, and standard deviation of each of the following random variables:
   (a) $Y = 2X + 3$;
   (b) $W = -3X$;
   (c) $U = (X-10)^2/16$.

5. The monthly starting salaries (X) for individuals holding a B.S. degree in management are assumed to be normally distributed with mean \$800 and standard deviation \$80. Find the following:
   (a) $P(X \geq \$900.)$;
   (b) $P(X \leq \$720.)$;
   (c) $P(\$640. \leq X \leq \$900.)$;
   (d) $P(X \leq \$1000.)$.

6. The protein content (X) of milo is assumed to be normally distributed with mean value 10% and standard deviation of 1.6%. Find the following:
   (a) $P(8.0\% \leq X)$;
   (b) $P(X \leq 15\%)$;
   (c) $P(10\% \leq X)$;
   (d) $P(9.2\% \leq X \leq 14.8\%)$.

7. The bagging operation used to bag More Pep dog food is such that the net weight (X) of each bag is normally distributed with standard deviation of .5 pounds about the set value of μ. What value of μ would you recommend for use on 25-pound bags of dog food if you want no more than 10% of the bags to contain less than the labeled 25 pounds?

8. Consider a light bulb production process that produces defective bulbs at the rate of 1%.
   (a) Find the exact (binomial) probability of 2 or fewer defectives in a random sample of 100 bulbs.
   *(b) Find the Poisson approximation for part (a).
   (c) Find the normal approximation for part (a).
   (d) Explain any discrepancies observed.

9. Consider a sample of size 100 from an electorate in which 36% are in favor of the new aid to higher education bill.
   (a) Find the binomial probability of at least 40 voters favoring the bill.
   (b) Find the normal approximation for part (a).
   *(c) Would the Poisson approximation for part (a) be appropriate? Why or why not?

10. Solve the following exercises:
   (a) $Z \sim N(0,1)$; $P(-1.50 \leq Z \leq 1.00) = ?$
   (b) $Z \sim N(0,1)$; $P(-.67 \leq Z) = ?$
   (c) $X \sim N(5,9)$; $P(-1.50 \leq X \leq 10.5) = ?$
   (d) $Z \sim N(0,1)$; $P(Z^2 \leq 3.8415) = ?$

*The asterisk indicates that the solution uses material from a starred section (in these cases, Section 4.7).

11. Solve the following exercises:
(a) $Y \sim \chi^2(6)$; $P(Y \geq c) = .25$; $c = ?$
(b) $Y \sim \chi^2(d)$; $P(Y \leq 19.3689) = .75$; $d = ?$
(c) $Z_i \sim N(0,1)$ for $i = 1, 2, \cdots, 20$;

$$Y = \sum_{i=1}^{20} Z_i^2.$$

$P(Y \geq 39.9968) = ?$ (Assume that Z_i are independent variables.)

12. (a) If $Z_i \sim N(0,1)$ and if the Z_i are independent, what are $E(Y)$ and $V(Y)$ when

$$Y = \sum_{i=1}^{8} Z_i^2?$$

(b) If $X_i \sim N(\mu,\sigma^2)$ and if the X_i are independent, what are $E(Y)$ and $V(Y)$ when

$$Y = \sum_{i=1}^{8} (X_i - \mu)^2/\sigma^2?$$

13. If 10% of all members of a specified human population are left-handed, find the probability that a random sample of 400 delegates to a convention contains 50 left-handed persons.

CHAPTER 6

Sampling Distribution Results

6.1 A LOOK BACK, THEN AHEAD

Chapters 4 and 5 present many potential models for empirical data. The types of data that can be modeled by the distributions studied vary widely. That is, the preceding models are appropriate for a wide variety of experimental data. Yet, when attempting to model any given set of data, we face two questions immediately. First, which family of distributions is appropriate for our data? Second, which specific member of that family should be used? The answers to these questions provide a specific model for the empirical data.

The selection of the appropriate family depends on what assumptions we make about our data. For example, if we accept either one of two possible answers to a question posed to a subject and assume that the responses for different subjects are independent, then the binomial family is the appropriate model for our experiment. For most of the situations considered in this text, the appropriate family of distributions is specified. Thus, our efforts are shifted to that of selecting a particular distribution from a designated family. Usually that requires only the specification of values for one or more parameters. Then the objective is to use any available data to help in the selection of best values for the parameters in the family of potential models for a data set. Estimation techniques which address that problem are discussed in Chapter 7.

To understand the underlying principles and philosophy of statistical estimation we need to explore the sampling properties of some potential estimators (statistics). Thus, we devote this chapter to a study of how some selected statistics vary from sample to sample. The results illustrated in this chapter serve as a foundation for the discussion of statistical inference which follows in subsequent chapters.

Since many of the statistics used for statistical inference depend on the *sum of sample observations* (or equivalently, on the sum of independent random variables), it is necessary that some basic results are introduced before proceeding to a study of sampling distributions. If $X_1, X_2, \ldots, X_n$ denote a set of independent random variables such that $E(X_i) = \mu_i$ and $V(X_i) = \sigma^2_i$, then the random variable

$$W = \sum_{i=1}^{n} a_i X_i$$

is such that

$$E(W) = \sum_{i=1}^{n} a_i \mu_i, \tag{6.1}$$

$$V(W) = \sum_{i=1}^{n} a^2_i \sigma^2_i. \tag{6.2}$$

Partial Proof: Equation (6.1) follows from the definition of the expected value and from Eq. (4.4) as shown below:

$$E(W) = E\{\Sigma a_i X_i\} = \Sigma E\{a_i X_i\} = \Sigma a_i E(X_i) = \Sigma a_i \mu_i.$$

By the independence of the X_i,

$$V(W) = V(\Sigma a_i X_i) = \Sigma V(a_i X_i).$$

Then by Eq. (4.5),

$$V(W) = \Sigma a_i^2 V(X_i) = \Sigma a_i^2 \sigma_i^2.$$

The result in Eq. (6.1) holds for dependent variables as well as independent variables, but independence is required for Eq. (6.2) to hold.

Another important result is the following. If the variables have independent normal distributions [i.e., $X_i \sim N(\mu_i, \sigma_i^2)$ $i = 1, 2, \ldots, n$], it follows that

$$W = \sum_{i=1}^{n} a_i X_i \sim N\left(\sum_{i=1}^{n} a_i \mu_i, \sum_{i=1}^{n} a_i^2 \sigma_i^2\right). \qquad (6.3)$$

Thus, in addition to the results for E(W) and V(W) as given in Eqs. (6.1) and (6.2), a linear combination of normal variables is also a normal variable. The result in Eq. (6.3) is usually proved as a theorem in elementary statistical theory textbooks (e.g., Hogg and Craig [1]).

Illustrations and applications of the preceding results are provided by the following developments and examples.

6.2 SAMPLING DISTRIBUTION OF $\overline{X}$

If random samples of size n are observed for a random variable X and if

$$\overline{X} = \sum_{i=1}^{n} X_i/n$$

is calculated for each sample, we expect to obtain a set of different $\overline{X}$ values. That is, we expect $\overline{X}$ to vary from sample to sample because it is a function of random observations on the variable X. The set of all possible $\overline{X}$ values (the sample space of $\overline{X}$) and their respective probabilities (or densities) constitute the *sampling distribution of* $\overline{X}$ and control the sample-to-sample variation observed in random sampling. The purpose of this section is to present results for the sampling distributions of $\overline{X}$. After the general results are stated, some examples for sampling discrete random variables are provided. The hypothetical distributions considered are such that the possible samples of various sizes can be enumerated. Those enumerations then yield the desired sampling distributions.

For convenience, we adopt the following notation:

$E(\overline{X}_n)$ = mean of sampling distribution of $\overline{X}$ based on a sample size of n = 1, 2, ..., .

$V(\overline{X}_n)$ = variance of sampling distribution of $\overline{X}$ based on a sample size of n = 1, 2, ..., .

One consequence of this notation is that $E(\overline{X}_1)$ and $V(\overline{X}_1)$ are the mean and variance (i.e., μ and σ^2) of the distribution being sampled. That result follows because observing $\overline{X}$ for a sample size n = 1 means that $\overline{X}$ is identical to the variable X. Thus, we use μ and σ^2 interchangeably with $E(\overline{X}_1)$ and $V(\overline{X}_1)$.

DISTRIBUTION OF $\overline{X}$ WHEN SAMPLING WITHOUT REPLACEMENT: In this section, we suppose that the variable being sampled consists of N distinct values and that the sampling is conducted without replacement.

THEOREM 6.1: If a random sample of size n, say, X_1, X_2, ..., X_n, is selected (without replacement) from the set of N values for a variable X such that $E(X) = \mu$ and $V(X) = \sigma^2$, and if

$$\overline{X}_n = \sum_{i=1}^{n} X_i/n$$

is calculated for that sample, then

$$E(\overline{X}_n) = \mu \text{ and } V(\overline{X}_n) = [\sigma^2/n][(N - n)/(N- 1)] \quad (6.4)$$

for n = 1, 2, ..., N.

Partial Proof: Since all observations in the random sample are on the same variable X, we have $\mu_i = \mu$ and

$a_i = 1/n$ for i = 1, 2, ..., n in Eq. (6.1).

Thus,

$$E(\overline{X}) = \sum_{i=1}^{n} \frac{1}{n}\mu = n(\frac{1}{n})\mu = \mu.$$

Since the sampling is completed without replacement, the observations in the sample are *dependent*. (For an example, see Case 1 in Section 2.5.) Thus, the result in Eq. (6.2) does not apply, and we accept the result for $V(\overline{X})$ in Eq. (6.4) without proof. However, that result is illustrated by the following example.

EXAMPLE 1: Suppose 4 ping pong balls (or poker chips) are marked with numbers 1,3,5,7. Then suppose those balls are placed in a box, and random samples of size $n(n = 2, 3,$ or $4)$ are selected without replacement from the box for the purpose of calculating $\bar{X}_n$. (Note: The described experiment for $n = 2, 3,$ or 4 may be conducted as a class exercise to demonstrate how $\bar{X}$ varies from sample to sample.) Find the sampling distribution of $\bar{X}_n$, $E(\bar{X}_n)$, and $V(\bar{X}_n)$ for 2, 3, and 4.

SOLUTION: The random variable, X, being sampled has the following probability distribution, mean, and variance:

x	P(x)
1	1/4
3	1/4
5	1/4
7	1/4
	1.0

$$\mu = \sum_x x\,P(x) = 1(1/4) + 3(1/4) + 5(1/4) + 7(1/4)$$
$$= 16/4 = 4.$$
$$\sigma^2 = \sum_x (x-\mu)^2\,P(x) = (-3)^2\,(1/4) + (-1)^2\,(1/4)$$
$$+ (1)^2\,(1/4) + (3)^2\,(1/4) = 20/4 = 5.$$

We remark that if $\bar{X}$ for samples of size $n = 1$ is considered, then X and $\bar{X}_1$ are identical variables, and $E(\bar{X}) = \mu = 4$ and $V(\bar{X}) = \sigma^2 = 5$.

For the sampling distribution of $\bar{X}_2$, we note that there are 6 equally likely samples of size $n = 2$ when random sampling (1, 3, 5, 7) without replacement. Those samples and their means, $\bar{X}_2$, are given in Table 6.1.

TABLE 6.1.
Shown are the possible samples of size n = 2 and their means for random samples (without replacement) from (1, 3, 5, 7).

Samples	(1, 3)	(1, 5)	(1, 7)	(3, 5)	(3, 7)	(5, 7)
$\bar{X}_2$	2	3	4	4	5	6

Since the 6 samples are equally likely events, we assign each of them the probability of 1/6. Then the sampling distribution of $\bar{X}_2$ is as given in Table 6.2. Figure 6.1 presents the graph of the $\bar{X}$ distribution.

TABLE 6.2.
Shown here is the distribution of $\overline{X}_2 (n = 2)$ from (1, 3, 5, 7) without replacement.

$\overline{x}_2$	$P(\overline{x}_2)$
2	1/6
3	1/6
4	2/6
5	1/6
6	1/6
	1.0

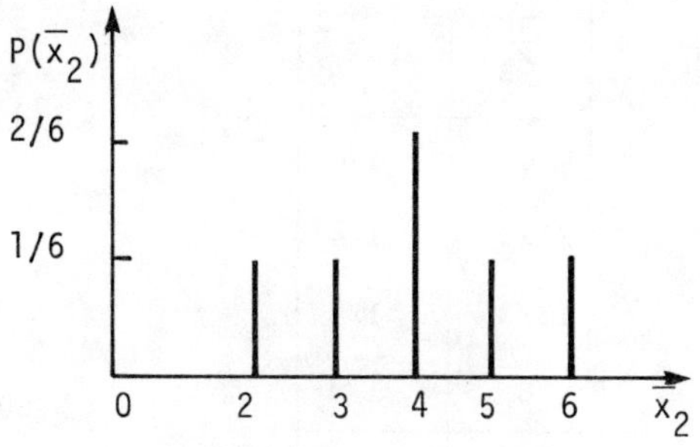

FIGURE 6.1. A graph of the distribution in Table 6.2 is shown.

From Table 6.2, it follows that

$$E(\overline{X}_2) = \Sigma\, \overline{x}_2\, P(\overline{x}_2) = [2 + 3 + 2(4) + 5 + 6]/6 = 4.$$

$$V(\overline{X}_2) = \Sigma [\overline{x}_2 - E(\overline{X}_2)]^2\, P(\overline{x}_2)$$

$$= [(-2)^2 + (-1)^2 + 2(0)^2 + 1^2 + 2^2]/6 = 5/3.$$

It is easy to verify that these results agree with Eq. (6.4) for $\mu = 4$, $\sigma^2 = 5$, $N = 4$, and $n = 2$.

In similar fashion, we can consider samples of size $n = 3$ from (1, 3, 5, 7) when sampling is without replacement. The results are given in Tables 6.3 and 6.4 and Figure 6.2.

TABLE 6.3
Given are the possible samples of size $n = 3$ and their means for random samples without replacement from (1, 3, 5, 7).

Samples	(1, 3, 5)	(1, 3, 7)	(1, 5, 7)	(3, 5, 7)
$\overline{X}_3$	3	11/3	13/3	5

TABLE 6.4.
Shown here is the distribution of $\overline{X}_3(n = 3)$ from (1, 3, 5, 7) without replacement.

$\overline{x}_3$	$P(\overline{x}_3)$
4	1/4
11/3	1/4
13/3	1/4
6	1/4
	1.0

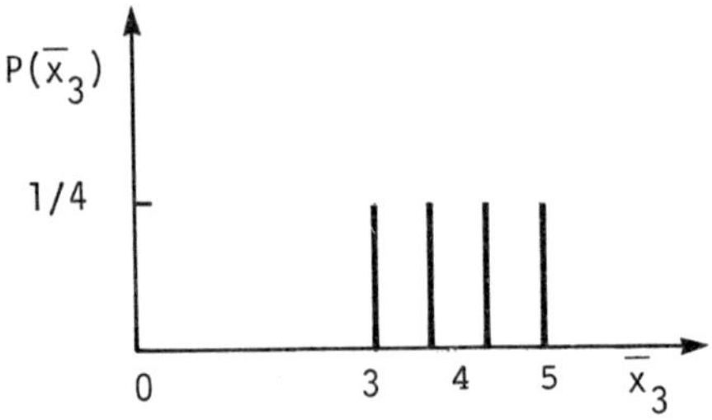

FIGURE 6.2. A graph of distribution in Table 6.4 is shown.

In the usual manner, we find that, for the distribution in Table 6.4,

$$E(\overline{X}_3) = (3 + 11/3 + 13/3 + 5)/4 = 4.$$

$$V(\overline{X}_3) = [(-1)^2 + (-1/3)^2 + (1/3)^2 + 1^2]/4 = 5/9.$$

Again, it is easy to verify that Eq. (6.4) holds for n = 3 in this example.

If we consider random samples of size n = 4 (without replacement) from (1, 3, 5, 7), there is only one possible sample, namely, (1, 3, 5, 7). So the distribution of $\overline{X}_4$ is as given in Table 6.5 and graphed in Figure 6.3.

TABLE 6.5.
Shown is the distribution of $\overline{X}_4$ (n = 4) from (1, 3, 5, 7) without replacement.

$\overline{x}_4$	$P(\overline{x}_4)$
4	1
	1.0

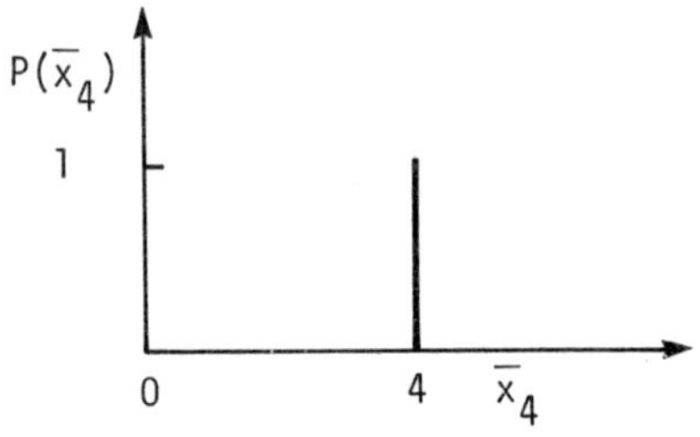

FIGURE 6.3. A graph of distribution in Table 6.5 is shown.

For Table 6.5, we find

$$E(\overline{X}_4) = 4 \cdot 1 = 4;$$

$$V(\overline{X}_4) = 0^2 \cdot 1 = 0.$$

We notice that these results agree with Eq. (6.4) for this example.

At this point it is worthwhile to summarize some of the implications of Theorem 6.1 by means of a few remarks.

REMARKS: When sampling without replacement from a variable that takes on N values, the only possible sizes of samples are n = 1, 2, 3, ..., N. In the preceding example, it is impossible to consider samples of size greater than 4.

The mean of each distribution of $\overline{X}_n$ in this sample is 4, which is the mean, μ, of the distribution being sampled. The variance of the distribution of $\overline{X}_n$ decreases as n increases. In fact, the variance is 0 when n = N. Table 6.6 summarizes the results for distributions of $\overline{X}_n$ for n = 1, 2, 3, 4 when (1, 3, 5, 7) is sampled without replacement.

TABLE 6.6.
Means and variances are shown for distributions of $\overline{X}_n$ from (1, 3, 5, 7) when sampling without replacement.

n	1	2	3	4
$E(\overline{X}_n)$	4	4	4	4
$V(\overline{X}_n)$	5	5/3	5/9	0

We neither wish nor need to determine the sampling distribution of $\overline{X}_n$ to find $E(\overline{X}_n)$ and $V(\overline{X}_n)$ for any variable that we want to sample. Since the results in Theorem 6.1 apply to sampling (without replacement) any variable that takes on a finite (N) number of values, we need only μ, σ^2, N, and n to use Eq. (6.4) in order to find $E(\overline{X}_n)$ and $V(\overline{X}_n)$ for n = 1, 2, ..., N. The results in Table 6.6 show that Eq. (6.4) holds for the preceding example. The following exercises provide further illustrations of Eq. (6.4).

In any real life situation, we desire information about model parameters from *one* sample. The techniques for extracting information from the sample are discussed in the following chapters. However, we should note here that our view of the population (as seen through the sample) may vary greatly and depends on the sample that we observe. For example, in samples of size 2 (Table 6.2), the observed $\overline{X}_2$ varies between 2 and 6.

EXERCISES

1. Use the formulas in Eq. (6.4) to verify the results in Table 6.6.
2. Find the sampling distribution for $\overline{X}$ based on samples of size 3 when (1, 2, 3, 4, 5, 6) is sampled without replacement. What is μ? What is σ^2? What is $E(\overline{X}_3)$? What is $V(\overline{X}_3)$?
3. Find $E(\overline{X}_5)$ and $V(\overline{X}_5)$ for samples selected without replacement from (1, 2, 3, 4, 5, 6, 7, 8, 9) without finding the sampling distribution of $\overline{X}$.

DISTRIBUTION OF $\overline{X}$ WHEN SAMPLING WITH REPLACEMENT: When sampling with replacement, the sample observations are independent. For an example, see Case 2 in Section 2.5. The purpose of this section is to present and illustrate properties of the sampling distributions for $\overline{X}$ when sampling with replacement is used.

THEOREM 6.2: If a random sample of size n, say, $X_1, X_2, \ldots, X_n$ is observed (with replacement) for a variable X for which $E(X) = \mu$ and $V(X) = \sigma^2$, and if

$$\overline{X}_n = \sum_{i=1}^{n} X_i/n$$

is calculated for that sample, then

$$E(\overline{X}_n) = \mu \quad \text{and } V(\overline{X}_n) = \sigma^2/n \tag{6.5}$$

for n = 1, 2, ...

Proof: Since $E(X_i) = \mu$ for i = 1, 2, ..., n, Eq. (6.1) gives

$$E(\overline{X}) = \sum_{i=1}^{n} \frac{1}{n} E(X_i) = \Sigma \frac{1}{n}\mu = n(\frac{1}{n})\mu = \mu.$$

Since the X_i are independent and $V(X_i) = \sigma^2$ for i = 1, 2, .., n, Eq. (6.2) gives

$$V(\overline{X}) = \Sigma\, V(X_i/n) = \Sigma\, [1/n^2]V(X_i)$$

$$= \Sigma\,(1/n^2)\sigma^2 = n(1/n^2)\sigma^2 = \sigma^2/n.$$

The results in Theorem 6.2 hold for all distributions. They apply to both discrete and continuous variables, but to illustrate them we use a discrete variable in the following example.

EXAMPLE 2: As in Example 1, suppose numbers 1, 3, 5, 7 are placed on 4 ping pong balls. Then suppose samples of size n (n = 2, 3, or 4) are taken (with replacement) from the balls to calculate $\overline{X}_n$. (The described experiment is suitable for a class exercise to illustrate sampling variations.) Find the sampling distribution of $\overline{X}_n$, $E(\overline{X}_n)$, and $V(\overline{X}_n)$ for n = 2, 3, and 4.

SOLUTION: The distribution for the random variable X being sampled is presented in Table 6.7 and graphed in Figure 6.4.

TABLE 6.7.
A uniform distribution on (1, 3, 5, 7) is shown.

x	P(x)
1	1/4
3	1/4
5	1/4
7	1/4
	1.0

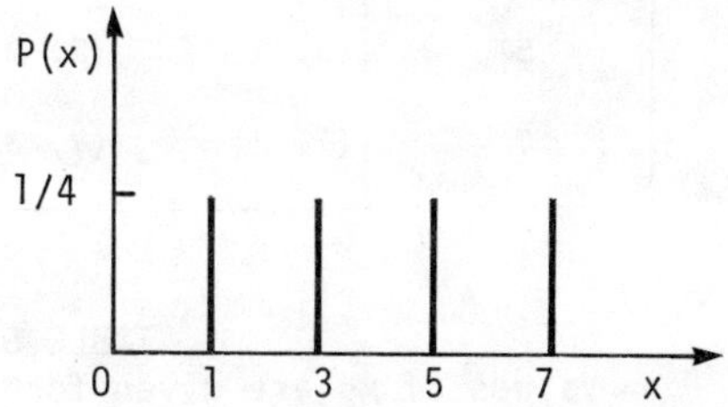

FIGURE 6.4. A graph of the X distribution in Table 6.7 is shown.

By applying the usual formulas, we find the mean and variance for the distribution of X to be

$$\mu = \sum_x xP(x) = 1(1/4) + 3(1/4) + 5(1/4) + 7(1/4) = 4.$$

$$\sigma^2 = \sum_x (x - \mu)^2\, p(x) = (1 - 4)^2\,(1/4) + (3 - 4)^2\,(1/4)$$

$$+ (5 - 4)^2\,(1/4) + (7 - 4)^2\,(1/4)$$

$$= 9(1/4) + 1(1/4) + 1(1/4) + 9(1/4) = 5.$$

These values, of course, are the same as the mean and variance obtained in the previous section and correspond to $E(\bar{X}_1)$ and $V(\bar{X}_1)$, respectively.

We now investigate the sampling distribution of $\bar{X}_2$ based on random samples of size 2 selected with replacement from the distribution of X. To do this we first obtain all possible samples of size 2. Second, we obtain the mean (i.e., $\bar{X}_2$) of each sample. Third, we find the probability distribution of the resulting sample means. We can best represent all samples of size 2 in the table form given in Table 6.8. Table 6.9 gives the values of $\bar{X}_2$ for samples of size 2.

TABLE 6.8.
Possible samples (with replacement) of size 2 from a uniform distribution on (1, 3, 5, 7) are given.

First sample element	Second sample element			
	1	3	5	7
1	(1, 1)	(1, 3)	(1, 5)	(1, 7)
3	(3, 1)	(3, 3)	(3, 5)	(3, 7)
5	(5, 1)	(5, 3)	(5, 5)	(5, 7)
7	(7, 1)	(7, 3)	(7, 5)	(7, 7)

TABLE 6.9.
Values of $\bar{x}_2$ are given for samples of size 2 from a uniform distribution on (1, 3, 5, 7).

First sample element	Second sample element			
	1	3	5	7
1	1	2	3	4
3	2	3	4	5
5	3	4	5	6
7	4	5	6	7

To find the probability distribution of $\bar{X}_2$ for samples of size 2, we observe the relative frequency of each value of $\overline{x}_2$ in Table 6.9. These relative frequencies show the distribution of $\overline{X}_2$ to be as presented in Table 6.10 and graphed in Figure 6.5.

TABLE 6.10.
Distribution of $\overline{X}_2$ is shown for sample size 2 from a uniform distribution on (1, 3, 5, 7).

$\overline{x}_2$	$P(\overline{x}_2)$
1	1/16
2	2/16
3	3/16
4	4/16
5	3/16
6	2/16
7	1/16
	1.0

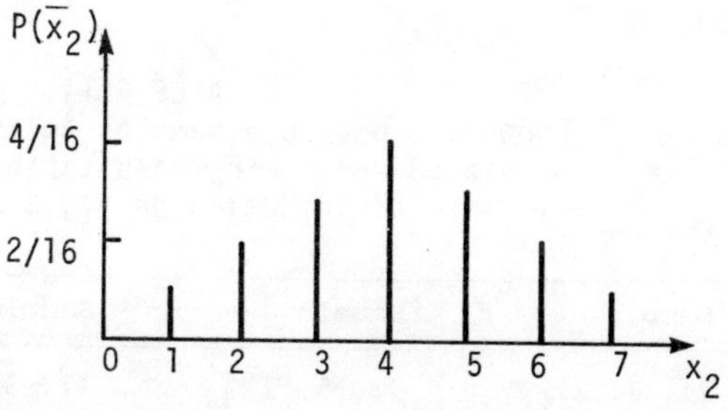

FIGURE 6.5. A graph is given for the distribution of $\overline{X}_2$.

Since the quantities in Table 6.10 define a probability distribution, it is reasonable to calculate its parameters. In particular, we calculate its mean and variance as follows:

$$E(\overline{X}_2) = \sum_{\overline{x}_2} \overline{x}_2\, P(\overline{x}_2) = 1(1/16) + 2(2/16) + 3(3/16) + 4(4/16) + 5(3/16) + 6(2/16) + 7(1/16) = 4.$$

$$V(\overline{X}_2) = \sum_{\overline{x}_2} [\overline{x}_2 - E(\overline{X}_2)]^2\, p(\overline{x}_2) = (1-4)^2\,(1/16) + (2-4)^2\,(2/16) + (3-4)^2\,(3/16) + (4-4)^2\,(4/16) + (5-4)^2\,(3/16) + (6-4)^2\,(2/16) + (7-4)^2\,(1/16) = 5/2.$$

Observe that the preceding results agree with those given by Eq. (6.5) for $\mu = 4$, $\sigma^2 = 5$, and $n = 2$.

We next repeat those steps followed for $n = 2$, except that we now consider samples of size $n = 3$. We can list the possible samples of size $n = 3$ by first taking a third sample element to be 1 for each of the 16 samples given in Table 6.8 and, next, by taking a third element to be 3 for each of the 16 possible samples of size 2. This can be done similarly for 5 and 7 as the third element of the sample. The 64 possible samples of size 3 are listed with their means, $\bar{x}_3$, in Table 6.11.

TABLE 6.11.
Shown are possible samples (with replacement) of size 3 and their means taken from a uniform distribution on (1, 3, 5, 7).

Sample	$\bar{x}$	Sample	$\bar{x}$	Sample	$\bar{x}$	Sample	$\bar{x}$
(1, 1, 1)	1	(3, 5, 1)		(1, 5, 7)		(3, 7, 5)	
(1, 1, 3)		(5, 1, 3)	3	(1, 7, 5)		(5, 3, 7)	
(1, 3, 1)	5/3	(5, 3, 1)		(5, 1, 7)		(5, 7, 3)	5
(3, 1, 1)		(3, 3, 3)		(5, 7, 1)		(7, 3, 5)	
(1, 1, 5)		(1, 3, 7)		(7, 1, 5)		(7, 5, 3)	
(1, 5, 1)		(1, 7, 3)		(7, 5, 1)	13/3	(5, 5, 5)	
(5, 1, 1)		(3, 1, 7)		(3, 3, 7)		(3, 7, 7)	
(1, 3, 3)	7/3	(3, 7, 1)		(3, 7, 3)		(7, 3, 7)	
(3, 1, 3)		(7, 1, 3)		(7, 3, 3)		(7, 7, 3)	17/3
(3, 3, 1)		(7, 3, 1)		(3, 5, 5)		(5, 5, 7)	
(1, 1, 7)		(1, 5, 5)	11/3	(5, 3, 5)		(5, 7, 5)	
(1, 7, 1)		(5, 1, 5)		(5, 5, 3)		(7, 5, 5)	
(7, 1, 1)		(5, 5, 1)		(1, 7, 7)		(5, 7, 7)	
(1, 3, 5)	3	(3, 3, 5)		(7, 1, 7)	5	(7, 5, 7)	19/3
(1, 5, 3)		(3, 5, 3)		(7, 7, 1)		(7, 7, 5)	
(3, 1, 5)		(5, 3, 3)		(3, 5, 7)		(7, 7, 7)	7

By summarizing the results in Table 6.11, we obtain the sampling distribution of $\bar{X}_3$ for samples (with replacement) of size 3 from a uniform distribution on

(1, 3, 5, 7). Table 6.12 gives that distribution, and Figure 6.6 presents its graph.

TABLE 6.12.
A distribution of $\overline{X}_3$ is given for samples of size 3 from a uniform distribution on (1, 3, 5, 7).

$\overline{x}_3$	$P(\overline{x}_3)$
1	1/64
5/3	3/64
7/3	6/64
3	10/64
11/3	12/64
13/3	12/64
5	10/64
17/3	6/64
19/3	3/64
7	1/64
	1.0

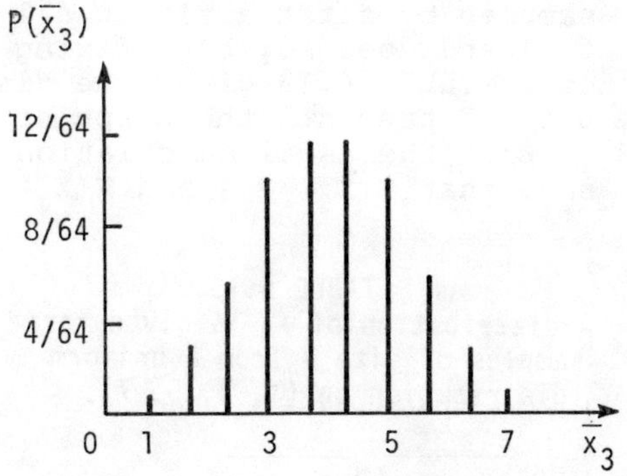

FIGURE 6.6. Presented here is a graph of the distribution of $\overline{X}_3$.

Again, it is reasonable to ask for the mean and variance of our sampling distribution of $\overline{X}_3$. These are

$$
\begin{aligned}
E(\overline{X}_3) = \sum_{\overline{x}_3} \overline{x}_3\, p(\overline{x}_3) &= 1(1/64) + (5/3)(3/64) \\
&+ (7/3)(6/64) + 3(10/64) + (11/3)(12/64) \\
&+ (13/3)(12/64) + 5(10/64) + (17/3)(6/64) \\
&+ (19/3)(3/64) + 7(1/64) \\
&= 4.
\end{aligned}
$$

$$V(\bar{X}_3) = \sum_{\bar{x}_3} [\bar{x}_3 - E(\bar{X}_3)]^2 \, p(\bar{x}_3) = [9(1/64) + (49/9)(3/64)$$
$$+ (25/9)(6/64) + 1(10/64) + (1/9)(12/64)] \cdot 2$$
$$= 5/3.$$

Here, too, the results verify Eq. (6.5) for $\mu = 4$, $\sigma^2 = 5$, and $n = 3$.

We consider those results after we present the sampling distribution of $\bar{X}$ for samples (with replacement) of size 4 from a uniform distribution on (1, 3, 5, 7). The interested student can easily list all 256 different samples by first affixing a 1 to the 64 samples in Table 6.11 and, second, by affixing a 3 to those 64 samples, etc. Table 6.13 gives the distribution of $\bar{X}_4$, and Figure 6.7 presents the graph of that distribution.

Both the usual calculation techniques and Eq. (6.5) show that $E(\bar{X}_4) = 4$ and $V(\bar{X}_4) = 5/4$.

TABLE 6.13.
A distribution of $\bar{X}_4$ is given for samples of size 4 from a uniform distribution on (1, 3, 5, 7).

$\bar{x}_4$	$P(\bar{x}_4)$
1	1/256
1.5	4/256
2.0	10/256
2.5	20/256
3.0	31/256
3.5	40/256
4.0	44/256
4.5	40/256
5.0	31/256
5.5	20/256
6.0	10/256
6.5	4/256
7.0	1/256
	1.0

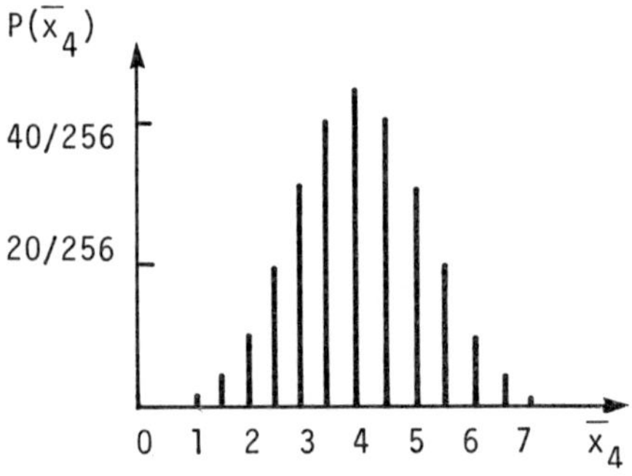

FIGURE 6.7. A graph is given for the distribution of $\bar{X}_4$.

The following remarks summarize some of the implications of Theorem 6.2 illustrated by the preceding example.

REMARKS: When sampling with replacement, samples of size n = 1, 2, 3, ... can be studied. That is, there is no upper limit on the sample size. So we could continue the foregoing procedure and study the sampling distribution of $\overline{X}_n$ for larger values of n.

We observe that, for n = 1, 2, 3, 4, the distribution of $\overline{X}_n$ in this example has mean 4, which is the mean of the distribution of X. That is, our sampling distribution of $\overline{X}_n$ is centered (located) at the parameter μ of the original uniform distribution. On the other hand, the variance of the $\overline{X}_n$ distribution decreases as n increases. However, it does not go to 0 as we observed when sampling without replacement. Table 6.14 summarizes the results for distributions of $\overline{X}_n$ for n = 1, 2, 3, 4 when a uniform distribution on (1, 3, 5, 7) is sampled with replacement.

TABLE 6.14.
Means and variances are given for the distribution of $\overline{X}_n$ when sampling (1, 3, 5, 7) with replacement.

n	1	2	3	4
$E(\overline{X}_n)$	4	4	4	4
$V(\overline{X}_n)$	5	5/2	5/3	5/4

The results in Eq. (6.5) are general in that they apply to any sampling situation (either discrete or continuous variables) where the sampling is conducted with replacement. Also, the results hold when sampling any variable that can take an infinity of values. With that in mind, let us compare the results in Eqs. (6.4) and (6.5). The only difference between the results in Eqs. (6.4) and (6.5) is the *finite correction factor* (N - n)/(N - 1) for the variance in Eq. (6.4). In sampling with replacement, N is effectively ∞. Thus, it follows that the correction factor is 1 for the present case.

Since, in later work, we discuss the use of *one* sample to deduce information about one or more parameters, we should note again that $\overline{X}_n$ for any given sample from (1, 3, 5, 7) can vary from 1 to 7. Yet, in viewing the distribution of $\overline{X}_n$ for increasing n, we find that those values of $\overline{X}_n$ near μ become more and more likely to occur.

This concentration of sample points is further demonstrated by the decreasing variance for increasing n.

EXERCISES

1. Use Eq. (6.5) to verify the results in Table 6.14.
2. (a) Find the sampling distribution of $\overline{X}_3$ for samples (with replacement) from the following distribution of X.

x	1	2	3
P(x)	1/3	1/3	1/3

   (b) Find μ and σ^2.
   (c) Find $E(\overline{X}_3)$ and $V(\overline{X}_3)$ both by definition and by Eq. (6.5).
3. (a) Find the sampling distribution of $\overline{X}_2$ for samples (with replacement) from the following distribution of X.

x	2	4	6
P(x)	.4	.2	.4

   (b) Graph the distribution of X and $\overline{X}_2$.
   (c) Find μ and σ^2.
   (d) Find $E(\overline{X}_2)$ and $V(\overline{X}_2)$ both by definition and by Eq. (6.5).
4. Suppose we select random samples of size 10 (with replacement) from the distribution in Table 6.10. Let $Y = \overline{X}_2$ be the variable being sampled. What is the mean value of

$$\overline{Y} = \sum_{i=1}^{10} Y_i/10?$$

   What is the variance of $\overline{Y}$?

DISTRIBUTION OF $\overline{X}$ WHEN SAMPLING A NORMAL DISTRIBUTION: The examples in the preceding section consider the sampling distribution of X when sampling the distribution of the discrete variable X. For the cases considered, it is possible to enumerate all possible samples and thus deduce the desired sampling distributions. However, when we sample a continuous variable, the results in Eq. (6.5) still apply, but it is no longer possible to enumerate the totality of samples. Thus, we rely on statistical theory to determine the desired sampling distribution of $\overline{X}$. Since the mathematics for deriving the distribution of $\overline{X}$ are beyond the scope of this text, the pertinent results for sampling normal random variables are stated in theorem form without proof. The theorem statement indicates the assumptions and conclusions for samples of size n.

THEOREM 6.3: If $X_1, X_2, \ldots, X_n$ is a random sample of size n selected from a *normal distribution* having mean μ and variance σ^2, then the sampling distribution of $\overline{X}$ is *normal* with mean μ and variance σ^2/n.

Theorem 6.3 is a special case of Eq. (6.3) with $a_i = 1/n$, $\mu_i = \mu$, and $\sigma^2_i = \sigma^2$ $(i = 1, 2, \ldots, n)$. Hogg and Craig's text [1] and most elementary statistical theory textbooks provide a proof for Theorem 6.3.

The important ideas contained in Theorem 6.3 are that the distribution of $\overline{X}$ *is normal* for *all* values of n, *has mean* μ, and *has variance* σ^2/n. Thus, as n increases, the distribution of $\overline{X}$ is normal and centered about μ for each n. Furthermore, the distribution of $\overline{X}$ becomes more concentrated (i.e., has higher probability of being contained in any given interval) about μ as n increases. Figure 6.8 illustrates that point.

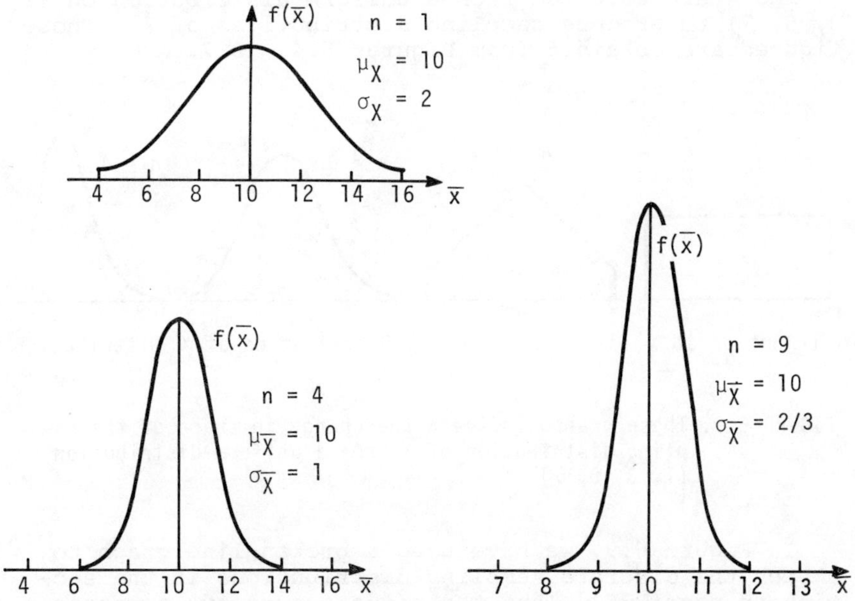

FIGURE 6.8. Sketches are shown of normal distributions for X, for $\overline{X}$ based on samples of size n = 4, and for $\overline{X}$ based on samples of size n = 9.

EXERCISES

1. If $X \sim N(10, 25)$, what is the distribution of $\bar{X}$ for sample size $n = 9$? Sketch the distributions of X and $\bar{X}$.
2. If $X \sim N(5, 9)$, compare the $P(2 \leq X \leq 8)$ with $P(2 \leq \bar{X} \leq 8)$ when $n = 4, 9$, and 16.

CENTRAL LIMIT THEOREM: The results in the preceding section suppose that the variable X being sampled is *normally* distributed. However, when we examine the results in the section on page 236, we find that the two general properties, $E(\bar{X}_n) = \mu$ and $V(\bar{X}_n) = \sigma^2/n$, stated in Eq. (6.5), hold when sampling a discrete uniform distribution on the points (1, 3, 5, 7). Furthermore, the results in the section on page 236 indicate that the appearance of the $\bar{X}$ distribution derived from sampling that discrete uniform distribution becomes normal (bell-shaped) as n increases. That is, the range of values for the variable $\bar{X}$ remains from *1 to 7* for $n = 2, 3, 4$. However, there are more possible values of $\bar{X}$ as n increases and the values of $\bar{X}$ near $\mu = 4$ become more likely to occur, while the values near either 1 or 7 have less likelihood of occurrence. Figure 6.9 illustrates the changes in shape that occur when samples of sizes 2, 3, and 4 are selected from a uniform distribution on (1, 3, 5, 7) to produce sampling distributions of $\bar{X}$. Those figures are obtained from Figures 6.4 - 6.7.

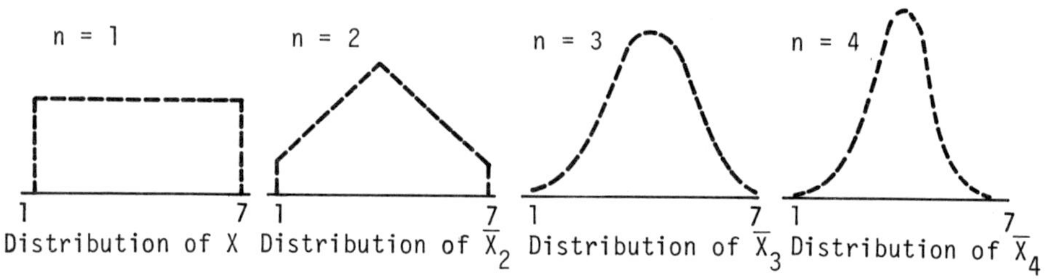

FIGURE 6.9. These graphs indicate the change in shape of the sampling distribution of $\bar{X}$ from a uniform distribution on (1, 3, 5, 7).

In Figure 6.9, we have used a broken-line graph to sketch the discrete sampling distributions in the section on page 236. That technique serves two purposes. First, it emphasizes the discreteness of the distribution. Second, it indicates that the individual probability spikes are becoming more dense between the limits 1 and 7 as n increases.

The foregoing sampling demonstration for a discrete uniform variable could be repeated using a continuous uniform variable, and the results would be similar, except that the broken curves in Figure 6.9 would be replaced by solid curves. In fact, the properties we have discussed hold for any distribution having finite values for μ and σ^2. Thus, the primary purpose of this section is to state (without proof) a theorem that indicates the general tendency for the distribution of $\overline{X}$ to approach a normal distribution. Actually there are many so-called central limit theorems that provide conditions under which the mean of random samples converge to a normal distribution. One such theorem follows.

THEOREM 6.4: (Central Limit Theorem): If $X_1, X_2, \ldots, X_n$ is a random sample from a population (distribution) having mean, μ, and variance, σ^2 (both finite), then

$$\overline{X} = \sum_{i=1}^{n} X_i/n$$

is approximately a normal variable with mean, μ, and variance, σ^2/n, for moderate to large values of n.

The interested reader can find a proof for Theorem 6.4 in any text on elementary statistical theory, such as that of Hogg and Craig [1].

Another way of stating the result in Theorem 6.4 is that if $(X_1, X_2, \ldots, X_n)$ is a random sample on a variable X such that $E(X) = \mu$ and $V(X) = \sigma^2$ (both finite), and if $Z \sim N(0, 1)$, then

$$\lim_{n\to\infty} P[(\overline{X} - \mu)/(\sigma/\sqrt{n}) \leq x] = P(Z \leq x). \qquad (6.6)$$

Equation (6.6) simply states that the cumulative distribution of $\overline{X}$ agrees with the cumulative distribution of Z (Table D2) for large values of n. Thus, Table D2 can be used to determine probabilities associated with $\overline{X}$.

Further consideration of Theorem 6.4 reveals some remarkable implications. First, the only condition for the convergence of $\overline{X}$ to a normal variable is the existence of a mean and a variance. This condition is met by most of the distributions we use in statistical applications. Notice that nothing is required as far as the form of the distribution of X is concerned. The variable X may be either discrete or continuous, symmetric or asymmetric, unimodal or multimodal, etc., and the conclusion holds as long as μ and σ^2 exist.

Second, it can be shown that the sampling distribution of $\overline{X}$ approaches normality very rapidly for nonnormal distributions. For distributions that are unimodal or even uniform, the approximation for probability calculations is quite good for n as small as 10. Certainly samples of sizes 25 to 30 yield excellent results for any distribution. The rate at which the distribution of $\overline{X}$ tends to that of a normal variable is illustrated by the results in Tables 6.10, 6.12, 6.13, and Figure 6.9.

Third, central limit theorems are the basis for using normal distributions to approximate other distributions. For example, the normal approximations for binomial and chi-square distributions, as discussed in the sections on pages 213 and 218, are justified by a central limit theorem similar to Theorem 6.4. In particular, a central limit theorem which states that

$$\sum_{i=1}^{n} X_i$$

(rather than $\overline{X}$) is approximately normally distributed with mean $n\mu$ and variance $n\sigma^2$ applies to those situations. Such a theorem justifies use of the above approximations because a binomial variable X is the sum of n independent variables which are either 0 or 1, and a chi-square variable Y is the sum of squares of independent N(0, 1) variables.

Thus, in conclusion, we emphasize that only knowledge of the values for the mean μ and variance σ^2 of the distribution being sampled is needed to treat the variable $\overline{X}$, which is based on a moderate size sample, as if it is a normal variable with mean μ and variance σ^2/n. This property is responsible to a large degree for the important role that normal distributions play in statistical analyses. Example 3 next illustrates the general ideas involved.

EXAMPLE 3: The average life of BAD 100-watt light bulbs is 1,000 hours, with a variance of 2,500 hours. What is the probability that the average life, $\overline{X}$, for a sample of 25 such light bulbs is greater than 1,010 hours?

SOLUTION: Although we have not assumed any distributional form for X, Theorem 6.4 implies that $\overline{X}$ is approximately $\sim N(1000, 2500/25)$. Therefore,

$$P(1010 \leq \overline{X}) \approx P[(1010 - 1000)/10 \leq Z] = P(1 \leq Z).$$

However, using Table D2 in Appendix D, we find that

$$P(1 \leq Z) = .5 - .3413 = .1587.$$

Thus, we expect to find that 15-16% of our test lots of 100-watt bulbs yield an average life greater than 1,010 hours.

EXERCISES

1. Suppose the distribution of X has an unknown form, but $\mu = 50$ and $\sigma^2 = 100$. What is the sampling distribution of $\overline{X}$ based on $n = 25$? What is $P(47 \leq \overline{X} \leq 53)$?

2. Name a situation in your field of study where a variable of interest is probably not normally distributed but where a central limit theorem could be used for calculating probabilities for $\overline{X}$.

CONCLUSION: By means of examples and discussions, we have explored some of the results for sampling various distributions and determining the sampling distribution of $\overline{X}$. First, we sampled a finite discrete population without replacement and illustrated formulas for expressing the mean and variance of $\overline{X}$. One limiting factor is that the variance of $\overline{X}$ is adjusted downward (even to 0) as n approaches the population size, N.

Second, we considered sampling with replacement and illustrated the results by means of a discrete uniform distribution. Theorem 6.2 gives specific formulas for expressing the mean and variance of $\overline{X}$. Example 2 illustrates those formulas, while exhibiting a tendency for the distribution of $\overline{X}$ to concentrate around μ as n increases. In the section on page 246, that tendency toward a symmetric unimodal distribution is summarized in a central limit theorem (Theorem 6.4). That theorem holds for any random variable, X, for which both μ and σ^2 are finite. Also, Theorem 6.3 states that when the variable X being sampled is a normal variable, the distribution of $\overline{X}$ is normal for all values of n. Thus, $\overline{X}$ is *approximately* a normal variable for any variable X having a finite mean and variance and is *exactly* a normal variable when X is a normal variable.

6.3 SAMPLING DISTRIBUTION OF $\hat{p}$

The procedures for obtaining the preceding sampling results for $\overline{X}$ can be duplicated to investigate the sampling properties of any statistic (function of sample data). As stated earlier, we desire the sampling properties of various statistics in order to study their usefulness in making inferences about unknown parameters. The study of $\overline{X}$ and its sampling distribution leads to inference about the mean, μ, for some population. In an analogous spirit, we now turn to the sampling distribution of the sample proportion

$$\hat{p} = X/n, \tag{6.7}$$

which contains information concerning the true value of p in a binomial distribution.

The binomial model defined by Eq. (4.15) is repeated here for reference.

$$P(x) = \binom{n}{x} p^x q^{n-x}, \quad x = 0, 1, 2, \ldots, n. \qquad (6.8)$$

Thus if we repeatedly sample a binomial variable, X, and for each sample we calculate $\hat{p} = X/n$, we can determine the sampling distribution of $\hat{p}$. It is clear that we observe one and only one value of $\hat{p}$ for each value of X. Furthermore, as X varies over the values 0, 1, 2, ..., n, then $\hat{p}$ varies over 0, 1/n, 2/n, ..., 1. Thus, the distribution of $\hat{p}$ is deduced from Eq. (6.8) as follows:

$$P(\hat{p}) = P(X/n = \hat{p}) = P(X = n\hat{p}), \ \hat{p} = 0, 1/n, 2/n, \ldots, 1.$$

Thus, to evaluate the probability distribution of $\hat{p}$, we use $x = n\hat{p}$ in Eq. (6.8) for the (n + 1) possible values of $\hat{p}$. Obviously, the probabilities assigned to the values of $\hat{p}$ by this method are exactly equal to the probabilities assigned to the (n + 1) values of X by Eq. (6.8). The only difference between the distribution of X and $\hat{p}$ is that the probabilities are assigned to different sample spaces. The following example helps to clarify that property.

EXAMPLE 4: Let X be a binomial variable for n = 4 and p = 1/3. Then the distributions of X and $\hat{p}$ are as follows:

x	P(x)
0	16/81
1	32/81
2	24/81
3	8/81
4	1/81
Total	1.0

$\hat{p}$	$P(\hat{p})$
0	16/81
1/4	32/81
1/2	24/81
3/4	8/81
1	1/81
Total	1.0

If the mean and variance for the two distributions in Example 2 are calculated, we find

$$E(X) = np = 4/3, \ V(X) = npq = 8/9,$$

$$E(\hat{p}) = 1/3, \ V(\hat{p}) = 2/36,$$

where 1/3 is p and 2/36 is pq/n = (1/3)(2/3)/4. These results hold in general. In fact, whenever we sample a binomial distribution and calculate $\hat{p}$, we have

$$E(\hat{p}) = \mu_{\hat{p}} = p, \tag{6.9}$$

$$V(\hat{p}) = \sigma^2_{\hat{p}} = pq/n. \tag{6.10}$$

Proof: To prove Eq. (6.9) we first recall that by Eq. (4.20),

$$\mu_X = np.$$

Then by Eq. (4.5) and $\hat{p} = X/n$,

$$\mu_{\hat{p}} = \mu_X/n = np/n = p.$$

To prove Eq. (6.10) we note that by Eq. (4.21),

$$\sigma^2_X = npq.$$

Then by Eq. (4.5) and the definition of $\hat{p}$,

$$\sigma^2_{\hat{p}} = \sigma^2_X/n^2 = npq/n^2 = pq/n.$$

Since any binomial variable X is the sum of n independent variables which are either 0 or 1, it follows that $\hat{p}$ is a sample mean. Thus, Theorem 6.4 applies, and we can state that $\hat{p}$ is approximately a normal variable with mean p and variance pq/n. So the distribution of $\hat{p}$ may be approximated by a normal distribution just as the distribution of a binomial variable X may be approximated by a normal model.

EXERCISES

1. Conduct the following experiment. Toss 4 coins 32 times, and record the distribution of $\hat{p}$. Compare the empirical results obtained with the expected results produced by 32 trials of Eq. (6.8) with n = 4 and p = 1/2.

   What are the mean and variance of the empirical results? How do the sample values compare with the mean and variance of the theoretical model produced by Eq. (6.8)? Would you expect the same empirical results for each new set of 32 trials?

2. Write out the sampling distribution of $\hat{p}$ for samples from a bi-

nomial distribution with n = 5 and p = .2. Find $E(\hat{p})$ and $V(\hat{p})$ both by Eqs. (6.9) and (6.10) and by direct computation.

3. Draw line graphs for the distributions of X and $\hat{p}$ in Example 4. How do the two graphs differ?

*6.4 SAMPLING DISTRIBUTION OF S^2

As a final example of a sampling distribution for a statistic, we consider the distribution of the variable S_n^2, which is defined as

$$S_n^2 = \sum_{i=1}^{n} (X_i - \overline{X})^2/(n - 1). \qquad (6.11)$$

The results are presented for two situations. First, samples of sizes 2, 3, and 4 are studied for the discrete uniform distributions presented in Example 2. That sampling is conducted with replacement. Second, the result for S_n^2 when the sample values are normally distributed is presented with illustrations.

SAMPLING DISTRIBUTION OF S_n^2 WHEN SAMPLING A DISCRETE UNIFORM VARIABLE WITH REPLACEMENT: In this section we refer to samples from a uniform distribution on (1, 3, 5, 7). (See Table 6.7.) The possible samples of size n = 2 are given in Table 6.8. By applying Eq. (6.11) to each sample in Table 6.8, the distribution of S_2^2 is found to be as given in Table 6.15.

TABLE 6.15. The distribution of S_2^2 is given for samples from a uniform distribution on (1, 3, 5, 7). (See Table 6.7.)

s_2^2	$P(s_2^2)$
0	4/16
2	6/16
8	4/16
18	2/16
	1.0

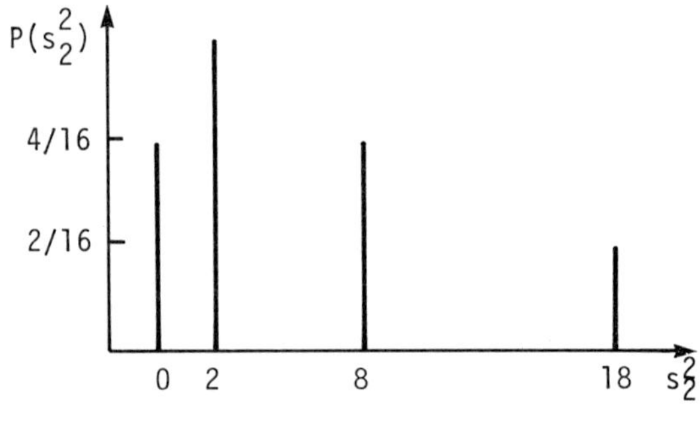

FIGURE 6.10. A graph is given for the distribution of S_2^2.

Let us now determine the mean and variance of the distribution of S_2^2. Again, by following the usual definitions for means and variances of variables, we find

$$E(S_2^2) = 0(4/16) + 2(6/16) + 8(4/16) + 18(2/16)$$

$$= 80/16 = 5.$$

$$V(S_2^2) = 25(4/16) + 9(6/16) + 9(4/16) + 169(2/16)$$

$$= 33.$$

Repeating the preceding steps for the 64 samples given in Table 6.11, we find the distribution of S_3^2 for samples of size n = 3 to be as given in Table 6.16.

TABLE 6.16.
The distribution of S_3^2 is given for samples from a uniform distribution on (1, 3, 5, 7).

s_3^2	$P(s_3^2)$
0	4/64
4/3	18/64
4	12/64
16/3	12/64
28/3	12/64
12	6/64
	1.0

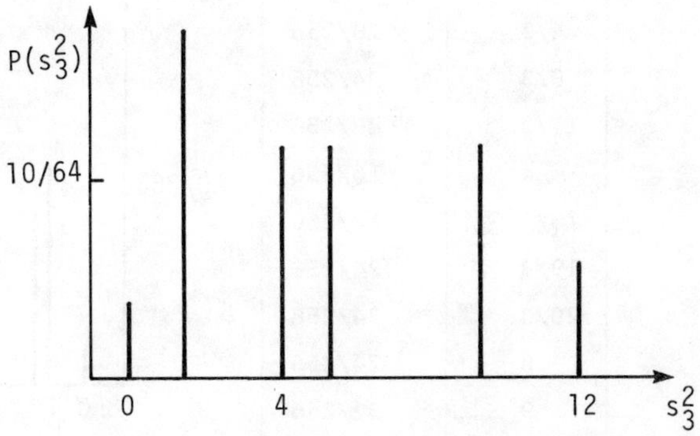

FIGURE 6.11. A graph is given for the distribution of S_3^2.

By the usual computations, we have

$$E(S_3^2) = 0(4/64) + (4/3)(18/64) + 4(12/64)$$

$$+ (16/3)(12/64) + (28/3)(12/64) + (12)(6/64)$$

$$= 5.$$

$$V(S_3^2) = (25)(4/64) + (121/9)(18/64) + (1)(12/64)$$
$$+ (1/9)(12/64) + (169/9)(12/64) + 49(6/64)$$
$$= 41/3.$$

After some tedious arithmetic, we can show that the probability distribution for S_4^2 based on samples of size 4 from a uniform distribution on (1, 3, 5, 7) is as given in Table 6.17.

TABLE 6.17.
The distribution of S_4^2 is given for samples from a uniform distribution on (1, 3, 5, 7).

s_4^2	$P(s_4^2)$
0	4/256
1	24/256
4/3	18/256
8/3	24/256
11/3	48/256
4	16/256
16/3	12/256
19/3	24/256
20/3	24/256
8	24/256
9	32/256
12	6/256
	1.0

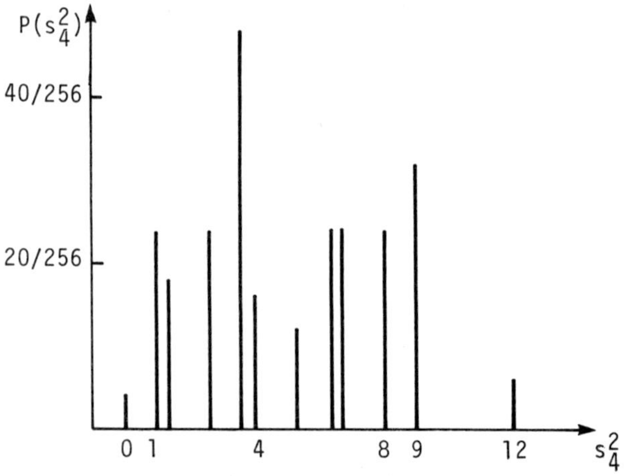

FIGURE 6.12. A graph is given for the distribution of S_4^2.

Following the usual calculations, we find

$$E(S_4^2) = [1/3(256)](3840) = 5.$$

$$V(S_4^2) = [1/9(256)](18,816) = 49/6.$$

Just as we summarized the results for the sampling

distributions of $\overline{X}$ (in Table 6.14), we now summarize the sampling results for S_n^2 in Table 6.18.

TABLE 6.18.
A summary of $E(S_n^2)$ and $V(S_n^2)$ is given for S_n^2 based on samples of sizes 2, 3, and 4.

Sample size n	2	3	4
$E(S_n^2)$	5	5	5
$V(S_n^2)$	33	41/3	49/6

At this point, we can now make several remarks to highlight the sampling properties of S_n^2 and to indicate similarities and differences that the sampling distributions of $\overline{X}_n$ and S_n^2 exhibit.

REMARKS:

1. Note that the mean of the sampling distribution of S_n^2 is σ^2 for each value of n = 2, 3, 4.
2. Next we observe that $V(S_n^2)$ is a decreasing function of n. It can be shown that as n increases without limit, $V(S_n^2)$ becomes small.
3. Contrast the graphs of the $\overline{X}_n$ distributions (in Figures 6.5 - 6.7) with those of the S_n^2 distributions (in Figures 6.10 - 6.12). First, the distributions of $\overline{X}_n$ are *symmetric* (as was the original uniform distribution) about the mean of X, μ. Also, the relative frequencies of the values of $\overline{X}_n$ farthest away from μ tend to decrease as n increases. That is, more and more of the $\overline{X}_n$ values are located near the mean of X even for small values of n. This tendency to smoothly and rapidly approach normality (i.e., tend toward the shape of a normal distribution with mean μ and variance σ^2/n) is not shared by the distribution of S_n^2. In fact, for small values of n (i.e., n = 2, 3, 4), there is a tendency for the distribution of S_n^2 to become erratic as n increases. It is worthy to note that the lack of smoothness in the distribution of S_n^2 does not prevent either the mean of S_n^2 from being σ^2 (the population variance) or $V(S_n^2)$ from decreasing as n increases.

4. Next, we observe that the statistic $S_n^2 = \Sigma(X_i - \overline{X})^2/(n - 1)$ is a squared function of observations and does not present the same symmetric and equally spaced set of sample points as given by $\overline{X}_n = \Sigma X_i/n$, a linear function of our observations.

5. Formulas similar to those presented for $\overline{X}_n$ in Eq. (6.5) can be derived for S_n^2. These formulas depend on σ^2 and $E(X - \mu)^4$ and are as follows:

$$E(S_n^2) = \sigma^2.$$

$$V(S_n^2) = \{E(X - \mu)^4 - [(n-3)/(n-1)]\sigma^4\}/n. \quad (6.12)$$

These formulas emphasize Remarks 1 and 2, but they can also be used to verify the values in Table 6.18. To illustrate the use of Eq. (6.12), we first calculate $E(X - \mu)^4$ for the uniform distribution in Table 6.7.

x	P(x)	$(x - \mu)^4$	$P(x)(x - \mu)^4$
1	1/4	81	81/4
3	1/4	1	1/4
5	1/4	1	1/4
7	1/4	81	81/4
	1		41

We have $\mu = 4$ and $\sigma^2 = 5$ for the distribution in Table 6.7, and from the table above, we find $E(X - \mu)^4 = 41$. Therefore, for n = 2,

$$V(S_2^2) = (41 + 25)/2 = 33.$$

That value agrees with the direct calculation of $V(S_2^2)$ above. Formulas (6.12) are derived in Kendall and Stuart [2].

SAMPLING DISTRIBUTION OF S_n^2 WHEN SAMPLING A NORMAL VARIABLE: In the preceding section, we observe some of the sampling properties of the sample variance S_n^2 for a discrete variable. The erratic behavior observed for that case does not occur when we sample a normal variable. Of course, the distribution of S_n^2 is continuous, and we can only study a finite number of samples to investigate its sampling distribution by the techniques of the preceding sections. Therefore, we state the result as a theorem with a partial proof. The interested reader can find a complete proof in any mathemetical statistics book, e.g., in Hogg and Craig [1].

THEOREM 6.5: If $(X_1, X_2, \ldots, X_n)$ is a random sample from a normal distribution with variance σ^2, and if

$$S_n^2 = \sum_{i=1}^{n} (X_i - \overline{X})^2/(n - 1),$$

then the sampling distribution of S_n^2 is such that $Y = (n - 1)S^2/\sigma^2$ is a chi-squared variable with $(n - 1)$ d.f. (degrees of freedom).

Partial Proof: In Section 5.5, we introduced the chi-square distribution as the sum of squares of independent $N(0, 1)$ variables. By using that result, we can motivate the result of Theorem 6.5. First, we write

$$\sum_{i=1}^{n} (X_i - \mu)^2 = \Sigma[(X_i - \overline{X}) + (\overline{X} - \mu)]^2$$

$$= \Sigma(X_i - \overline{X})^2 + n(\overline{X} - \mu)^2.$$

Second, we rearrange the terms above and divide each term by σ^2 to find

$$\Sigma(X_i - \overline{X})^2/\sigma^2 = \Sigma[(X_i - \mu)/\sigma]^2 - [(\overline{X} - \mu)^2/(\sigma^2/n)].$$

Third, $Z_i = (X_i - \mu)/\sigma \sim N(0, 1)$ and $(\overline{X} - \mu)/(\sigma/\sqrt{n}) \sim N(0, 1)$. So $\Sigma[(X_i - \mu)/\sigma]^2 \sim \chi^2(n)$ and $(\overline{X} - \mu)^2/(\sigma^2/n) \sim \chi^2(1)$. It can be shown that $\Sigma(X_i - \overline{X})^2$ and $\overline{X}$ are independent variables. That independence, plus the fact that the sum of two independent chi-square variables is a chi-square variable, leads to

$$\Sigma(X_i - \overline{X})^2/\sigma^2 = [(n - 1)S_n^2]/\sigma^2 \sim \chi^2(n - 1).$$

The conclusion of Theorem 6.5 implies that the density function of S_n^2 is given by

$$f(s_n^2) = \frac{1}{\Gamma[(n-1)/2]} [(n - 1)/2\sigma^2]^{(n-1)/2} (s_n^2)^{[(n-1)/2] - 1}$$

$$\cdot\ e^{-[(n-1)s_n^2]/2\sigma^2},\ s_n^2 > 0. \tag{6.13}$$

$$= 0 \text{ otherwise,}$$

where $\Gamma[(n-1)/2]$ is the gamma function defined by Eq. (5.41).

Figure 6.13 illustrates how the shape of $f(s_n^2)$ varies with n and σ^2. Notice that as n increases, the asymmetric distribution tends to flatten out and become more symmetric in appearance.

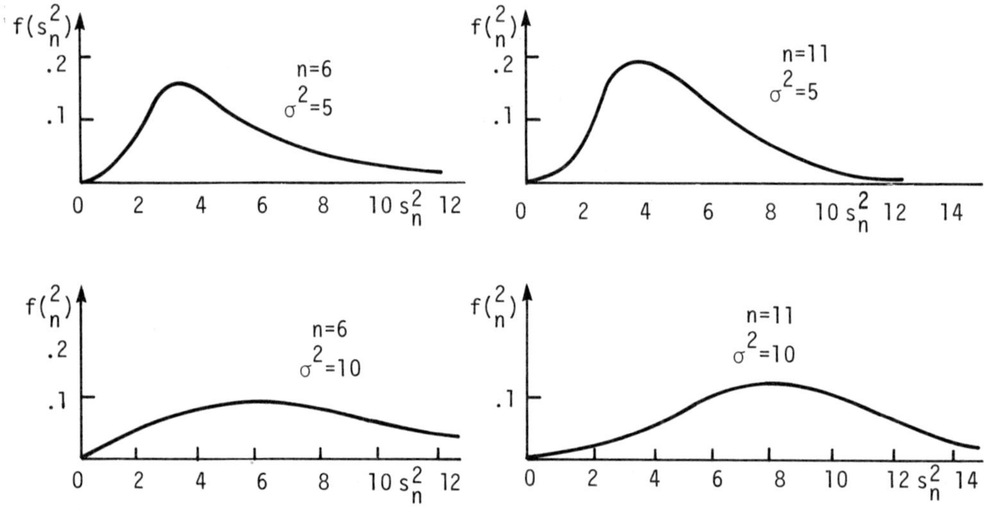

FIGURE 6.13. Illustrations of $f(s_n^2)$ are shown for selected values of n and σ^2.

By using the properties of a chi-square variable as presented in Eq. (5.42), we find

$$E(S_n^2) = \sigma^2,$$

$$V(S_n^2) = 2\sigma^4/(n - 1). \tag{6.14}$$

Proof: From Eq. (5.42),

$$E[\chi^2(n-1)] = (n-1) \text{ and}$$

$$V[\chi^2(n-1)] = 2(n-1).$$

Therefore, Theorem 6.5 implies that

$$E[(n-1)S_n^2/\sigma^2] = (n-1) \text{ and}$$

$$V[(n-1)S_n^2/\sigma^2] = 2(n-1). \text{ Hence,}$$

$$E(S_n^2) = (n-1)[\sigma^2/(n-1)] = \sigma^2$$

and

$$V(S_n^2) = 2(n-1)[\sigma^4/(n-1)^2] = 2\sigma^4/(n-1).$$

The application of Eq. (6.14) to the distributions in Figure 6.13 yields the results in Table 6.19.

TABLE 6.19.
Some applications of Eq. (6.14) are given.

σ^2	n	
	6	11
5	$E(S^2) = 5$ $V(S^2) = 10$	$E(S^2) = 5$ $V(S^2) = 5$
10	$E(S^2) = 10$ $V(S^2) = 40$	$E(S^2) = 10$ $V(S^2) = 20$

These examples help to illustrate that the distribution of S_n^2 has mean σ^2 for all values of n, but doubling σ^2 quadruples the variance of S_n^2 for any fixed n. Further, the variance of S_n^2 for fixed σ^2 decreases with n.

In conclusion, we summarize some of the sampling results for the variables (statistics), $\overline{X}_n$ and S_n^2, when the distribution being sampled is $N(\mu,\sigma^2)$. First, $\overline{X}_n \sim N(\mu,\sigma^2/n)$ and $[(n-1)S_n^2/\sigma^2] \sim \chi^2(n-1)$. Second, it can be shown that those distributions for $\overline{X}_n$ and S_n^2 are independent. Thus, if we have a number of independent samples from a normal distribution and plot the points $(\overline{x}_n, s_n^2)$ for each sample on a graph with $\overline{x}_n$ as the horizontal axis and s_n^2 as the vertical axis, we can expect to find a random (shotgun) pattern about the point (μ, σ^2). This property is true for normal samples only.

EXERCISES

1. Find the sampling distribution of S_n^2 for samples of size 2, 3 when sampling (1, 3, 5, 7) without replacement. (See Tables 6.1 and 6.3 for a listing of the possible samples.)
2. Find the mean and variance for each distribution of S_n^2 in Exercise 1.
3. Consider the distribution of X to be as follows:

x	1	2	3
P(x)	.3	.4	.3

   (a) Find μ, σ^2, and $E(X - \mu)^4$.
   (b) Use the results for (a) in Eq. (6.12) to find the $E(S^2)$ and $V(S^2)$ for samples of size 10.
   (c) Why can we not use σ^2 in Eq. (6.14) to determine $E(S^2)$ and $V(S^2)$?
4. (a) Find the $E(S^2)$ and $V(S^2)$ for samples of size 10 when $X \sim N(8, 9)$.
   (b) Sketch the distributions of X and S^2 for part (a).

6.5 SUMMARY

In this chapter we have attempted through the use of three commonly used statistics, $\bar{X}$, $\hat{p}$, and S^2, to illustrate some properties of sampling distributions of statistics. We should note that the *location* and *variance* of these distributions have been important in our studies. Furthermore, we have shown how a statistic (i.e., either $\bar{X}$, $\hat{p}$, or S^2) may vary from sample to sample. That variation must be understood in order for us to appreciate and comprehend the basic ideas of estimation theory presented in the next chapter.

To elaborate on this property briefly, we first remark that, in practice, we are sampling (i.e., taking or receiving observations from) a distribution for which both the form and parameters are unknown. Our problem is to use the observations (n in number) both to select a representative family of distributions and to estimate the necessary parameters to select a particular member of that family as a model for our data. Now all of this inference is being made on the basis of *one* sample which is assumed to have been selected at random from all possible samples of that size.

At this point, we ask the reader to refer to the 16 samples of size 2 given in Table 6.8. Imagine that one of those 16 samples is selected at random and that, on the basis of those observations, we are going to make some announcement about the magnitude of μ (the unknown mean of the unknown distribution being sampled). We note that samples (1, 1) and (7, 7) are possible samples and that each should occur about 1/16 of the time under random sampling. If the *one* selection should yield either of those extreme samples (i.e., samples that yield extreme values of $\overline{X}$), then the impression we have concerning the magnitude of μ will be quite different from that derived from one of the 10 samples that yield $\overline{X}$ (and thus an estimate of μ) of 3, 4, or 5.

This discussion perhaps makes more apparent the role played in inference by variation from sample to sample. Also, we hope that the illustrations of this chapter indicate the need for studying the sampling distributions of statistics in order to assess their effectiveness as estimators for unknown parameters.

6.6 REFERENCES

[1] Hogg, R. V., and Craig, A. T. 1970. *Introduction to mathematical statistics*. New York: Macmillan.

[2] Kendall, M. G., and Stuart, A. 1969. *The advanced theory of statistics*. Vol. 1. New York: Hafner Pub. Co., pp. 276-277.

6.7 REVIEW EXERCISES

1. Let X be a variable having the following distribution:

x	1	2	3
P(x)	.4	.2	.4

   (a) Find μ and σ^2.
   (b) Find the sampling distribution of $\overline{X}$ for n = 3 (with replacement). (Note: The 27 possible samples are not equally likely because 1, 2, and 3 are not equally likely.)
   (c) What is the approximate distribution of $\overline{X}$ for n = 30?
   (d) $P(1.7 \leq \overline{X} \leq 2.3) = ?$ (Use n = 30.)

2. (a) Find the sampling distribution of S^2 for n = 3 (with replacement) for the distribution in Exercise 1.
   (b) Find the mean and variance of S^2 by direct calculation and by Eq. (6.12).

3. As a class project, select 100 samples of size 10 from Table 3.1. Calculate $\overline{X}$ and S^2 for each sample, and form a distribution for $\overline{X}$ and S^2.

(a) Describe the shape of the resulting distributions.
(b) Find the mean and variance for the distributions.

4. As a class project, select 100 samples of size 10 from Table 3.2. Calculate $\overline{X}$ and S^2 for each sample, and form a distribution for $\overline{X}$ and S^2.
(a) Describe the shape of the resulting distributions.
(b) Find the mean and variance for the distributions.

CHAPTER 7

Estimation of Parameters

7.1 INTRODUCTION TO STATISTICAL INFERENCE

In various places in the preceding chapters, we have made overtures toward defining what we mean by statistical inference, but no specific definitions were provided. Therefore, we now pull the preceding concepts together so that explicit definitions for some of the basic ideas in statistical inference can be given.

At the beginning of this study of statistics, we stated that a statistical investigation begins with the statement of one or more questions concerning some phenomenon. Some data are then collected in the form of a sample to be used in answering those questions. Frequently, the questions are either stated or can be stated in terms of the parameters of one of the models discussed in Chapters 4 and 5. For example, any one of the following questions could be the basis for a statistical investigation:

(1) What proportion of the students at MYU are registered to vote in the next election?

(2) What is the average age of students at MYU who are enrolled in the basic statistics course this year?

(3) What is the average grade point average (GPA) of students at MYU who are enrolled in the basic statistics course this year?

If we suppose that the true proportion of students at MYU who are registered to vote is p, then question (1) reduces to the following: What is the value of p? Similarly, using μ_A as the true average age of MYU students enrolled in the basic statistics course and μ_G as the true average GPA of those same students permits us to interpret questions (2) and (3) as follows: What is the value of μ_A? What is the value of μ_G?

By raising such questions as these, we seem to be *asking for information concerning the values* of the quantities, p, μ_A, and μ_G. Yet there are other ways in which we may wish to study those quantities. For example, we might raise the following questions: Is p = .7? Is $\mu_A = 20$? Is $\mu_G = 2.9$? Stating questions in this way shifts the emphasis to *asking whether or not specific values* for the quantities, p, μ_A, and μ_G are reasonable. In a simplified way, these two sets of questions illustrate the two types of statistical inference, which are: (1) *estimation of parameters* and (2) *tests of hypotheses*.

Problems of *estimation* arise frequently in statistical analyses because the motivation for conducting many statistical studies is that of securing information about the unknown value of one or more parameters. That is, the reason for the study is to *estimate* one or more parameters. We may be interested in estimating the percentage of voters who favor Candidate R in the upcoming election. We may be interested in estimating the percentage of tranfer students who drop out of school during their first semester at MYU. An estimation for the average life-time (mileage) for a radial tire may be desired. Or we may wish to estimate the average increase (or decrease) in mileage for radial tires over belted tires.

The presentation of techniques for obtaining answers to the preceding questions and other questions of the estimation type is the purpose of this chapter. Discussion of the hypothesis-testing type questions begins in Chapter 8. So we conclude this introduction with a brief discussion of the philosophy of statistical inference.

In the theoretical world, we are able to observe and study all possible samples of a given size, n, from a designated distribution (see Chapter 6). However, in the real (experimental) world, we observe only *one* of the many possible samples and use that sample to answer questions about the distribution (or population or model) being studied. Pictorially we can represent the situation as shown in Figure 7.1. There is the full distribution represented by the larger set (circle) A and the sample represented by the smaller set (circle) B. Notice that B is a subset of A. In Chapter 6, the reasoning we used followed from the full distribution A to a specific sample B. That type of reasoning from a general situation to a specific situation is called *deductive reasoning*. However, the reasoning needed for *statistical inference* follows from the specific sample B to the full distribution A, and that type of reasoning is called *inductive reasoning*.

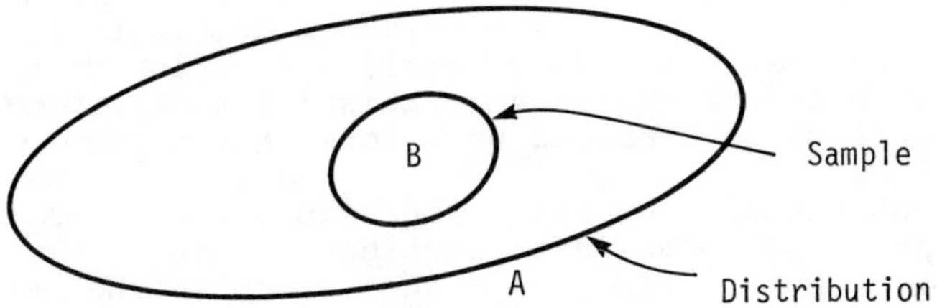

FIGURE 7.1. A pictorial representation is given of a distribution A and a sample B.

In light of the preceding discussion, one final point should be made before we proceed to specific estimation methods. When reaching decisions by inductive reasoning, we must always be aware that our view of the full distribution is limited to one sample. Also, as pointed out in Chapter 6, the impression of the distribution presented by a sample varies greatly from one sample to another. Thus, we *cannot* make the inductive step (leap) from our specific example to the full distribution with certainty. So a part of statistical inference is to make the inductive step (inference) with specified or controlled risk. The meaning and methods of the *controlled-risk inductive reasoning*, which is called *statistical inference*, are presented in the subsequent sections of this and the following chapters. Finally, the only way to make inference without risk is to study the full distribution (population) by means of a census. Since that procedure is usually either impossible or impractical or both, we proceed with the techniques for making judgments about a distribution on the basis of a sample.

7.2 POINT ESTIMATION

In this section, we assume that our sample (X_1, X_2, ..., X_n) may be modeled by a specified family of probability distributions. However, in order to select a particular model from the designated family, values for one or more parameters are needed. The objective of *point estimation* techniques is to use the sample to provide the best value (one number) for each unknown parameter. Now the definition of best varies from source to source and requires more sophisticated mathematics than we assume for complete justification, so we simply define two of the properties we want for our estimators and proceed with illustrations. To define those properties we need to establish some convenient notation.

Let Θ denote any parameter that we wish to estimate. Let t denote a statistic that, when evaluated for a given sample $(X_1, X_2, \ldots, X_n)$, provides a number to estimate Θ. Note that t is, in general, a function of the sample. We call the function (statistic) t an *estimator* for Θ and the value produced by t for a given sample an *estimate* of Θ.

For illustration, suppose Θ represents the parameter μ for a family of normal distributions. Then $t_1 = \overline{X}$ is an estimator for μ. Also, $t_2 = X_{md}$, which is the median, is an estimator for μ. That is, for any given sample, the evaluation of either $\overline{X}$ or X_{md} of the sample provides a number that estimates μ. We could consider other measures of location as estimators for μ, but we limit our discussion to the two most common statistics.

Now any estimator, t, is a function of the variables in the sample and thus varies from sample to sample. That is to say, t has a sampling distribution. So when we want to compare two estimators, t_1 and t_2, we compare them by comparing properties of their sampling distributions. As has been the case throughout this text, we use two characteristics when comparing distributions. These characteristics are the means, $E(t_1)$ and $E(t_2)$, and the variances, $V(t_1)$ and $V(t_2)$. Now the important question becomes the following: What are the desirable properties for the mean and variance of an estimator, t? In other words, how can the means and variances for the sampling distributions of two estimators, t_1 and t_2, be used to select the better estimator? Answers for this type of question are contained in the concepts of *unbiasedness* and *small variance* found in Definition 1 and the subsequent discussion.

DEFINITION 1: An estimator, t, for a parameter, Θ, is said to be *unbiased* if $E(t) = \Theta$.

Thus, an unbiased estimator is one which on the average is equal to the parameter being estimated. If $E(t) \neq \Theta$, we say that t is a *biased* estimator. Now it is possible to have two estimators, t_1 and t_2, which are both unbiased. If that is the case, we compare their variances in order to make our selection. Since the variance of t is an indication of how the values of t are concentrated around E(t), we prefer an estimator with *small variance*. Hence, when considering two estimators, t_1

and t_2, for Θ such that $E(t_1) = E(t_2) = \Theta$ and $V(t_1) < V(t_2)$, we choose t_1 because its sampling distribution tends to exhibit a higher concentration near the true value, Θ. In some cases, it is possible to show mathematically that an estimator, t, is such that $E(t) = \Theta$ *and* $V(t)$ is less than or equal to the variance of any other unbiased estimator of Θ. Such an estimator is called a *minimum-variance unbiased* estimator of Θ, and that is one definition of a best estimator. The following example illustrates this discussion.

EXAMPLE 1: To compare the sample mean, $\overline{X}$, and the sample median, X_{md}, as estimators for μ, we consider the samples of size 3 presented in Table 6.11. Consider the sampling distributions of $\overline{X}$ and X_{md}, and select the better estimator for $\mu(=4)$, the mean of a uniform distribution of (1, 3, 5, 7).

SOLUTION: The sampling distribution of $\overline{X}$ is available from Table 6.12, while the sampling distribution of X_{md} can be obtained from the samples in Table 6.11. Those distributions are as follows:

$\overline{x}$	$P(\overline{x})$
1	1/64
5/3	3/64
7/3	6/64
3	10/64
11/3	12/64
13/3	12/64
5	10/64
17/3	6/64
19/3	3/64
7	1/64
	1.0

x_{md}	$P(x_{md})$
1	10/64
3	22/64
5	22/64
7	10/64
	1.0

The calculations following Table 6.12 give $E(\overline{X}) = 4$ and $V(\overline{X}) = 5/3$. Using Eqs. (4.2) and (4.3), we find $E(X_{md}) = 4$ and $V(X_{md}) = 7/2$. Therefore, both $\overline{X}$ and X_{md} are unbiased estimators of μ, but $\overline{X}$ has smaller variance than X_{md}. Hence, on the basis of unbiasedness, there are no grounds for choosing one statistic (estimator) over the other, but the desire for smaller sampling variance leads us to choose $\overline{X}$ over X_{md}.

The preceding example illustrates the method by which we choose one estimator over another by considering the two properties of unbiasedness and small variance. Similar results for continuous variables can be obtained by using mathematical arguments beyond the scope of this text, so the derivations for those results are of no special interest to us. Our concern is simply that the means and variances of our statistics (estimators) are available for comparisons. Thus, in the following subsections, we present the usual point estimators for the parameters in binomial, normal, and Poisson models, with their means and variances.

The use of data collected for one or more of the following experiments can make the techniques more understandable.

BINOMIAL EXPERIMENTS:

(a) Ask 30 students to classify their class standing as either junior or above or sophomore or below. Let X = the number of junior or above responses.
(b) Ask 30 coeds to classify their hair color as either blonde or not blonde. Let X = the number of blonde responses.
(c) Ask 30 students to classify their political affiliation as either independent or not independent. Let X = the number of independent responses.

NORMAL EXPERIMENTS:

(a) Select a random sample of size 9 from the standard normal deviates in Table D3.
(b) Obtain the weights (to the nearest pound) for 25 male (or female) students.

*POISSON EXPERIMENTS:

(a) Count the number of cars passing a given intersection per minute for 20 1-minute intervals.
(b) Count the number of people entering a given door per minute for 20 1-minute intervals.

EXERCISES

1. Show that $\overline{X}$ for samples of size 3 is unbiased for μ in a uniform distribution on 1, 2, and 3. (See Exercise 2 on page 244.) Show that X_{md} (n=3) is also unbiased for μ in the same uniform distribution. How do $V(\overline{X})$ and $V(X_{md})$ compare?

2. If we select random samples of size 2 from the following distribution, is $\bar{X}$ unbiased for μ?

x	2	4	6
P(x)	.4	.2	.4

(See Exercise 3 on page 244.)

3. Suppose t_1 and t_2 are two estimators for a parameter Θ, and suppose they have means and variances as indicated. Which estimator is preferred? (Give reasons for your answers.)
   (a) $E(t_1) = \Theta$, $E(t_2) = \Theta$, $V(t_1) = 10$, and $V(t_2) = 5$.
   (b) $E(t_1) = \Theta + 1$, $E(t_2) = \Theta$, $V(t_1) = 6$, and $V(t_2) = 4$.
   (c) $E(t_1) = \Theta + .5$, $E(t_2) = \Theta$, $V(t_1) = 4$, and $V(t_2) = 8$.
   (d) $E(t_1) = \Theta - .5$, $E(t_2) = \Theta + .5$, $V(t_1) = 3$, and $V(t_2) = 5$.

POINT ESTIMATOR FOR BINOMIAL DISTRIBUTIONS: In binomial models (distributions), there is one parameter, p, which represents that proportion of the population exhibiting a designated characteristic that, for convenience, can be called success. Thus, we desire a statistic based on a sample that provides one number to estimate, p. The most common estimator of p is

$$\hat{p} = X/n, \tag{7.1}$$

where X denotes the number of successes in n independent trials. That is, we use the sample proportion of successes, $\hat{p}$, to estimate the corresponding population proportion, p. As shown in Section 6.3,

$$E(\hat{p}) = p, \qquad V(\hat{p}) = pq/n. \tag{7.2}$$

Thus, we have an estimator of p that is unbiased and has variance that decreases as the sample size increases. Hence, whenever economic conditions do not limit the sample size, we can obtain a more precise estimate of p by increasing the number of observations in the sample. We also observe that since $0 < p < 1$ and $q = 1 - p$, it follows that $pq \leq 1/4$. Therefore,

$$V(\hat{p}) \leq 1/4n \text{ for all } 0 < p < 1. \tag{7.3}$$

For example, the variance of the estimator, $\hat{p}$, based on 100 observations is $\leq .0025$. Thus, the standard deviation of $\hat{p}$ for $n = 100$ is $\leq .05$. Using Chebyshev's inequality [Equation (3.23)], we can expect at least 90% of the $\hat{p}$ values to fall within $3(.05) = .15$ of the true, but unknown, value of p. Now if that type of precision is not appropriate for the purposes at hand, a larger sample is needed. More comments about precision are made in Section 7.3 for the discussion of interval estimates.

EXAMPLE 2: A sample of 30 coeds yields 8 who classify their hair color as blonde. Find an estimate for the proportion of blondes in the population from which the sample was obtained.

SOLUTION: Using $\hat{p}$ as our estimator, we estimate the proportion of blondes in the population to be

$$\hat{p} = 8/30 = .267.$$

Notice that the variance of the estimator, $\hat{p}$, depends on the true value of p. Since we do not know the value of p in practice (otherwise we would not try to estimate it), we do not know the value of $V(\hat{p})$. However, we can estimate $V(\hat{p})$ by using $\hat{p}$ and $\hat{q} = (1 - \hat{p})$ in place of p and q in Eq. (7.2). In particular,

$$\widehat{V(\hat{p})} = \hat{p}\hat{q}/n. \tag{7.4}$$

The use of the caret (^) over either a symbol or an expression denotes an estimator (or estimate) for the symbol or expression.

EXAMPLE 3: A sample of 30 students who were entering the Student Union produced 12 students who classified themselves as juniors or above. Find an estimator for p, the true proportion of students who are juniors or above, and estimate the variance of the estimate.

SOLUTION: First, the estimate of p is

$$\hat{p} = 12/30 = .4.$$

Second, we use $\hat{p}$ to estimate $V(\hat{p})$ by

$$\widehat{V(\hat{p})} = [(.4)(.6)]/30 = .008.$$

EXERCISES

1. A sample of 50 individuals includes 6 people who consider themselves to be left-handed. Find an estimate for the proportion of left-handed individuals in the population being sampled.

2. Estimate the proportion of men who wear shoe size 10 or larger when a sample of 100 men shows 23 with shoe size 10 or larger.

3. A sample of 40 bolts produced by machine A includes 2 defective bolts. Estimate the defective rate for machine A.

4. An opinion poll shows that 58 of 100 students questioned favor Mr. Senator for re-election. What percentage of the student

body do you predict favors Mr. Senator? Find an estimate for the variance of your estimate. Is your estimate appropriate for Mr. Senator's home state?

5. If a coin is tossed 100 times to estimate the value of $p = P(H)$, which of the following values of $\hat{p} = \#(H)/100$ would you consider to be evidence that $p \neq .5$? $\hat{p}$ = .55, .35, .59, .73, .50, .47, .10. (Indicate your method of classification.)

6. Find point estimates for p in the binomial experiments outlined on page 268 by means of any data you have available.

POINT ESTIMATION FOR ONE NORMAL DISTRIBUTION: When considering a normal distribution as a model for empirical data, we must specify (estimate) two parameters, μ and σ^2. The most common point estimators of these parameters are

$$\bar{X} = \sum_{i=1}^{n} X_i/n \qquad \text{for an estimator of } \mu,$$

$$S^2 = \sum_{i=1}^{n} (X_i - \bar{X})^2/(n-1) \qquad \text{for an estimator of } \sigma^2. \tag{7.5}$$

By using $\bar{X}$ and S^2 to estimate μ and σ^2, respectively, we are selecting a normal model with the same mean and variance as the sample being modeled.

When sampling a normal distribution with mean, μ, and variance, σ^2, Theorem 6.3 and Eq. (6.14) provide

$$E(\bar{X}) = \mu, \qquad V(\bar{X}) = \sigma^2/n, \tag{7.6}$$

$$E(S^2) = \sigma^2, \qquad V(S^2) = 2\sigma^4/(n-1). \tag{7.7}$$

Thus, both $\bar{X}$ and S^2 are unbiased estimators of the respective parameters, μ and σ^2. Furthermore, it can be shown that $\bar{X}$ is a minimum variance estimator of μ. That is, we cannot find any unbiased estimator of μ that has smaller sampling variance than $\bar{X}$. In particular, the sample median, X_{md}, is such that

$$E(X_{md}) = \mu \text{ and } V(X_{md}) = (\pi/2)(\sigma^2/n). \tag{7.8}$$

Thus, the median X_{md} based on a sample of size n exhibits about 1.57 times as much sampling variance as $\bar{X}$ for the same size sample. This property is analogous to that illustrated by Example 1 on page 267. In practical terms, it simply means that the estimate of μ obtained by using $\bar{X}$ is more precise than the estimate of μ obtained by using X_{md} for the same sample. In particular, the standard deviation of X_{md} is about 1.25 times as large as the standard deviation of $\overline{X}$. Thus, where we might expect the range of $\overline{X}$ values to span $6\sigma/\sqrt{n}$ units, the range of X_{md} values might be expected to span $6(1.25)\sigma/\sqrt{n} = 7.5\ \sigma/\sqrt{n}$ units.

If σ^2 is unknown, we can estimate the variance of $\overline{X}$, $V(\overline{X})$, by the estimator,

$$\widehat{V(\overline{X})} = S^2/n. \tag{7.9}$$

Since S^2 is an unbiased estimator of σ^2, S^2/n is an unbiased estimator of $V(\overline{X}) = \sigma^2/n$.

EXAMPLE 4: The following 9 observations have been obtained from a normal distribution: 1.03, -1.30, -.66, 1.35, -.43, -.01, -2.51, -.15, -.13. Find estimates for μ, σ^2, and $V(\overline{X})$.

SOLUTION: First, we calculate Σx_i and Σx_i^2 to be $\Sigma x_i = -2.81$ and $\Sigma x_i^2 = 11.5335$. Second,

$$\bar{x} = -2.81/9 = -.312,$$

$$\begin{aligned} s^2 &= \{11.5335 - [(-2.81)^2/9]\}/8 \\ &= (11.5335 - .8773)/8 \qquad \text{using Eq. (3.19)} \\ &= 1.332. \end{aligned}$$

Then, the estimate of $V(\overline{X})$ is

$$\widehat{V(\overline{X})} = 1.332/9 = .148.$$

REMARK: The preceding sample was taken from a $N(0, 1)$ distribution. Thus, $\bar{x} = -.312$ is an estimate of $\mu = 0$; $s^2 = 1.332$ is an estimate of $\sigma^2 = 1$; and $s^2/n = .148$ is an estimate of $\sigma^2/n = 1/9$ ($\approx .111$).

The foregoing results pertain to normal distributions, but frequently we are interested in estimating either the mean, μ, or the variance, σ^2, or both for a distribution that is not necessarily normal. In such a case, $\overline{X}$ and S^2 are still unbiased estimators of μ and σ^2 as long as μ and σ^2 are finite. Also, the variance of $\overline{X}$ is σ^2/n, and the distribution of $\overline{X}$ becomes normal as n increases, as demonstrated by the central limit theorem (Theorem 6.4). The variance of S^2 no longer has the simple form of Eq. (7.7) but instead is given by Eq. (6.12).

EXAMPLE 5: A sample of 10 men out of a total of 50 who are attending a short course at MYU exhibits the following ages: 27, 32, 40, 21, 26, 19, 45, 36, 29, 38. Find an estimate for the mean and standard deviation of the age of the men attending the short course.

SOLUTION: First,

$$\Sigma x_i = 313, \quad \Sigma x_i^2 = 10437.$$

Second,

$$\bar{x} = 313/10 = 31.3,$$

$$s^2 = [10437 - (313)^2/10]/9$$

$$= 640.1/9 = 71.12.$$

Thus, we estimate the average age of a participant, μ, to be 31.3 years, and we estimate σ, the standard deviation of the participants' ages, to be $s = \sqrt{71.12} = 8.43$ years.

POINT ESTIMATION FOR TWO INDEPENDENT, NORMAL DISTRIBUTIONS: Frequently statistical studies involve comparisons of two independent, normal distributions. For example, a consumer agency may want to compare the performance of two different automobile tires. So an experiment is conducted by studying two independent samples of tires. The data collected are $(X_1, X_2, \ldots, X_{n_1})$ and $(Y_1, Y_2, \ldots, Y_{n_2})$ for radial and belted tires, respectively. Suppose μ_1 is the true but unknown average mileage for radial tires, and μ_2 is the true but unknown mileage for belted tires. Then one problem of interest is the estimation of $\mu_1 - \mu_2$.

First we suppose that the two samples are independent. Second, we suppose that the variables X and Y are distributed such that $E(X) = \mu_1$, $V(X) = \sigma_1^2$, $E(Y) = \mu_2$, and $V(Y) = \sigma_2^2$. Third, we observe the two independent samples of size n_1 and n_2, respectively. Fourth, we calculate

$$\bar{X} = \sum_{i=1}^{n_1} X_i/n_1$$

and

$$\bar{Y} = \sum_{i=1}^{n_2} Y_i/n_2$$

and use

$$D = \bar{X} - \bar{Y} \tag{7.10}$$

as an estimator for $\mu_1 - \mu_2$.

Since D is the difference of two independent normal variables, Eq. (6.3) gives

$$E(D) = \mu_1 - \mu_2,$$

$$V(D) = \sigma_1^2/n_1 + \sigma_2^2/n_2. \qquad (7.11)$$

So D is an unbiased estimator of the difference, $\mu_1 - \mu_2$.

The problem of estimating V(D) must be considered in two cases. First, there is the case when $\sigma_1^2 = \sigma_2^2$. Second, there is the case when $\sigma_1^2 \neq \sigma_2^2$. The following results developed for those two cases are used in Section 7.3.

When $\sigma_1^2 = \sigma_2^2$, we let σ^2 denote the common value and write

$$V(D) = \sigma^2/n_1 + \sigma^2/n_2 = \sigma^2(1/n_1 + 1/n_2).$$

In this case, our problem reduces to one of combining the information from the two independent samples into one estimate for σ^2. We accomplish that feat by *pooling* the sums of squares from the two samples and defining the *pooled estimator* for σ^2 to be

$$S_p^2 = \frac{\left[\sum_{i=1}^{n_1} (X_i - \bar{X})^2 + \sum_{i=1}^{n_2} (Y_i - \bar{Y})^2\right]}{[(n_1 - 1) + (n_2 - 1)]}. \qquad (7.12)$$

Notice that the numerator of S_p^2 is just the sum of squared deviations of the elements in each sample from the respective sample means. Furthermore, the denominator is the sum of the degrees of freedom for the two sums in the numerator. Thus, S_p^2 is the *pooled* sum of squared deviations divided by the *pooled* degrees of freedom. This approach seems reasonable in that the variation in each sample is estimating the same σ^2.

When $\sigma_1^2 \neq \sigma_2^2$, the sample sums of squared deviations are estimating different values. So it does not make sense to combine the two samples as done previously. Thus, we define

$$S_1^2 = \left[\sum_{i=1}^{n_1} (X_i - \bar{X})^2\right]/(n_1 - 1)$$

and

$$S_2^2 = [\sum_{i=1}^{n_2} (Y_i - \overline{Y})^2]/(n_2 - 1) \qquad (7.13)$$

as estimators for σ_1^2 and σ_2^2, respectively. That is, we use the two sample variances to estimate the two parameters, σ_1^2 and σ_2^2.

It can be shown that the estimators S_p^2, S_1^2, and S_2^2 are unbiased for estimating σ^2, σ_1^2, and σ_2^2, respectively. So we can summarize the previous comments by saying that an unbiased estimator for V(D) is

$$\hat{V}(D) = \begin{cases} S_p^2 (1/n_1 + 1/n_2) & \text{for } \sigma_1^2 = \sigma_2^2, \\ S_1^2/n_1 + S_2^2/n_2 & \text{for } \sigma_1^2 \neq \sigma_2^2. \end{cases} \qquad (7.14)$$

EXAMPLE 6: A sample of 10 radial tires is tested for mileage and produces $\overline{x} = 35{,}000$ with $s_1^2 = 900{,}000$. Similarly, a test of 15 belted tires provides $\overline{y} = 33{,}500$ and $s_2^2 = 1{,}000{,}000$. Find estimates of $\mu_1 - \mu_2$ and V(D), where $D = \overline{X} - \overline{Y}$.

SOLUTION: For any data set having values $\overline{x}$ and $\overline{y}$ for the variables $\overline{X}$ and $\overline{Y}$, Eq. (7.10) states that the best estimate of $\mu_1 - \mu_2$ is $d = \overline{x} - \overline{y}$ or $d = 35{,}000 - 33{,}500 = 1{,}500$ miles.

Now to estimate V(D) we consider two cases.

Case 1: Assume that $\sigma_1^2 = \sigma_2^2$. To use Eq. (7.14) we first calculate

$$\Sigma(x_i - \overline{x})^2 = (n_1 - 1)s_1^2 = 9(900{,}000) = 8{,}100{,}000,$$

$$\Sigma(y_i - \overline{y})^2 = (n_2 - 1)s_2^2 = 14(1{,}000{,}000) = 14{,}000{,}000.$$

Second, we find that

$$s_p^2 = [(8{,}100{,}000) + (14{,}000{,}000)]/(9 + 14)$$

$$= 960{,}869.56.$$

Third, Eq. (7.14) gives

$$\hat{V}(D) = (960{,}869.56)(1/10 + 1/15)$$

$$= 160{,}144.92.$$

Case 2: Assume that $\sigma_1^2 \neq \sigma_2^2$. Using Eq. (7.14) plus the given information, we find that

$$V(D) = 900{,}000/10 + 1{,}000{,}000/15$$
$$= 90{,}000 + 66{,}666{,}67$$
$$= 156{,}666.67.$$

POINT ESTIMATION FOR TWO DEPENDENT, NORMAL DISTRIBUTIONS: Another type of analysis that uses two samples is what we might call a before-after study. That is, measurements for each member in a group of subjects are taken *before* a treatment is administered and again *after* the treatment occurs in order to observe its effect. The preceding discussion is no longer valid because the two samples are dependent in that both measurements pertain to the same subjects. The usual interest in such a study is an estimate for the *mean change* that occurs between the before and after measurements.

To be explicit in our discussion, we suppose there are n subjects, each of which yields a *before-score*, X, and an *after-score*, Y. Thus, we want to estimate the mean change, μ_c, between the before and after observations. Let $C_i = Y_i - X_i$ denote the change for the ith subject. Then the best estimator for μ_c is

$$\overline{C} = \sum_{i=1}^{n} C_i/n. \qquad (7.15)$$

If we use σ_c^2 to represent the variance of the change produced by the treatment, then

$$s_c^2 = \sum_{i=1}^{n} (C_i - \overline{C})^2/(n-1) \qquad (7.16)$$

is an unbiased estimator for σ_c^2. We remark that $\overline{C} = \overline{Y} - \overline{X}$ because C_i is simply the difference between Y_i and X_i, but $\sigma_c^2 \neq \sigma_y^2 + \sigma_x^2$ because the samples are dependent.

EXAMPLE 7: The heart beat rates for heart patients are recorded before (X) and after (Y) they are administered a drug to slow their heart beat rates. Use the following data to estimate μ_c and σ_c^2:

Patient #	1	2	3	4	5	6	7	8	9	10
x	95	86	85	92	90	87	80	83	89	91
y	88	84	78	85	86	79	78	80	85	86

SOLUTION: First, we calculate $c_i = y_i - x_i$ to be -7, -2, -7, -7, -4, -8, -2, -3, -4, -5. Then $\Sigma c_i = -49$ and $\Sigma c_i^2 = 285$. Second, we estimate μ_c by $\bar{c} = -49/10 = -4.9$. Third, we estimate σ_c^2 by

$$s_c^2 = \{285 - [(-49)^2/10]\}/9 = 44.9/9$$
$$= 4.99.$$

EXERCISES

1. Assume that the heights of 6th grade boys are normally distributed with unknown mean, μ, and unknown variance, σ^2. A sample of 9 boys shows heights of 63.6, 65.1, 59.7, 60.5, 60.1, 57.0, 61.1, 58.9, and 61.5 inches.
   (a) Find a point estimate for μ.
   (b) Find a point estimate for σ^2.
   (c) Find a point estimate for the variance of the point estimate for μ.
2. The Placement Office at MYU selects a random sample of 25 bachelor's degree candidates to estimate the average starting salary per pay period, μ, for all individuals receiving a bachelor's degree at MYU this year. The sample produces $\bar{x}$ = \$650 and $s^2 = 4900$. How would you interpret such data for the Placement Office?

3. Two classes of elementary statistics are being taught by different methods of instruction. Group 1 is experiencing the common lecture-recitation instruction, while group 2 is experiencing a new computer-assisted instruction approach. At the end of this course, all students take a common achievement test for the course content, and the results are as follows:

$$n_1 = 30, \quad \bar{x}_1 = 85, \quad s_1^2 = 45,$$

$$n_2 = 25, \quad \bar{x}_2 = 80, \quad s_2^2 = 50.$$

   (a) What is your estimate of the true difference, $\mu_1 - \mu_2$?

   (b) Find an estimate for the variance of the answer to (a) when:
   (i) we assume that the two classes share a common variance
   (ii) we assume that the two classes exhibit different variances.

4. The before and after reading rates in words per minute for a sample of 5 students who have completed a speed-reading course are as follows:

Student	1	2	3	4	5
Before	350	570	760	620	480
After	540	690	840	750	630

(a) Find an estimate for μ_C, the true change in the reading rate.
(b) Find an estimate for σ_C^2, the variance of the change in the reading rate.
(c) What interpretation can be made for the answers in (a) and (b)?

5. Find point estimates for μ and σ^2 in the normal experiments on page 268 by using any data you have collected.

***POINT ESTIMATION FOR POISSON DISTRIBUTIONS:** In Poisson models we are concerned with one parameter, λ, which is both the mean and variance of the model. Thus, it might appear that either the sample mean, $\overline{X}$, or the sample variance, S^2, could serve as an estimator for λ. Furthermore, both of these estimators are unbiased. However, when we consider small sampling variance and ease of calculation, we choose

$$\overline{X} = \sum_{i=1}^{n} X_i/n \tag{7.17}$$

as the best estimator for λ. It follows from Theorem 6.2 that

$$E(\overline{X}) = \lambda \text{ and } V(\overline{X}) = \lambda/n. \tag{7.18}$$

Also, we can estimate the variance of $\overline{X}$ by

$$\hat{V}(\overline{X}) = \overline{X}/n. \tag{7.19}$$

Notice that in addition to being an unbiased estimator of λ, $\overline{X}$ has a variance that decreases at the rate of $1/n$ as the sample size increases.

Thus, we can select a value for λ to determine a particular Poisson model for empirical data by using the observations to calculate $\overline{X}$.

EXAMPLE 8: The following 10 numbers represent the number of telephone calls received per day by Mr. Jones for the last 2 weeks: 5, 5, 4, 2, 2, 4, 1, 2, 2, 3. If a Poisson model is assumed to be an appropriate distribution for describing the number of calls Mr. Jones receives, what value of λ should we use in determining the model? [Notice that the estimate of λ you use is also the estimated number (or mean number) of calls Mr. Jones receives per day.]

SOLUTION: The best estimate of λ is $\bar{x} = 30/10 = 3$. Therefore, the model for the number of incoming calls per day, X, would be modeled by

$$P(X=x) = e^{-3}\, 3^x/x!,\ x = 0, 1, 2, \ldots .$$

EXERCISES

1. A local hospital reports the number of births for each day. A sample of 10 from the daily reports produces the following data: 0, 3, 2, 1, 0, 4, 2, 2, 1, 2.
   (a) Estimate λ for using a Poisson model for X, the number of births per day at the hospital.
   (b) Estimate the variance of the answer in (a).
   (c) Estimate the probability that there will be 4 births next Saturday.
2. The $ Savings & Loan Company has found the following number of new accounts per month for a sample of 5 randomly selected months: 9, 8, 12, 7, 13. What Poisson model would you recommend for modeling the variable X, which denotes the number of new accounts per month?
3. Estimate λ for any of the Poisson data you have collected by using the experiments on page 268.

7.3 AN INTRODUCTION TO INTERVAL ESTIMATION

In the preceding discussion, we have been concerned with the problem of finding a *best number* for estimating an unknown parameter. Yet, in every case, the parameter (p, μ, σ^2, or λ) being estimated could take on any one of an infinity of numbers. For example, $0 < p < 1$, $-\infty < \mu < \infty$, $0 < \sigma^2 < \infty$, and $0 < \lambda < \infty$. Thus, the probability that we observe a sample statistic that agrees with the true, but unknown, value of a parameter is essentially 0. That is, we cannot assess any probability of being correct when using a point estimate to provide a parameter value.

Thus, we now direct our attention toward methods by which we can determine the end points for an interval which has a stated probability of containing the true value of a parameter. Such intervals are called *confidence intervals* and provide *interval estimates* for unknown parameters.

To be more explicit suppose we let $X = (X_1, X_2, \ldots, X_n)$ denote the random sample of variables and let $x = (x_1, x_2, \ldots, x_n)$ denote a set of observations for these variables. Then suppose we seek a confidence interval for a parameter, Θ. Next we want to determine two functions, say, L(X) and U(X), which depend on the sample and are such that

$$P[L(X) \leq \Theta \leq U(X)] = 1 - \alpha, \qquad (7.20)$$

where α is such that $0 < \alpha < 1$. Furthermore, α denotes the probability that Θ is *not* in the interval, $[L(X), U(X)]$. It should be noted that the values of $L(X)$ and $U(X)$ change with α, as well as with the sample, X.

Equation (7.20) is a probability statement that pertains to the random variable, X (the sample). However, once the sample is observed, say, $x = (x_1, x_2, \ldots, x_n)$, and we evaluate the functions, $L(X)$ and $U(X)$, to obtain the numbers, $L(x)$ and $U(x)$, then the true, but unknown value of Θ will either lie inside the interval, $[L(x), U(x)]$, or it will lie outside the interval. After the sample is observed, there is no more randomness involved. That is, *the interval*, $[L(x), U(x)]$, that we obtain for a given sample either includes the true value of Θ or excludes the true value of Θ. Thus, we call the interval, $[L(x), U(x)]$, a $(1 - \alpha)$-level $[100(1 - \alpha)\%]$ *confidence interval* for Θ and write

$$CI_{1-\alpha}: \quad L(x) \leq \Theta \leq U(x) \qquad (7.21)$$

to indicate that the functions, L and U, are determined so that *prior* to taking our sample there is a probability of $(1 - \alpha)$ that the resulting interval contains the true value of Θ.

The quantity, α, denotes the risk that an experimenter is willing to take in finding a confidence interval which *does not* include the true parameter value. The most widely used values of α are .10, .05, and .01, which represent a 1-in-10, 1-in-20, and 1-in-100 chance of being wrong, respectively. Note that we choose $\alpha = .50$ and obtain a 50% confidence interval, but this is rarely done because approximately 50% of such intervals do not include the true parameter value.

Another way of interpreting a $(1 - \alpha)$-level confidence interval is to consider repeated samples of size n from a designated distribution. Then calculate the interval, $[L(x), U(x)]$, for each sample. By using repeated sampling, we can expect about $(100\alpha)\%$ of the resulting intervals to *exclude* the true value of Θ and the remaining $[100(1 - \alpha)]\%$ of the intervals to *include* Θ. The following class experiment helps to illustrate these concepts.

CLASS EXPERIMENT: Select a random sample of size 9 from the N(0,1) elements in Table D3, say, $z_1, z_2, \ldots, z_9$. Transform these data to N(10, 36) data by using $x_i = 6z_i + 10$ [see Eq. (5.23)]. Now we have a normal sample of size 9 with $\mu = 10$ and $\sigma = 6$. Thus, if we calculate $\bar{x} = \Sigma x_i/9$, then we have a normal observation with $\mu = 10$ and

$\sigma_{\overline{X}} = \sigma/\sqrt{9} = 2$. For each sample, calculate the interval, $[L(x), U(x)]$, where

$$L(x) = \overline{x} - 1.96\sigma_{\overline{X}} = \overline{x} - 3.92,$$

$$U(x) = \overline{x} + 1.96\sigma_{\overline{X}} = \overline{x} + 3.92.$$

Repeat the experiment until a set of 20 intervals is obtained. Count the number of intervals that include the true value of μ, namely, $\mu = 10$. If possible, obtain several sets of 20 intervals.

REMARK: The preceding experiment takes advantage of the classroom situation in which we *know* the parameter values being estimated. So we can determine the number of intervals including that value and the number of intervals excluding that value. However, we must remain aware that in practice we observe only one sample and determine only one confidence interval for making inference about the parameter in question. Furthermore, we do not know whether our interval includes or excludes the true parameter value. We only know that prior to observing the sample there was a $(1 - \alpha)$ probability that the resulting interval would include the true parameter value.

The usual procedure for obtaining a confidence interval is to find a *pivotal quantity* that has these two characteristics: (1) it is a function of the sample points and the parameter for which the confidence interval is desired, and (2) it has a known sampling distribution. At this point, we are not going to attempt to define exactly what is meant by a pivotal quantity because the subsequent material provides examples that both exhibit pivotal quantities and also demonstrate their use in finding confidence intervals.

Now that the general concepts of confidence intervals are presented, we turn to the problem of determining the functions, $L(X)$ and $U(X)$, for the parameters of a normal model.

7.4 CONFIDENCE INTERVALS FOR STUDYING NORMAL DISTRIBUTIONS

In this section, we consider the problem of finding confidence intervals for the parameters involved when studying one or more normal distributions. In particular, we present confidence intervals for μ and σ^2 when sampling one normal distribution; $\mu_1 - \mu_2$ and σ_1^2/σ_2^2 when sampling two normal distributions independently; and μ_C and σ_C^2 when studying paired observations from two dependent, normal distributions.

INTERVAL ESTIMATION FOR μ AND σ^2: Frequently we use a random sample from a normal distribution to provide confidence intervals for the parameters, μ and σ^2. The development of confidence intervals for these parameters is presented in this section by considering three distinct problems. First, we present an interval for μ when σ^2 is known. Second, an interval for μ is developed when σ^2 is unknown. Third, an interval for σ^2 is given.

For reference in this section, we assume that a random sample $(X_1, X_2, \ldots, X_n)$ is taken from a $N(\mu,\sigma^2)$ distribution. Also,

$$\overline{X} = \Sigma X_i/n \text{ and } S^2 = \sum_{i=1}^{n} (X_i - \overline{X})^2/(n - 1) \qquad (7.22)$$

are the sample mean and variance, respectively.

CONFIDENCE INTERVALS FOR μ WHEN σ^2 IS KNOWN: From Theorem 6.3, we have $\overline{X} \sim N(\mu, \sigma^2/n)$. Thus,

$$Z = (\overline{X} - \mu)/(\sigma/\sqrt{n}) = (\overline{X} - \mu)/\sigma_{\overline{X}} \qquad (7.23)$$

has a $N(0, 1)$ distribution. For the present case, Z is a function of the sample through $\overline{X}$; Z is also a function of μ, the parameter for which a confidence interval is desired. The other factors, σ and $\sqrt{n}$, are known constants. Thus, Z satisfies the properties previously outlined for a *pivotal quantity*.

Now if we define $z(\alpha)$ to be that value of Z such that

$$P[Z > z(\alpha)] = \alpha \qquad (7.24)$$

for $0 < \alpha < .5$, then we can call $z(\alpha)$ the upper α-level $[(100\alpha)\%]$ value of Z, a standard normal variable. By the symmetry of Z and by Eq. 24, we have $P[Z < -z(\alpha)] = \alpha$ and

$$P[-z(\alpha/2) \leq Z \leq z(\alpha/2)] = 1 - \alpha.$$

In accord with the previous terminology, we can call $-z(\alpha)$ the lower α-level value of Z and refer to $\pm z(\alpha/2)$ as the two-sided, α-level values of Z. Figure 7.2 illustrates these ideas and probability statements. Table 7.1 provides the most widely used values of $z(\alpha)$. These $z(\alpha)$ values and $z(\alpha)$ for other choices of α may be obtained from Appendix D, Table D2 by interpolation and by the methods found in the section on page 199.

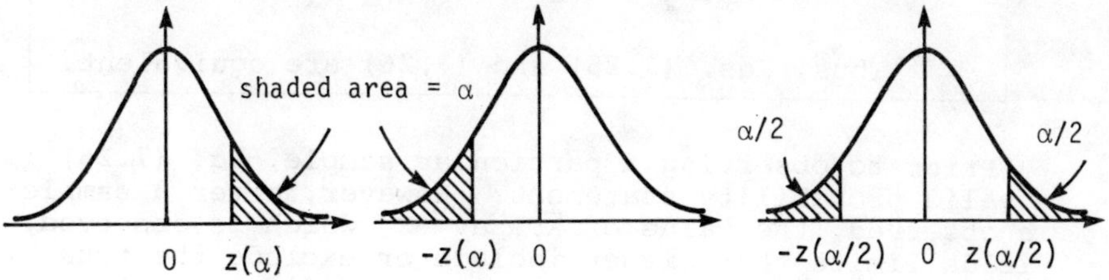

FIGURE 7.2. Illustrations of upper, lower, and two-sided, α-level values of z.

TABLE 7.1.
For a select set of α values, z(α) is shown here.

α	.10	.05	.025	.01	.005	.001	.0005
z(α)	1.282	1.645	1.96	2.326	2.576	3.09	3.291

In practice, experimenters usually specify a risk, α, which they are willing to assume in finding a confidence interval that omits the true value of μ. Then for given values of n, σ^2, and α, we can select $z(\alpha/2)$ from Table 7.1 so that

$$P[-z(\alpha/2) \le Z \le z(\alpha/2)] = 1 - \alpha.$$

Next, by using Eq. (7.23) to substitute for Z, we find that

$$P[-z(\alpha/2) \le (\overline{X} - \mu)/\sigma_{\overline{X}} \le z(\alpha/2)] = 1 - \alpha. \qquad (7.25)$$

The inequalities in Eq. (7.25) can be rearranged to give

$$P[\overline{X} - z(\alpha/2)\sigma_{\overline{X}} \le \mu \le \overline{X} + z(\alpha/2)\sigma_{\overline{X}}] = 1 - \alpha. \qquad (7.26)$$

Proof: By using Eqs. (5.25)-(5.27), the following inequalities become equivalent:

$$-z(\alpha/2) \le (\overline{X} - \mu)/\sigma_{\overline{X}} \le z(\alpha/2),$$

$$-\sigma_{\overline{X}}\, z(\alpha/2) \le \overline{X} - \mu \le \sigma_{\overline{X}}\, z(\alpha/2),$$

$$-\sigma_{\overline{X}}\, z(\alpha/2) \le \mu - \overline{X} \le \sigma_{\overline{X}}\, z(\alpha/2),$$

$$\overline{X} - z(\alpha/2)\sigma_{\overline{X}} \leq \mu \leq \overline{X} + z(\alpha/2)\sigma_{\overline{X}}.$$

Thus, Eqs. (7.25) and (7.26) are equivalent.

Prior to observing a particular sample, Eq. (7.26) is a valid probability statement. However, after a sample is observed, the value of $\overline{X}$, say $\overline{x}$, which is observed, yields limits that either include or exclude the true value of μ. Thus, we can write the confidence interval produced by a particular sample as follows:

$$CI_{1-\alpha}: \quad \overline{x} - z(\alpha/2)\sigma_{\overline{X}} \leq \mu \leq \overline{x} + z(\alpha/2)\sigma_{\overline{X}}, \tag{7.27}$$

where

$$L(x) = \overline{x} - z(\alpha/2)\sigma_{\overline{X}} \text{ and } U(x) = \overline{x} + z(\alpha/2)\sigma_{\overline{X}}$$

in the symbolism used in the introductory remarks of Section 7.3. The interval, $CI_{1-\alpha}$, is called a $(1-\alpha)$-*level, central confidence interval for* μ. Also, we sometimes refer to the preceding interval as a two-sided confidence interval, in contrast to the one-sided confidence intervals to be discussed following some examples and interpretation remarks.

EXAMPLE 9: Consider the following 9 numbers to be one trial of the class experiment outlined in the introductory remarks of Section 7.3: 16.6, 11.7, 19.6, -.3, 6.6, 5.3, 8.6, 7.0, 7.4. Find a 95% central confidence interval for μ.

SOLUTION: First of all, when $\alpha = .05$, we find that $z(.025) = 1.96$ in Table 7.1. Second, $\sigma^2 = 36$ is given and $n = 9$, so $\sigma_{\overline{X}} = 6/\sqrt{9} = 2$. Third, $\overline{x} = 82.5/9 = 9.17$. Fourth,

$$L(\overline{x}) = \overline{x} - z(\alpha/2)\sigma_{\overline{X}} = 9.17 - 1.96(2) = 5.25,$$

$$U(\overline{x}) = \overline{x} + z(\alpha/2)\sigma_{\overline{X}} = 9.17 + 1.96(2) = 13.09.$$

Thus, the desired CI is

$$CI_{.95}: \quad 5.25 \leq \mu \leq 13.09.$$

REMARK: You recall that $\mu = 10$ for the class experiment used in Example 9. Thus, the confidence interval we obtained did, in fact, include the true value of μ. However, in a real-world situation, we would not know μ, but we would conclude that its true value is somewhere between 5.25 and 13.09, with 95% confidence. The exact meaning of this statement is explained in the following discussion.

After the repeated sampling described for the class experiment outlined in Section 7.3 and illustrated by Example 9, we expect about 19 confidence intervals to include 10 (the true value of μ) for every interval that excludes 10. Table 7.2 presents 5 sets of 20 confidence intervals for the class experiment as outlined. Compare any available classroom results with those in Table 7.2. Notice that the intervals which do not include 10 are preceded by an asterisk. Furthermore, sets 1, 2, and 5 each have one interval that excludes 10, set 3 has two intervals that exclude 10, and set 4 has no intervals that exclude 10. Thus, if we pool the 5 sets of intervals together, we have 95 intervals that include 10 and 5 intervals that exclude 10. Although we do not expect exactly 5 intervals in every set of 100 intervals to exclude μ when $\alpha = .05$, we do expect the ratio of the number of intervals including μ to the number of intervals excluding μ to be about 19 to 1. (This is based on 19/20 = 95%.)

TABLE 7.2.
Five sets of 20 95% confidence intervals for μ are shown based on samples of size 9 from a N(10, 36) distribution.

Set 1		Set 2		Set 3		Set 4		Set 5	
L(x)	U(x)	L(x)	U(x)	L(x)	U(x)	L(x)	U(x)	L(x)	U(x)
9.13	16.97	7.73	15.57	6.74	14.59	5.17	13.01	4.89	12.73
4.38	12.22	4.67	12.51	8.01	15.85	7.32	15.16	3.00	10.84
4.55	12.39	6.67	14.51	3.62	11.46	4.26	12.10	*1.11	8.95
5.76	13.60	*1.41	9.25	7.36	15.20	5.72	13.56	7.79	15.63
6.97	14.81	7.08	14.92	*1.72	9.56	6.26	14.10	7.26	15.10
8.18	16.02	4.02	11.86	3.00	10.84	7.51	15.35	5.38	13.22
7.24	15.08	4.98	12.82	6.44	14.28	8.12	15.96	7.81	15.65
7.74	15.59	6.74	14.58	6.55	14.39	6.81	14.65	6.36	14.20
6.20	14.04	9.51	17.35	5.91	13.75	5.55	13.39	8.53	16.37
5.99	13.84	3.17	11.01	5.08	12.92	5.92	13.76	7.07	14.91
5.97	13.81	4.46	12.30	5.42	13.26	7.84	15.68	6.89	14.73
*10.16	18.00	5.59	13.43	4.16	12.00	4.08	11.92	9.87	17.71
6.99	14.83	7.19	15.03	6.05	13.89	6.05	13.89	7.20	15.04
7.84	15.68	9.66	17.50	4.15	11.99	5.61	13.45	4.45	12.29

TABLE 7.2 . Continued.

Set 1		Set 2		Set 3		Set 4		Set 5	
L(x)	U(x)	L(x)	U(x)	L(x)	U(x)	L(x)	U(x)	L(x)	U(x)
4.41	12.25	7.70	15.55	5.38	13.22	9.23	17.07	5.83	13.67
3.48	11.32	5.80	13.64	7.57	15.41	4.19	12.03	5.41	13.25
5.02	12.86	5.67	13.51	8.18	16.02	6.11	13.95	6.55	14.39
8.78	16.62	5.65	13.49	*1.65	9.49	6.61	14.45	3.22	11.06
7.55	15.39	5.44	13.28	5.24	13.08	8.44	16.28	5.98	13.82
3.74	11.58	7.52	15.36	2.69	10.53	6.41	14.25	7.77	15.61

The meaning of confidence intervals for repeated samples can be illustrated by representing each interval with a line joining its endpoints. Figure 7.3 graphs the first 10 intervals in Set 3. Note that only one interval does not include the true value of $\mu = 10$.

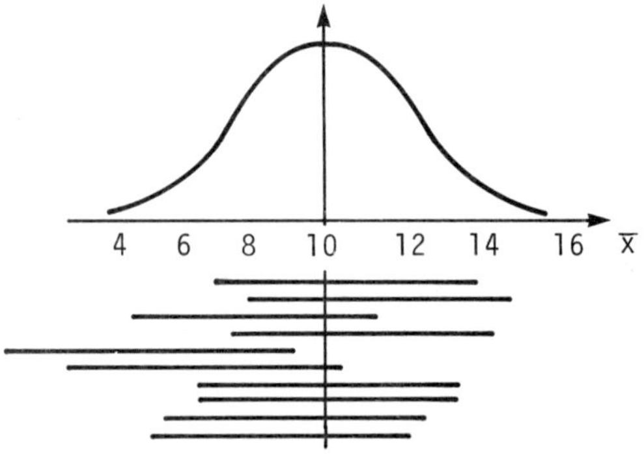

FIGURE 7.3. Presented are the sampling distribution of $\overline{X} \sim N(10,4)$ and the first 10 confidence intervals in Set 3 of Table 7.2.

Frequently only an upper limit or lower limit for μ is needed to answer the particular question of concern. That is, the question of concern can best be answered by a one-sided confidence interval rather than a two-sided interval such as Eq. (7.27). When this is the case, arguments similar to the one for deriving Eq. (7.27) show that

$$LCI_{1-\alpha} : \bar{X} - z(\alpha)\sigma_{\bar{X}} \leq \mu \tag{7.28}$$

and

$$UCI_{1-\alpha} : \mu \leq \bar{X} + z(\alpha)\sigma_{\bar{X}} \tag{7.29}$$

provide the desired lower and upper α-level confidence limits for μ. In a trivial way, the intervals in Eqs. (7.28) and (7.29) are two-sided with $U(x) = \infty$ for Eq. (7.28) and $L(x) = -\infty$ for Eq. (7.29). Thus, the full α risk, as specified by the experimenter, is taken at one end of the normal distribution of $\bar{X}$ for Eqs. (7.28) and (7.29). Whereas, α is equally divided between the two ends for a two-sided CI.

EXAMPLE 10: An automatic machine fills boxes with More Go, a breakfast cereal. Let the variable, X, denote the weight of a box of cereal. If the standard deviation of the weight of a box of cereal is .5 ounces, and if a random sample of 16 boxes shows that $\bar{x} = 12.1$ ounces, find a lower limit (95%) for μ, the true net weight of each box.

SOLUTION: First, for a one-sided interval with $\alpha = .05$, we use Table 7.1 to obtain $z(.05) = 1.645$. Second,

$$\sigma_{\bar{X}} = \sigma/\sqrt{n} = .5/\sqrt{16} = .125.$$

Third,

$$\bar{x} - z(\alpha)\sigma_{\bar{X}} = 12.1 - 1.645(.125) = 11.89.$$

Fourth,

$$LCI_{.95} : 11.89 \leq \mu.$$

Thus, we are 95% confident that the true average weight per box is at least 11.89 ounces.

The confidence intervals for μ, as presented in Eqs. (7.27)-(7.29), require knowledge of σ^2. That is, the value of σ (or σ^2) is needed in order to calculate the confidence interval endpoints. In many experimental situations, the preceding techniques are not applicable because σ^2 is unknown. The next section develops techniques for such cases.

CONFIDENCE INTERVALS FOR μ WHEN σ^2 IS UNKNOWN: The approach to this problem is similar to that used in the previous section. That is, we seek a pivotal quantity that yields confidence intervals for μ when σ^2 is unknown. If S^2, as defined in Eq. (7.5), is used as an estimator for σ^2, then it seems natural to use S as an estimator for σ and to consider

$$t = (\bar{X} - \mu)/(S/\sqrt{n}) = (\bar{X} - \mu)/S_{\bar{X}} \qquad (7.30)$$

as a pivotal quantity. Clearly, the quantity, t, depends on the sample data through $\bar{X}$, S, and n. Furthermore, it involves the parameter, μ, for which a CI is desired. Thus, the only remaining required property for t to be considered a pivotal quantity is that its sampling distribution be known so that α-level values of t can be obtained. W. S. Gosset (a brewery employee who published under the pseudonym, Student) studied the sampling distribution of t in 1908. He found that the distribution was a function of one parameter, say, k, which is called the degrees of freedom. For the present case, $k = n - 1$. Thus, we say that t has a Student's t-distribution with $(n - 1)$ degrees of freedom. Furthermore, selected α-level values of t are provided in Table E1. The use of that table is illustrated following the presentation of the probability density function for t.

The variable, t, as defined in Eq. (7.30), is a continuous random variable, and its probability density function is

$$f(t) = \{\Gamma[(k+1)/2]/[\Gamma(k/2)\sqrt{\pi k}\,]\} \cdot (1 + t^2/k)^{-(k+1)/2}, \quad -\infty < t < \infty, \qquad (7.31)$$

where $\Gamma(a)$ is defined by Eq. (5.41) and k (the degrees of freedom) is the only parameter.

The density, f(t), is symmetric about 0 and looks very much like a N(0,1) distribution as shown in Figure 7.4. In fact, as k gets large, f(t) approaches a N(0,1) density. For $k > 2$, the $E(t) = 0$ and $V(t) = k/(k - 2)$.

Whereas, the α-level values of Z, $z(\alpha)$ as given in Table 7.1 depend only on α, the α-level values of t, say, $t(k, \alpha)$, depend on both k and α. Thus, we define

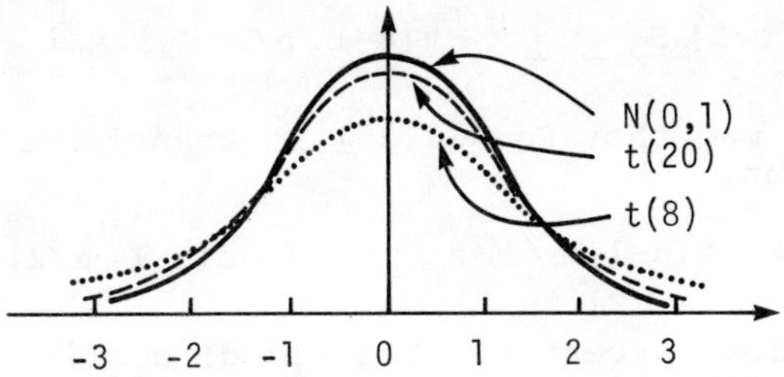

FIGURE 7.4. A graphic comparison of a N(0,1) density and t densities with k = 8 and 20.

$t(k, \alpha)$ to be that value of a t variable having k degrees of freedom for which

$$P[t > t(k, \alpha)] = \alpha. \tag{7.32}$$

Table E1 provides a wide selection of critical values, $t(k, \alpha)$, for selected values of k and α. The next example illustrates the use of Table E1.

EXAMPLE 11: From Table E1, we find the following α-level values for various t distributions:

(a) $t(10, .05) = 1.812$ (b) $t(10, .95) = -1.812$

(c) $t(24, .025) = 2.064$ (d) $t(\infty, .05) = 1.645.$

REMARKS: First, the values in Table E1 are all positive, but part (b) illustrates that $t(k,\alpha) = -t(k,1-\alpha)$ by symmetry. Second, part (d) illustrates that $t(\infty, \alpha) = z(\alpha)$.

Now that we have a pivotal quantity, Eq. (7.30), for which we can obtain α-level values from Table E1, the next step is to transform

$$P[-t(n-1,\alpha/2) \le t = (\bar{X} - \mu)/S_{\bar{X}} \le t(n-1, \alpha/2)] = 1 - \alpha \tag{7.33}$$

into

$$P[\overline{X} - t(n-1, \alpha/2)\, S_{\overline{X}} \le \mu \le \overline{X} + t(n-1, \alpha/2)\, S_{\overline{X}}] = 1 - \alpha \quad (7.34)$$

in the same way that Eq. (7.25) is transformed into Eq. (7.26). Thus,

$$CI_{1-\alpha} : \overline{x} - t(n-1, \alpha/2)\, s_{\overline{x}} \le \mu \le \overline{x} + t(n-1, \alpha/2)\, s_{\overline{x}} \quad (7.35)$$

is a $(1-\alpha)$-level, central (two-sided) confidence interval for μ when σ^2 is unknown. In a similar way, we can use Eq. (7.32) to obtain the one-sided confidence intervals,

$$LCI_{1-\alpha} : \overline{x} - t(n-1, \alpha)\, s_{\overline{x}} \le \mu, \quad (7.36)$$

$$UCI_{1-\alpha} : \mu \le \overline{x} + t(n-1, \alpha)\, s_{\overline{x}}, \quad (7.37)$$

to serve as lower and upper limits for μ when σ^2 is unknown. Example 12 illustrates the use of Eqs. (7.35) and (7.36).

EXAMPLE 12: A random sample of 9 boxes of breakfast cereal shows that $\overline{x} = 11.6$ ounces, with $s^2 = .36$. (a) Find a 95% central confidence interval for μ. (b) Find a 99% lower limit for μ.

SOLUTION: (a) First, $s_{\overline{x}} = \sqrt{.36}/\sqrt{9} = .2$, and $t(8, .025) = 2.306$. Then

$$11.6 \pm 2.306\,(.2) = 11.6 \pm .4612.$$

Next, $CI_{.95}$: $11.1388 \le \mu \le 12.0612$.
(b) We have $t(8, .01) = 2.896$, and $11.6 - 2.896(.2) = 11.6 - .5792$. Then $LCI_{.99} = 11.0208 \le \mu$.

REMARKS: As before, the interpretation of the confidence intervals is as follows. We are 95% confident that the true net weight of a box of cereal, μ, is between 11.1388 and 12.0612 ounces. We are 99% confident that the true net weight is greater than 11.0208 ounces.

Now that confidence intervals for μ are available when σ^2 is either known or unknown, we continue by developing confidence interval techniques for the variance of a normal distribution, σ^2.

***CONFIDENCE INTERVALS FOR σ^2:** When studying some phenomenon of interest, experimenters usually desire inference about the unknown variability present, as well as inference about the unknown mean. To do that we proceed to develop confidence intervals about σ^2 to supplement the point estimation by S^2 in Eq. (7.5). We again seek a pivotal quantity for use in forming probability statements that can be inverted to provide the desired confidence intervals. *Theorem 6.5* states that

$$Y = (n-1)S^2/\sigma^2 = \sum_{i=1}^{n} (X_i - \overline{X})^2/\sigma^2 \tag{7.38}$$

is a chi-square variable with (n - 1) degrees of freedom. We note that Y depends only on the sample sum of squares, $\Sigma(X_i - \overline{X})^2$, and on σ^2, the parameter for which a CI is desired. The chi-square probability density function is given in Eq. (5.40), and Table E2 provides α-level values of Y for selected values of α. Thus, Y satisfies the requirements of a pivotal quantity.

Section 5.5 presents several properties for χ^2 variables, and Eq. (5.43) defines the α-level value of a χ^2 variable, say Y, with d degrees of freedom to be $\chi^2(d, \alpha)$, which satisfies

$$P[Y \geq \chi^2(d, \alpha)] = \alpha.$$

We note that the asymmetry of a χ^2 distribution makes it necessary for Table E2 to include α-level values for α near 1, as well as near 0. This is a change from both the $z(\alpha)$ and $t(k, \alpha)$, which exhibited symmetry. Since Examples 22 and 23 on page 223 illustrate the use of Table E2 for determining $\chi^2(d, \alpha)$ values for given d and α, we turn now to the development of a two-sided confidence interval for σ^2.

First, we have

$$P[\chi^2(n-1, 1-\alpha/2) \leq Y \leq \chi^2(n-1, \alpha/2)] = 1 - \alpha,$$

which, after substituting for Y by Eq. (7.38), gives

$$P[\chi^2(n-1, 1-\alpha/2) \leq \Sigma(X_i-\overline{X})^2/\sigma^2 \leq \chi^2(n-1, \alpha/2)] = 1 - \alpha. \tag{7.39}$$

By taking reciprocals and multiplying by $\Sigma(X_i - \overline{X})^2$, we obtain

$$P[\Sigma(X_i-X)^2/\chi^2(n-1, \alpha/2) \leq \sigma^2 \leq \Sigma(X_i-\overline{X})^2/\chi^2(n-1, 1-\alpha/2)] = 1 - \alpha. \tag{7.40}$$

Proof: We first recall for positive quantities, a and b, that $a \le b$ implies that $1/a \ge 1/b$. Thus,

$$\chi^2(n-1,\ 1-\alpha/2) \le \Sigma(X_i-\bar{X})^2/\sigma^2 \le \chi^2(n-1,\ \alpha/2),$$

$$1/\chi^2(n-1,\ 1-\alpha/2) \ge \sigma^2/\Sigma(X_i-\bar{X})^2$$

$$\ge 1/\chi^2(n-1,\ \alpha/2),$$

and

$$\Sigma(X_i-\bar{X})^2/\chi^2(n-1,\ \alpha/2) \le \sigma^2$$

$$\le \Sigma(X_i-\bar{X})^2/\chi^2(n-1,\ 1-\alpha/2)$$

are all equivalent, and so Eq. (7.40) is established.

For any sample data, $x_1, x_2, \ldots, x_n$, the two-sided, $(1-\alpha)$-level confidence interval for σ^2 is

$$CI_{1-\alpha}:\ \Sigma(x_i-\bar{x})^2/\chi^2(n-1,\ \alpha/2) \le \sigma^2$$

$$\le \Sigma(x_i-\bar{x})^2/\chi^2(n-1,\ 1-\alpha/2), \tag{7.41}$$

where $\Sigma(x_i-\bar{x})^2$ can be replaced by $(n-1)s^2$ when s^2 is available. Similarly, we have $(1-\alpha)$-level lower and upper limits for σ^2 given by

$$LCI_{1-\alpha}:\ \Sigma(x_i-\bar{x})^2/\chi^2(n-1,\alpha) \le \sigma^2, \tag{7.42}$$

$$UCI_{1-\alpha}:\ \sigma^2 \le \Sigma(x_i-\bar{x})^2/\chi^2(n-1,\ 1-\alpha). \tag{7.43}$$

The following example illustrates the use of Eqs. (7.41) and (7.43).

EXAMPLE 13: Use the 9 boxes of breakfast cereal in Example 12 with $s^2 = .36$ to answer the following questions. (a) Find a two-sided 95% confidence interval for σ^2. (b) Find a 95% upper limit for σ^2.

SOLUTION: (a) Table E2 gives $\chi^2(8,\ .025) = 17.53$ and $\chi^2(8,\ .975) = 2.18$. Then $\Sigma(x_i-\bar{x})^2 = 8(.36) = 2.88$. Next, $2.88/17.53 = .164$ and $2.88/2.18 = 1.32$. Thus,

$CI_{.95}$: $.164 < \sigma^2 < 1.32$.
(b) First, $\chi^2(8, .95) = 2.73$. Second, $2.88/2.73 = 1.05$. Thus, $UCI_{.95}$: $\sigma^2 \leq 1.05$.

REMARKS: The appropriate conclusions are as follows. We are 95% confident that the variance of the net weights of cereal boxes, σ^2, is between .164 and 1.32. We are 95% confident that σ^2 is less than or equal to 1.05.

The two-sided interval for σ^2, as given in Eq. (7.41), is not a central interval in that S^2, which is the point estimate of σ^2, is not the center of the CI. That property is a consequence of the asymmetry of the χ^2 distribution.

PROPERTIES OF CONFIDENCE INTERVAL LENGTH: Ideally, when seeking a confidence interval for an unknown parameter, we desire a *short* interval, which is associated with a small α. That is, a narrow interval with high confidence of containing the true parameter value is considered to be a good interval. Thus, we conclude this section with a few comments about the factors affecting the *length* of a confidence interval. The length of a confidence interval is denoted by L and defined by

$$L = U(X) - L(X), \tag{7.44}$$

where U(X) and L(X) are the upper and lower confidence interval limits, respectively. For the confidence interval in Eq. (7.27),

$$L = [\bar{x} + z(\alpha/2)\sigma_{\bar{x}}] - [\bar{x} - z(\alpha/2)\sigma_{\bar{x}}] = 2z(\alpha/2)\sigma_{\bar{x}}. \tag{7.45}$$

Since $\sigma_{\bar{x}} = \sigma/\sqrt{n}$, we see that the value of L is changed by any change in either α or n. In particular, an *increase in* α produces a decrease in $z(\alpha/2)$ and yields a *decrease in L*. Similarly, an *increase in n* yields a *decrease in L*. Therefore, whenever it is economically feasible to take larger samples, the length of a confidence interval can be controlled by changing n.

EXAMPLE 14: If the standard deviation for net weights of cereal boxes is $\sigma = .5$ ounces, how large a sample is needed to give a 90% confidence interval for μ, which has $L \leq .25$ ounces?

SOLUTION: We obtain $z(.05) = 1.645$ from Table 7.1. Then the constraints specified require that

$$L = 2(1.645)(.5)/\sqrt{n} \leq .25.$$

Therefore, $\sqrt{n} \geq 2(1.645)(.5)/.25$ or $n \geq [4(1.645)]^2 = 43.3$. Hence, any sample of size 44 or larger provides a CI that meets the specifications.

In a similar fashion, the values of α and n change the length of the confidence intervals in Eqs. (7.35) and (7.41). However, the lengths of those intervals are affected by two additional properties not present in Eq. (7.45). First, the lengths of Eqs. (7.35) and (7.41) are changed by the data through $S_{\overline{X}}$ and $\Sigma(X_i - \overline{X})^2$, respectively. Second, the dependence on n is not a simple division by $\sqrt{n}$ because the values, $t(n - 1, \alpha/2)$, $\chi^2(n - 1, \alpha/2)$, and $\chi^2(n - 1, 1 - \alpha/2)$, change with n. Thus, any procedure for determining L in those cases is a trial-and-error technique for different choices of n.

EXERCISES

1. Find the following critical values from the appropriate table:
   (a) $z(.025)$;
   (b) $z(.95)$;
   (c) $t(15, .10)$;
   (d) $t(10, .90)$;
   (e) $t(\infty, .95)$;
   (f) $\chi^2(5, .90)$;
   (g) $\chi^2(8, .025)$;
   (h) $\chi^2(20, .975)$.

2. Suppose the variable, X, is normally distributed with $\sigma = 10$. Find the indicated confidence intervals and limits for μ when the sampling situations summarized below are given.
   (a) $CI_{1-\alpha}$ for $n = 9$, $\overline{x} = 5$, and $\alpha = .10$.
   (b) $CI_{1-\alpha}$ for $n = 9$, $\overline{x} = 5$, and $\alpha = .05$.
   (c) $CI_{1-\alpha}$ for $n = 25$, $\overline{x} = 5$, and $\alpha = .05$.
   (d) $LCI_{1-\alpha}$ for $n = 16$, $\overline{x} = 10$, and $\alpha = .01$.
   (e) $UCI_{1-\alpha}$ for $n = 16$, $\overline{x} = 8$, and $\alpha = .005$.
   (f) $UCI_{1-\alpha}$ for $n = 25$, $\overline{x} = -5$, and $\alpha = .10$.

3. Suppose the variable, X, is normally distributed with σ unknown. Find the indicated confidence intervals and limits for μ when the sampling situations summarized below are given.
   (a) $CI_{1-\alpha}$ for $n = 16$, $\overline{x} = 5$, $s^2 = 25$, and $\alpha = .05$.
   (b) $CI_{1-\alpha}$ for $n = 25$, $\overline{x} = 5$, $s = 5$, and $\alpha = .05$.
   (c) $CI_{1-\alpha}$ for $n = 9$, $\overline{x} = -3$, $s = 4$, and $\alpha = .01$.
   (d) $LCI_{1-\alpha}$ for $n = 16$, $\overline{x} = 7$, $s^2 = 9$, and $\alpha = .10$.
   (e) $LCI_{1-\alpha}$ for $n = 4$, $\overline{x} = 10$, $s = 5$, and $\alpha = .01$.
   (f) $UCI_{1-\alpha}$ for $n = 25$, $\overline{x} = -8$, $s = 7$, and $\alpha = .05$.

*4. Let X be a normal variable with σ^2 unknown. Find the indicated confidence intervals and limits for σ^2 under the following assumed sampling situations.
   (a) $CI_{.95}$ for $n = 12$ and $s^2 = 30$.
   (b) $CI_{1-\alpha}$ for $n = 5$, $\Sigma(x_i-\overline{x})^2 = 15$, and $\alpha = .10$.
   (c) $CI_{1-\alpha}$ for $n = 7$, $s = 5$, and $\alpha = .05$.
   (d) $LCI_{1-\alpha}$ for $n = 10$, $s^2 = 3$, and $\alpha = .01$.
   (e) $UCI_{.99}$ for $n = 11$ and $\Sigma(x_i-\overline{x})^2 = 10$.

5. The Long-Mileage tire is advertised as being good for at least 40,000 miles of use. A random sample of 16 tires gives $\overline{x}$ = 38,000 and s = 4,000. Calculate $UCI_{.95}$ and decide whether or not the sample evidence is sufficient to question the advertisement.

6. Find $CI_{.95}$ for the average weights, μ, of 6th-grade boys if a random sample of 25 produces $\overline{x}$ = 100 pounds and s = 15 pounds.

7. Assume that the starting salaries for individuals holding a bachelor's degree in a given field are normally distributed. A random sample of 9 yields starting monthly salaries of $650, $800, $750, $675, $700, $825, $850, $775, $725. Find $CI_{.90}$ for μ, the true average starting salary. *Find $CI_{.95}$ for σ^2, the variance of starting salaries.

8. Is it possible for a 95% confidence interval to be shorter than a 90% confidence interval for μ? Defend your answer.

9. What size sample is needed to obtain a $CI_{.99}$ for μ, which is no longer than 5 units when $\sigma^2 = 100$? $\sigma^2 = 25$?

10. Identify one or more problems in an area of interest, and explain how the techniques of this section could be used to answer questions concerning μ and σ^2.

CONFIDENCE INTERVALS FOR $\mu_1 - \mu_2$ AND σ_1^2/σ_2^2 WHEN SAMPLING TWO INDEPENDENT, NORMAL DISTRIBUTIONS: The purpose of some statistical investigations is to compare either the averages or the variances for two different populations. For example, the purpose of a statistical study may be to compare either the *average life* and/or the *variation in lives* for two brands of tires. Another example could be an instructor who teaches two similar classes and compares the *average score* and/or the *variation in scores* for the two classes. This section presents confidence intervals for three types of problems of interest when studying two independent, normal samples.

First we find confidence intervals and limits for the difference of means, $\mu_1 - \mu_2$, when both σ_1^2 and σ_2^2 are known. Second, confidence intervals and limits for $\mu_1 - \mu_2$ are presented when we assume that a common variance, $\sigma_1^2 = \sigma_2^2 = \sigma^2$, is unknown. Third, confidence intervals and limits for the variance ratio, σ_1^2/σ_2^2, are illustrated.

Throughout this section, we assume that two independent samples, $(X_1, X_2, \ldots, X_{n1})$ and $(Y_1, Y_2, \ldots, Y_{n2})$, are available where $X_i \sim N(\mu_1, \sigma_1^2)$ and $Y_i \sim N(\mu_2, \sigma_2^2)$. Furthermore, X, Y, S_1^2, and S_2^2 are the usual statistics denoting the sample means and variances [see Eqs. (7.10) and (7.13)].

CONFIDENCE INTERVALS FOR μ_1 μ_2 WHEN BOTH σ_1^2 AND σ_2^2 ARE KNOWN: Since the two samples being studied are assumed to be from independent, normal distributions, it follows from Eq. (6.3) that

$$D = (\bar{X} - \bar{Y}) \sim N(\mu_1 - \mu_2, \sigma_1^2/n_1 + \sigma_2^2/n_2).$$

Thus, if σ_1^2 and σ_2^2 are known, we can use the pivotal quantity,

$$Z = [D - (\mu_1 - \mu_2)]/(\sigma_{\bar{x} - \bar{y}}) \sim N(0, 1), \qquad (7.46)$$

where

$$\sigma_{\bar{x} - \bar{y}} = \sqrt{\sigma_1^2/n_1 + \sigma_2^2/n_2}$$

to develop confidence intervals for the difference of means, $\mu_1 - \mu_2$. Then arguments identical to those used to motivate and present Eqs. (7.27)-(7.29) provide the following $(1 - \alpha)$-level confidence intervals and limits for $(\mu_1 - \mu_2)$:

$$CI_{1-\alpha}: \quad (\bar{x} - \bar{y}) - z(\alpha/2)\sigma_{\bar{x} - \bar{y}} \le (\mu_1 - \mu_2) \le (\bar{x} - \bar{y}) + z(\alpha/2)\sigma_{\bar{x} - \bar{y}}. \qquad (7.47)$$

$$LCI_{1-\alpha}: \quad (\bar{x} - \bar{y}) - z(\alpha)\sigma_{\bar{x} - \bar{y}} \le (\mu_1 - \mu_2). \qquad (7.48)$$

$$UCI_{1-\alpha}: \quad (\mu_1 - \mu_2) \le (\bar{x} - \bar{y}) + z(\alpha)\sigma_{\bar{x} - \bar{y}}. \qquad (7.49)$$

The next example illustrates the use of Eq. (7.47) and provides an interpretation of the resulting CI.

EXAMPLE 15: A random sample of the weights (X) for 20 6th-grade boys exhibited $\bar{x} = 100$ pounds, while a sample of the weights (Y) of 25 6th-grade girls exhibited $\bar{y} = 85$ pounds. If we suppose the variance of weights for boys is $\sigma_1^2 = 200$ and the variance of weights for girls is $\sigma_2^2 = 150$, find a $CI_{.95}$ for $\mu_1 - \mu_2$. Is it reasonable to assume that the average weight (μ_1) for 6th-grade boys is greater than the average weight (μ_2) for 6th-grade girls?

SOLUTION: First, $z(.025) = 1.96$ as given in Table 7.1. Then

$$\bar{x} - \bar{y} = 100 - 85 = 15,$$

$$\sigma_{\bar{x} - \bar{y}} = \sqrt{200/20 + 150/25} = \sqrt{16} = 4,$$

$$z(.025)\sigma_{\bar{x} - \bar{y}} = 1.96(4) = 7.84.$$

Thus, 15 ± 7.84 gives

$$CI_{.95}: \quad 7.16 \le (\mu_1 - \mu_2) \le 22.84.$$

Since the point 0, which implies $\mu_1 = \mu_2$, is not in the interval, and since, in fact, the lower limit of the $CI_{.95}$ is positive, we are quite confident that μ_1 is greater than μ_2.

CONFIDENCE INTERVALS FOR $\mu_1 - \mu_2$ WHEN σ_1^2 AND σ_2^2 ARE UNKNOWN BUT EQUAL: When the two variances, σ_1^2 and σ_2^2, are equal to some common value, say, σ^2, then

$$D = (\bar{X} - \bar{Y}) \sim N[\mu_1 - \mu_2, \sigma^2(1/n_1 + 1/n_2)]$$

and

$$[D - (\mu_1 - \mu_2)]/\sigma\sqrt{(1/n_1 + 1/n_2)} \sim N(0, 1).$$

However, if σ^2 is unknown, we pool the sums of squared deviations from both samples in order to obtain the estimator, S_p^2, as defined in Eq. (7.12). Then we use S_p to estimate the unknown value of σ. Now S_p^2 is associated with $(n_1 + n_2 - 2)$ degrees of freedom, and with that it can be shown that the pivotal quantity

$$t = [D - (\mu_1 - \mu_2)]/S_p\sqrt{(1/n_1 + 1/n_2)} \sim t(n_1 + n_2 - 2). \quad (7.50)$$

Thus, we can follow the logic used to present Eqs. (7.35)-(7.37) to obtain the following $(1 - \alpha)$-level confidence intervals for $\mu_1 - \mu_2$ when σ_1^2 and σ_2^2 are unknown but equal.

$$CI_{1-\alpha}: \quad (\bar{x} - \bar{y}) - t(n_1 + n_2 - 2, \alpha/2)\, s_p\sqrt{1/n_1 + 1/n_2} \le (\mu_1 - \mu_2) \le (\bar{x} - \bar{y}) + t(n_1 + n_2 - 2, \alpha/2) \cdot s_p\sqrt{1/n_1 + 1/n_2}. \quad (7.51)$$

$$LCI_{1-\alpha}: \quad (\bar{x} - \bar{y}) - t(n_1 + n_2 - 2, \alpha)\, s_p\sqrt{1/n_1 + 1/n_2} \le (\mu_1 - \mu_2). \quad (7.52)$$

$$UCI_{1-\alpha}: \quad (\mu_1 - \mu_2) \le (\bar{x} - \bar{y}) + t(n_1 + n_2 - 2, \alpha) \cdot s_p\sqrt{1/n_1 + 1/n_2}. \quad (7.53)$$

The following example illustrates Eq. (7.51).

EXAMPLE 16: Consider again the problem of comparing the average weights of 6th-grade boys and girls. Furthermore, suppose we are willing to assume that the two groups share a common but unknown variance, σ^2. The samples show $n_1 = 16$, $n_2 = 26$, $\bar{x} = 100$, $\bar{y} = 85$, $s_1^2 = 200$, $s_2^2 = 180$. Find a $CI_{.90}$ for $\mu_1 - \mu_2$.

SOLUTION: First, $t(40, .05) = 1.684$ is obtained from Table E2, and $\bar{x} - \bar{y} = 100 - 85 = 15$. Then Eq. (7.12) gives

$$s_p^2 = [15(200) + 25(180)]/(16 + 26 - 2) = 7500/40 = 187.5.$$

Next,

$$\begin{aligned} s_p\sqrt{1/16 + 1/26} &= \sqrt{(187.5)(1/16 + 1/26)} \\ &= \sqrt{187.5[42/(16)(26)]} = \sqrt{7875/416} \\ &= \sqrt{18.93} = 4.35. \end{aligned}$$

Finally, $15 \pm (1.684)(4.35)$ gives

$$CI_{.90}: \quad 7.67 \le (\mu_1 - \mu_2) \le 22.33.$$

Since the point, 0, is not an element of the CI, we conclude that $\mu_1 = \mu_2$ is not acceptable. In particular, we conclude that μ_1 is greater than μ_2 because the lower limit in $\mu_1 - \mu_2$ is greater than 0.

SOME REMARKS CONCERNING TWO INDEPENDENT, NORMAL SAMPLES: In the preceding work, we have not addressed the problem of determining confidence intervals for $\mu_1 - \mu_2$ when both variances are unknown and assumed to be unequal. A natural way to attack the problem might by to use S_1^2 and S_2^2 to estimate σ_1^2 and σ_2^2, respectively. With these substitutions, we could consider the quantity.

$$[D - (\mu_1 - \mu_2)]/\sqrt{S_1^2/n_1 + S_2^2/n_2}$$

for use in determining confidence intervals for $(\mu_1 - \mu_2)$. A problem arises in that the above quantity *does not* have a t-distribution when $n_1 \neq n_2$. Furthermore, it does not have a convenient distribution for determining α-level values for use in calculating confidence intervals. Thus, any procedure we consider will include some approximation. Approximate techniques for the problem are discussed under the term, Behrens-Fisher problem, in Snedecor [1], Fisher and Yates [4], and in other statistical books.

Since S_1^2 and S_2^2 approach σ_1^2 and σ_2^2, respectively, as n_1 and n_2 increase, a good, large sample (approximate) procedure for the Behrens-Fisher problem is to use Eqs. (7.47)-(7.49) with σ_1^2 and σ_2^2 replaced by S_1^2 and S_2^2. Those approximations should be sufficient for practical purposes when *both* n_1 and n_2 are greater than 30.

CONFIDENCE INTERVALS FOR THE RATIO OF VARIANCES, σ_1^2/σ_2^2: The particular form in which we compare two parameters is frequently a result of the way in which those parameters occur in the pivotal quantity being used. For example, the preceding sections present confidence intervals for $\mu_1 - \mu_2$ to compare means because the Z-term in Eq. (7.46) and the t-term in Eq. (7.50) are pivotal quantities involving the difference, $\mu_1 - \mu_2$. However, as we direct our attention toward the development of confidence intervals for comparing two variances, the pivotal quantity used involves the variance ratio, σ_1^2/σ_2^2. To study that variance ratio we must introduce a new probability model and learn to read a table that provides α-level values for that model.

Suppose S_1^2 and S_2^2 are independent estimators for the two variances, σ_1^2 and σ_2^2, respectively. Then consider the pivotal quantity

$$F = (S_1^2/\sigma_1^2)/(S_2^2/\sigma_2^2) = (\sigma_2^2\ S_1^2)/(\sigma_1^2\ S_2^2). \qquad (7.54)$$

The sampling distribution of the quantity F was derived by Snedecor [1] when using some earlier work by Fisher [2]. Snedecor chose F to designate the variable in order to acknowledge Fisher's contributions.

As previously indicated, we suppose the two samples are of sizes n_1 and n_2. Thus, S_1^2 and S_2^2 are associated with $(n_1 - 1)$ and $(n_2 - 1)$ degrees of freedom, respectively. Now S_1^2 and S_2^2 are the only variables in the F ratio, and it can be shown that the sampling distribution of F depends on two parameters, $k_1 = n_1 - 1$, the degrees of freedom for S_1^2, and $k_2 = n_2 - 1$, the degrees of freedom for S_2^2. In terms of those parameters, the density of F is

$$f(F) = \{\Gamma[(k_1 + k_2)/2]/[\Gamma(k_1/2)\Gamma(k_2/2)]\}(k_1/k_2)^{k_1/2}$$
$$F^{(k_1-2)/2}\ (1 + k_1F/k_2)^{-(k_1 + k_2)/2},\ F > 0, \qquad (7.55)$$

where $\Gamma(a)$ is defined by Eq. (5.41). The F density given by Eq. (7.55) is a skewed distribution such that

$$E(F) = k_2/(k_2 - 2) \text{ if } k_2 > 2,$$
$$V(F) = 2k_2^2(k_1 + k_2 - 2)/k_1(k_2 - 2)^2(k_2 - 4) \text{ if } k_2 > 4. \quad (7.56)$$

Figure 7.5 presents a sketch of an F density with $k_1 = 5$ and $k_2 = 7$.

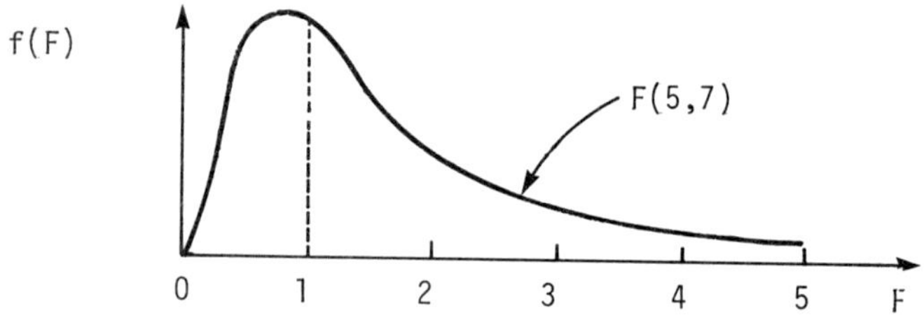

FIGURE 7.5. A graph of an F density with $k_1 = 5$ and $k_2 = 7$.

Table E3 provides α-level values for a select set of α, k_1, and k_2. To use those tables we define the α-level value of an F variable with parameters k_1 and k_2 [i.e., $F \sim F(k_1, k_2)$] as $F(k_1, k_2, \alpha)$, where

$$P[F > F(k_1, k_2, \alpha)] = \alpha. \quad (7.57)$$

Only upper-tail critical points are provided in Table E3 because the lower-tail points ($\alpha > .5$) can be obtained by using the following identity:

$$F(k_1, k_2, \alpha) = 1/F(k_2, k_1, 1 - \alpha). \quad (7.58)$$

Proof: The justification of Eq. (7.58) relies on the fact that if

$$F \sim F(k_1, k_2),$$

then

$$1/F \sim F(k_2, k_1).$$

This follows directly from inspection of Eq. (7.54) because the reciprocal of F just interchanges the roles of samples 1 and 2. Thus, if $F(k_1, k_2, \alpha)$ is such that

$$P[(F > F(k_1, k_2, \alpha)] = \alpha,$$

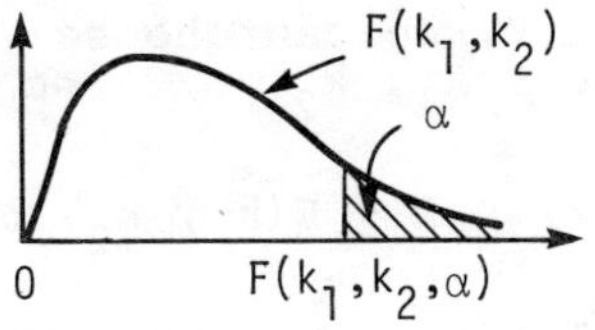

it follows that

$$P[1/F < 1/F(k_1, k_2, \alpha)] = \alpha.$$

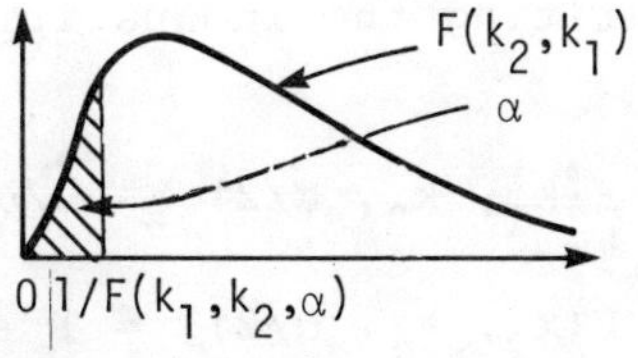

However, by Eq. (7.57),

$$P[1/F > F(k_2, k_1, 1 - \alpha)] = 1 - \alpha.$$

Therefore, we must have

$$1/F(k_1, k_2, \alpha) = F(k_2, k_1, 1 - \alpha),$$

which, by reciprocals, is equivalent to Eq. (7.58).

Example 17 demonstrates the reading of Table E3 and the use of Eq. (7.58).

EXAMPLE 17: Use Table E3 to find: (a) $F(3, 4, .05)$; (b) $F(6, 2, .01)$; (c) $F(4, 7, .95)$.

SOLUTION: (a) Entering Table E3 with $\alpha = .05$, $k_1 = 3$, and $k_2 = 4$, we read directly that $F(3, 4, .05) = 6.59$.
(b) Entering Table E3 with $\alpha = .01$, $k_1 = 6$, and $k_2 = 2$, we read directly that $F(6, 2, .01) = 99.33$.
(c) From Eq. (7.58), we have $F(4, 7, .95) = 1/F(7, 4, .05)$. From Table E3 with $\alpha = .05$, $k_1 = 7$, and $k_2 = 4$,

we find that $F(7, 4, .05) = 6.09$. Therefore, $F(4, 7, .95) = 1/6.09 = .164$.

Now that Table E3 is available to supply the necessary α-level values of F, we can choose $F(k_1, k_2, 1 - \alpha/2)$ and $F(k_1, k_2, \alpha/2)$ so that

$$P[F(k_1, k_2, 1 - \alpha/2) \leq F \leq F(k_1, k_2, \alpha/2)] = 1 - \alpha.$$

Then by using Eq. (7.54) to substitute for F, we find that

$$P[F(k_1, k_2, 1 - \alpha/2) \leq (\sigma_2^2 S_1^2)/(\sigma_1^2 S_2^2) \leq F(k_1, k_2, \alpha/2)] = 1 - \alpha. \tag{7.59}$$

By algebraically rewriting the inequalities in Eq. (7.59), we obtain

$$P[(S_1^2/S_2^2)/F(k_1, k_2, \alpha/2) \leq \sigma_1^2/\sigma_2^2 \leq (S_1^2/S_2^2)\, F(k_2, k_1, \alpha/2)] = 1 - \alpha. \tag{7.60}$$

Proof: Taking reciprocals and reversing the order give

$$1/F(k_1, k_2, \alpha/2) \leq (\sigma_1^2/\sigma_2^2)(S_2^2/S_1^2) \leq 1/F(k_1, k_2, 1 - \alpha/2)$$

for the inequality in Eq. (7.59). Multiplying by S_1^2/S_2^2 and using Eq. (7.58) to replace $1/F(k_1, k_2, 1 - \alpha/2)$ by $F(k_2, k_1, \alpha/2)$ give

$$(S_1^2/S_2^2)/F(k_1, k_2, \alpha/2) \leq \sigma_1^2/\sigma_2^2 \leq (S_1^2/S_2^2) \cdot F(k_2, k_1, \alpha/2).$$

This completes the justification for Eq. (7.60).

Thus, the desired $(1-\alpha)$-level confidence intervals and limits are as follows:

$$\text{CI}_{1-\alpha}\text{:}\quad (s_1^2/s_2^2)/F(k_1,\ k_2,\ \alpha/2) \le \sigma_1^2/\sigma_2^2$$

$$\le (s_1^2/s_2^2)F(k_2,\ k_1,\ \alpha/2). \qquad (7.61)$$

$$\text{LCI}_{1-\alpha}\text{:}\quad (s_1^2/s_2^2)/F(k_1,\ k_2,\ \alpha) \le \sigma_1^2/\sigma_2^2. \qquad (7.62)$$

$$\text{UCI}_{1-\alpha}\text{:}\quad \sigma_1^2/\sigma_2^2 \le (s_1^2/s_2^2)F(k_2,\ k_1,\ \alpha). \qquad (7.63)$$

Equation (7.61) is illustrated by the next example.

EXAMPLE 18: A manufacturer of breakfast cereal has added a new machine to fill boxes. To study the variability in the net weights of boxes filled by the new machine as opposed to the old machine, the weights of the contents are measured for 25 boxes from the old machine and 16 boxes from the new machine. The data show that $s_0^2 = .36$ and $s_N^2 = .30$. Find a 95% confidence interval for the ratio, σ_0^2/σ_N^2. Is there any reason to doubt that the variances of the net weights for the two machines are equal?

SOLUTION: First, using $\alpha = .025$, $k_0 = 25\text{-}1 = 24$, and $k_N = 16\text{-}1 = 15$ permits us to find $F(24,\ 15,\ .025) = 2.70$ and $F(15,\ 24,\ .025) = 2.44$ from Table E3. Then

$$(s_0^2/s_N^2) = (.36/.30) = 1.2.$$

Next, $1.2/2.70 = .44$ and $1.2(2.44) = 2.93$ give

$$\text{CI}_{.95}\text{:}\quad .44 \le \sigma_0^2/\sigma_N^2 \le 2.93.$$

Since the value 1 is included in the interval, we have no reason to doubt that $\sigma_0^2 = \sigma_N^2$.

REMARKS: If the upper limit of the confidence interval in Eq. (7.61) is less than 1, then we are confident that $\sigma_1^2/\ _2^2 < 1$, or equivalently we believe that $\sigma_1^2 < \sigma_2^2$. Similarly, when the lower limit of Eq. (7.61) is greater than 1, we believe that $\sigma_1^2/\sigma_2^2 > 1$ and conclude that $\sigma_1^2 > \sigma_2^2$. Thus, when 1 is included in the interval produced by Eq. (7.61), we have no reason to doubt that $\sigma_1^2 = \sigma_2^2$.

EXERCISES

1. Find the following critical values for the indicated F-distributions:
   (a) $F(4, 5, .05)$;
   (b) $F(10, 9, .025)$;
   (c) $F(6, 10, .95)$;
   (d) $F(7, 8, .99)$;
   (e) $F(20, 30, .25)$.

2. Find the indicated confidence intervals and limits for $\mu_1 - \mu_2$:
   (a) $CI_{.99}$: $n_1 = 10$, $n_2 = 12$, $\sigma_1^2 = 40$, $\sigma_2^2 = 60$, $\bar{x}_1 = 15$, $\bar{x}_2 = 12$.
   (b) $LCI_{.95}$ and $UCI_{.99}$ for the data in (a).
   (c) $CI_{.95}$: $n_1 = 9$, $n_2 = 16$, $s_p^2 = 144$, $\bar{x}_1 = 28$, $x_2 = 36$.
   (d) $LCI_{.99}$ and $UCI_{.95}$ for the data in (c).
   (e) $CI_{.90}$: $n_1 = 9$, $n_2 = 16$, $s_1^2 = 110$, $s_2^2 = 90$, $\bar{x}_1 = 75$, $\bar{x}_2 = 80$. (Assume $\sigma_1^2 = \sigma_2^2$.)

*3. Find the indicated confidence intervals and limits for σ_1^2/σ_2^2:
   (a) $CI_{.95}$: $n_1 = 16$, $n_2 = 20$, $s_1^2 = 16$, $s_2^2 = 25$.
   (b) $LCI_{.95}$: $n_1 = 25$, $n_2 = 15$, $s_1^2 = 4000$, $s_2^2 = 1000$.
   (c) $UCI_{.975}$: $n_1 = 41$, $n_2 = 16$, $\Sigma(x_{1i} - \bar{x}_1)^2 - 1600$, $\Sigma(x_{2i} - \bar{x}_2)^2 = 500$. [The two samples are $(x_{11}, x_{12}, \ldots, x_{1n_1})$ and $(_{21}, x_{22}, \ldots, x_{2n_2}]$.

4. An instructor of a basic statistics course prepares two different tests for the two sections of his class. It is reasonable to assume that the two classes are random samples of basic statistics students and that the scores for the two exams have the same variance. The data for the 1 p.m. class are $n_1 = 31$, $\bar{x}_1 = 75$, $s_1^2 = 105$. The data for the 2 p.m. class are $n_2 = 26$, $\bar{x}_2 = 81$, $s_2^2 = 95$. Find a $CI_{.95}$ for $\mu_1 - \mu_2$. Is there any reason to believe that one test is easier than the other test?

*5. Use the data in Exercise 4 to find a confidence interval for σ_1^2/σ_2^2. Is the instructor's assumption of equal variance reasonable?

*6. From the definition of a median, the median of any F-distribution is $F(k_1, k_2, .5)$. Use Eq. (7.58) to show that the median for any F-distribution with $k_1 = k_2 = k$ is 1.

*7. Identify a problem in a field of interest for which the techniques of this section could be used to answer questions concerning $(\mu_1 - \mu_2)$ or (σ_1^2/σ_2^2).

CONFIDENCE INTERVALS FOR μ_c AND σ_c^2 WHEN STUDYING TWO DEPENDENT, NORMAL SAMPLES: In this discussion, we are considering dependent samples where the two samples are derived from the same individuals. One general class of such experiments is the before-and-after treatment comparison. Thus, we have a set of measurements (X) for n individuals prior to some treatment and a similar set of measurements (Y) for the *same* n individuals after the treatment. Since the two sets of data are obtained from the same individuals, we can eliminate the differences between individuals by considering only the *change*, $C_i = Y_i - X_i$. Then the fact that one individual enters the experiment at a different level from another individual is immaterial, since we base our conclusions only on the individual *changes*. Thus, we have reduced the two samples of paired (X_i, Y_i) observations to one sample (C_i) from the viewpoint of analysis.

Since X_i and Y_i are assumed to be normally distributed variables, it follows that C_i is normally distributed with mean, μ_C, and variance, σ_C^2. Now the unbiased estimators of μ_C and σ_C^2 are $\overline{C}$ and S_C^2, as defined in Eqs. (7.15) and (7.16). Thus, the problem of finding confidence intervals for μ_C and σ_C^2 is identical to that of the section, *Interval estimation for* μ *and* σ^2 on page 282. That is, we can use the pivotal quantities,

$$Z = (\overline{C} - \mu_C)/\sigma_{\overline{C}} \sim N(0, 1), \tag{7.64}$$

$$t = (\overline{C} - \mu_C)/S_{\overline{C}} \sim t(n-1), \tag{7.65}$$

$$\chi^2 = \sum_{i=1}^{n} (C_i - \overline{C})^2/\sigma_C^2 \sim \chi^2(n-1), \tag{7.66}$$

where $\sigma_{\overline{C}} = \sigma_C/\sqrt{n}$ and $S_{\overline{C}} = S_C/\sqrt{n}$, in the same way that Eqs. (7.23), (7.30), and (7.38) are used in the section just cited on page 282.

Therefore, the $(1-\alpha)$-level confidence intervals and limits for μ_C are

$$\begin{aligned} &CI_{1-\alpha}: \ \overline{c}-z(\alpha/2)\sigma_{\overline{c}} \le \mu_C \le \overline{c}+z(\alpha/2)\sigma_{\overline{c}} \\ &LCI_{1-\alpha}: \ \overline{c}-z(\alpha)\sigma_{\overline{c}} \le \mu_C \\ &UCI_{1-\alpha}: \ \mu_C \le \overline{c}+z(\alpha)\sigma_{\overline{c}} \end{aligned} \tag{7.67}$$

when σ_C^2 *is known, and*

$$\begin{aligned} &CI_{1-\alpha}: \ \overline{c} - t(n-1,\alpha/2)s_{\overline{c}} \le \mu_C \le \overline{c} + t(n-1,\alpha/2)s_{\overline{c}} \\ &LCI_{1-\alpha}: \ \overline{c} - t(n-1,\alpha)s_{\overline{c}} \le \mu_C \\ &UCI_{1-\alpha}: \ \mu_C \le \overline{c} + t(n-1,\alpha)s_{\overline{c}} \end{aligned} \tag{7.68}$$

when σ_C^2 *is unknown.* Similarly, the $(1-\alpha)$-level confidence intervals and limits for σ_C^2 are

$$\text{CI}_{1-\alpha}: \quad \Sigma(c_i - \bar{c})^2/\chi^2(n-1,\alpha/2) \le \sigma_c^2$$
$$\le \Sigma(c_i - \bar{c})^2/\chi^2(n-1,1-\alpha/2)$$

$$\text{LCI}_{1-\alpha}: \quad \Sigma(c_i - \bar{c})^2/\chi^2(n-1,\alpha) \le \sigma_c^2 \qquad (7.69)$$

$$\text{UCI}_{1-\alpha}: \quad \sigma_c^2 \le \Sigma(c_i - \bar{c})^2/\chi^2(n-1,1-\alpha).$$

EXAMPLE 19: A new additive for gasoline is supposed to increase mileage. A sample of 9 cars is selected for the test. Each test car is driven at a constant rate to determine the miles per gallon of gasoline without the additive (X) and for one gallon of gasoline with the additive (Y). The data are as follows:

Car #	1	2	3	4	5	6	7	8	9
x_i	15.5	14.2	17.3	16.7	12.8	15.9	14.7	16.1	15.4
y_i	16.8	15.2	17.9	17.1	14.7	16.6	15.8	17.3	16.2

(a) Find a 90% confidence interval for μ_c, the true change due to the additive.
(b) Find a 95% confidence interval for σ_c^2, the variance of the change due to the additive.
(c) Does the additive have any effect on gas mileage?

SOLUTION: (a) First, we calculate the differences, $c_i = y_i - x_i$, to be: (1.3, 1.0, .6, .4, 1.9, .7, 1.1, 1.2, .8). Then $\bar{c} = 9.0/9 = 1.0$ and $\Sigma(c_i - \bar{c})^2 = 1.60$. Next, $t(8,.05) = 1.86$, $s_c^2 = 1.60/8 = .20$, and $s_c = \sqrt{.20} = .45$. Finally, $\bar{c} \pm 1.86 s_c/\sqrt{9}$ in Eq. (7.68) gives
$\text{CI}_{.90}$: $.721 \le \mu_c \le 1.279$.
(b) First, $\chi^2(8, .975) = 2.18$ and $\chi^2(8, .025) = 17.53$. Second, using $\Sigma(c_i - \bar{c})^2 = 1.60$ in Eq. (7.69) gives

$$\text{CI}_{.95}: \quad .091 \le \sigma_c^2 \le .734.$$

(c) Part (a) indicates that we are 90% confident that μ_c is between .721 and 1.279 miles per gallon. Since the entire interval is above 0, we conclude that the additive probably increases the mileage performance of gasoline.

REMARK: In the preceding example, we have made two different inferences ($CI_{.90}$ for μ_c and $CI_{.95}$ for σ_c^2) from the same set of data. Technically, we should have an independent set of data for each inference. That is, we should use two different sets of data to answer questions (a) and (b).

EXERCISES

1. Consider the data in Example 19, and find a 95% lower limit for μ_c.
2. Find a 95% upper limit for σ_c^2 by using the data in Example 19.
3. Identify one or more situations in your field of interest where the variable of interest is the change produced by some treatment. (Treatment here is a very general term to include any before-after type measurements.)
4. To study the average change in weight for 3rd-grade boys during the school year, the weights for 16 boys were recorded in September (X) and in May (Y). The results are as follows:

Boy #	1	2	3	4	5	6	7	8	9	10	11	12	13	14	15	16
x_i	52	48	51	66	71	66	66	74	92	64	83	75	75	61	71	56
y_i	55	51	56	75	76	68	72	79	93	68	86	81	78	65	77	64

   Find a 95% confidence interval for μ_c.

5. Use the data in Exercise 4 to obtain a confidence interval for σ_c^2. (Use $\alpha = .01$.)
6. Twenty-five students are given two achievement tests, between which they are given a self-study refresher course. The results show $\bar{c} = 3$ and $s_c^2 = 16$. Find a 99% confidence interval for μ_c. Is there any evidence to support the claim by the developer of the refresher course that users' achievement will increase by at least 2 points?
7. Use the data for change in heart beat rate produced by a new drug as presented in Example 7 on page 276 to find a 90% confidence interval for μ_c. The pertinent data are $n = 10$, $\bar{c} = -4.9$, and $s_c^2 = 4.99$. Interpret your results.
8. Refer again to the data in Exercise 7 to find a 90% confidence interval for σ_c^2, the variance of change in heart beat rate produced by the drug being tested.

7.5 INTERVAL ESTIMATION WHEN SAMPLING BINOMIAL DISTRIBUTIONS

The pupose of this section is to present methods for setting confidence intervals for p, the true probability of a so-called success in a binomial model. First, we present the Clopper-Pearson intervals, which depend on the F distribution defined by Eq. (7.55). Second, we present an approximate method, using the normal approximation to the binomial.

Bowker and Lieberman [3] present the following formulas for calculating the Clopper-Pearson $(1-\alpha)$-level confidence interval and limits for p when a sample size n shows X successes.

$$CI_{1-\alpha}: \quad \frac{x}{x+(n-x+1)F[2(n-x+1),2x,\alpha/2]} \leq p \leq \frac{(x+1)F[2(x+1),2(n-x),\alpha/2]}{(n-x)+(x+1)F[2(x+1),2(n-x),\alpha/2]}. \qquad (7.70)$$

$$LCI_{1-\alpha}: \quad \frac{x}{x+(n-x+1)F[2(n-x+1),2x,\alpha]} \leq p. \qquad (7.71)$$

$$UCI_{1-\alpha}: \quad p \leq \frac{(x+1)F[2(x+1),2(n-x),\alpha]}{(n-x)+(x+1)F[2(x+1),2(n-x),\alpha]}. \qquad (7.72)$$

Equations (7.70)-(7.72) can be used to find confidence intervals for any choice of values for α and n, provided the necessary α-level F values are available. Since binomial samples for certain values of n occur frequently in practice, and since Table E3 does not cover all values of degrees of freedom for F, Tables B3, B4, and B5 are provided for a selected set of n values and for α = .10, .05, and .01, respectively. The limits, L(x) and U(x), as presented in those tables, have been calculated by Eq. (7.70) and expressed as percentages. Graphs of L(x) and U(x), again as presented in Tables B3, B4, and B5 are provided by Pearson and Hartley [5]. Before giving an illustrative example, we note that the Clopper-Pearson intervals and limits are exact in the sense that the true confidence coefficient is always greater than or equal to $(1 - \alpha)$.

EXAMPLE 20: Find a 95% confidence interval for p, the true proportion of voters who favor Candidate A when a random sample of 20 voters yield 11 for A.

SOLUTION: Using the Eq. (7.70) method, F(20, 22, .025) = 2.39 and F(24, 18, .025) = 2.50. Then

$$L(x) = 11/[11 + 10(2.39)] = .315,$$

$$U(x) = 12(2.50)/[9 + 12(2.50)] = .769.$$

Therefore, $CI_{.95}$: $.315 \leq p \leq .769$.

Using the Table B4 method and entering Table B4 for $n = 20$ and $x = 11$, we find that $L(x) = 32$ and $U(x) = 77$. Therefore,

$$CI_{.95}: \ .32 \leq p \leq .77.$$

REMARK: The decimal points are omitted in Table B4, so each confidence limit in the table should be divided by 100.

For $n = 100$, 200, and 500, Tables B3, B4, and B5 are entered with the fraction x/n rather than x. This is a matter of convenience, but the accuracy provided should be enough for most purposes. Furthermore, when $x/n > .50$, we enter Tables B3, B4, and B5 with $1-(x/n)$ and use $L(x) = 100 - U(x)$ and $U(x) = 100 - L(x)$. These results follow directly from the symmetry of binomial models when $p = .5$.

EXAMPLE 21: Suppose a sample of $n = 100$ for Example 20 produces $x = 55$. Find a 95% confidence interval for p.

SOLUTION: First, $x/n = .55$. We enter Table B4 with $n = 100$ and $x/n = .45$ to find $L(x) = 35$ and $U(x) = 55$. Thus, $100 - 55 = 45$ and $100 - 35 = 65$ give

$$CI_{.95}: \ .45 \leq p \leq .65.$$

Note that we may use Tables B3, B4, and B5 for obtaining either $LCI_{1-\alpha}$ or $UCI_{1-\alpha}$ for $\alpha = .05$, .025, and .005 by using either $L(x)$ for $LCI_{1-\alpha}$ or $U(x)$ for $UCI_{1-\alpha}$ from the tables for $\alpha = .10$, .05, and .01, respectively.

By using the fact from Section 6.3 that $\hat{p}$ is approximately a normal variable with mean, p, and variance, pq/n, it follows that

$$Z = (\hat{p} - p)/\sqrt{\hat{p}\hat{q}/n}, \tag{7.73}$$

where $\hat{p} = x/n$, and $\hat{q} = (n-x)/n$ is approximately a $N(0,1)$ variable. Then, by following steps identical to those used to derive Eqs. (7.27)-(7.29), the approximate $(1-\alpha)$-level confidence interval and limits are

$$CI_{1-\alpha}: \ \hat{p} - z(\alpha/2)\sqrt{\hat{p}\hat{q}/n} \leq p \leq \hat{p} + z(\alpha/2)\sqrt{\hat{p}\hat{q}/n} \tag{7.74}$$

$$LCI_{1-\alpha}: \ \hat{p} - z(\alpha)\sqrt{\hat{p}\hat{q}/n} \leq p \tag{7.75}$$

$$UCI_{1-\alpha}: \ p \leq \hat{p} + z(\alpha)\sqrt{\hat{p}\hat{q}/n}. \tag{7.76}$$

Since the normal approximation for a binomial model is best when p is near .5, it follows that the intervals provided by Eqs. (7.74)-(7.76) are best for p near .5 and not so good for p near either 0 or 1. Finally, one other comment should be made regarding the approximate intervals. Observe that the intervals in Tables B3, B4, and B5 are symmetric about $\hat{p}$ only for p near .5, but the approximate intervals are symmetric for all values of $\hat{p}$. The next examples help to illustrate these ideas.

EXAMPLE 22: Use Eq. (7.74) to find a 95% confidence interval for p when the data are as in Example 21.

SOLUTION: First, $z(\alpha/2) = 1.96$. Next, $\hat{p} = .55$ and

$$\sqrt{\hat{p}\hat{q}/n} = \sqrt{(.55)(.45)/100} = .0497.$$

Then $\hat{p} \pm 1.96(.0497)$ gives $L(x) = .55 - .098$ and $U(x) = .55 + .097$. Thus,

$$CI_{.95}: \quad .452 \leq p \leq .648.$$

Note that this interval agrees to 2 decimal places with the result in Example 21.

EXAMPLE 23: Compare the 95% confidence intervals provided by Table B4 and Eq. (7.74) for a study yielding $x = 5$ for $n = 20$.

SOLUTION: From Table B4 for $n = 20$ and $x = 5$,

$$CI_{.95}: \quad .09 \leq p \leq .49.$$

By Eq. (7.74), $L(x)$ and $U(x)$ are $.25 \pm 1.96\sqrt{(.25)(.75)/20}$. Thus,

$$CI_{.95}: \quad .06 \leq p \leq .44.$$

Notice that the two intervals have comparable lengths but that the symmetry of the normal approximations forces the approximate interval to be to the left of the table interval. The degree of distortion increases as $\hat{p}$ deviates from .5, but for any fixed value of $\hat{p}$ the approximate interval approaches the table interval as n increases.

EXERCISES

1. Use Tables B3, B4, and B5 to provide confidence intervals and limits for p when given the following data:
   (a) $CI_{.90}$ for n = 20 and x = 12.
   (b) $CI_{.95}$ for n = 100 and x = 25.
   (c) $CI_{.99}$ for n = 100 and x = 60.
   (d) $LCI_{.95}$ for n = 50 and x = 40.
   (e) $UCI_{.975}$ for n = 20 and x = 10.
2. Use Eqs. (7.74)-(7.76) to solve Exercise 1.
3. Use Eqs. (7.70)-(7.72) to obtain confidence intervals and limits for the following situations:
   (a) $CI_{.95}$ for n = 15 and x = 6.
   (b) $CI_{.90}$ for n = 21 and x = 12.
   (c) $LCI_{.95}$ for n = 10 and x = 8.
   (d) $UCI_{.95}$ for n = 17 and x = 9.
4. A random sample of 30 students at MYU produced 16 who said they were classified as juniors or above. Find a 90% confidence interval for p, the true proportion of students who are juniors or above. Could the available data have been used to find a confidence interval for the proportion of students who are not classified as juniors or above?
5. Find confidence intervals for p based on any data you have collected for the binomial experiments proposed in Section 7.2. (If you have not already collected some data for these or similar experiments, we urge you to do so.)
6. A random sample of 50 registered voters in a given precinct shows 20 who are registered as Republicans. Find a $CI_{.99}$ for the true proportion of Republicans among registered voters in the designated precinct.
7. The manufacturers of long-mileage tires claim that their tires last for at least 30,000 miles. In a test of 100 tires, 4 failed before 30,000 miles. Is there any reason to doubt the manufacturers' claim?
8. In a test of 100 60-watt light bulbs, 5 failed before the 1,000-hour mark. Find a $CI_{.99}$ for the proportion of bulbs that will burn at least 1,000 hours.

7.6 SUMMARY OF ESTIMATION TECHNIQUES

The key results for both point estimation and interval estimation are summarized here for easy reference. The beginning page number for the section in the text discussing each main topic is indicated in parentheses.

POINT ESTIMATORS:

I. Binomial models (page 269)
Point estimator of p is $\hat{p} = X/n$.

II. Normal models (page 271)
- A. One normal distribution (page 271)
  1. Point estimator for μ is $\overline{X} = \Sigma X_i/n$.
  2. Point estimator for σ^2 is $S^2 = \Sigma (X_i - \overline{x})^2/(n - 1)$.
- B. Two independent, normal distributions (page 273)
  1. Point estimator for $\mu_1 - \mu_2$ is $D = \overline{X} - \overline{Y}$.
  2. Point estimator for common variance, σ^2, is $S_p^2 = [\Sigma(X_i - \overline{X})^2 + \Sigma(Y_i - \overline{Y})^2]/(n_1 + n_2 - 2)$.
- C. Two dependent, normal distributions (page 276)
  1. Point estimator for μ_c is $\overline{C}$, where $C_i = X_i - Y_i$.
  2. Point estimator for σ_c^2 is $S_c^2 = \Sigma(C_i - \overline{C})^2/(n - 1)$.

III. Poisson models (page 278)
Point estimator for $\lambda = \overline{X}$

INTERVAL ESTIMATORS:

I. Normal models, (page 281)
- A. $X \sim N(\mu, \sigma^2)$ (page 281)
  1. Confidence intervals for μ
     - a. With σ^2 known, the pivotal quantity, $Z = (\overline{X} - \mu)/\sigma_{\overline{X}}$, where $\sigma^2_{\overline{X}} = \sigma^2/n$ (page 282)
       - (i) $CI_{1-\alpha}$: $\overline{x} - z(\alpha/2)\sigma_{\overline{x}} \le \mu \le \overline{x} + z(\alpha/2)\sigma_{\overline{x}}$.
       - (ii) $LCI_{1-\alpha}$: $\overline{x} - z(\alpha)\sigma_{\overline{x}} < \mu$.
       - (iii) $UCI_{1-\alpha}$: $\mu < \overline{x} + z(\alpha)\sigma_{\overline{x}}$.
     - b. With σ^2 unknown, the pivotal quantity, $t = (\overline{X} - \mu)/S_{\overline{X}}$, where $S^2_{\overline{X}} = S^2/n$ (page 288)
       - (i) $CI_{1-\alpha}$: $\overline{x} - t(n - 1, \alpha/2)\ S_{\overline{x}} \le \mu \le \overline{x} + t(n-1, \alpha/2)\ S_{\overline{x}}$.
       - (ii) $LCI_{1-\alpha}$: $\overline{x} - t(n - 1, \alpha)\ S_{\overline{x}} < \mu$.
       - (iii) $UCI_{1>\alpha}$: $\mu \le \overline{x} + t(n - 1, \alpha)\ S_{\overline{x}}$

  *2. Confidence interval for σ^2 with the pivotal quantity, $Y = \Sigma(X_i - \overline{X})^2/\sigma^2$ (page 290)
     - a. $CI_{1-\alpha}$: $\Sigma(X_i - \overline{X})^2/\chi^2(n - 1, \alpha/2) \le \sigma^2 \le \Sigma(X_i - \overline{x})^2/\chi^2(n - 1, 1 - \alpha/2)$.
     - b. $LCI_{1-\alpha}$: $\Sigma(X_i - \overline{X})^2/\chi^2(n - 1, \alpha) \le \sigma^2$.
     - c. $UCI_{1-\alpha}$: $\sigma^2 \le \Sigma(X_i - \overline{X})^2/\chi^2(n - 1, 1 - \alpha)$.

- B. $X \sim N(\mu_1, \sigma_1^2)$, $Y \sim N(\mu_2, \sigma_2^2)$, and X and Y are independent (page 295).

1. Confidence intervals for $\mu_1 - \mu_2$

   a. With σ_1^2 and σ_2^2 known, the pivotal quantity, $Z = [D - (\mu_1 - \mu_2)]/\sqrt{\sigma_1^2/n_1 + \sigma_2^2/n_2}$ (page 295).

   (i) $CI_{1-\alpha}$: $D - z(\alpha/2)\sqrt{\sigma_1^2/n_1 + \sigma_2^2/n_2} \le \mu_1 - \mu_2 \le D - z(\alpha/2)\sqrt{\sigma_1^2/n_1 + \sigma_2^2/n_2}$.

   (ii) $LCI_{1-\alpha}$: $D - z(\alpha)\sqrt{\sigma_1^2/n_1 + \sigma_2^2/n_2} \le \mu_1 - \mu_2$.

   (iii) $UCI_{1-\alpha}$: $\mu_1 - \mu_2 \le D + z(\alpha)\sqrt{\sigma_1^2/n_1 + \sigma_2^2/n_2}$.

   b. With $\sigma_1^2 = \sigma_2^2 = \sigma^2$ unknown, the pivotal quantity, $t = [D - (\mu_1 - \mu_2)]/\sqrt{s_p^2(1/n_1 + 1/n_2)}$ (page 297)

   (i) $CI_{1-\alpha}$: $D \pm t(n_1 + n_2 - 2, \alpha/2)\sqrt{s_p^2(1/n_1 + 1/n_2)}$.

   (ii) $LCI_{1-\alpha}$: $D - t(n_1 + n_2 - 2, \alpha)\sqrt{s_p^2(1/n_1 + 1/n_2)}$.

   (iii) $UCI_{1-\alpha}$: $D + t(n_1 + n_2 - 2, \alpha)\sqrt{s_p^2(1/n_1 + 1/n_2)}$.

*2. Confidence intervals for σ_1^2/σ_2^2 with the pivotal quantity, $F = s_1^2\sigma_2^2/s_2^2\sigma_1^2$ (page 299)

   a. $CI_{1-\alpha}$: $(S_1^2/S_2^2)/F(n_1 - 1, n_2 - 1, \alpha/2) \le \sigma_1^2/\sigma_2^2 \le (S_1^2/S_2^2)\ F(n_2 - 1, n_1 - 1, \alpha/2)$.

   b. $CI_{1-\alpha}$: $(S_1^2/S_2^2)/F(n_1 - 1, n_2 - 1, \alpha) \le \sigma_1^2/\sigma_2^2$.

   c. $UCI_{1-\alpha}$: $\sigma_1^2/\sigma_2^2 \le (S_1^2/S_2^2)\ F(n_2 - 1, n_1 - 1, \alpha)$.

*C. Sampling two dependent, normal variables (paired data), $C = X - Y \sim N(\mu_c, \sigma_c^2)$ (page 304)
Apply results in A by replacing $\overline{x}$ with $\overline{c}$, s^2 by s_c^2, μ by μ_c, and σ^2 by σ_c^2.

II. Binomial models (page 308)

A. Clopper-Pearson intervals

1. $CI_{1-\alpha}$: $\dfrac{x}{x + (n - x + 1)F[2(n - x + 1),\ 2x,\ \alpha/2]} \leq p \leq \dfrac{(x + 1)F[2(x + 1),\ 2(n - x),\ \alpha/2]}{(n - x) + (x + 1)F[2(x + 1),\ 2(n - x),\ \alpha/2]}$

2. $LCI_{1-\alpha}$: $\dfrac{x}{x + (n - x + 1)F[2(n - x + 1),\ 2x,\ \alpha]} \leq p.$

3. $UCI_{1-\alpha}$: $p \leq \dfrac{(x + 1)F[2(x+1),\ 2(n - x),\ \alpha]}{(n-x) + (x + 1)F[2(x + 1),\ 2(n - x),\ \alpha]}.$

B. Approximate intervals

1. $CI_{1-\alpha}$: $\hat{p} - z(\alpha/2)\sqrt{\hat{p}\hat{q}/n} \leq p \leq \hat{p} + z(\alpha/2)\sqrt{\hat{p}\hat{q}/n}.$
2. $LCI_{1-\alpha}$: $\hat{p} - z(\alpha)\sqrt{\hat{p}\hat{q}/n} \leq p.$
3. $UCI_{1-\alpha}$: $p \leq \hat{p} + z(\alpha)\sqrt{\hat{p}\hat{q}/n}.$

7.7 REFERENCES

[1] Snedecor, G. W. 1956. *Statistical methods.* 5th ed. Ames, Iowa: The Iowa State College Press.

[2] Fisher, R. A. 1942. On a distribution yielding the error functions of several well-known statistics. In *Proceeding International Math. Congress.* Vol.II. Toronto, pp. 805-813.

[3] Bowker, A. H., and Lieberman, G. J. 1959. *Engineering statistics.* Engelwood Cliffs, N. J.: Prentice-Hall, p. 373.

[4] Fisher, R. A., and Yates, F. 1963. *Statistical tables for biological, agricultural, and medical research.* 6th ed. New York: Hafner Pub. Co.

[5] Pearson, E. S., and Hartley, H. O., eds. 1966. *Biometrika tables for statisticians.* 3rd. ed. Cambridge: The University Press. 1966.

7.8 REVIEW EXERCISES

1. The Bakery Department at MYU has acquired a new loaf machine which is set to prepare one-pound loaves of bread. In order to compare the performance of the new machine, 25 loaves are prepared, and their weights (X) are compared with the weights (Y) for 25 loaves prepared by the machine currently in use.

The results are as follows: $\bar{x} = .98$ pounds, $\bar{y} = 1.01$ pounds, $s_x^2 = .0004$, $s_y^2 = .0009$.

*(a) Is it reasonable to assume $\sigma_x^2 = \sigma_y^2$?

(b) Assume $\sigma_x^2 = \sigma_y^2$, and find a 99% confidence interval for $\mu_x - \mu_y$.

(c) Are the loaves produced by the new machine significantly lighter than those produced by the current machine?

*2. Suppose an independent study of the loaf machines in Exercise 1 shows $n_y = 25$, $s_y^2 = .0005$, $n_x = 17$, and $s_x^2 = .00010$. Find a 95% confidence interval for σ_y^2/σ_x^2. What response to part (a) of Exercise 1 follows from your answer?

*3. Nine individuals with a weight problem have been on a 2-week weight reduction program. The individuals are weighed at the beginning and end of the program. Then the weight change for each individual is recorded. The data are as follows: $\bar{c} = 10$ pounds and $s_c = 3$ pounds.

Find a $LCI_{.95}$ for μ_c. Do you doubt the claim that people participating in the program can expect to lose at least 9 pounds in 2 weeks?

4. A survey of 30 coeds at MYU reveals 10 who could be classified as blondes. Give a point estimate for p, the true proportion of blondes at MYU. Find an estimate of the variance for your estimate of p. Find a $CI_{.90}$ for p.

5. The Fair Trade Auto Sales Company finds that the average cost for preparation of a new car for delivery is $25 for a sample of 20 cars. The same sample yields $s^2 = 25$. What is your best prediction of the cost for preparing the next new car for delivery? How would you estimate the variance of your prediction? Find a $CI_{.95}$ for the true, but unknown, average cost of new car preparation.

6. A random sample of 100 voters shows 60 who support the incumbent governor for re-election. Is the governor's re-election campaign in trouble?

7. Find 95% confidence intervals for p based on the following combinations of (n, x): (20, 6); (50, 15); (100, 30); and (500, 150). What is the effect of increasing n on the length of a confidence interval for p?

8. Find confidence intervals for p based on n = 100 and x = 40 and the following choices of α: .10, .05, and .01. What is the effect of increasing α on the length of a confidence interval for p?

9. If we suppose the aptitude scores for management graduates from MYU are normally distributed with $\sigma^2 = 36$, what is the smallest value of n that will produce a 95% confidence interval of no more than 3 units?

10. A new exhaust emission control unit for automobiles is tested on 20 cars of which 3 did not show at least a 75% reduction in exhaust pollution. Find a point estimate for p, the proportion of cars which show an exhaust pollution reduction of at least 75%. Find a $CI_{.99}$ for p.

CHAPTER 8

Tests of Statistical Hypotheses

8.1 INTRODUCTION, PHILOSOPHY, AND DEFINITIONS

In the preceding chapter, we studied that part of statistical inference called estimation, where our primary concern was the use of sample evidence to obtain information about the value of various unknown parameters. That view can be contrasted with the part of statistical inference called *testing statistical hypotheses*, where sample evidence is used to either *support* (*accept*) or *refute* (*reject*) one or more assumptions made concerning the distribution (model) being studied. The purpose of this section is to present the basic terminology and concepts of hypothesis testing. These ideas are illustrated by examples in Section 8.2.

Frequently the assumptions made about a distribution (model) being studied can be summarized in terms of the parameters involved. For example, we can assume that the variable X is uniformly distributed [see Eq. (5.6)] between $\Theta - .5$ and $\Theta + .5$. Then we can express our assumptions about the location of the variable X in terms of Θ. For reference and illustrations, we graph the distribution of X in Figure 8.1.

The set of all assumptions made about a model is called the *null hypothesis* and is denoted by H_0. For the

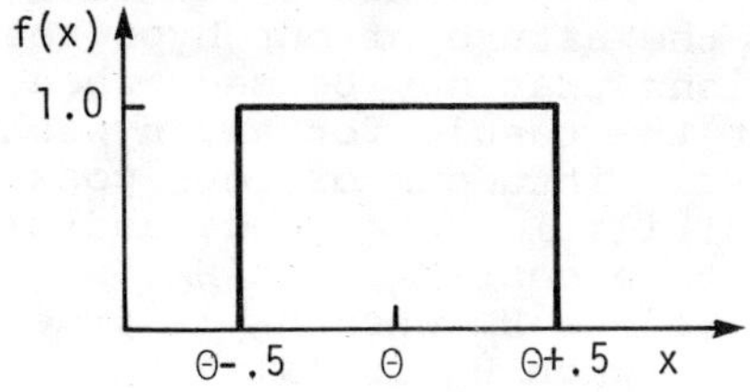

$$f(x) = \begin{cases} 1, & \Theta - .5 < x < \Theta + .5 \\ 0, & \text{elsewhere.} \end{cases}$$

FIGURE 8.1. A graph and density function are shown for a uniform variable, X.

uniform variable X in Figure 8.1, the null hypothesis H_o: $\Theta = 2$ implies that we believe the uniform distribution is centered at 2. Other null hypotheses might be as follows:

(i) H_o: $\mu = 25{,}000$ to express the belief that the average life of a given automobile tire is 25,000 miles.

(ii) H_o: $\sigma^2 = 50$ to express the belief that the variance of weights in pounds of 5th-grade boys is 50.

(iii) H_o: $p = .60$ to express the belief that 60% of the state voters support the governor in his re-election bid.

In testing H_o, sample data are used to either *reject* H_o or *not reject* H_o. Thus, hypothesis testing is a *two-decision* procedure. We note that the second decision is technically that of not rejecting H_o. However, we frequently refer to it as accepting H_o. That reference is for convenience only in that no amount of data can prove that H_o is true (acceptable). Data can only fail to reject H_o.

A *statistical test* for H_o is a rule designating those sample values for which H_o is to be rejected. In other words, a statistical test partitions the sample space, S, of a test statistic into two disjoint regions as follows. Let R denote those points in S for which H_o is to be rejected. Let A denote those points in S for which H_o is to be accepted. Clearly, $R \cup A = S$ and $R \cap A = \emptyset$.

When we state a hypothesis, H_0, there are two possibilities. Either H_0 *is true* or H_0 *is false*. By combining these two possibilities for the nature of our hypothesis with the two possible decisions that may be made when testing H_0, we conclude that the result for any hypothesis testing situation must fall into one of four possibilities. Two of these possibilities are good, and two are bad. First, it is *good* to accept H_0 when H_0 is true. Second, it is *bad* to reject H_0 when H_0 is true. Third, it is *good* to reject H_0 when H_0 is false. Fourth, it is *bad* to accept H_0 when H_0 is false. We assign names and probabilities to those four situations for the sake of discussion.

When H_0 *is true* and the decision is to *reject* H_0, we say that a *type I* error is committed. We use α to denote the probability of a type I error. Thus,

$$\alpha = P(\text{type I error}) = P(H_0 \text{ is rejected} | H_0 \text{ is true}). \quad (8.1)$$

We also call α the *level of significance* for testing the null hypothesis, H_0.

When H_0 *is false*, and the decision is to *accept* H_0, we say that a *type II* error is committed. We use β to denote the probability of a type II error. Thus,

$$\beta = P(\text{type II error}) = P(H_0 \text{ is accepted} | H_0 \text{ is false}). \quad (8.2)$$

The quantities, α and β, represent the probabilities for the two possible types of errors that can result from testing a hypothesis, H_0. The complements of these events are desirable in that we want to accept H_0 when H_0 is true and we want to reject H_0 when H_0 is false. In recognition of the positive nature of such situations, we introduce the following names. We call the probability that H_0 *is accepted* when H_0 *is true* the *level of confidence* for the test. The probability that H_0 *is rejected* when H_0 *is false* is called the *power* of the test. Thus,

$$1 - \alpha = \text{level of confidence} = P(H_0 \text{ is accepted} | H_0 \text{ is true}), \quad (8.3)$$

$$1 - \beta = \text{power} = P(H_0 \text{ is rejected} | H_0 \text{ is false}). \quad (8.4)$$

The preceding results are summarized in Table 8.1.

In Table 8.1, the double vertical line is used to emphasize that on any particular test we are either on the right side or on the left side of the table. That is, either H_0 is true or H_0 is false. Thus, on any one test of a hypothesis, H_0, we can either commit an error of the first type (type I) or an error of the second type (type II), but we cannot commit an error of both types

TABLE 8.1
An outline of the four possibilities that exist when testing a null hypothesis, H_0.

Decision	Nature of hypothesis, H_0	
	H_0 is true	H_0 is false
Accept H_0	Level of confidence $= 1 - \alpha$	P(type II error) $= \beta$
Reject H_0	Level of significance = P (type I error) $= \alpha$	Power $= 1 - \beta$

on the same test. That exclusiveness property is supported by the complementary nature of the probabilities [α and $(1-\alpha)$] and [β and $(1-\beta)$].

A question could be raised at this point. What happens when H_0 is false? Obviously, we would not knowingly test a false hypothesis in practice, but we can never be sure whether or not the hypothesis being tested is true. Thus, we state a second hypothesis, H_1, to describe an *alternative situation* to that stated in H_0. H_1 is called an *alternative hypothesis*. The alternative hypothesis *is not tested*, but it determines the type of test (rejection region) used to test H_0. That property is illustrated in Section 8.2 and subsequent chapters.

Ideally, we want to keep α and β small, preferably 0, for every hypothesis testing problem, but the three quantities, α, β, and n (the sample size) are interdependent so that fixing values for any two of them determines the value for the third one. For most of the inference considered in this text, we assume a fixed sample size n and specify the level of significance, α. Then we choose a statistical test which, for those choices of α and n, minimizes the value of β (or equivalently, maximizes the power). To do that, the choice of test depends on H_1, as illustrated by later examples.

We conclude this section with some remarks concerning the formulation of statistical hypotheses. In view of the preceding remark that we usually fix α and choose a test to minimize β, it is important that we state as H_0 that hypothesis which designates the more serious error to be of type I. Then we can choose α as small as necessary to provide the level of protection that we consider appropriate. For illustration, suppose we are given the assignment of testing the safety of a new miracle drug, Cure-all. Consider the following pair of hypotheses:

$$H_{o1}: \text{ drug is safe;}$$
$$H_{o2}: \text{ drug is harmful.} \tag{8.5}$$

Which hypothesis should be designated H_o, and which hypothesis should be designated H_1? To answer that question, we must consider the errors involved with each choice and determine which of these errors is the more serious one. First, the pair of hypotheses,

$$H_o: \text{ drug is safe; } H_1: \text{ drug is harmful,} \tag{8.6}$$

implies that the two types of errors are as follows:

Type I. Reject H_o when H_o is true. Thus, we conclude that a safe drug is harmful.

Type II. Accept H_o when H_o is false. Thus, we conclude that a harmful drug is safe.

Second, the pair of hypotheses,

$$H_o: \text{ drug is harmful; } H_1: \text{ drug is safe,} \tag{8.7}$$

reverses the preceding errors and gives the following:

Type I. Conclude that a harmful drug is safe.

Type II. Conclude that a safe drug is harmful.

Since concluding that a harmful drug is safe has more serious consequences than concluding that a safe drug is harmful, we choose to test the null hypothesis, $H_o = H_{o2}$, against the alternative hypothesis, $H_1 = H_{o1}$, as stated in Eq. (8.7). In this way, we can choose α very small to provide adequate protection against the release of a dangerous drug. More illustrations of statistical hypotheses are given in the following examples.

EXAMPLE 1: Mr. Senator is conducting a campaign for re-election and is concerned about the proportion of voters who support his campaign. Provide hypotheses, H_o and H_1, for answering his question if it is a two-person contest.

SOLUTION: Since any proportion, p, greater than .5 spells victory, we choose to test the following pair of hypotheses:

$$H_o: \ p \leq .5,$$
$$H_1: \ p > .5.$$

Mr. Senator can then specify the risk, α, he is willing to accept when concluding that $p > .5$, when in reality $p \leq .5$ (i.e., H_o is true). Then a rejection of H_o at his designated value of α gives him a high degree of confidence that his bid for re-election is headed for success.

EXAMPLE 2: A guidance counselor believes that the average aptitude score, μ, for incoming freshmen is 80. What hypotheses, H_o and H_1, should be tested to check his belief?

SOLUTION: Since the counselor does not indicate a direct interest such as that the average score is at least 80, or that the average score is at most 80, we choose a two-sided, alternative hypothesis. That is, we state the pair of hypotheses to be

$$H_o: \quad \mu = 80,$$
$$H_1: \quad \mu \neq 80.$$

By testing $\mu = 80$ against the two-sided alternative $\mu \neq 80$, the counselor's test has the opportunity of detecting a shift from 80, regardless of whether the change is to lower or higher values.

EXERCISES

1. An entomologist claims that a new insecticide is at least 90% effective. Suggest a null hypothesis, H_o, and an appropriate alternative hypothesis, H_1, for the claim. Give reasons for your choices.

2. The etomologist in Exercise 1 also claims that the new insecticide is ecologically safe. What hypotheses, H_o and H_1, should be used to check his claim?

*3. The makers of More-Go gasoline additive claim that their product causes a mean change (μ_c) in gasoline mileage of at least 2 m.p.g. (miles per gallon). How should we begin to check their assertion?

4. Which of the following pairs of hypotheses would you prefer for checking whether or not the net contents of boxes of Zippo cereal contain 14 ounces of cereal?
   (a) H_o: $\mu = 14$, H_1: $\mu \neq 14$.
   (b) H_o: $\mu \leq 14$, H_1: $\mu > 14$.
   (c) H_o: $\mu \geq 14$, H_1: $\mu < 14$.
   Give reasons for your answer by considering the type I and type II errors involved with each pair of hypotheses.

5. In checking the mileage performance of automobile tires, a testing agency tests the hypothesis H_o: $\mu \leq 30{,}000$ against H_1: $\mu > 30{,}000$. What is a type I error for the test? What is a type II error for the test?

6. Newman [1] presents a classic hypothesis testing situation as discussed by the late Sir Ronald A. Fisher. A lady claims that by taste alone, she can tell which component, milk or tea, was first added to a cup containing a mixture of milk and tea. Identify the type I and type II errors involved with the following two pairs of hypotheses:

   H_{o1}: The lady can distinguish whether milk or tea was first added to the cup.
   H_{11}: The lady cannot distinguish whether milk or tea was first added to the cup.
   H_{o2}: The lady cannot distinguish -------.
   H_{12}: The lady can distinguish -------.

   Which pair of hypotheses should we test to check the claim? Which pair of hypotheses would the lady prefer that we test?

8.2 SELECTION OF STATISTICAL TESTS FOR UNIFORM AND BINOMIAL MODELS

In the preceding section, we defined a number of terms and considered various statistical hypotheses that we may want to test, but we did not address the problem of selecting a *statistical test* for a specified hypothesis, H_o. Thus, the purpose of this section is to illustrate by means of two extensive examples (uniform models and binomial models) how we select statistical tests. Our presentation includes the calculation of α and β for specific hypotheses and provides further insight for the interpretation of those quantities.

We suppose that the study of some phenomenon has resulted in the collection of a sample of data. Those data are then reduced to a *test statistic*, t. The selection of a statistical test thus becomes one of selecting a subset, say R, of the sample space of t, say S, such that $t \in R$ leads to the decision of rejecting H_o, while $t \notin R$ leads to the decision of accepting H_o. Since the sample space S for a test statistic is usually a set of real numbers, we can graphically represent the testing situation as shown in Figure 8.2.

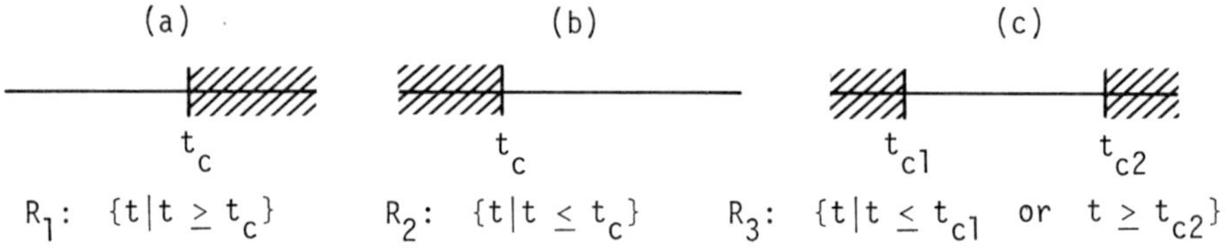

FIGURE 8.2. Graphic representation is given for three different rejection regions.

In Figure 8.2, we use t_c to denote a *critical point* for the test statistic, t. Thus, t_c is a boundary point for the rejection region (critical region). Parts (a) and (b) in Figure 8.2 depict one-sided rejection regions, while part (c) shows a two-sided region. Correspondingly, any test defined by a one-sided rejection region is called a *one-sided* (one-tailed) *test*, while a test defined by a two-sided rejection region is called a *two-sided* (two-tailed) *test*.

Before discussing a specific hypothesis testing example, we list the sequence of steps involved in any statistical test.

Step 1. Indicate the null and alternative hypotheses, H_o and H_1.
Step 2. Specify α, the probability of a type I error.
Step 3. State the rejection region, R, in terms of the test statistic, t.
Step 4. Collect the sample data.
Step 5. Calculate the test statistic, t.
Step 6. If $t \in R$, reject H_o; otherwise, accept H_o.
Step 7. State a conclusion by interpreting the decision reached in step 6.

We emphasize that steps 1, 2, and 3 occur *before* the sample is observed. That is, the experimenter has some question in mind that is formulated into a statistical hypothesis, H_o. An alternative hypothesis is then stated, and the acceptable risk for a type I error is specified. Those steps are the responsibility of the experimenter with the help of the statistician. Then the statistician uses H_o, H_1, and α to select a statistical test by selecting a test statistic and providing the rejection region for it. Following these preliminaries, the data are collected, and the test is completed. The next two sections illustrate the testing procedure for two types of models.

UNIFORM MODELS: We consider the uniform model in Figure 8.1 for our first example. Thus, the density function is

$$f(x) = \begin{cases} 1, & \Theta - .5 < x < \Theta + .5, \\ 0, & \text{otherwise}, \end{cases} \tag{8.8}$$

and the distribution function is

$$F(x) = \begin{cases} 0, & x \le \Theta - .5, \\ x - \Theta + .5, & \Theta - .5 < x \le \Theta + .5, \\ 1, & \Theta + .5 \le x. \end{cases} \tag{8.9}$$

Note that both f(x) and F(x) depend on Θ, so we need only specify a value for Θ to determine particular density and distribution functions. Therefore, we can hypothesize a particular model by hypothesizing a value for Θ. That property is evident in the subsequent experiment and examples.

EXPERIMENT: Select 100 2-digit, random numbers from Table A2. Divide each number by 100 to obtain a random number between 0 and 1. Use the first 50 numbers as a random sample from Eq. (8.8) with $\Theta = .50$. Add .50 to each of the second set of 50 numbers to generate a random sample of 50 from Eq. (8.8) with $\Theta = 1.00$. Use $t = x$ as a test statistic for testing H_o: $\Theta = .50$ against H_1: $\Theta = 1.00$ by using the (test) rejection region R: $\{x | x \geq .90\}$. Count the number of elements included in R for each set of 50 to determine estimates for α and $(1 - \beta)$. Repeat the experiment 4 times, or compare your data with the results of your classmates.

The test statistic for testing H_o against H_1 in the foregoing experiment is $t = x$ when the one sample point, x, is from the uniform density determined by H_o. Then the density of the test statistic is

$$f(x) = 1, \quad 0 < x < 1.$$

Thus, the true value of α is given by

$$\alpha = P(x \geq .9 | \Theta = .5) = .10.$$

Under H_1, the density being sampled is

$$f(x) = 1, \quad .5 < x < 1.5.$$

Hence,

$$\begin{aligned}(1 - \beta) &= P(\text{rejecting } H_o | \Theta = 1) \\ &= P(x \geq .9 | \Theta = 1) = .6.\end{aligned}$$

Clearly we do not expect to observe these exact values for each set of data generated by the experiment, but the results should fluctuate around those values. Table 8.2 provides 4 data sets for the experiment. The asterisks indicate those data points for which H_o is rejected.

TABLE 8.2.
Four data sets of 50 observations each are given from Eq. (8.8) with Θ=.5 and Θ=1. Asterisks denote rejection of H_0: Θ=.5 when using the rejection region R: $\{x | x \geq .9\}$.

No.	Set 1		Set 2		Set 3		Set 4	
	Θ=.5	Θ=1	Θ=.5	Θ=1	Θ=.5	Θ=1	Θ=.5	Θ=1
1	.15	*1.44	.42	.73	.04	.50	.35	*1.09
2	.46	.82	.69	.69	.45	*1.44	*.98	.83
3	.80	*1.29	.18	*1.24	.54	.61	.48	*1.19
4	.09	*1.40	.73	*1.25	.54	.58	.28	*1.03
5	*.91	*1.39	.77	.69	.21	*1.01	*.99	.83
6	.85	*1.34	.56	*1.15	.19	* .97	.55	* .98
7	.52	* .99	.54	*1.46	.80	.55	.17	.73
8	.56	.65	.86	.58	*.95	*1.25	.63	.52
9	.17	*1.16	.32	.57	.18	*1.26	.01	.85
10	.30	*1.34	.75	* .91	.51	*1.31	.47	*1.05
11	.71	.77	.13	*1.23	.13	*1.19	.44	*1.22
12	.11	.87	.35	*1.49	.31	*1.30	.88	*1.40
13	.46	*1.04	.51	* .93	.62	*1.01	.10	.74
14	.33	*1.44	.84	* .94	.47	* .99	.82	*1.08
15	*.95	*1.37	.64	*1.36	.60	.58	.75	*1.09
16	.30	.73	.85	* .96	*.96	.84	.48	.88
17	.21	*1.35	.04	*1.02	.85	*1.12	.83	.74
18	.76	*1.03	.83	*1.02	.05	.64	.14	* .99
19	.19	* .98	.62	.54	*.91	.75	.39	*1.06
20	*.98	*1.29	.41	.50	.54	.52	.64	* .90
21	*.99	.82	.62	.87	.26	.88	.71	*1.43
22	.64	*1.33	.80	*1.18	*.96	*1.12	.68	* .96
23	.66	*1.23	.13	.67	*.94	*1.04	.07	*1.41
24	.36	*1.47	.06	.80	.38	*1.44	.26	.82
25	.06	* .90	.37	.52	.11	*1.33	.28	.88
26	*.97	*1.47	.07	*1.40	.56	.60	.64	.83
27	.41	*1.44	.20	*1.24	.13	*1.49	.31	.57
28	*.93	* .90	.34	.56	.26	*1.45	.67	*1.47
29	.73	*1.25	.64	.76	.45	.51	.87	.70
30	.01	.69	.36	* .95	.41	*1.46	.71	*1.48
31	.77	*1.30	.52	.81	.87	.82	.51	* .97
32	.20	*1.16	.78	*1.31	.81	*1.11	.00	*1.46
33	.69	*1.43	.68	*1.12	*.93	.61	.44	.67
34	.87	* .95	.27	*1.39	.34	.70	.24	.55
35	.89	* .92	.39	.87	.11	*1.25	.47	.66
36	.24	*1.44	.38	*1.46	.14	*1.05	*.99	.57
37	.24	*1.13	.82	*1.48	.67	*1.47	.28	*1.04
38	.33	*1.09	.61	.89	.83	*1.47	.27	.60
39	.53	.80	*.97	.55	.59	.52	.35	.54
40	*.94	*1.12	*.90	.50	.13	* .90	.51	*1.00
41	.12	*1.18	.55	.55	*.94	*1.19	.51	.78
42	.73	.60	.34	*1.42	.36	*1.12	.86	*1.43
43	.86	.61	.35	*1.10	.28	*1.06	*.95	* .91
44	.66	*1.38	*.99	* .93	.15	*1.36	.01	*1.29

TABLE 8.2. (Continued).

No.	Set 1		Set 2		Set 3		Set 4	
	Θ=.5	Θ=1	Θ=.5	Θ=1	Θ=.5	Θ=1	Θ=.5	Θ=1
45	.33	.88	.29	*1.43	.29	* .92	.68	*1.16
46	.51	.71	.26	*1.10	.41	.71	.68	.54
47	.77	*1.42	.83	*1.35	.12	.81	.43	*1.28
48	.24	*1.34	.59	.75	*.96	.63	*.94	.64
49	.70	*1.16	*.92	*1.29	.88	*1.40	.54	.62
50	.10	.52	.01	*1.01	.17	*1.03	.40	.58

When H_o: $\Theta = .5$ is true, we expect about 10% (5 out of 50) of the observations to fall in the rejection region, R. The actual numbers of elements for the 4 sample data sets in Table 8.2 are 7, 4, 8, and 5 for the sets 1, 2, 3, and 4, respectively. Thus, we obtain 4 estimates of $\alpha = .10$ to be .14, .08, .16, and .10. Similarly, we expect 60% (30 out of 50) of the observations to fall in R when H_o is false and $\Theta = 1$. The actual numbers are 37, 30, 31, and 26 for sets 1, 2, 3, and 4, respectively. Those numbers provide estimates of .74, .60, .62, and .52 as estimates for $1 - \beta = .60$.

The remainder of this section is concerned with the presentation of test procedures (rejection regions) when using one observation, x, as the test statistic, t, for testing H_o: $\Theta = \Theta_o$ when x has the uniform density given in Eq. (8.8). The results are motivated and illustrated by examples.

EXAMPLE 3: We want to test the null hypothesis, H_o: $\Theta = 1$, against the alternative hypothesis, H_1: $\Theta = 1.25$, for the uniform density given in Eq. (8.8). The value of α is .05, and the decision is based on one observation, x. Find a statistical test (rejection region) for that situation. Calculate β and $1-\beta$ (power) for the test.

SOLUTION: On the basis of one observation, x, we want to either reject or accept H_o: $\Theta = 1$. Thus, we use $t = x$ as the test statistic for selecting one of the two possible decisions. We proceed by following the steps outlined in Section 8.2.

Step 1: H_o: $\Theta = 1$,

H_1: $\Theta = 1.25$.

Step 2: $\alpha = .05$.

Step 3: R: $\{x | x \geq 1.45\}$.

Step 4: A random observation from the distribution shows that $x = 1.37$.

Step 5: $x = 1.37$.
Step 6: Since $x = 1.37 \notin R$, we accept H_o.
Step 7: We do not have sufficient evidence to conclude that $\Theta = 1$ should be rejected in favor of $\Theta = 1.25$.

REMARKS: To show that the rejection region (statistical test), R, as defined by step 3 is a 5% ($\alpha = .05$) test, we must show that

$$P(x \varepsilon R | \Theta = 1) = .05.$$

If $\Theta = 1$, then the distribution being sampled is

$$f(x) = 1, \quad .5 < x < 1.5.$$

Therefore, $P(x \varepsilon R | \Theta = 1) = P(1.45 \leq x \leq 1.50 | \Theta = 1)$

$= (.05)(1) = .05$. (See Figure 8.3.)

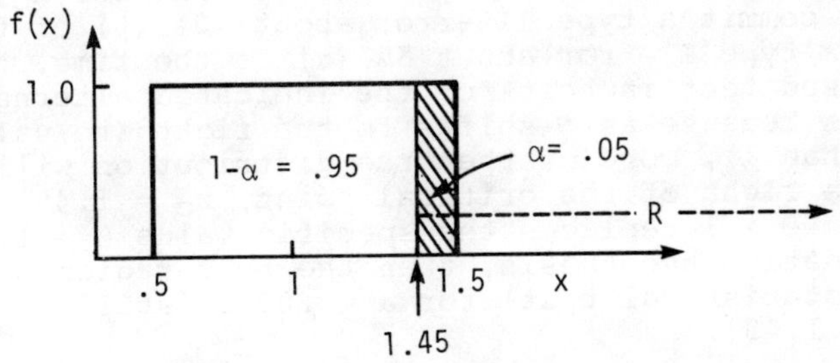

FIGURE 8.3. A graph of the rejection region, $R = \{x | x \geq 1.45\}$, is shown.

Clearly, there are many regions, R, such that $\alpha = .05$. In fact, any interval $a \leq x \leq b$, such that $a \geq .5$, and $b \leq 1.5$, and $b - a = .05$, provides an α-level (5%) statistical test of H_o. Then the question becomes that of asking, why choose $R = \{x | x \geq 1.45\}$? That question is answered by considering β. If H_o is false and, in reality, $\Theta = 1.25$ (rather than 1 as stated for H_o), then the distribution being sampled is

$$f(x) = 1, \quad .75 < x < 1.75,$$

and the power $(1 - \beta)$ is given by

$$1 - \beta = P(R|\Theta = 1.25) = P(1.45 \leq x \leq 1.75)$$

$$= .30(1) = .30. \text{ (See Figure 8.4.)}$$

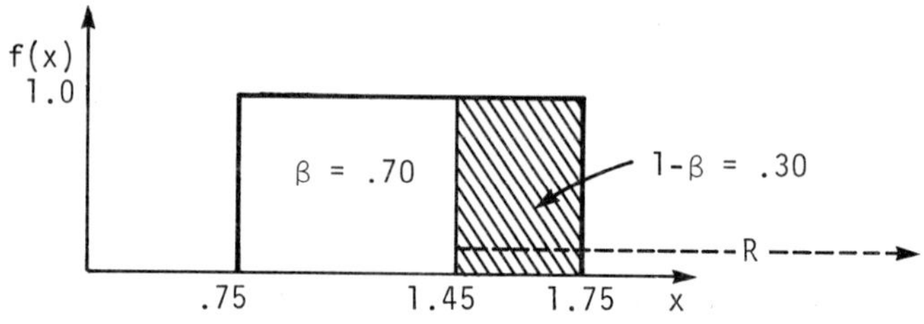

FIGURE 8.4. A graph of β and $1 - \beta$ is given when $R = \{x|x \geq 1.45\}$ and $\Theta = 1.25$.

Observe that by using the indicated test for H_o, we expect to commit a type II error about 70% (β) of the time and a type I error about 5% (α) of the time. Yet the proposed test is best for the indicated alternative hypothesis because as Θ shifts to the right (to values greater than 1), more of the true distribution will occur to the right of the critical point, $x_c = 1.45$.

If H_1: $\Theta > 1$ replaces the specific value $\Theta = 1.25$ as the alternative hypothesis, then the best region of rejection (statistical test) for $\alpha = .05$ is still R: $\{x|x \geq 1.45\}$.

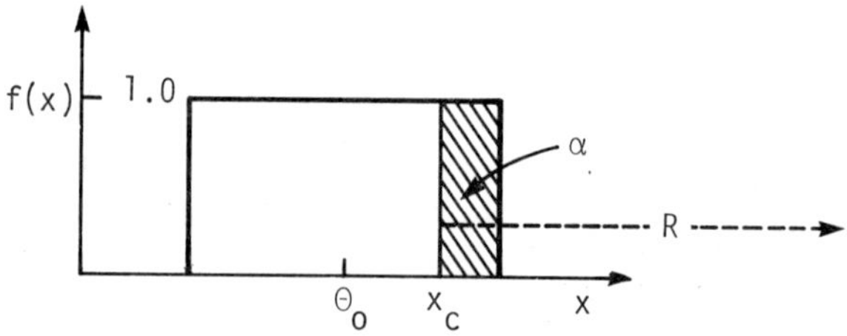

FIGURE 8.5. A graph of the best α-level test is presented for testing H_o: $\Theta = \Theta_o$ against H_1: $\Theta > \Theta_o$.

In general, the best α-level test for testing H_o: $\Theta = \Theta_o$ against H_1: $\Theta > \Theta_o$ is given by

$$R: \quad \{x \mid x \geq x_c\},$$

where x_c is that point which excludes $(100\alpha)\%$ of the X distribution's upper tail. Figure 8.5 illustrates that concept. By selecting our statistical test as outlined in Figure 8.5, we obtain the most powerful test available for the indicated pair of hypotheses.

Sometimes it is of interest to evaluate the power $(1-\beta)$ for each point Θ in the set of values designated by H_1. If H_1: $\Theta > \Theta_o$ is the alternative hypothesis, then the graph that plots Θ against

$$(1 - \beta) = P(x \geq x_c \mid \Theta)$$

is called the *power curve* (power function) for the statistical test defined by R: $\{x \mid x \geq x_c\}$ and based on the test statistic, x. Figure 8.6 illustrates the calculation of a power curve for testing H_o: $\Theta = \Theta_o$ against H_1: $\Theta > \Theta_o$ by using R: $\{x \mid x \geq x_c\}$.

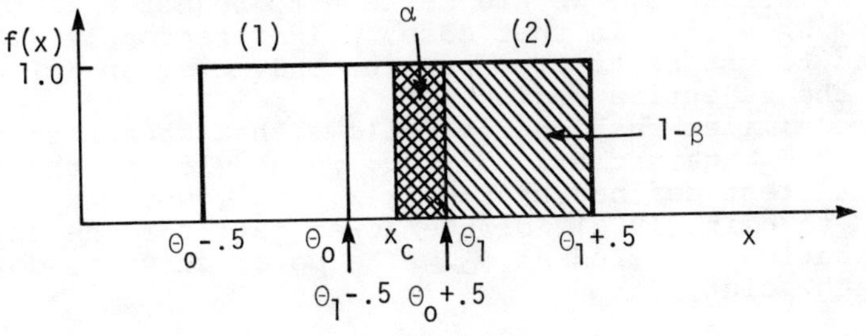

FIGURE 8.6. Graphic calculation of the power, $(1 - \beta)$, is given for $\Theta = \Theta_1$.

From Figure 8.6, it is easy to see that for every point, $\Theta > \Theta_o$, the power is

$$(1 - \beta) = \begin{cases} [(\Theta + .5) - x_c](1) = \Theta + .5 - x_c, \\ \qquad\qquad \Theta_o < \Theta < x_c + .5 \\ 1, \quad x_c + .5 \leq \Theta. \end{cases} \tag{8.10}$$

Thus, the power function is given by Eq. (8.10) and may be graphed as in Figure 8.7.

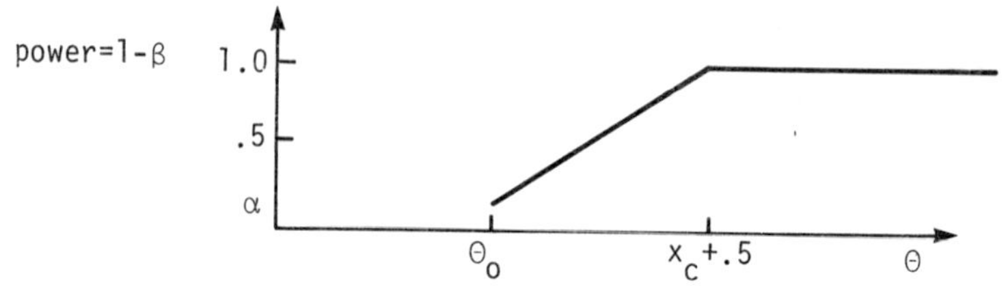

FIGURE 8.7. A graph of the power curve for using R: $\{x|x \geq x_c\}$ is given for testing H_0: $\Theta = \Theta_0$ against H_1: $\Theta > \Theta_0$.

Notice that the power function starts at α when $\Theta = \Theta_0$ and increases to 1 at $\Theta = x_c + .5$. We can imagine that rectangle (2) in Figure 8.6 starts by coinciding with rectangle (1) and then moves to the right. As rectangle (2) moves right, more of its area falls into the rejection region, R: $\{x|x \geq x_c\}$. That is, the probability of rejecting H_0: $\Theta = \Theta_0$, when the true value of Θ is $\Theta_1 > \Theta_0$, increases as Θ_1 shifts further away from Θ_0. Finally, we are certain to reject H_0: $\Theta = \Theta_0$ if $\Theta_1 \geq x_c + .5$. In that case, all of rectangle (2) is located to the right of x_c so the only overlap (if any) is in the rejection region.

In a similar fashion, it follows that when testing H_0: $\Theta = \Theta_0$ against H_1: $\Theta < \Theta_0$, we should use the statistical test defined by R: $\{x|x \leq x_c\}$, where x_c is chosen to omit (100α)% of the lower tail for the uniform distribution centered at Θ_0. The power curve is defined for each point, $\Theta < \Theta_0$, by

$$(1 - \beta) = \begin{cases} x_c - (\Theta - .5), & x_c - .5 < \Theta < \Theta_0 \\ 1, & \Theta \leq x_c - .5 \end{cases} \tag{8.11}$$

and graphed in Figure 8.8.

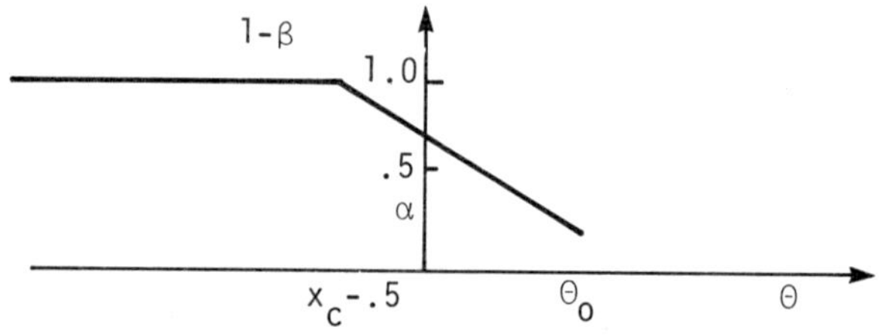

FIGURE 8.8. A graph of the power curve for using R: $\{x|x \leq x_c\}$ is given to test H_0: $\Theta = \Theta_0$ against H_1: $\Theta < \Theta_0$.

EXAMPLE 4: If we use R: $\{x|x \leq 1.6\}$ to test $H_o:\Theta = 2$ against H_1: $\Theta = 1.5$, what are α and β when the test statistic, x, has the uniform density given in Eq. (8.8)?

SOLUTION: We first graph the null distribution (1) and the alternative distribution (2) as shown.

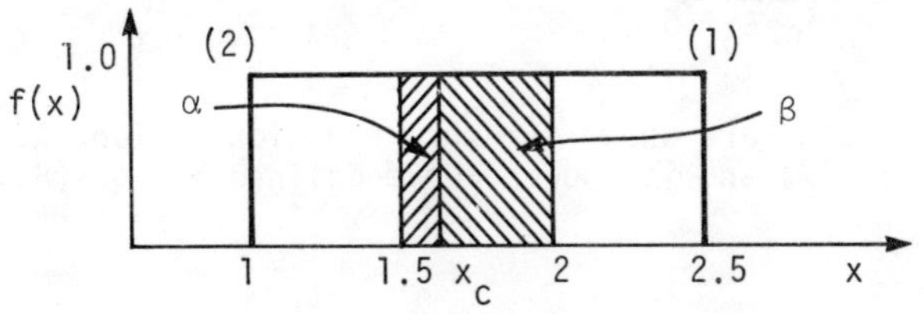

Then $\alpha = (1.6 - 1.5)(1) = .1$ and $\beta = (2.0 - 1.6)(1) = .4$. Thus, R: $\{x|x \leq x_c\}$ defines a 10% test for H_o: $\Theta = 2.0$ against H_1: $\Theta = 1.5$ with power of $1 - .4 = .6$. That is, when using the rejection region, R, we expect to reject H_o about 10% of the time when it is true, and we expect to accept H_o about 40% of the time when H_o is false. Notice that we can change α and β by shifting the critical point, x_c. However, any decrease in α implies an increase in β and vice versa.

The preceding developments illustrate statistical tests for testing one-sided, alternative hypotheses. To complete the discussion we conclude this section by presenting a two-sided testing procedure.

When testing H_o: $\Theta = \Theta_o$ against H_1: $\Theta \neq \Theta_o$, we need an α-level test that rejects H_o whenever Θ gets either too large or too small. Thus, we select a two-sided rejection region

$$R: \quad \{x|x \leq x_{c1} \quad \text{or} \quad x \geq x_{c2}\}, \tag{8.12}$$

where x_{c1} is such that $P(x \leq x_{c1}|\Theta = \Theta_o) = \alpha/2$, and x_{c2} is such that $P(x \geq x_{c2}|\Theta = \Theta_o) = \alpha/2$. The power curve for the test defined by R is defined for each point $\Theta \neq \Theta_o$ by

$$(1 - \beta) = \begin{cases} 1 & , \Theta \leq x_{c1} - .5 \text{ or } \Theta \geq x_{c2} + .5, \\ \Theta + .5 - x_{c2} & , \Theta_o + \alpha/2 < \Theta < x_{c2} + .5, \\ x_{c1} - (\Theta - .5) & , x_{c1} - .5 < \Theta \leq \Theta_o - \alpha/2, \\ \alpha & , \Theta_o - \alpha/2 \leq \Theta \leq \Theta_o + \alpha/2, \end{cases} \tag{8.13}$$

and graphed in Figure 8.9.

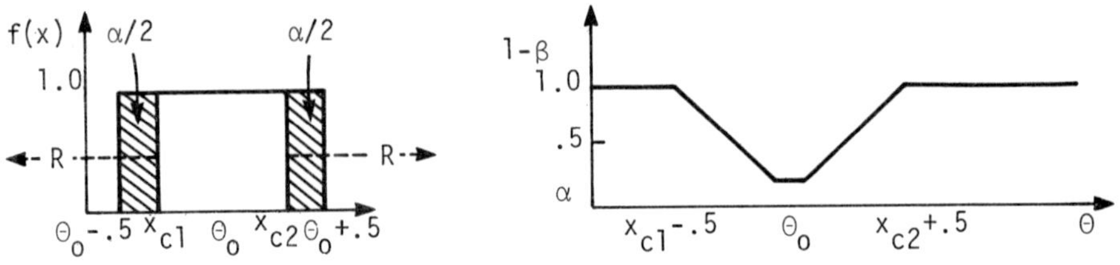

FIGURE 8.9. Graphs are shown for the rejection region, R, in Eq. (8.12) and its power curve defined by Eq. (8.13).

From the power curve defined by Eq. (8.13) and graphed in Figure 8.9, it is apparent that R defines the desired type of test. In particular, we have good probability of detecting a shift in the value of Θ whether the shift is to larger or smaller values. That fact is shown graphically by the lines that increase as Θ moves away from Θ_o. The values on the power curve for each value of Θ represent the amount of a uniform distribution on the interval $(\Theta - .5, \Theta + .5)$, which falls in the rejection region, R. The flat part of $(1 - \beta)$ occurs because, until Θ shifts a distance of $\alpha/2$ away from Θ_o, part of each end of the distribution falls into a part of R.

EXAMPLE 5: Find R to define a 5% test for testing H_o: $\Theta = 2$ against H_1: $\Theta \neq 2$ for the uniform density in Eq. (8.8).

SOLUTION: Graphically we want to determine x_{c1} and x_{c2} as shown in Figure 8.9 so that $P(x \leq x_{c1}) = .025$ and $P(x \geq x_{c2}) = .025$ when $\Theta = 2$. Therefore, we require $x_{c1} - 1.5 = .025$ and $2.5 - x_{c2} = .025$, which in turn implies that $x_{c1} = 1.525$ and $x_{c2} = 2.475$. Thus,

$$R: \{x | x \leq 1.525 \quad \text{or} \quad x \geq 2.475\}.$$

More illustrations of the preceding techniques are developed in the following exercises.

EXERCISES

1. Use one observation, x, from the uniform density on the interval $(\Theta - .5, \Theta + .5)$ [see Eq. (8.8)] to test H_o: $\Theta = 1$ against H_1: $\Theta = .60$.

(a) Find R to define a 10% test.
(b) Find β and the power of the test in (a).
(c) Find α and β for the test defined by R: $\{x|x \leq .65\}$.

2. For the situation in Exercise 1, graph the power curve for testing H_o: $\Theta = 1$ against H_1: $\Theta < 1$ with R: $\{x|x \leq .65\}$.

3. Consider the uniform density

$$f(x) = 1/\Theta, \quad 0 < x < \Theta.$$

(a) Graph f(x) for $\theta = 1$ and $\Theta = 2$.
(b) Find α and β when using R: $\{x|x \geq .95\}$ to test H_o: $\Theta = 1$ against H_1: $\Theta = 2$ (test statistic, x).
(c) Graph the power curve for part (b) when H_1: $\Theta > 1$.

4. Determine sample estimates for α and β by using any data that you collected for the uniform experiment on page 324.

5. Complete the 7 steps for using one observation from

$$f(x) = 1, \quad \Theta - .5 < x < \Theta + .5$$

to test H_o: $\Theta = 1$ against H_1: $\Theta > 1$ with $\alpha = .05$ (using $x = .73$).

6. Repeat Exercise 5, using $x = 1.48$ as the observation.

7. Complete the following statistical tests where x is one observation from f(x) as defined in Exercise 5.
(a) H_o: $\Theta = 2$, H_1: $\Theta \neq 2$, $\alpha = .05$, $x = 1.52$.
(b) H_o: $\Theta = 2$, H_1: $\Theta = 3$, $\alpha = .10$, $x = 2.40$.
(c) H_o: $\Theta = 2$, H_1: $\Theta < 2$, $\alpha = .10$, $x = 1.55$.

8. Graph the power curve for parts (a) and (c) of Exercise 7.

9. Find β for part (b) of Exercise 7.

10. All of the discussion in this section has been based on a sample of size 1 by using the test statistic, $t = x$. Now we can make inference concerning Θ for a uniform density by using a larger sample. To do so we need only have a test statistic with a known distribution so that we can calculate quantities such as α and β. By referring to Example 3 on page 189, we see that the sum of two uniform variables, from the density in Eq. (8.8), with $\Theta = .5$, has a triangular density between 0 and 2. Similarly, if $\Theta = 1$, then $t = X_1 + X_2$ has a triangular density between 1 and 3. Let R: $\{t|t \geq 1.5\}$ define a test for testing H_o: $\Theta = .5$ against H_1: $\Theta = 1$ in the density in Eq. (8.8), where the test statistic is $t = x_1 + x_2$. Find α and β for the test.

*11. Refer to the situation in Exercise 10, and graph the power curve for the test defined by R when H_o: $\Theta = .5$ and H_1: $\Theta > .5$.

BINOMIAL MODELS: While working with the uniform models of the preceding section, the determination of α and β [or $(1 - \beta)$] required only the calculation of areas for certain rectangles. Also, it was possible to find critical points, x_c, designating rejection regions that were of any specified size α. That is, the continuity of the uniform model being studied made it possible to specify α and determine the appropriate critical region of that size. However, when we consider a discrete distribution such as a binomial model, the sample space of the variable is the set of integers, say 0, 1, 2, ..., n. Thus, it is impossible to choose a subset of the sample space as a rejection region to meet any arbitrary value of α. The subsequent discussion and examples emphasize the differences between testing hypotheses with continuous variables such as the uniform models of the preceding section and testing hypotheses with discrete variables such as a binomial variable.

To begin the development we suppose that our interest is in testing the null hypothesis, H_o: $p = .5$, for a binomial model. Further suppose that a sample of size $n = 20$ is to be used in testing H_o. The test statistic is to be X, the number of successes observed in the random sample of 20 trials. To obtain personal data for use in the discussions of this section, one or more of the following experiments can be performed.

EXPERIMENT 1: Select a political candidate for office, and ask 20 randomly selected people whether or not they support his candidacy. Let p be the true unknown proportion of the population being sampled that supports the candidate. Observe X, the number of respondents supporting the candidate.

EXPERIMENT 2: Select some political office-holder, and question 20 randomly selected people as to the adequacy of his job performance. Let p be the true proportion who believe the official is doing an adequate job. Observe X, the number of respondents who believe the official's performance is adequate.

EXPERIMENT 3: Select a question of local (school, community, county, or state) interest, and solicit responses from 20 randomly selected individuals. Observe X, the number of yes responses received. Use p to denote the proportion of yes responses in the population being studied.

EXPERIMENT 4: Flip a coin 20 times, and observe X, the number of heads that result. Let p denote the true probability of a head on each toss.

In selecting a test for the hypothesis, H_0: $p = .5$, for data coming from any of the foregoing binomial experiments, the following types of rejection regions are appropriate for the indicated alternative hypotheses:

$$R: \{x \mid x \geq x_c\} \quad \text{for} \quad H_1: \; p > .5.$$

$$R: \{x \mid x \leq x_c\} \quad \text{for} \quad H_1: \; p < .5. \tag{8.14}$$

$$R: \{x \mid x \leq x_{c1} \text{ or } x \geq x_{c2}\} \quad \text{for} \quad H_1: p \neq .5.$$

In each of the above rejection regions, the critical points, x_c, x_{c1}, and x_{c2}, are integer values. The size of the test, α, depends on the critical point or points used in R. In particular, $\alpha = P(x \varepsilon R \mid p = .5)$, where $X \sim b(20, .5)$. So that with R_1: $\{x \mid x \geq 14\}$, we have $\alpha = P(x \geq 14 \mid p = .5)$ for a binomial variable with $n = 20$. In Table B1, we use $n = 20$ and $p = .5$ to find

$$\alpha = P(x \geq 14 \mid p = .5) = .0577.$$

Similarly, R_2: $\{x \mid x \leq 6\}$ yields

$$\alpha = P(x \leq 6 \mid p = .5) = .0577,$$

and R_3: $\{x \mid x \leq 5 \quad \text{or} \quad x \geq 15\}$ yields

$$\alpha = P(x \leq 5 \mid p = .5) + P(x \geq 15 \mid p = .5)$$

$$= .0207 + .0207 = .0414.$$

The three preceding rejection regions, R_1, R_2, and R_3, are examples of approximate 5% level rejection regions for the types illustrated in Eq. (8.14).

Now we present Table 8.3 to indicate the limited choice of α values for testing H_0: $p = .5$ when using samples of size $n = 20$ and rejection regions of the type R: $\{x \mid x \leq x_c\}$. The α values are read directly from Table B1 and show that the possible choices are not unlimited, as was the case for uniform models in the preceding section.

TABLE 8.3.
Some possible α levels are shown for testing H_0: $p = .5$ for $n = 20$ and R: $\{x \mid x \leq x_c\}$.

x_c	0	1	2	3	4	5	6	7	8
α	.0000	.0000	.0002	.0013	.0059	.0207	.0577	.1316	.2517

Tables similar to Table 8.3 could be made to show the possible choices for α when using R: $\{x|x \geq x_c\}$ or R: $\{x|x \leq x_{c1}$ or $x \geq x_{c2}\}$ to test H_0: $p = .5$. We note also that other rejection regions (e.g., $x \leq 9$, etc.) could be added to Table 8.3, but, for most practical purposes, we choose tests having α no larger than .10. Some further remarks concerning the choice of α are given in Section 8.3.

EXAMPLE 6: Find α when a sample of size $n = 20$ and R: $\{x|x \leq 3\}$ are being used to test H_0: $p = .3$, where p is the proportion of housewives who have tried the latest static-free laundry softener in their dryers.

SOLUTION: When entering Table B1 with $n = 20$ and $p = .3$, we find

$$\alpha = P(x \leq 3|p = .3) = .1071.$$

Thus, the outlined test is a 10.7% test. Using the designated R yields the only test which has α approximately equal to 10%.

EXAMPLE 7: Suppose the marketing specialist in charge of obtaining consumer reaction to the laundry product in Example 6 questions a sample of 50 users and classifies the responses as favorable and not favorable. Let X denote the number of favorable reactions observed. Find a rejection region with $\alpha \approx .05$ for testing H_0: $p = .6$ against H_1: $p > .6$, where p is the true proportion of favorable reactions among users.

SOLUTION: When entering Table B1 with $n = 50$ and $p = .6$, we find

$$P(x \geq 36|p = .6) = .0540.$$

Therefore, R: $\{x|x \geq 36\}$ is a rejection region with $\alpha \approx .05$.

REMARKS: For the preceding discussion and examples, the primary concern is to determine the probability of a type I error, α. Notice that to determine α, we use the value of p specified by the null hypothesis and the given rejection region, R. However, to find a rejection region, R, for which α is approximately as specified, we use the alternative hypothesis, H_1, to determine whether a one-sided or a two-sided R is desired. For a one-sided region, we find the critical point, x_c, which omits approximately (100α)% of the binomial distribution having p as given by H_0. However, for a two-sided region, we seek two critical points, x_{c1} and x_{c2}, such that x_{c1} and x_{c2} omit approximately $[100(\alpha/2)]$% from the lower and upper tails, respectively, of a binomial distribution with p as given by H_o. In summary, discussions of α and R depend on the value of p as specified by H_o and depend on whether H_1 is one- or two-sided. However, when we consider the determination either of the power, $(1 - \beta)$, of a test or of β, the probability of a type II error, then H_1 must specify one value of p as an alternative to the value specified by H_o. The following examples illustrate the techniques for calculating $(1 - \beta)$ when testing hypotheses about p for binomial models.

EXAMPLE 8: A sample of size 50 is to be used to test H_o: $p = .3$ against H_1: $p = .5$ with R: $\{x|x \geq 21\}$. Find α and $(1 - \beta)$.

SOLUTION: To find α we need the probability that X, a binomial variable with $n = 50$ and $p = .3$, is greater than or equal to 21. Thus, by Table B1 we find

$$\alpha = P(X \geq 21|p = .3) = .0478.$$

Similarly, by definition of $(1 - \beta)$ and Table B1, we have

$$(1 - \beta) = P(X \geq 21|p = .5) = .8987.$$

In summary, a sample of size 50 and R: $\{x|x \geq 21\}$ provide a test of H_o: $p = .3$ against H_1: $p = .5$, which has $\alpha = .0478$ and $(1 - \beta) = .8987$. Thus, we have approximately a 5% risk for a type I error and approximately a 10% risk for a type II error.

EXAMPLE 9: Find a rejection region, R, for testing H_o: $p = .5$ against H_1: $p = .4$ such that $\alpha \approx .05$. Calculate the power of the test. Assume $n = 50$.

SOLUTION: By examining Table B1 for $n = 50$ and $p = .5$, we find that

$$P(X \leq 19|p = .5) = .0595.$$

So we select R: $\{x|x \leq 19\}$ as the desired rejection region because the alternate value of p is less than .5, and $\alpha = .0595$ is as close as possible to the designated .05-level test. Again we refer to Table B1 with $n = 50$ and $p = .4$ to find

$$(1 - \beta) = P(X \leq 19|p = .4) = .4465.$$

So the test defined by R has an α risk of .0595 and power equal to .4465.

The two preceding examples help to illustrate how the power, $(1 - \beta)$, of a test depends on H_1. In fact, the power is very sensitive to shifts in the true value of p. In Example 8, the alternative value (p = .5) of p is .2 larger than the null value (p = .3) of p, and the power of an approximate 5% test is about .90. When the difference between the alternate value (p = .4) and the null value (p = .5) of p is reduced to .1 in Example 9, however, the power of an approximate 5% test is only .45. That characteristic of power is also apparent in the work with power curves for the uniform models in the section on page 323. That is, as the true value of a parameter (either θ or p) varies away from the value specified by H_o, the power increases.

We have considered the power for a one-point alternative hypothesis when testing H_o: $p = p_o$, and those techniques can be used to evaluate the power curve for testing H_o: $p = p_o$ against any appropriate alternative hypothesis, say H_1: $p < p_o$, H_1: $p \neq p_o$. To find the exact power curve we would need to evaluate the probability of the rejection region for every possible value of p under the alternative hypothesis, H_1. However, we can get a good approximation by using Table B1 to find the power for a selected set of p values and sketch a curve to follow those points. The next example illustrates this idea.

EXAMPLE 10: Consider the coin-tossing Experiment 4 at the beginning of this section. Let p denote the true probability of a head on any toss. Use R: $\{x \mid x \leq 5$ or $x \geq 15\}$ and a sample of size 20.
(a) Graph the power function for the test of H_o: $p = .5$ against H_1: $p \neq .5$.
(b) If a sample of 20 tosses of a coin shows 12 heads, is there any reason to suspect that the coin is not fair?

SOLUTION: (a) For the indicated choices of p, we obtain the values of $(1 - \beta)$ as given from Table B1 with n = 20.

p	.1	.2	.3	.4	.5	.6	.7	.8	.9
$1 - \beta$	.9887	.8042	.4165	.1272	.0414	.1272	.4165	.8042	.9887

Now we can graph $(1 - \beta)$ against p and sketch a curve between those choices of p to approximate the power curve.

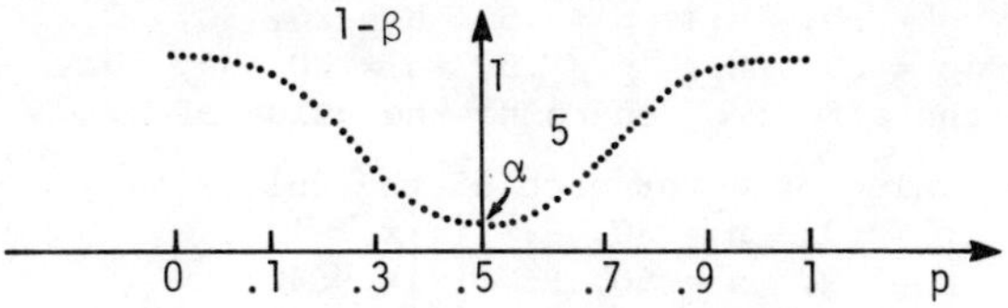

Shown is a graph of the power curve for R: $\{x \mid x \leq 5 \text{ or } x \geq 15\}$.

(b) To test the fairness of the coin we conduct the following statistical test:

H_0: $p = .5$,

H_1: $p \neq .5$,

R: $\{x \mid x \leq 5 \quad \text{or} \quad x \geq 15\}$ with $\alpha = .0414$.

The test statistic is $x = 12$. Decision: Accept H_0 because $12 \notin R$. Conclusion: We have no reason to suspect that the coin is biased.

If a more complete set of binomial distributions than provided by Table B1 is needed for calculating power functions or selecting rejection regions, reference should be made to either the National Bureau of Standards [2] or Romig [3].

EXERCISES

1. Find R to define an approximate 5% test for H_0: $p = .5$ against H_1: $p > .5$ for Experiment 1 at the beginning of the section on page 334. Sketch the power curve for the test.

2. Use data you have collected for Experiment 1 and Exercise 1 to assess the candidate's chances for election.

3. State appropriate null and alternative hypotheses for the experiments outlined at the beginning of the section on page 334. Find approximate 5% tests for each experimental situation. Use any available data to test your hypotheses.

4. A bent coin is tossed 50 times and shows 31 heads. Is there any reason to doubt that the probability of a head occurring on any toss of the coin is .5? (Use $\alpha \approx .10$.)

5. Find rejection regions approximately of size 5% for the following testing situations:

5. (a) H_0: $p = .5$, H_1: $p > .5$, $n = 20, 30, 50$.
   (b) H_0: $p = .5$, H_1: $p < .5$, $n = 20, 30, 50$.
   (c) H_1: $p = .5$, H_1: $p \neq .5$, $n = 20, 30, 50$.
   What is the effect of changing the value of n?

6. Find the value of α for each of the following situations:
   (a) H_0: $p = .3$, $n = 30$, R: $\{x \mid x \geq 13\}$.
   (b) H_0: $p = .9$, $n = 50$, R: $\{x \mid x \leq 40\}$.
   (c) H_0: $p = .7$, $n = 20$, R: $\{x \mid x \leq 9 \text{ or } x \geq 18\}$.

7. Find the power for each of the following tests:
   (a) H_0: $p = .4$, H_1: $p = .6$, $n = 50$, R: $\{x \mid x \geq 26\}$.
   (b) H_0: $p = .4$, H_1: $p = .6$, $n = 30$, R: $\{x \mid x \geq 17\}$.
   (c) H_0: $p = .4$, H_1: $p = .6$, $n = 20$, R: $\{x \mid x \geq 12\}$.
   What is the effect of changing n?

8. An insecticide is advertised as being at least 90% effective. When applied to a sample of 50 insects, it kills 41. Is there any reason to doubt the advertisement? (Use $\alpha \approx .01$.)

9. Mr. Representative finds that a sample of 50 voters yields 28 who favor his re-election. Is it reasonable to believe that at least 50% of the voters favor his re-election? (Use $\alpha \approx .05$.)

10. A marketing specialist believes that at least 60% of all customers prefer the new 12-pound box of Clean-All detergent to the old 10-pound box. If a sample of 30 customers shows only 15 who prefer the new, larger box, should we doubt the specialist's claim? (Use $\alpha \approx .10$.)

11. Identify one or more questions in your field that could be answered by the hypothesis testing techniques of this section. Collect data and perform the tests wherever possible.

8.3 CONCLUSIONS AND A LOOK AHEAD

This chapter presents numerous definitions and ideas concerning the general techniques of hypothesis testing. In summary, we begin by stating a null hypothesis, H_0, to be tested. Then we continue by specifying an appropriate, alternative hypothesis, H_1. Next, an acceptable value of α is designated, and that leads to a rejection region, R. A test statistic, t, is then evaluated and compared with the rejection region. If $t \in r$, we reject H_0; otherwise, we accept H_0. Finally, appropriate conclusions are stated.

One part of the procedure that needs further clarification is the choice of α. We emphasize again that choosing α is the experimenter's responsibility. It represents the risk the experimenter is willing to assume for rejecting H_0 when it is true. Usually the experimenter has some investment in the null hypothesis, H_0, and, therefore, is eager to protect it from being rejected when it is true. The investment in H_0 may be in terms of a warranty for a product, the performance of a product, the voter's choice for a political office, the cost of adjusting machines, or any number of other items. To reflect that interest in H_O, the experimenter usually chooses α to be .01, .05, or .10 to reflect a willingness to take a 1-in-100, 1-in-20, or 1-in-10 chance of wrongly rejecting H_O. Thus, the most frequent choices of α are .01, .05, and .10. Yet there are situations, such as testing a new drug for harmful effects, when a much smaller risk is required. So in the remaining chapters, we may occasionally run a test with α = .001 or even smaller values of α. We stress that the choice of α is arbitrary, but most of the tables of critical values used in the following chapters are for the more popular values. Finally, we remark that when using a discrete test statistic, such as X for the binomial models in the section on page 334, there is a limited number of choices of exact α values.

Now that we have motivated and illustrated the basic methods of testing statistical hypotheses, the next step is to present testing techniques for specific problems of interest. In such cases, we present the methods of analysis used for testing a variety of hypotheses, but we do not justify the results. Complete justification requires rather sophisticated mathematics and is better left for the more mathematical courses in the theory of statistics. The testing procedures are divided into three chapters to emphasize the different types of experimental problems being solved.

EXERCISES

1. Specify an experimental situation for which you believe an α = .001 might be required for testing a hypothesis.
2. Identify several experimental situations that might be investigated by tests of statistical hypotheses. What value of α would be appropriate for each situation?

8.4 REFERENCES

[1] Newman, J. R. 1960. *The world of mathematics*. Vol. 3, New York: Simon and Schuster, pp. 1512-1521.

[2] National Bureau of Standards. 1949. Applied Math. Series No. 6. *Tables of the binomial probability distribution*. Washington, D.C.: U.S. Govt. Printing Office.

[3] Romig, H. G. 1953. *50-100 binomial tables*. New York: Wiley.

8.5 REVIEW EXERCISES

1. Explain in your own words the meaning of the following terms:
   (a) null hypothesis
   (b) alternative hypothesis
   (c) type I error
   (d) type II error
   (e) power
   (f) statistical test
   (g) test statistic
   (h) rejection region
   (i) α and $(1 - \alpha)$
   (j) β and $(1 - \beta)$.

2. Assume that one observation, x, is obtained from the uniform density,

$$f(x) = 1, \quad \Theta - .5 < x < \Theta + .5.$$

   (a) Graph f(x) for $\Theta = 2$ and $\Theta = 2.5$.
   (b) Find a 5% rejection region for testing H_o: $\Theta = 2$ against H_1: $\Theta > 2$.
   (c) If $x = 2.47$, test H_o: $\Theta = 2$ vs. H_1: $\Theta = 2.5$ (with $\alpha = .05$).
   (d) What is the power of your test in part (c)?
   (e) Indicate α, $(1 - \alpha)$, β, and $(1 - \beta)$ on the graph for part (a).
   (f) If $x = 2.47$, test H_o: $\Theta = 2$ vs. H_1: $\Theta \neq 2$ (with $\alpha = .05$).

3. The Ready-To-Go Auto Agency claims that no more than 10% of their used cars are returned for repairs during the first week (7 days) following delivery. If a sample of 20 customers results in 4 who returned their purchases within the first week, is there reason to doubt the agency's claim? (Use $\alpha \approx .05$.)

4. The inspector on the production line of 60-watt light bulbs suspects that something is out of adjustment and that more than the usual 1% defective bulbs are being produced. A sample of 50 bulbs is tested, and three fail. Use a 10% (approximately) test to see whether or not the data support the inspector's suspicion.

5. Determine α and $(1 - \beta)$ whenever possible for the following tests on binomial models:
   (a) H_o: $p = .4$, H_1: $p = .7$, $n = 30$, R: $\{x | x \geq 16\}$.
   (b) H_o: $p = .5$, H_1: $p \neq .5$, $n = 50$, R: $\{x | x \leq 17 \text{ or } x \geq 33\}$.
   (c) H_o: $p = .9$, H_1: $p = .8$, $n = 50$, R: $\{x | x \leq 40\}$.

6. Determine appropriate rejection regions, R, for the following tests on binomial models:
   (a) H_o: $p = .1$, H_1: $p > .1$, $n = 30$, $\alpha \approx .10$.
   (b) H_o: $p = .5$, H_1: $p \neq .5$, $n = 10$, $\alpha \approx .10$.
   (c) H_o: $p = .9$, H_1: $p < .9$, $n = 50$, $\alpha \approx .10$.

CHAPTER 9

Chi-Square Test Procedures

9.1 INTRODUCTION

The basic philosophy for testing statistical hypotheses is presented in the preceding chapter. Briefly, a statistical test uses a *test statistic* to either reject H_O or not reject H_O. The particular decision is determined by comparing a calculated value of the test statistic with a tabled percentile value obtained from the sampling distribution of the statistic. Thus, to build on the previous material, we must specify test statistics and their sampling distributions for specific hypotheses. Of course, the particular hypothesis being tested depends on the type of question being answered. However, there are many situations for which the hypothesis of interest can be tested by a test statistic with a sampling distribution that is approximately a chi-square density, as presented in Section 5.5. Therefore, the purpose of this chapter is to introduce a set of chi-square procedures that are useful in answering questions such as the following:

1. Is the proportion of voters favoring additional support to higher education equal to .60?
2. Is the proportion of college women who smoke equal to the proportion of college men who smoke?

3. Are the two variables, political philosophy and party affiliation, independent?
4. Are the data in a given sample normally distributed?

For each situation introduced, a test statistic is presented which is approximately a chi-square variable with specified degrees of freedom (d.f.). Then the chi-square percentile values in Table E2 are used to state the necessary rejection regions.

9.2 TESTS OF HYPOTHESES FOR BINOMIAL MODELS

TESTS OF H_0: $p = p_0$ FOR ONE BINOMIAL MODEL: Section 8.2 presents exact test procedures for testing whether or not the parameter, p, in a binomial model is a specific value, say p_0. Those exact tests require extensive binomial tables (i.e., tables for all combinations of n and p) which are not always available. So we present an approximate procedure based on a chi-square test statistic. Our interest is in testing H_0: $p = p_0$ against H_1: $p \neq p_0$.

Suppose n independent observations of a binomial variable yield X successes. Under H_0, $E(X) = np_0$ and $V(X) = np_0(1 - p_0)$. Furthermore, the normal approximation to the binomial in the section on page 213 implies that

$$Z = (X - np_0)/\sqrt{np_0(1 - p_0)}$$

is approximately a N(0, 1) variable. Therefore, Eq. (5.38) shows that

$$Y = Z^2 = (X - np_0)^2/np_0(1 - p_0) \qquad (9.1)$$

is approximately a $\chi^2(1)$ variable. Since any departure of X from its expected value, np_0, causes an increase in Y, the appropriate α-level rejection region for testing H_0 against H_1 is

$$R: \ \{y \mid y \geq \chi^2(1, \alpha)\}, \qquad (9.2)$$

where $\chi^2(1, \alpha)$ is obtained from Table E2.

Before providing illustrative examples of the testing procedures, we need to specify some binomial experiments through which data can be collected. Any data obtained can then be compared to the data presented in the subsequent examples.

EXPERIMENT 1: Flip a coin 20 times, and record X, the number of heads observed. Hypothesize a value for p, the true probability of heads appearing on the coin tossed.

EXPERIMENT 2: Roll a die 24 times, and record X, the number of times the event 1 or 2 occurs. Hypothesize a value for p, the true probability for the event designated.

EXPERIMENT 3: Select some topic of interest (e.g., a political contest, an athletic event, a community project, a campus question, etc.), and record the number of people, X, whose response is yes to your question about the topic when 20 people are questioned. Hypothesize a value of p for the true proportion of people in the designated population who would, if given the opportunity, respond yes.

Table 9.1 presents results for three independent repetitions of Experiments 1 and 2. These results are used in the subsequent examples.

TABLE 9.1.
Three sets of independent data are given for Experiments 1 and 2.

Experiment 1				Experiment 2			
Outcome	Set 1	Set 2	Set 3	Outcome	Set 1	Set 2	Set 3
Heads	9	7	9	(1,2)	6	6	10
Tails	11	13	11	(3,4,5,6)	18	18	14
Total	20	20	20	Total	24	24	24

EXAMPLE 1: Use $\alpha = .05$ and data set 1 for Experiment 1 in Table 9.1 to test whether or not the coin being used is fair.

SOLUTION:

$$H_o: \; p = .5,$$

$$H_1: \; p \neq .5,$$

$$\alpha = .05.$$

R: $\{y | y \geq 3.84\}$, where $\chi^2(1, .05) = 3.84$ is obtained from Table E2. Using H_o, we find $E(X) = 20(.5) = 10$. Thus,

$$y = (9 - 10)^2/(20)(.5)(.5) = .2.$$

Since $.2 < 3.84$, we accept H_0 and conclude that the coin is fair.

EXAMPLE 2: Use $\alpha = .10$ and data set 1 for Experiment 2 in Table 9.1 to test whether or not the probability of observing a 1 or 2 is 1/3.

SOLUTION:

$$H_0: \ p = 1/3,$$

$$H_1: \ p \neq 1/3,$$

$$\alpha = .10.$$

R: $\{y | y \geq 2.71\}$, where $\chi^2(1, .10) = 2.71$ is obtained from Table E2. Using H_0, we find $E(X) = 24(1/3) = 8$. Thus,

$$y = (6 - 8)^2/(24)(1/3)(2/3) = .75.$$

Since $.75 < 2.71$, we accept H_0 and conclude that 1/3 is a reasonable value for p.

EXAMPLE 3: A random sample of 50 students produces 18 who say they agree with the editorial position of the *MYU Campus News*. Is there any reason to doubt that the true proportion, p, of students having opinions similar to those of the editorial staff is .5? (Use $\alpha = .05$.)

SOLUTION:

$$H_0: \ p = .5,$$

$$H_1: \ p \neq .5,$$

$$\alpha = .05.$$

R: $\{y | y \geq 3.84\}$. Under H_0, $E(X) = 25$. Therefore,

$$y = (18 - 25)^2/(50)(.5)(.5) = 98/25 = 3.92.$$

Since $3.92 > 3.84$, we reject H_0. The reason for y being too large is that there is a shortage of those who agree (i.e., $X < 25$). So we conclude that fewer than 50% of the students agree with the editorial view of the *MYU Campus News*.

EXERCISES

1. Combine the 3 sets of coin-tossing data in Table 9.1 to obtain a sample of 60 tosses. Use the combined sample to test H_0: $p = .5$ against H_1: $p \neq .5$ with $\alpha = .05$. Is there any reason to suspect that the coin is biased?

2. Test the fairness of any coin you have used to generate data for the coin-tossing experiment (Experiment 1).

3. Combine the 3 sets of die-rolling data in Table 9.1 to obtain a sample of size 72. Use that data to test H_o: $p = 1/3$ against H_1: $p \neq 1/3$ with $\alpha = .10$.

4. Use any data you have collected for Experiment 2 to test whether or not the event 1 or 2 occurs with probability 1/3 on the die used.

5. Formulate hypotheses and conduct tests for any other binomial data you have collected.

6. A sample of 28 coeds at MYU are asked to classify their hair color as blonde or not blonde. If 7 designated themselves as blonde, is there any reason to doubt the hypothesis that H_o: $p = .40$, where p denotes the proportion of blonde coeds on campus? (Use $\alpha = .10$.)

7. A sample of 100 students is questioned to test the hypothesis that 10% of all students at MYU are left-handed. The sample includes 17 left-handed individuals. Test the hypothesis, and draw appropriate conclusions.

8. In Exercise 7, find the limits L and U for X, the number of left-handed individuals in a sample of 100, such that $X \leq L$ or $X \geq U$ leads to the rejection of H_o: $p = .10$, but $L < X < U$ is acceptable.

TESTS OF H_o: $p_1 = p_2 = \ldots = p_c$ FOR c BINOMIAL MODELS: There are situations in which the primary purpose of a statistical investigation is to ascertain whether or not a set of independent binomial samples is coming from binomial models with the same value for the parameter, p. For example, suppose that p denotes the proportion of students in a particular college who voted in the last election, and we wish to test whether or not p is the same for a set of c colleges. To investigate the problem we observe a random sample of students from each college and classify each student as either a voter or a nonvoter. Then we use X_i to denote the number of voters in the sample of n_i students interviewed from the ith college for $i = 1, 2, \ldots, c$. We assume that the X_i are independent variables such that $X_i \sim b(n_i, p_i)$, 1, 2, ..., c. Then we want to test

$$H_o: p_1 = p_2 = \ldots = p_c$$

against H_1: Some of the p_i's are unequal.

The data for any experiment of the type just described are $X_1, X_2, \ldots, X_c$. Since H_o does not provide a number (value) for the common proportion, say p_o, we must use the data to estimate p_o. When H_o is true, the best estimate of p_o is provided by pooling all of the samples together to obtain

$$\hat{p}_o = \sum_{i=1}^{c} X_i / \sum_{i=1}^{c} n_i .$$

Then the expected value and variance of each X_i are estimated by $n_i\hat{p}_o$ and $n_i\hat{p}_o(1 - \hat{p}_o)$, respectively. Thus, it again follows that

$$(X_i - n_i\hat{p}_o)^2/n_i\hat{p}_o(1 - \hat{p}_o)$$

is approximately a $\chi^2(1)$ variable. We note that the definition of $\hat{p}_o$ implies that

$$\sum_{i=1}^{c} (X_i - n_i\hat{p}_o) = 0.$$

The consequence of this restriction is that

$$Y = \sum_{i=1}^{c} (X_i - n_i\hat{p}_o)^2/n_i\hat{p}_o(1 - \hat{p}_o) \qquad (9.3)$$

is approximately a chi-square variable with (c - 1) d. f. rather than c d.f., as might be expected from Eq. (5.38). Therefore, the appropriate α-level rejection region for testing H_o against H_1 is

$$R: \{y|y \geq \chi^2(c - 1, \alpha)\}, \qquad (9.4)$$

where $\chi^2(c - 1, \alpha)$ is obtained from Table E2.

EXAMPLE 4: Suppose the following data represent the number of voters in random samples from four different colleges. Test H_o: $p_1 = p_2 = p_3 = p_4$ against H_1: Some of the p_i's are unequal. Use $\alpha = .05$.

College	1	2	3	4
n_i	20	50	20	10
x_i	10	30	5	5

SOLUTION: Since c = 4, R: $\{y \mid y \geq \chi^2(3, .05) = 7.81\}$. By pooling the data together, we find

$$\hat{p}_o = 50/100 = .5.$$

Then $E(X_1) = 10$, $E(X_2) = 25$, $E(X_3) = 10$, and $E(X_4) = 5$. Therefore,

$$y = (10 - 10)^2/5 + (30 - 25)^2/12.5$$

$$+ (5 - 10)^2/5 + (5 - 5)^2/2.5$$

$$= 0 + 2 + 5 + 0 = 7.$$

Since $y = 7 < 7.81$, we accept H_o and conclude that the true proportion of voters is the same for the four colleges.

EXERCISES

1. A random sample of 25 women reveals 5 who smoke, while a random sample of 50 men shows 16 smokers. Test H_o: $p_w = p_m$.
2. Use the data in Table 9.1 to test whether or not the proportion of heads is the same for the 3 sets of coin tosses. H_o: $p_1 = p_2 = p_3$.
3. Use the data in Table 9.1 to test whether or not the proportion of #(1, 2) is the same for the 3 sets of dice rolls. H_o: $p_1 = p_2 = p_3$.
4. Two samples of 100 flies are sprayed with two different insecticides. The first insecticide kills 75 flies, while the second insecticide kills 95 flies. Test H_o: $p_1 = p_2$ with $\alpha = .10$.

9.3 TESTS OF HYPOTHESES FOR MULTINOMIAL MODELS

Section 4.5 introduces a multinomial model as a natural extension of a binomial model. In those models, we suppose that the outcomes from n independent trials of an experiment are classified into k categories. Furthermore, we suppose that p_i, i = 1, 2, ..., k denotes the probability of any trial producing an outcome that falls into the ith category. Then we let X_i, i = 1, 2, ..., k denote the number of outcomes from n trials observed in the ith category. Clearly,

$$\sum_{i=1}^{k} X_i = n$$

and

$$\sum_{i=1}^{k} p_i = 1.$$

With the foregoing notation and model description, our concern is to test

$$H_o: \ p_i = p_{io}, \ i = 1, 2, \ldots, k$$

against

$$H_1: \ p_i \neq p_{io},$$

for at least one value of $i = 1, 2, \ldots, k$. For the values of p_i specified in H_o, we have

$$E(X_i) = np_{io}, \ i = 1, 2, \ldots k.$$

In 1900, Karl Pearson, a British mathematician and statistician, studied the problem we have outlined and showed that the statistic

$$Y = \sum_{i=1}^{k} (X_i - np_{io})^2/np_{io} \qquad (9.5)$$

is approximately a chi-square variable with $(k - 1)$ d. f.

Notice that the test statistic in Eq. (9.5) is similar to the earlier chi-square statistics in Eqs. (9.1) and (9.3) in that Y becomes large as the discrepancies between X_i and $E(X_i)$ increase in either direction. Thus, the appropriate α-level test for testing H_o against H_1 is defined by the rejection region

$$R: \ \{y | y \geq \chi^2(k - 1, \alpha)\}. \qquad (9.6)$$

Data from the following multinomial experiments are used in the subsequent examples and exercises to illustrate the use of Eqs. (9.5) and (9.6).

EXPERIMENT 1: Flip 2 coins simultaneously 24 times, and record X_1 to be the number of times 2 heads occur, X_2 to be the number of times 1 head and 1 tail occur, and X_3 to be the number of times 2 tails occur. Suggest an appropriate hypothesis, H_o, to be tested when the coins are assumed to be fair.

EXPERIMENT 2: Roll a die 36 times, and record X_1 to be the number of times 1 is observed, X_2 to be the number of times 2 or 3 is observed, and X_3 to be the number of times 4, 5, or 6 is observed. Suggest an appropriate H_o when the die is assumed to be fair.

EXPERIMENT 3: Select a sample of 30 coeds, and record X_1 = #(blondes), X_2 = #(brunettes), X_3 = #(redheads), X_4 = #(other than blonde, brunette, or redhead).

Three data sets for Experiments 1 and 2 are given in Table 9.2. These data are used in the subsequent examples and exercises and can be compared to any personal data available.

TABLE 9.2.
Three sets of independent data are given for Experiments 1 and 2.

Experiment 1				Experiment 2			
Outcome	Set 1	Set 2	Set 3	Outcome	Set 1	Set 2	Set 3
(H,H)	5	6	4	(1)	2	3	9
(H,T) or (T,H)	12	9	14	(2 or 3)	16	15	12
(T,T)	7	9	6	(4,5, or 6)	18	18	15
Total	24	24	24	Total	36	36	36

EXAMPLE 5: Use $\alpha = .05$ and data set 3 for Experiment 1 in Table 9.2 to test the hypothesis, H_o: $p_1 = 1/4$, $p_2 = 1/2$, $p_3 = 1/4$.

SOLUTION: H_o: $p_1 = 1/4$, $p_2 = 1/2$, $p_3 = 1/4$.

H_1: At least one p_i is not as stated in H_o.

$\alpha = .05$.

R: $\{y | y \geq 5.99\}$, where 5.99 is $\chi^2(2, .05)$, as given in Table E2.

Under H_o, $E(X_1) = (1/4)(24) = 6$, $E(X_2) = (1/2)(24) = 12$, and $E(X_3) = (1/4)(24) = 6$. Therefore,

$$y = (4 - 6)^2/6 + (14 - 12)^2/12 + (6 - 6)^2/6 = 1.00.$$

Since $1.00 < 5.99$, we have no reason to doubt the probabilities specified by H_o.

EXAMPLE 6: Use $\alpha = .10$ and data set 3 for Experiment 2 in Table 9.2 to test the hypothesis, H_o: $p_1 = 1/6$, $p_2 = 1/3$, $p_3 = 1/2$.

SOLUTION: H_o: $p_1 = 1/6$, $p_2 = 1/3$, $p_3 = 1/2$.

H_1: At least one p_i is not as specified by H_o.

$\alpha = .10$.

R: $\{y | y \geq 4.61\}$, where 4.61 is $\chi^2(2, .10)$, as given in Table E2.

Under H_o, $E(X_1) = (1/6)(36) = 6$, $E(X_2) = (1/3)(36) = 12$, and $E(X_3) = (1/2)(36) = 18$. Therefore,

$$y = (9 - 6)^2/6 + (12 - 12)^2/12 + (15 - 18)^2/18 = 2.00.$$

Since $2.00 < 4.61$, there is no reason to doubt the probabilities given by H_o.

EXERCISES

1. Pool the 3 data sets for Experiment 1 in Table 9.2 to test H_o: $p_1 = .25$, $p_2 = .50$, $p_3 = .25$. (Use $\alpha = .10$.)
2. Use any data you have collected for Experiment 1 to test H_o: $p_1 = .25$, $p_2 = .50$, $p_3 = .25$. (State α.)
3. Pool the 3 data sets for Experiment 2 in Table 9.2 to test H_o: $p_1 = 1/6$, $p_2 = 1/3$, $p_3 = 1/2$. (Use $\alpha = .05$.)
4. Use any data you have collected for Experiment 2 to test H_o: $p_1 = 1/6$, $p_2 = 1/3$, $p_3 = 1/2$. (State α.)
5. Three candidates, A, B, and C, are opponents for the office of senator. A sample of 100 voters shows 45 for A, 25 for B, and 30 for C. If $\alpha = .05$, is there any reason to doubt the hypothesis, H_o: $p_A = .40$, $p_B = .40$, and $p_C = .20$?
6. State and test appropriate hypotheses for any data you have collected for Experiment 3 at the beginning of this section.
7. A die is rolled 72 times and shows 7 ones, 15 twos, 9 threes, 10 fours, 16 fives, and 15 sixes. Is there any reason to doubt that the 6 faces are equally likely to occur on any given roll of the die?
8. It has been hypothesized that the percentages of coeds at MYU who are either blonde, brunette, or neither blonde nor brunette are 40, 30, and 30, respectively. If a sample of 28 coeds shows 7 blondes, 16 brunettes, and 5 other than blonde or brunette, are the above percentages acceptable?

9. (a) Write a random sequence of 50 digits using the 10 digits, 0, 1, 2, ..., 9. Form a frequency distribution of the digits.
   (b) Select a set of 50 random digits from Table A2. Form a frequency distribution of the results.
   (c) Calculate the chi-square test statistics, y_a and y_b, for the data in parts (a) and (b), and test H_o: $p_1 = .1$, $p_2 = .1$, ..., $p_{10} = .1$ for both sets of data.

10. As a class exercise, have each student report the values of y_a and y_b for Exercise 9. Do the two sets of data appear to be alike? Compare the means and variances for the two data sets.

11. From the data in Exercise 10, count the number of pairs of observations (y_a, y_b) for which $y_a < y_b$, and denote that number by X. Then let n denote the number of pairs of observations (students), and use n - X as the number of pairs (y_a, y_b) for which $y_a \geq y_b$. Use those data to test H_o: $p = .5$, where p is the probability that $y_a < y_b$. Is there any reason to suspect that the random digits from the table are more random than those written by the class?

12. Since a multinomial model with k = 2 is a binomial model, we compare the test statistics given in Eqs. (9.1) and (9.5). For k = 2, Eq. (9.5) reduces to

$$Y_1 = (X_1 - np_{10})^2/np_{10} + [(n - X_1) - n(1 - p_{10})]^2/n(1 - p_{10})$$

because $X_1 + X_2 = n$ and $p_{10} + p_{20} = 1$. In the same notation, Eq. (9.1) is

$$Y_2 = (X_1 - np_{10})^2/np_{10}(1 - p_{10}).$$

Show that the two tests are equivalent when k = 2 by showing $Y_1 = Y_2$.

9.4 CONTINGENCY TABLES

In this section we introduce another situation for which inference can be made by using a test statistic which is approximately a χ^2 variable. To illustrate the general ideas involved, let us investigate a hair coloring manufacturer's claim that blondes have more fun. Our study begins by formulating a more precise statistical hypothesis to be tested. For example, we can test H_o against the indicated H_1 as follows:

H_o: Fun status is independent of hair color.

H_1: Fun status is dependent on hair color.

If we reject H_0 in favor of H_1 because the proportion of blondes having more fun is higher than the proportion of nonblondes having more fun, then the manufacturer's claim appears to be reasonable. Otherwise, we doubt the claim. Before collecting data to be used in testing H_0, we must establish categories (values) for the two variables, fun status and hair color. Since the claim is made for blondes only, it is reasonable to include two categories, blonde and not blonde, for hair color. Now categories for fun status are a bit more nebulous, but suppose we agree to use the categories, more fun, average fun, and less fun. By observing those rules of play, we can now collect some empirical data from the following experiment.

EXPERIMENT: Determine the hair color and fun status for 50 college women, and record the frequencies for the 6 cells in the two-way table below.

Hair color	Fun status		
	More fun	Average fun	Less fun
Blonde			
Not blonde			

Note: The sum of the frequencies in the 6 cells is 50. Thus, the 50 women are classified with respect to the categories of both variables simultaneously.

There are numerous problems that fall into the format of the preceding example. A *political scientist* may wish to test whether or not political attitudes are independent of party affiliation. A *marketing specialist* may wish to test whether or not the volume of sales for a product is independent of the type of packaging used. An *entomologist* may wish to test whether or not the insect damage is independent of variety of corn. An *educator* may wish to test whether or not the verbal and quantitative scores of students are independent. A *miller* may wish to test whether or not the mixability of an additive in a ration is independent of the particle size of the additive. A *psychologist* may wish to test whether or not a child's aggression classification is independent of the hours of weekly television violence viewed. A *labor relations specialist* may wish to test whether or not attitudes toward unionization are independent of the rank of college faculty members. *These are only a few illustrations of studies where a set of n items or individuals is simultaneously classified with respect to the categories of two variables.* So the next step is to establish some general notation for representing and analyzing data of that type.

To begin we denote one variable as the *row variable* and use $R_1, R_2, \ldots, R_r$ to denote the possible categories for that variable. The other variable is called the *column variable*, and we use $C_1, C_2, \ldots, C_c$ to denote its categories. Notice that the categories of a variable may be discrete numbers, intervals of real numbers, qualitative values such as colors, or any variety of other possible classes.

After the variables and the respective categories are designated, a sample of n items is observed and classified into the $r \times c$ different combinations of row and column cells. Those cells are illustrated by the two-way table shown in Figure 9.1. In that figure, n_{ij} is the number of items that fall into categories, R_i and C_i, with respect to the two variables being studied. Also, $n_{i.}$ (read as n, subscript i dot) and $n_{.j}$ (read as n, subscript dot j) denote the marginal frequencies of R_i and C_j, respectively. Thus, there are n_{ij} items exhibiting both properties, R_i *and* C_j; there are $n_{i.}$ items exhibiting the property, R_i, when considering the row variable alone; and there are $n_{.j}$ items exhibiting the property, C_j, when considering the column variable alone. Notice that $n_{i.} = n_{i1} + n_{i2} + \ldots + n_{ic}$, $i = 1, 2, \ldots, r$

Categories of row variable	Categories of column variable				Row total
	C_1	C_2	...	C_c	
R_1	n_{11}	n_{12}	...	n_{1c}	$n_{1.}$
R_2	n_{21}	n_{22}	...	n_{2c}	$n_{2.}$
⋮	⋮	⋮		⋮	
R_r	n_{r1}	n_{r2}	...	n_{rc}	$n_{r.}$
Column total	$n_{.1}$	$n_{.2}$	...	$n_{.c}$	n

FIGURE 9.1. A contingency table for 2 variables shows the joint frequencies, n_{ij}, and marginal frequencies, $n_{i.}$ and $n_{.j}$.

and $n_{.j} = n_{1j} + n_{2j} + \ldots + n_{rj}$, $j = 1, 2, \ldots, c$. Therefore, a dot is used to indicate a summation over the row index, i, or the column index, j. That is, the marginal frequencies, $n_{i.}$ and $n_{.j}$, are the total frequencies for the ith row and jth column, respectively. The fact that

$$\sum_{j=1}^{c} n_{.j} = n \quad \text{and} \quad \sum_{i=1}^{r} n_{i.} = n$$

indicates that the same n items have been simultaneously classified with respect to two different variables.

The null hypothesis, H_o, of interest when studying contingency tables, is usually a statement that the classifications with respect to the row and column variables are independent. Or in terms of probability concepts as introduced in Chapter 2, we assume our sample observations in the contingency table are independent. Thus, we expect our probability estimator, n_{ij}/n, for $P(R_iC_j)$ to be approximately equal to the product of the marginal probability estimators, $n_{i.}/n$ and $n_{.j}/n$, for $P(R_i)$ and $P(C_j)$, respectively. In other words, Eq. (2.8) implies that

$$P(R_iC_j) = P(R_i)P(C_j)$$

if R_i and C_j are independent. So we expect

$$n_{ij}/n \approx (n_{i.}/n)(n_{.j}/n) \tag{9.7}$$

when the classification variables for rows and columns are independent. By multiplying each side of Eq. (9.7) by n, we see that when H_o is true,

$$n_{ij} \approx (n_{i.})(n_{.j})/n.$$

Therefore, we define the expected frequency, $E_{ij} = E(n_{ij})$, for the ith row and jth column to be

$$E_{ij} = (n_{i.})(n_{.j})/n. \tag{9.8}$$

Then in the format of Eq. (9.5), we define the test statistic,

$$Y = \sum_{i=1}^{r} \sum_{j=1}^{c} (n_{ij} - E_{ij})^2/E_{ij}, \tag{9.9}$$

where the n_{ij} are the observed cell frequencies, and the E_{ij} are the expected frequencies given by Eq. (9.8). It can be shown that when the row and column variables are independent (i.e., H_O is true), the test statistic, Y, in Eq. (9.9) is approximately a χ^2 variable with $[(r-1)\cdot(c-1)]$ d.f. Details of this result are provided in mathematical statistics books such as Freund [1]. Therefore, the α-level test of H_O against H_1 is defined by the rejection region,

$$R: \ \{y | y \geq \chi^2 [(r-1)\cdot(c-1), \alpha]\}. \qquad (9.10)$$

The next example illustrates the use of Eqs. (9.9) and (9.10).

EXAMPLE 7: The following hair color - fun status data were obtained by interviewing 110 coeds.

Hair color	Fun status			Total
	More fun	Average fun	Less fun	
Blonde	25	15	2	42
Not blonde	30	32	6	68
Total	55	47	8	110

Use the data to test the following H_O against the stated H_1:

H_o: Fun status and hair color are independent variables.

H_1: Fun status and hair color are dependent variables.

SOLUTION: If we choose $\alpha = .05$ and use $r = 2$ and $c = 3$ to give $(r-1)\cdot(c-1) = 2$ d.f. for the test statistic, then the rejection region is

$$R: \ \{y | y \geq 5.99\},$$

where $\chi^2(2, .05) = 5.99$ is taken from Table E2. Next we calculate the expected frequencies:

$E_{11} = (42)(55)/110 = 21$, $\quad E_{12} = (42)(47)/110 \approx 18$,

$E_{13} = (42)(8)/110 \approx 3$, $\quad E_{21} = (68)(55)/110 = 34$,

$E_{22} = (68)(47)/110 \approx 29$, $\quad E_{23} = (68)(8)/110 \approx 5$.

Then the calculated value for the test statistic is

$$y = (25 - 21)^2/21 + (15 - 18)^2/18 + (2 - 3)^2/3$$
$$+ (30 - 34)^2/34 + (32 - 29)^2/29 + (6 - 5)^2/5$$
$$= 2.57.$$

Since $y = 2.57 < 5.99$, we accept H_0 and conclude that fun status and hair color are *independent* variables.

EXERCISES

1. The data used in Example 7 were originally recorded in the following table:

Hair color	Fun status		
	More fun	Average fun	Less fun
Blonde	25	15	2
Brunette	14	12	3
Redhead	5	6	1
Other	11	14	2

Use $\alpha = .05$ to test H_0: Fun status and hair color are independent variables.

2. Choose an appropriate α, and use any data you have collected for the fun status and hair color experiment outlined at the beginning of Section 9.4.

3. One hundred voters have been classified with respect to their party affiliation and their political philosophy. The results are as follows:

Party affiliation	Political philosophy			Total
	Conservative	Liberal	Other	
Democrat	20	25	5	50
Republican	18	19	3	40
Other	2	6	2	10

Use $\alpha = .05$ to test whether or not party affiliation and political philosophy are independent variables.

4. Berger and Oswald [2] present the following table of data from a study of eye movements during dreams:

Eye movements	Dream content	
	Active	Passive
Active	42	16
Passive	8	23

Are dream content and eye movement independent variables? (Choose any appropriate α.)

5. A study of university faculty members has been made to investigate their attitudes toward collective bargaining. The results are classified with respect to academic rank and attitudes toward collective bargaining in the following table:

Academic rank	Attitude toward collective bargaining		
	Favored	Opposed	Undecided
Asst. prof.	40	20	10
Assoc. prof.	30	20	10
Professor	40	10	10

Are the two classifications, academic rank and collective bargaining attitude, independent? (Use $\alpha = .05$.)

6. An entomologist inspects 100 plots of corn and classifies each plot relative to variety and insect damage. The results are as follows:

Variety	Insect damage		
	0 - 20%	21 - 75%	76 - 100%
#742	10	25	5
#360	35	20	5

Is there any reason to prefer either #742 or #360? Give reasons for your answer.

7. An educator wants to know whether or not test performance is independent of sex for college students. A sample of 100 students gives the following two-way classification:

Sex	Test performance		
	0 - 60%	61 - 80%	81 - 100%
Female	5	15	20
Male	10	25	25

Use $\alpha = .10$ to help the educator answer his question.

8. Professor A. Dale Allen conducted a mail survey of the faculties of the three state colleges and the three state universities in Kansas to study attitudes toward collective bargaining. The 262 responses have been classified with respect to the two variables, rank and join bargaining organization if legal. The data are as follows:

Rank	Join bargaining organization if legal		
	Yes	No	No opinion
Instructor	34	10	8
Asst. prof.	61	9	8
Assoc. prof.	39	15	17
Professor	33	17	11

Are the two variables independent? Interpret your analysis. (Use $\alpha = .05$.)

9. King [3] presented two photographs of a coed to 106 subjects and asked them to select their favorite. The only difference between the two photographs was that one had the coed looking to the left, and the other had the coed looking to the right. The 106 subjects were also classified as left-lookers or right-lookers by their conjugate lateral eye movements (CLEM) while responding to questions. The data are as follows:

Selected photo	CLEM classification	
	Left-lookers	Right-lookers
Eyes to left	31	14
Eyes to right	20	41

Is there any reason to suspect that left-lookers prefer the photo having eyes to the left or that right-lookers prefer the photo having eyes to the right? Give reasons for your answers. (Use $\alpha = .05$.)

9.5 GOODNESS OF FIT PROCEDURES

In Chapters 4 and 5, several probability distributions are presented. Those distributions are frequently proposed as models for empirical frequency distributions. When using one of those models for an experimental situation, it is of interest to determine whether or not the model *fits* the observed data. One method for assessing the fit of a model is to use a chi-square statistic which compares observed frequencies with expected frequencies in a method similar to that used for Eq. (9.5). The procedure is outlined next.

First, the null hypothesis states the assumptions being made about the proposed model. Some possible null hypotheses are as follows: (1) H_0: X is uniformly distributed on the integers 1, 2, 3, 4. (2) H_0: X is binomially distributed with n = 5 and p = .5. (3) H_0: $X \sim N(10, 9)$. (4) H_0: $X \sim N(\mu, \sigma^2)$, where μ and σ^2 are unspecified. Note that hypotheses (1) - (3) completely specify the model by giving *the type of model* and *the values of any parameters involved*. Hypothesis (4), however, gives only the type of model (normal).

Second, the k categories into which observations of the variable are to be classified are specified. The resulting distribution yields observed frequencies, f_i, $i = 1, 2, \ldots, k$, for the k categories. If the variable is discrete, the categories are usually the possible values or sample space of the variable. For a continuous variable, the categories are selected intervals of the variable which form a partition of the sample space. We note that the intervals for a continuous variable need not have equal lengths, and that for discrete variables the tail events are sometimes combined. In this way, it is possible to obtain categories that have larger expected frequencies. Further comments on the use of chi-square procedures and the need for avoiding categories that have very small expected frequencies are delayed until Section 9.6.

Third, we calculate the probabilities, p_i, i = 1, 2, ..., k, associated with the k categories of the variable. If H_o completely specifies the model, then the techniques of Chapters 4 and 5 are used to obtain the desired probabilities. When H_o provides only the type of model and leaves parameters unspecified, we first use the data and the techniques of Chapter 7 to calculate any unspecified parameters before applying the probability techniques of Chapters 4 and 5. We note that the degrees of freedom for the test statistic depend on whether or not H_o completely specifies the model.

Finally, the model probabilities, p_i, are used to calculate expected frequencies, $E_i = E(f_i)$, for the k categories of the variable. Those values are compared with the f_i values to test H_o. The test statistic given below in Eq. (9.11) is such that wide discrepancies between observed frequencies and those expected under the hypothesized model lead to a rejection of H_o and hence a rejection of the proposed model.

Now to formalize the discussion we suppose that the sample space of a variable, X, is partitioned into k categories. Let $p_1, p_2, \ldots, p_k$ denote the probabilities for those categories when H_o is true. When n independent values of X are observed, then the expected frequencies for the k categories are

$$E_i = np_i, \quad i = 1, 2, \ldots, k.$$

If the observed frequencies are denoted by f_i, i = 1, 2, ..., k, then the test statistic is

$$Y = \sum_{i=1}^{k} (f_i - np_i)^2/np_i. \tag{9.11}$$

It can be shown that Y is approximately a chi-square variable with (k-m-1) d.f., where m is the number of parameters unspecified by H_o. Thus, Y is approximately a chi-square variable with (k-1) d.f. when H_o completely specifies the model.

The next three subsections illustrate the technique for uniform, binomial, and normal models. Of course, the methods can be extended to any other models that are of interest to researchers.

UNIFORM MODELS: In Section 9.3, a procedure is given for testing H_o: $p_i = p_{io}$, i = 1, 2, ..., k. One special case of that hypothesis is

$$H_o: \; p_i = 1/k, \quad i = 1, 2, \ldots, k. \tag{9.12}$$

Expression (9.12) simply states that all categories are equally likely to occur. Thus, if the k categories are the points for a discrete sample space, then a *discrete uniform model* is hypothesized. Or if the k categories are equal-length intervals, then a *continuous uniform model* is proposed. To illustrate the concepts involved, we consider two different experiments for investigating the randomness of numbers in a random digit table.

EXPERIMENT 1: Select 50 digits from Table A2, and record the frequency of the 10 digits, 0, 1, 2, ..., 9.

EXPERIMENT 2: Select 50 3-digit numbers from Table A2. Divide each number by 1000, and record the frequencies for the following 10 classes: 0 - .1, .1 - .2, ..., .9 - .1.0. We suppose that each lower class limit is a member of the class, but we exclude the upper limit from the class.

In Experiment 1, we observe frequencies, f_i, for the 10 digits, 0, 1, 2, ..., 9. If the digits are random, a discrete uniform model should describe the observed frequencies. The use of Eq. (9.11) to test whether or not a uniform model is appropriate for a given set of data is illustrated by the next example.

EXAMPLE 8: A sample of 50 digits from Table A2 gives the following data:

Digit	0	1	2	3	4	5	6	7	8	9
Frequency	4	8	8	4	10	3	2	2	4	5

Is there any reason to doubt that the digits are uniformly distributed?

SOLUTION: The hypotheses are:

$$H_o\text{: } p_i = .1,\ i = 1, 2, \ldots, 10,$$

$$H_1\text{: Not all } p_i = .1.$$

Using $\alpha = .10$, the rejection region is

$$R\text{: } \{y \mid y \geq \chi^2\ (9, .10) = 14.68\}.$$

The expected frequencies are

$$E_i = np_{io} = 50(.1) = 5,\ i = 1, 2, \ldots, 10.$$

The test statistic is

$$y = \sum_{i=1}^{10} (f_i - 5)^2/5$$

$$= [(-1)^2 + 3^2 + 3^2 + (-1)^2 + 5^2 + (-2)^2 + (-3)^2$$

$$+(-3)^2 + (-1)^2 + 0^2]/5$$

$$= 68/5 = 13.6.$$

Since $y = 13.6 < 14.68$, we accept H_0. Thus, there is no reason to reject the uniform model.

In Experiment 2, we consider 3-digit decimals between 0 and 1. If the numbers in Table A2 are random, then a continuous uniform model between 0 and 1 should describe the observed frequencies. The probabilities for uniform models are the areas over the class intervals, 0 - .1, .1 - .2, ..., .9 - 1.0. Thus, just as for the previous example, $p_i = .1$, $i = 1, 2, \ldots, 10$. The full procedure is illustrated by the next example.

EXAMPLE 9: A sample of 50 3-digit numbers has been selected from Table A2 and divided by 1000. The resulting decimals give the following frequency distribution:

Classes	(0-.1)	(.1-.2)	(.2-.3)	(.3-.4)	(.4-.5)
Frequency	4	4	5	8	6

Classes	(.5-.6)	(.6-.7)	(.7-.8)	(.8-.9)	(.9-1.0)
Frequency	6	4	4	5	4

Is there any reason to doubt that the decimals are uniformly distributed on the interval (0, 1)?

SOLUTION: The hypotheses are:

H_o: The numbers are uniformly distributed on (0, 1).

H_1: The numbers are not uniformly distributed on (0, 1).

With $\alpha = .10$, the rejection region is

$$R: \{y | y \geq \chi^2(9, .10) = 14.68\}.$$

Since the 10 classes have equal length, the probability of each class is .1. Therefore, the expected frequencies are

$$E_i = 50(.1) = 5.$$

The test statistic is

$$Y = \sum_{i=1}^{10} (f_i - 5)^2/5$$

$$= [(-1)^2 + (-1)^2 + 0^2 + 3^2 + 1^2 + 1^2$$

$$+(-1)^2 + (-1)^2 + 0^2 + (-1)^2]/5$$

$$= 16/5 = 3.2.$$

Since $y = 3.2 < 14.68$, we accept H_0 and conclude that the continuous uniform model on (0, 1) provides a good fit for our data.

EXERCISES

1. An asymmetric block of wood is cut, and the 6 sides are numbered 1, 2, 3, 4, 5, 6. The die is rolled 36 times and yields the following data:

Number	1	2	3	4	5	6
Frequency	2	4	6	8	4	12

Is there any reason to doubt that the die is fair when $\alpha = .10$?

2. Four coins are tossed 40 times, and X, the number of heads, is observed for each toss. The data are as tabled.

x	0	1	2	3	4
f	5	7	17	8	3

Is there any reason to believe that X is not uniformly distributed on the integers 0, 1, 2, 3, 4? (Use $\alpha = .05$.) If you reject the null hypothesis, indicate any discrepancies between the observations and the expected frequencies.

3. Complete chi-square goodness of fit tests for any data you have collected for the experiments described at the beginning of this section.

4. The count of the number of cars per minute passing a given intersection is denoted by X and recorded for 30 independent, 1-minute periods. Do the following data cast doubt on a uniform model for X?

x	0	1	2	3	4
f	4	8	9	6	3

BINOMIAL MODELS: When surveying public opinion or studying product quality, we frequently desire to determine whether or not a binomial distribution provides an adequate model for the variable X, which denotes the number of successes in a sample of n observations. Let N denote the number of binomial samples observed. That is, we assume N realizations of a binomial variable X, which has parameter, p, and sample size, n. The N observations are classified in the categories 0, 1, 2, ..., n. The probabilities associated with those classes are given by the probability function (4.15), which is

$$p_i = P(X = i) = \binom{n}{i} p^i (1 - p)^{n-i}, \quad i = 0, 1, 2, \ldots, n. \qquad (9.13)$$

If a value of the parameter, p, is specified by the null hypothesis, the expected frequencies are

$$E_i = Np_i, \quad i = 0, 1, 2, \ldots, n. \qquad (9.14)$$

If no value of p is given by H_o, we first use the data to estimate p by

$$\hat{p} = \sum_{i=o}^{n} i\, f_i/Nn, \qquad (9.15)$$

where f_i is the number of samples showing exactly i successes. Equation (9.15) is nothing more than the total number of successes in the N binomial samples,

$$\sum_{i=o}^{n} i\, f_i,$$

divided by the total number of binomial trials, Nn. So Eq. (9.15) is a pooled estimate of p based on all N samples, which reduces to the usual estimate, (7.1), for p when only one sample is available. Thus, when p is not specified by H_o, Eq. (9.15) is used to estimate p. The estimate, say $\hat{p}$, is used in Eq. (9.13) to estimate the probabilities, p_i. Let $\hat{p}_i$ denote the estimate of p_i. Then the estimated expected frequencies are

$$\hat{E}_i = N\, \hat{p}_i, \quad i = 0, 1, 2, \ldots, n. \qquad (9.16)$$

The test statistics and rejection regions are

$$Y = \sum_{i=o}^{n} (f_i - Np_i)^2/Np_i; \quad R: \ \{y \mid y \geq \chi^2\ (n, \alpha)\} \qquad (9.17)$$

when p is specified by H_o, and

$$Y = \sum_{i=o}^{n} (f_i - N\, \hat{p}_i)^2/N\, \hat{p}_i; \quad R: \ \{y \mid y \geq \chi^2(n-1, \alpha)\} \qquad (9.18)$$

when p is unspecified by H_0 and estimated by Eq. (9.15). Since there are $(n+1)$ categories for the variable X, the d.f. for the test statistics are $(n+1) - 1 = n$ when H_0 completely specifies the model [i.e., Eq. (9.17)] and $(n+1) - 1 - 1 = n - 1$ when the data are used in Eq. (9.15) to estimate one parameter, p [i.e., Eq. (9.18)].

Data from the following experiments can be collected to illustrate the methods used.

EXPERIMENT 1: Flip 4 ($n = 4$) coins 64 ($N = 64$) times, and record X, the number of heads occurring on each trial.

EXPERIMENT 2: Select 2 ($n = 2$) cards (with replacement) from a deck of 52 playing cards, and record X, the number of spades observed. Repeat the experiment 64 ($N=64$) times.

EXPERIMENT 3: Roll 3 ($n = 3$) dice 81 ($N = 81$) times, record X, the number of dice showing either 1 or 2 on each roll (i.e., $X = \#\{1, 2\}$ on the roll of 3 dice), and the sample space of X is 0, 1, 2, 3.

EXPERIMENT 4: Select 50 ($N = 50$) groups of 4 ($n = 4$) coeds, and record X, the number of blondes in each group. The following examples provide insight into the use of the goodness of fit tests provided by Eqs. (9.17) and (9.18).

EXAMPLE 10: The following data are from Experiment 3.

x	0	1	2	3
f	21	39	17	4

Use those data to test the hypothesis H_0: $X \sim b(3, 1/3)$.

SOLUTION: The hypotheses are:

$$H_0: \quad X \sim b(3, 1/3),$$

$$H_1: \quad X \text{ is not} \sim b(3, 1/3).$$

Since $p = 1/3$ is specified by H_0, the rejection region for $\alpha = .05$ is

$$R: \quad \{y \mid y \geq \chi^2(3, .05) = 7.81\}.$$

Using Eq. (9.13) with $n = 3$ and $p = 1/3$ gives $p_0 = 8/27$, $p_1 = 12/27$, $p_2 = 6/27$, and $p_3 = 1/27$. Then Eq. (9.14) gives the expected frequencies, $E_0 = 81(8/27) = 24$, $E_1 = 81(12/27) = 36$, $E_2 = 81(6/27) = 18$, and $E_3 = 81(1/27) = 3$.

Next Eq. (9.17) gives

$$y = (-3)^2/24 + 3^2/36 + (-1)^2/18 + 1^2/3 = 1.014.$$

Since $y = 1.014 < 7.81$, we accept the binomial model expressed for the given data.

EXAMPLE 11: Suppose 50 groups of 4 coeds (Experiment 4) are observed, and X, the number of blondes in each group, is recorded. The data are as follows:

x	0	1	2	3	4
f	4	20	20	4	2

Is there any reason to suspect that X is not a binomial variable? (Use $\alpha = .05$.)

SOLUTION: The hypotheses are as follows:

H_o: $X \sim b(4, p)$ where p is unspecified,

H_1: X is not a binomial variable.

Since p is not specified by H_o, we use Eq. (9.15) to estimate p by

$$\hat{p} = [0(4) + 1(20) + 2(20) + 3(4) + 4(2)]/50(4) = .4.$$

The rejection region for $\alpha = .05$ is

$$R: \{y | y \geq \chi^2 (3, .05) = 7.81\}.$$

By using .4 for p in Eq. (9.13), or by entering Table B1 with $n = 4$ and $p = .4$, we find

$$\hat{p}_0 = 81/625, \hat{p}_1 = 216/625, \hat{p}_2 = 216/625,$$

$$\hat{p}_3 = 96/625, \hat{p}_4 = 16/625.$$

Then Eq. (9.16) gives the following estimated, expected frequencies:

$$\hat{E}_0 = 50(81/625) \approx 6.5, \quad \hat{E}_1 = 50(216/625) \approx 17.3,$$

$$\hat{E}_2 = 50(216/625) \approx 17.3, \quad \hat{E}_3 = 50(96/625) \approx 7.7,$$

$$\hat{E}_4 = 50(16/625) \approx 1.3.$$

The test statistic as defined by Eq. (9.18) is

$$y = (-2.5)^2/6.5 + (2.7)^2/17.3 + (2.7)^2/17.3 + (-3.7)^2/7.7 + (.7)^2/1.3 = 3.96.$$

Since $y = 3.96 < 7.81$, a binomial distribution is acceptable for modeling the number of blondes in random samples of size 4.

The techniques of this section can be extended to test the fit of multinomial models when we observe variables having more than two values. We use Eq. (4.28) to evaluate the probability for each value of the vector of multinomial variables. Then those probabilities are multiplied by N, the number of samples observed, to obtain the expected frequencies. These expected frequencies are compared to the observed frequencies by calculating a chi-square statistic similar to the one defined in Eq. (9.17). In those cases where H_0 does not specify all of the parameters for the model, the data can be used to estimate the parameters in a manner analogous to that used for the binomial.

EXERCISES

1. Use any data you have collected for Experiment 1 to test H_o: $X \sim b(4, .5)$.
2. Use any data you have collected for Experiment 2 to test H_o: $X \sim b(2, .25)$.
3. Use any data you have collected for Experiment 3 to test H_o: $X \sim b(3, 1/3)$.
4. Use any data you have collected for Experiment 4 to test H_o: $X \sim b(4, p)$, where p is unspecified.
5. In Exercise 2 on page 366, we rejected a uniform model for the following coin-tossing data:

x	0	1	2	3	4
f	5	7	17	8	3

   Use those data to test H_o: $X \sim b(4, .5)$ with $\alpha = .05$.

6. Select some phenomenon which is of interest to you and which can be studied by means of a binomial variable. Collect some data by means of an experiment, and test the fit of a binomial model for your data.

7. Forty samples of 3 students are asked to classify their political affiliation as either Republican or not Republican. The variable, X, denotes the number of Republicans in each sample of 3, and the results are as follows:

x	0	1	2	3
f	9	11	11	9

Is there any reason to reject H_0: $X \sim b(3, .5)$ with $\alpha = .05$? If so, can you offer any explanation for the discrepancies which lead to the decision to reject H_0?

8. Let A = {1, 2, 3} for the roll of a rough cut die. Let X denote the number of times A occurs on 3 independent rolls of the die. Suppose 40 observations on X are made, and the results are as follows:

x	0	1	2	3
f	14	15	8	3

(a) Test H_0: $X \sim b(3, .5)$, where $\alpha = .05$.
(b) Test H_0: $X \sim b(3, p)$, where p is unspecified and $\alpha = .05$.
(c) Compare the decisions for parts (a) and (b), and give reasons for any differences.

NORMAL MODELS: When studying a continuous variable X, we frequently want to know whether or not it is normally distributed. The question is important because many of the classical inference techniques assume normally distributed data and are not appropriate otherwise. Thus, we devote this section to procedures for testing whether or not a normal distribution is appropriate for modeling observations on some variable, X.

When questioning the normality of a variable, there are four different questions that can be raised. The questions differ in regard to the assumptions being made about the variable under study, and they can be stated in hypothesis form as follows:

(1) H_0: $X \sim N(\mu, \sigma^2)$, where both μ and σ^2 are unspecified.

(2) H_0: $X \sim N(\mu_0, \sigma^2)$, where μ is assumed to be some given value, μ_0, but σ^2 is unspecified.

(3) H_0: $X \sim N(\mu, \sigma_0^2)$, where μ is unspecified, but σ^2 is assumed to be some given value, σ_0^2.

(4) H_0: $X \sim N(\mu_0, \sigma_0^2)$, where μ and σ^2 are assumed to be the given values, μ_0 and σ_0^2, respectively.

Note that only H_o for (4) completely specifies the model being fit to the data. The null hypothesis for (2) and (3) requires that one parameter be estimated before continuing the goodness of fit test. Under (1), H_o only specifies that the distribution of X is normal, so estimates for both μ and σ^2 are required before completing a check of the fit. However, we must point out that when H_o for (1) is rejected, we conclude that the variable, X, is not normally distributed. However, when H_o for (4) is rejected, it may be because X is not normally distributed, *or* the mean of X is not μ_o, *or* the variance of X is not σ_o^2, *or* any combination of these three factors.

Whenever H_o for (1), (2), or (3) is being tested, the estimation techniques of Chapter 7 are used to estimate the parameter or parameters unspecified by H_o. In particular, the sample mean, $\overline{X}$, and the sample variance, S^2, as defined in Eq. (7.5), are used to estimate μ and σ^2, respectively.

To apply a chi-square goodness of fit test to assess whether or not a normal model is appropriate for a given data set of n observations we first group the data into k classes. For the convenience of discussion, we suppose the end-points of the k classes are

$$-\infty = a_0 < a_1 < a_2 < \cdots < a_{k-1} < a_k = \infty$$

and that the observed frequencies are denoted by $f_1, f_2, \ldots, f_k$. Now the probability that an observation of X falls in the ith class is given by

$$p_i = P(a_{i-1} \le X \le a_i) = P[(a_{i-1} - \mu)/\sigma \le Z \le (a_i - \mu)/\sigma], \quad i = 1, 2, \ldots, k, \tag{9.19}$$

where $X \sim N(\mu, \sigma^2)$ and $Z \sim N(0, 1)$. Thus, the necessary probabilities, p_i, can be obtained from Table D2 by the methods of the section on page 199. Whenever either μ or σ^2 or both are unspecified by H_o, the estimate or estimates provided by $\overline{X}$ or S^2 or both [see Eq. (7.5)] are used in Eq. (9.19) to determine the p_i values. Once the p_i values are available, the expected frequencies for comparison with the observed frequencies are

$$E_i = np_i, \quad i = 1, 2, \ldots, k.$$

The actual comparison between the f_i and E_i values is provided by the test statistic,

$$Y = \sum_{i=1}^{k} (f_i - E_i)^2/E_i, \tag{9.20}$$

for which the α-level rejection regions are

$$R: \quad \{y \mid y \geq \chi^2(k - 1, \alpha)\} \text{ for } H_o \text{ under (1)},$$

$$R: \quad \{y \mid y \geq \chi^2(k - 2, \alpha)\} \text{ for } H_o \text{ under either (2) or (3)},$$

$$R: \quad \{y \mid y \geq \chi^2(k - 3, \alpha)\} \text{ for } H_o \text{ under (4)}.$$

Note that we lose 1 d.f. for each parameter being estimated. That property is intuitively pleasing because when the data are used to estimate parameters for the model, we expect to get a better fit (that is, we expect smaller differences between the observed and model frequencies). In particular, when both $\overline{X}$ and S^2 are used in a test, the p_i values are based on a normal distribution with the same mean and variance as the data being modeled. So we expect that distribution to provide a better fit to our data than a distribution with parameter values that differ from the data statistics.

The following examples are presented to illustrate the previous comments and are based on the 100 grades for an elementary statistics class presented in Table 9.3.

TABLE 9.3.
The frequency distribution of 100 grades in an elementary statistics class is shown.

Score interval	Frequency
45 - 55	7
55 - 65	18
65 - 75	35
75 - 85	28
85 - 95	12
Total	100

EXAMPLE 12: Use the data in Table 9.3 to test whether or not a normal distribution with $\mu = 70$ and $\sigma = 10$ is an appropriate model for the statistics scores being analyzed. (Use $\alpha = .05$.)

SOLUTION: The hypotheses are

$$H_o: X \sim N(70, 100),$$

$$H_1: X \text{ is not } \sim N(70, 100).$$

The rejection region is

$$R: \{y|y \geq \chi^2(4, .05) = 9.49\}.$$

Equation (9.19) and Table D2 give

$$p_1 = P(Z \leq -1.5) = .0668,$$

$$p_2 = P(-1.5 \leq Z \leq -.5) = .2417,$$

$$p_3 = p(-.5 \leq Z \leq .5) = .3830,$$

$$p_4 = p(.5 \leq Z \leq 1.5) = .2417,$$

$$p_5 = p(1.5 \leq Z) = .0668.$$

(For these calculations, we act as if the lower limit, 45, and the upper limit, 95, are $-\infty$ and $+\infty$, respectively.) Therefore, the expected frequencies are $E_1 = 6.7$, $E_2 = 24.2$, $E_3 = 38.3$, $E_4 = 24.2$, and $E_5 = 6.7$.

The test statistic is

$$\begin{aligned} y = {} & (7-6.7)^2/6.7 + (18-24.2)^2/24.2 \\ & +(35-38.3)^2/38.3 + (28-24.2)^2/24.2 \\ & + (12-6.7)^2/6.7 \\ = {} & 6.67. \end{aligned}$$

Since $y = 6.67 < 9.49$, we accept the proposed normal model.

EXAMPLE 13: Test whether or not a normal distribution with unspecified mean, μ, and unspecified variance, σ^2, is an appropriate model for the statistics scores in Table 9.3. (Use $\alpha = .05$.)

SOLUTION: The hypotheses are

$$H_o: X \sim N(\mu, \sigma^2), \text{ with both parameters unspecified,}$$

$$H_1: X \text{ is not } \sim N(\mu, \sigma^2).$$

The methods of Chapter 3 for grouped data and for Eq. (7.5) provide the estimates of μ and σ to be $\overline{x} = 72$ and

and s = 10.9. Since the two parameters are being estimated, the rejection region is

$$R: \quad \{y | y \geq \chi^2(2, .05) = 5.99\}.$$

Using the estimators, $\bar{x}$ and s, for μ and σ in Eq. (9.19), as well as Table D2, gives

$p_1 = P(Z \leq -1.56) = .0594$, $p_2 = P(-1.56 \leq Z \leq -.64) = .2017$, $p_3 = P(-.64 \leq Z \leq .28) = .3492$, $p_4 = P(.28 \leq Z \leq 1.19) = .2727$, and $p_5 = P(1.19 \leq Z) = .1170$.

Therefore, the expected frequencies are $E_1 = 5.9$, $E_2 = 20.2$, $E_3 = 34.9$, $E_4 = 27.3$, and $E_5 = 11.7$.

The test statistic is

$$\begin{aligned} y &= (+1.1)^2/5.9 + (-2.2)^2/20.2 + (+.1)^2/34.9 \\ &\quad + (+.7)^2/27.3 + (+.3)^2/11.7 \\ &= .47. \end{aligned}$$

Since y = .47 < 5.99, we accept the hypothesis that the scores are normally distributed.

Before concluding this section with some exercises to provide additional practice in applying goodness of fit tests for normal models, we draw attention to an earlier comment about H_o for (1) and H_o for (4) as stated at the beginning of this section. In Example 12, the normal model being fit to the data has mean 70, while the mean for the data is 72. That difference causes larger discrepancies between the observed and expected frequencies for Example 12 than for Example 13, where the model uses 72 as its mean value. That property has also been exhibited for binomial models in Exercise 8.

EXERCISES

1. Use the data in Table 9.3 to test H_o: $X \sim N(\mu, 100)$, where μ is unspecified. (Use $\alpha = .05$.) How does your decision compare with those in Examples 12 and 13?

2. Select a random sample of 100 from Table D3, and form a distribution with classes less than -2, -2 to -1, -1 to 0, 0 to 1, 1 to 2, and over 2. Use the data to test H_o: $X \sim N(0, 1)$. (Use $\alpha = .10$.)

3. Let X denote the weekly consumption of hamburger (in pounds) for 100 families. Use the following data to test H_o: $X \sim N(3.5, 1)$ with $\alpha = .05$.

Pounds of hamburger	Number of families
Less than 2	9
2 - 3	18
3 - 4	48
4 - 5	23
Over 5	2

4. Use the procedures of this section to test the suitability of normal models for any data you collected for the experiments in Section 5.4.

5. Select some phenomenon of interest to you and for which a continuous variable, X, can be measured. Collect sample data on X, and test H_o: $X \sim N(\mu, \sigma^2)$. (Specify appropriate values for μ and σ^2 in H_o whenever possible.)

6. The statistics scores, X, for 100 students give the following distribution:

Class	45 - 55	55 - 65	65 - 75	75 - 85	85 - 95
f	2	13	35	40	10

Use $\alpha = .05$ to test the following hypotheses:

(a) H_o: $X \sim N(70, 100)$,
(b) H_o: $X \sim N(\mu, 100)$, where μ is unspecified,
(c) H_o: $X \sim N(\mu, \sigma^2)$, where both μ and σ^2 are unspecified.

9.6 SUMMARY AND A WORD OF CAUTION

This chapter presents some widely used inference techniques having a test statistic that is approximately a chi-square variable. The varied methods are all similar in that the data are recorded as observed frequencies, f_i, for a specific set of categories or cells. Also, the null hypothesis provides information for calculating the expected frequencies, E_i, for all categories. Then the approximate chi-square test statistic is of the form

$$Y = \sum_{i=1}^{k} (f_i - E_i)^2/E_i$$

and leads to the rejection of H_0 when Y becomes larger than a critical chi-square value obtained from Table E2. Under that general framework, we can test hypotheses concerning the following concepts: (1) the parameter of a binomial population; (2) the equality of parameters for two or more binomial models; (3) the parameters of a multinomial population; (4) the independence or classifications of elements with respect to two variables; (5) goodness of fit tests for any hypothesized distribution, e.g., uniform, binomial, and normal.

Now we conclude with a word of caution. Since each chi-square statistic, Y, is a sum of *positive* terms, it is sometimes possible for one term to make a large enough contribution to Y to lead to rejection. That may be the case when E_i is very small for a given category. For example, suppose the first category shows one observation, and $E_1 = .1$. The contribution to Y by that term is

$$(1 - .1)^2/.1 = 8.1.$$

This simply means that the occurrence of one rare (unlikely) event may make a large contribution to Y. To avoid difficulties of this type and to assume that the chi-square approximation for Y is good, a common rule-of-thumb is to require each E_i to be at least 5. This requirement is usually met by one of two methods. One approach is to increase the number of observations, n, until the expected frequency for each category is 5 or more. Since this approach is not always feasible, a more common technique is to combine (group) categories until a new set of categories, each of which exhibits $E_i \geq 5$, is obtained. Finally, we repeat that it is only the occurrence of rare (unexpected) events that causes any difficulty. For example, if $f_i = 0$ when $E_i = .1$, the contribution to Y by the ith category is only .1. Thus, it is often possible to use all categories of interest and not encounter any difficulties due to small values of E_i.

9.7 REFERENCES

[1] Freund, J. E. 1971. *Mathematical statistics*. 2nd ed. Englewood Cliffs, N. J.: Prentice-Hall, p.336.

[2] Berger, R. J., and Oswald, I. 1962. Eye movements during active and passive dreams. *Science* 137: 601.

[3] King, A. S. 1972. Pupil size, eye direction, and message appeal: some preliminary findings. *Journal of Marketing* 36: 57.

[4] Fisher, R. A. 1963. *Statistical methods for research workers*. 13th ed. New York: Hafner, p. 55.

[5] Lancaster, H. O. 1969. *The chi-squared distribution*. New York: Wiley.

9.8 REVIEW EXERCISES

1. When studying a characteristic carried by a single gene, the progeny produced by a single mating is of three types: homozygous dominant (FF), heterozygous (Ff), homozygous recessive (ff). According to the Mendelian theory of genetics, the probabilities associated with the three types of progeny are: 1/4 type (FF), 1/2 type (Ff), and 1/4 type (ff). Observed are 100 progeny which show 30 to be of type (FF), 40 of type (Ff), and 30 of type (ff). Does the Mendelian theory provide an acceptable model for the data? (Use $\alpha = .05$.)

2. A sample of 50 residents produces 30 who are in favor of purchasing an open area for development of a city park. If p denotes the true proportion of residents in favor of the purchase, is H_0: $p = .75$ an acceptable hypothesis? (Use $\alpha = .10$.)

3. The weights, X, of 50 men who have attended a short course for milling employees are as follows:

   180, 173, 173, 179, 168, 180, 162, 173, 167, 159, 164, 155,

   162, 179, 166, 157, 172, 168, 142, 170, 155, 184, 173, 178,

   168, 177, 170, 182, 178, 176, 174, 132, 182, 179, 162, 161,

   182, 186, 158, 154, 164, 169, 163, 155, 163, 161, 177, 177,

   173, 167.

   (a) Form a frequency distribution for x, and find $\bar{x}$ and s^2.
   (b) Test H_0: $X \sim N(175, 100)$. (Use $\alpha = .10$.)
   (c) Test H_0: $X \sim N(\mu, \sigma^2)$ when both parameters are unspecified. (Use $\alpha = .10$.)
   (d) Compare the results for parts (b) and (c).

4. A reflection meter is used to classify the brightness of 100 white shirts as dull, bright, or very bright. The same 100 shirts are classified with respect to the following laundry detergents used for cleaning: White Light, Bright White, or Brand X. Use the following data to test the hypothesis, H_0: Brightness is independent of detergent. (Use $\alpha = .05$.)

Detergent	Brightness		
	Dull	Bright	Very bright
White Light	5	10	15
Bright Light	10	15	15
Brand X	5	5	20

5. By testing a sample of 5 from each lot, 100 shipments of transistors have been inspected. The number of defectives in each sample is called X, and the frequencies for the possible values of X are as follows:

x	0	1	2	3	4	5
f	80	10	5	2	2	1

(a) Is $X \sim b(5, .1)$ an acceptable model for the data?
(b) Is $X \sim b(5, p)$ an acceptable model for X?

Note: You may want to pool together some classes that have small expected frequencies.

*6. Fisher [4] provides the following observed and expected frequencies for X, the number of deaths per year per army corps. There are 200 observations (10 army corps observed for 20 years), and the expected frequencies are calculated by using a Poisson model with $\lambda = .61$. Since .61 is an estimate of λ from the data, your test statistic should be compared with $\chi^2(3, .05)$ when $\alpha = .05$.

x	0	1	2	3	4
O_i	109	65	22	3	1
E_i	108.7	66.3	20.2	4.1	.6

Is there any reason to doubt that a Poisson model provides a good fit for the variable X?

CHAPTER 10

One-Sample Testing Procedures For Location and Dispersion

10.1 INTRODUCTION

Chapter 3 presents techniques for summarizing data by calculating the descriptive statistics, $\bar{X}$ and S^2, to represent the mean (location) and variance (dispersion) of a sample. Similar methods are included in Chapters 4 and 5 to determine μ and σ^2, which represent the mean (location) and variance (dispersion) of a probability model. In particular, we recall that μ and σ^2 are the parameters of a normal probability model. Thus, it was natural, in Chapter 7, to use the statistics, $\bar{X}$ and S^2, to provide point estimators and confidence intervals for the parameters, μ and σ^2. The relationship between those statistics and parameters when studying normal models is also basic to the hypothesis testing techniques concerning μ and σ^2 as presented in this chapter. Therefore, many similarities between the concepts of hypothesis testing and confidence interval estimation exist and are noted throughout this presentation.

Since normal models are not the only models for which we wish to test whether or not the location and dispersion are equal to some specified values, we present a selected set of nonparametric procedures, in addition to the classical methods for normal models. The nonparametric techniques are used whenever the experimenter cannot or does not wish to assume that his data are normally distributed. Thus, the classical procedures are appropriate when we assume that the data, $(X_1, X_2, \ldots, X_n)$, are a random sample from a normal distribution, while the nonparametric procedures require only that the data are a random sample from some unspecified distribution. For the purpose of discussion, the chapter is divided into three major parts. In Section 10.2, tests for the location of one sample are presented. Section 10.3 discusses tests for inference about the change in location occurring between before and after experiments. Then Section 10.4 treats the methods for testing hypotheses about the dispersion present in a model. Sample data for practice in using the techniques of this chapter can be obtained from the experiments described in Section 5.4 or one of the following experiments.

EXPERIMENT 1: Draw a line on a blank sheet of paper, and ask 25 individuals to guess its length, X.

EXPERIMENT 2: Draw two lines having the same length (one vertical and one horizontal) on a blank sheet of paper, and ask 25 individuals to guess X, the length of the vertical line, and Y, the length of the horizontal line.

EXPERIMENT 3: Obtain the weights of 25 male or female college students.

10.2 TESTS OF LOCATION FOR ONE SAMPLE

In this section, we consider some techniques for testing whether or not the location of a distribution is some specified value, μ_o. The procedures in the following section are appropriate for normal data such that the central limit theorem can be assumed to apply. If we are not willing to make those assumptions but still suppose that the variable is symmetric about a point, μ_o, then we can use the nonparametric procedure in the section on page 388.

CLASSICAL Z- AND t-TESTS FOR H_o: $\mu = \mu_o$: In this section we suppose that the data are a random sample of n observations, $(X_1, X_2, \ldots, X_n)$, from a normal distribution. The procedures are also appropriate when n is large enough to assume that $\overline{X}$ is approximately a normal variable by a central limit theorem. Our objective is to test the null hypothesis,

$$H_o: \mu = \mu_o, \tag{10.1}$$

against any appropriate alternative hypothesis. To achieve that objective the problem is divided into two cases. The first case assumes that σ^2, the variance of the distribution under study, is known; while the second case uses S^2 to estimate σ^2 when it is unknown.

Our discussion begins by referring to Chapters 6 and 7, which demonstrate some sampling properties for $\overline{X}$ and give reasons why $\overline{X}$ is a good estimator of μ, the mean of a distribution. Those results indicate that the sample mean, $\overline{X}$, is a useful statistic for making decisions about the unknown mean, μ, of a variable, X. In particular, when $\overline{X}$ and μ_o differ by a large value, we should reject H_o and conclude that the variable, X, under study does not have a mean of μ_o. As for previous testing methods, however, we need to find a test statistic with a known distribution when H_o is true so that α-level rejection regions can be found and used in making a decision about H_o. That is, a test statistic and its sampling distribution are used to determine when a difference between $\overline{X}$ and μ_o is large.

When σ^2 is known, the desired test statistic is

$$Z = (\overline{X} - \mu_o)/(\sigma/\sqrt{n}), \tag{10.2}$$

where Z has a N(0, 1) distribution when H_o is true. Thus, the α-level rejection regions (tests) are defined as follows:

$$R: \quad \{z \mid |z| > z(\alpha/2)\} \tag{10.3}$$

when the alternative hypothesis is H_1: $\mu \neq \mu_o$;

$$R: \quad \{z \mid z < -z(\alpha)\} \tag{10.4}$$

when the alternative hypothesis is H_1: $\mu < \mu_o$; and

$$R: \quad \{z \mid z > z(\alpha)\} \tag{10.5}$$

when the alternative hypothesis is H_1: $\mu > \mu_o$. These tests are valid when $z(\alpha/2)$ and $z(\alpha)$ are, respectively, the $(\alpha/2)$ and α upper-tail probability points of the N(0, 1) distribution as defined and used in the section on page 282 (see Table 7.1). The following example illustrates the method for (10.4).

EXAMPLE 1: The breakfast cereal, Crispy Crunch, is sold in boxes reporting a net weight of 12 ounces. If a sample of 9 boxes shows $\overline{x}$ = 11.6 ounces, is there any reason to doubt that H_o: $\mu = 12$ when $\sigma = .4$?

SOLUTION: Since our primary concern is with boxes containing less than 12 ounces, we decide to test H_o against the following one-sided H_1:

$$H_o: \quad \mu = 12,$$

$$H_1: \quad \mu < 12.$$

We choose $\alpha = .05$. Since $z(.05) = 1.645$, the rejection region is

$$R: \{z | z < -1.645\}.$$

The test statistic is

$$z = (11.6 - 12)/(.4/\sqrt{9}) = -.4(3)/.4 = -3.$$

Since $z = -3 < -1.645$, we reject H_o and conclude that the true value of μ is less than 12. Thus, we should stop the process and adjust the boxing machine.

The previous arguments also provide some insight into the relationship between hypothesis testing and interval estimation. Note that the pivotal quantity, Z, given in Eq. (7.23) and used to derive the confidence intervals and limits given in Eqs. (7.27)-(7.29) is similar to test statistic, Z, given in Eq. (10.2) and used to define the tests presented in (10.3)-(10.5). In fact, it is easy to show algebraically that any $\overline{X}$ leading to the rejection of H_o in (10.3) at some level of α produces a $CI_{1-\alpha}$ which does not include μ_o. Thus, the confidence interval in (7.27) and the hypothesis test in (10.3) yield the same conclusion. It can be shown that the limits in Eqs. (7.28) and (7.29) are algebraically analogous to (10.5) and (10.4), respectively.

Yet it is important to notice that the two concepts (interval estimation and hypothesis testing) are philosophically different. In particular, the primary purpose of a confidence interval is to find an interval of reasonable values for a parameter; while the primary purpose of a hypothesis test is to determine whether or not one specific value of a parameter is reasonable. So, in conclusion, we see that both procedures rely on the same statistic, $\overline{X}$, and its sampling distribution, but that the two procedures are used to answer different questions.

Now when σ^2 is unknown, but estimated by the sample variance, S^2, the test statistic is

$$t = (\overline{X} - \mu_o)/(S/\sqrt{n}), \tag{10.6}$$

where t has a t-distribution with (n - 1) degrees of freedom when H_o is true. So the α-level rejection regions (tests) are as follows:

$$R: \{t | |t| > t(n - 1, \alpha/2)\} \tag{10.7}$$

for testing H_o against H_1: $\mu \neq \mu_o$,

$$R: \{t | t < -t(n - 1, \alpha)\} \tag{10.8}$$

for testing H_o against H_1: $\mu < \mu_o$, and

$$R: \{t | t > t(n - 1, \alpha)\} \tag{10.9}$$

for testing H_o against H_1: $\mu > \mu_o$. These tests are valid when $t(n - 1, \alpha/2)$ and $t(n - 1, \alpha)$ are, respectively, the $\alpha/2$ and α upper-tail points of the $t(n - 1)$ distribution defined and used in the section on page 288. (See also Table E1.) The following example illustrates the technique.

EXAMPLE 2: If, in Example 1, σ^2 is unknown and $s^2 = .25$ is calculated from the sample, is there any reason to doubt that $\mu = 12$?

SOLUTION: The hypotheses are

$$H_o: \mu = 12,$$

$$H_1: \mu < 12.$$

With $\alpha = .05$ and $n = 9$, the critical value, $t(8, .05) = 1.86$, provides the rejection region,

$$R: \{t | t < - 1.86\}.$$

The test statistic value is

$$t = (11.6 - 12)/(.5/\sqrt{9}) = (.4)(3)/.5 = -2.4.$$

Since $t = -2.4 < -1.86$, we reject H_o and reach the same conclusion as that for Example 1.

Since the pivotal quantity, t, given in Eq. (7.30) is similar to the test statistic, t, given in Eq. (10.6), it follows that the confidence interval and limits in Eqs. (7.35) - (7.37) are analogous to the tests defined by (10.7), (10.9), and (10.8), respectively. Therefore, we again observe a close relationship between interval estimation and hypothesis testing.

We next present an alternate method for stating the rejection regions for the test statistics, z and t. First, for the two-sided region (10.3), we observe that $z \in R$ if and only if $z > z(\alpha/2)$ or $z < - z(\alpha/2)$. Thus,

$$|z| > z(\alpha/2)$$

implies that either

$$z = (\bar{x} - \mu_o)/(\sigma/\sqrt{n}) > z(\alpha/2)$$

or

$$z = (\bar{x} - \mu_o)/(\sigma/\sqrt{n}) < - z(\alpha/2).$$

Hence, $z \in R$ implies that either

$$\bar{x} > \mu_o + z(\alpha/2)\sigma/\sqrt{n} \tag{10.10}$$

or

$$\bar{x} < \mu_o - z(\alpha/2)\sigma/\sqrt{n}. \tag{10.11}$$

By rewriting (10.3) in the two parts given in (10.10) and (10.11), it is obvious that we reject H_o: $\mu = \mu_o$ when $\bar{x}$ deviates from the hypothesis value, μ_o. Rejection of H_o occurs when $\bar{x}$ is either too far above μ_o (too large) or too far below μ_o (too small). Furthermore, the smallest discrepancy between $\bar{x}$ and μ_o leading to the rejection of H_o is $z(\alpha/2)\sigma/\sqrt{n}$. Thus, the critical difference is a function of both α and n. Whether we use R as given in (10.3) or the pair of equivalent results (10.10) and (10.11) to test the H_o: $\mu = \mu_o$ against H_1: $\mu \neq \mu_o$ is a matter of personal preference. Similarly, the one-sided rejection regions, (10.4) and (10.5), are equivalent to

$$\bar{x} < \mu_o - z(\alpha)\sigma/\sqrt{n} \tag{10.12}$$

and

$$\bar{x} > \mu_o + z(\alpha)\sigma/\sqrt{n}, \tag{10.13}$$

respectively. In a similar way, the rejection regions (10.7)-(10.9) can be expressed in terms of $\bar{x}$ instead of t.

To support the earlier comments relating confidence intervals and rejection regions, we note that (10.10) and (10.11) are equivalent to $\mu_o < \bar{x} - z(\alpha/2)\sigma/\sqrt{n}$ and $\mu_o > \bar{x} + z(\alpha/2)\sigma/\sqrt{n}$, respectively. Thus, $Z \in R$ is equivalent to μ_o falling outside a $(1 - \alpha)$ confidence interval for μ based on $\bar{x}$. Similarly, (10.12) is equivalent to $\bar{x} + z(\alpha)\ \sigma/\sqrt{n} < \mu_o$. Hence, $Z \in R$ for (10.4) is equivalent to μ_o exceeding the $UCI_{1-\alpha}$ determined by $\bar{x}$. Also, $Z \in R$ for (10.5) is equivalent to μ_o being less than the $LCI_{1-\alpha}$ determined by $\bar{x}$.

We conclude this section with a brief discussion of power for the Z-test. In Eq. (8.4), the power of a test, $1 - \beta$, is defined to be the probability of rejecting a false H_o. If H_o is false, the true value of μ must differ from μ_o. Yet in order to calculate the power of a test, we need the true value of μ, say μ_T. Then in terms of the alternate forms, (10.10) and (10.11), for the rejection region, (10.3), we find that

$$
\begin{aligned}
1 - \beta &= P[\bar{X} < \mu_o - z(\alpha/2)\ \sigma/\sqrt{n} \mid \mu = \mu_T] \\
&\quad + P[\bar{X} > \mu_o + z(\alpha/2)\ \sigma/\sqrt{n} \mid \mu = \mu_T] \\
&= P[Z < (\mu_o - \mu_T)/(\sigma/\sqrt{n}) - z(\alpha/2)] \\
&\quad + P[Z > (\mu_o - \mu_T)/(\sigma/\sqrt{n}) + z(\alpha/2)]
\end{aligned}
\tag{10.14}
$$

where $Z \sim N(0, 1)$.

In a similar way, the use of (10.12) leads to

$$1 - \beta = P[Z < (\mu_0 - \mu_T)/(\sigma/\sqrt{n}) - z(\alpha)] \tag{10.15}$$

for the test defined in Eq. (10.4), and the use of (10.13) gives

$$1 - \beta = P[Z > (\mu_o - \mu_T)/(\sigma/\sqrt{n}) + z(\alpha)] \tag{10.16}$$

for the test defined by (10.5) when $Z \sim N(0, 1)$. Notice that the power of a test depends on μ_T. In particular, the power increases as the difference between μ_O and μ_T increases and as long as that difference is consistent with the alternative hypothesis. For example, the power function in Eq. (10.15) increases as long as μ_T decreases and remains less than μ_O; while the power function in (10.16) increases as long as μ_T increases and remains greater than μ_O. Clearly, the power function in Eq. (10.14) increases as $|\mu_O - \mu_T|$ increases. Thus, Eq. (10.14) increases whenever μ_T shifts in either direction away from μ_O. The next example illustrates the use of Eq. (10.15) and some of the indicated properties.

EXAMPLE 3: Find the power of the test used in Example 1 when $\mu_T = 11.8$, 11.5, and 11.0. Notice that $n = 9$, $\sigma = .4$, $\mu_o = 12$, and $\alpha = .05$.

SOLUTION: Since H_1: $\mu < 12$, we use Eq. (10.15) to calculate the power. For $\mu_T = 11.8$.,

$$
\begin{aligned}
1 - \beta &= P[Z < (12 - 11.8)/(.4/\sqrt{9}) - 1.645] \\
&= P(Z < -.145) \\
&= .44236 \text{ (see Table D2).}
\end{aligned}
$$

For $\mu_T = 11.5$,

$$
\begin{aligned}
1 - \beta &= P[Z < (12 - 11.5)/(.4/\sqrt{9}) - 1.645] \\
&= P(Z < 2.105) \\
&= .98236 \text{ (see Table D2).}
\end{aligned}
$$

For $\mu_T = 11.0$,

$$1 - \beta = P[Z < (12 - 11)/(.4/\sqrt{9}) - 1.645]$$

$$= P(Z < 5.855)$$

$$= 1.0000 \text{ (see Table D2).}$$

Notice that the test detects a shift of .2 pounds about 44% of the time, but a shift of .5 pounds is detected about 98% of the time. This illustrates the fact that the test becomes more powerful as μ_T decreases and goes farther away from μ_o.

EXERCISES

1. Select a sample of 16 men on campus, and obtain their weights. Test H_o: $\mu = 170$ against H_1: $\mu \neq 170$. Use $\alpha = .05$.
   (a) Assume $\sigma = 15$.
   (b) Estimate σ with S.
2. The Fair Tread tire is advertised as a 25,000-mile tire. If a sample of 25 tires provides $\bar{x} = 24{,}000$ and $s = 2000$, is there any reason to reject H_o: $\mu = 25{,}000$? (Use $\alpha = .10$.) Repeat the test when it is known that $\sigma = 2000$.
3. The following weights have been obtained by weighing the contents of 9 10-pound bags of sugar: 10.1, 9.8, 10.1, 9.9, 10.0, 9.7, 9.9, 10.0, 9.8. Is H_o: $\mu = 10$ an acceptable hypothesis when $\alpha = .05$? Use H_1: $\mu < 10$.
4. A sample of 16 scores for students in a basic statistics course gives $\bar{x} = 74$ and $s = 8$. Is there any reason to doubt that the average score for all students in this course is 80? (Use $\alpha = .01$.)
5. A total of 25 coeds shows an average shoe size of 8 with $s = .8$. Do these data contradict a shoe manufacturer's claim that the average shoe size for young women is 7.5? (Choose any appropriate α.)
6. Measure the length of the line used in Experiment 1 at the beginning of this chapter. If the length is μ_o, test H_o: $\mu = \mu_o$ against H_o: $\mu \neq \mu_o$, where μ is the mean guess. (Use $\alpha = .05$.)
7. Test H_o: $\mu = 180$ against H_1: $\mu \neq 180$ for Experiment 3 at the beginning of this chapter. (If female weights are observed, replace 180 by 120.) (Use $\alpha = .10$.)
8. Use H_1: $\mu \neq 25{,}000$ and $\sigma = 2000$ in Exercise 2, and calculate the power of the test when $\mu_T = 24{,}500$, 24,800, and 25,200.
9. Find the power of the Z-test associated with Exercise 3 when $\sigma = .6$ and $\mu_T = 9.9$, 9.8, 9.7, 9.6, 9.5, 9.4, 9.3, 9.2, 9.1, and 9.0. Graph your results on a chart with $(1 - \beta)$ = power on the vertical axis and μ_T on the horizontal axis. If we

join the points on the graph with a smooth curve, we have approximated the power function of the test.

10. Repeat Example 3 with α = .025, and discuss what effect (if any) changing α has on the power of a test.

A NONPARAMETRIC PROCEDURE: The procedures in the preceding section can be shown to have certain optimal properties (e.g., maximum power for testing against one-sided alternatives) when it is reasonable to assume that $\overline{X}$ is a normal variable. There are situations in statistical analyses where the variable being studied cannot be assumed to be a normal variable and where the sample size is not large enough to ensure that normality of $\overline{X}$ follows from a central limit theorem (Theorem 6.4). In that case, the methods of the section on page 381 no longer exhibit the desirable optimal properties and are, in fact, inappropriate. Thus, we turn now to the presentation of a method that does not require $\overline{X}$ to be a normal variable. The procedure requires only that the data are a random sample from a continuous symmetric distribution. Note that the assumption of symmetry is a much weaker requirement than the assumption of normality.

Procedures that do not assume a specific parametric model, such as a normal distribution, are called *nonparametric methods*. Conover [1] provides a comprehensive presentation of many nonparametric procedures for statistical analyses. Many of these procedures rely on the ranks of the data rather than on the actual observations. Some of these ranking procedures are presented in this text to be used whenever the necessary assumptions for parametric procedures are not reasonable.

To begin our discussion we define *ranks* for a set of data. The first step in ranking data is to order the observations. That ordered data set is usually presented from the smallest observation to the largest observation. The methods used to assign the ranks 1, 2, ..., n to the n observations vary with the particular procedure being used. That is, assigning ranks to a set of data requires only the assignment of the numbers 1, 2, ..., n to the ordered sample according to some rule, but the rule may vary with the method being studied.

EXAMPLE 4: Rank the following data from smallest to largest: 8, 10, -5, 6, 3, -1, 2.

SOLUTION: Ordered data: -5, -1, 2, 3, 6, 8, 10. Ranks: 1, 2, 3, 4, 5, 6, 7. Thus, rank 1 is assigned to the smallest data point, -5, rank 2 is assigned to the next smallest value, -1, ..., and rank 7 is assigned to the largest value, 10.

Some of the nonparametric procedures we consider require the use of *signed ranks*. If the data being ranked exhibit both negative and positive values, we first rank the absolute value of the data. That is, the signs of the data are ignored during the assignment of ranks. Then the signed ranks are obtained by associating the sign of each data value with the rank assigned to its absolute value. *We use R_i to denote the signed rank of the ith observation. When using signed ranks, zeros are omitted from consideration.*

EXAMPLE 5: Find the signed ranks for the following data: 10, -4, 2, -7, 5, 6, -9.

SOLUTION: The necessary operations can be completed line by line as follows:

Data, X_i: 10, -4, 2, -7, 5, 6, -9

$|X_i|$: 10, 4, 2, 7, 5, 6, 9

Ordered $|X_i|$: 2, 4, 5, 6, 7, 9, 10

Absolute ranks $|R_i|$: 1, 2, 3, 4, 5, 6, 7

Signed ranks, R_i: 1, -2, 3, 4, -5, -6, 7.

Statistical procedures using signed ranks as the basic data for analysis require test statistics that are a function of the R_i. Iman [2] presents a modification of the Wilcoxon [3] signed rank procedure for testing H_o: $\mu = \mu_o$ or, equivalently, H_o: median = μ_o. Let (x_1, x_2. ..., x_n) be a random sample of observations on a continuous symmetric variable, X. Use R_i, i = 1, 2, ..., n to denote the signed ranks for ($x_1 - \mu_o$, $x_2 - \mu_o$, ..., $x_n - \mu_o$). Then the mean signed rank is

$$\bar{R} = \sum_{i=1}^{k} R_i/k, \qquad (10.17)$$

where *k is the number of nonzero* terms being ranked. Iman suggests using the test statistic,

$$Y = (Z/2)\{1 + \sqrt{(k-1)/(k - Z^2)}\}, \qquad (10.18)$$

where

$$Z = \bar{R}/[(k + 1)(2k + 1)/6k]^{1/2}, \qquad (10.19)$$

in which k is the number of nonzero terms being ranked. The α-level upper-tail value for the test statistic, Y, is

$$y(k, \alpha) = [t(k - 1, \alpha) + t(\infty, \alpha)]/2, \qquad (10.20)$$

where $t(i, \alpha)$ is the α-level value of a t-statistic with i degrees of freedom (Table E1). For convenience, Table F2 presents $y(k, \alpha)$ for some of the popular values of α and a selection of k values. Table E1 and Eq. (10.20) can be used to calculate other upper-tail values of Y as needed.

The rejection regions for testing H_0: $\mu = \mu_0$ are

$$R: \{y \mid |y| > y(k, \alpha/2)\} \qquad (10.21)$$

when testing H_0 against H_1: $\mu \neq \mu_0$,

$$R: \{y \mid y > y(k, \alpha)\} \qquad (10.22)$$

when testing H_0 against H_1: $\mu > \mu_0$, and

$$R: \{y \mid y < - y(k, \alpha)\} \qquad (10.23)$$

when testing H_0 against H_1: $\mu < \mu_0$.

Even though we assume a continuous variable, round-off error can produce equal values of X. These values are called *ties*. To apply rank procedures a method for assigning ranks to ties is needed. In this text, we use the *average rank* technique for resolving ties that can be described as follows. If s values of X are tied for ranks, R_{i+1}, R_{i+2}, ..., R_{i+s}, each of the s-tied values is assigned a rank of

$$(\sum_{j=1}^{s} R_{i+j})/s.$$

The next example illustrates the technique.

EXAMPLE 6: Rank the following ages of professional football players: 29, 29, 26, 26, 25, 29.

SOLUTION: Data: 29, 29, 26, 26, 25, 29.

Ordered sample: 25, 26, 26, 29, 29, 29.

Ranks: 1, 2.5, 2.5, 5, 5, 5.

REMARK: Since the two 26's are competing for ranks 2 and 3, we assign a rank of $(2 + 3)/2 = 2.5$ to each 26. Similarly, each 29 is assigned a rank of $(4 + 5 + 6)/3 = 5$.

The next example illustrates the modified Wilcoxon procedure.

EXAMPLE 7: The ages of 10 professional football players are as follows: 24, 23, 34, 26, 24, 30, 26, 29, 37, 27. Is H_o: $\mu = 30$, where μ is the average age, an acceptable hypothesis? (Use $\alpha = .05$.)

SOLUTION: The hypotheses are

$$H_o: \ \mu = 30,$$

$$H_1: \ \mu \neq 30.$$

Since $\alpha = .05$ and $k = 9$ [one observation is 30 (μ_o) so $k = 10 - 1 = 9$], we use Table F2 to find $y(9, .025) = 2.13$. Therefore, the rejection region is

$$R: \ \{y \mid |y| > 2.13\}.$$

x_i data: 24, 23, 34, 26, 24, 30, 26, 29, 37, 27.

$x_i - 30$: −6, −7, 4, −4, −6, 0, −4, −1, 7, −3.

Ordered $|x_i - 30|$: 0, 1, 3, 4, 4, 4, 6, 6, 7, 7.

Absolute ranks, $|R_i|$: 1, 2, 4, 4, 4, 6.5, 6.5, 8.5, 8.5.

Signed ranks, R_i: −1, −2, −4, −4, 4, −6.5, −6.5, −8.5, 8.5.

Therefore,

$$\begin{aligned} \overline{R} &= [(-1) + (-2) + (-4) + (-4) + 4 + (-6.5) + (-6.5) \\ &\quad + (-8.5) + 8.5]/9 \\ &= -20/9 = -7/3 \end{aligned}$$

and

$$\begin{aligned} z &= (-20/9)/[(10)(19)/54]^{1/2} \\ &= -1.18. \end{aligned}$$

Using $z = -1.18$ in (10.18) gives

$$\begin{aligned} y &= (-1.18/2)\ \{1 + \sqrt{8/(9 - 1.39)}\} \\ &= -1.21. \end{aligned}$$

Since $|y| = 1.21 < 2.13$, we accept H_o and conclude that 30 is an acceptable average age for professional football players.

The following exercises provide practice in using the modified Wilcoxon procedure.

EXERCISES

1. Brand X movie floodlights are guaranteed to burn for 6 hours. To produce bulbs so that most of them exceed the guaranteed 6 hours, the production manager strives for an average life of 8 hours. A sample of 10 bulbs shows 7.3, 8.5, 3.2, 6.8, 12.7, 9.4, 8.9, 6.1, 8.3, 8.6. Use $\alpha = .10$ to test H_o: $\mu = 8$ against H_1: $\mu \neq 8$.

2. A sample of 8 players from a professional football team gives weights of 214, 250, 176, 226, 200, 185, 250, and 178. Is it reasonable to believe that the average weight of men on the team is 230 pounds? (Use $\alpha = .05$.)

3. Six randomly selected exam papers from a basic statistics class show scores of 62, 76, 81, 80, 72, 88. Is H_o: $\mu = 75$ acceptable against H_1: $\mu > 75$? (Use $\alpha = .05$.)

4. The weights of 9 10-pound bags of sugar are 10.1, 9.8, 10.1, 9.9, 10.0, 9.7, 9.9, 10.0, 9.8. Test H_o: $\mu = 10$ against H_1: $\mu \neq 10$ with $\alpha = .05$. Compare your analysis with that in Exercise 3 on page 387.

5. Test H_o: $\mu = \mu_o$ against H_1: $\mu \neq \mu_o$, where μ is the mean of the guesses and μ_o is the measured length of the line in Experiment 1 at the beginning of this chapter.

6. Test H_o: $\mu = 180$ (or H_o: $\mu = 120$ for females) against H_1: $\mu \neq 180$ (or H_1: $\mu \neq 120$) for Experiment 3 at the beginning of this chapter.

*10.3 PAIRED SAMPLE TESTS

In this section, we consider testing procedures for studying the average change, μ_c, in before-after experiments as introduced in the sections beginning on pages 276 and 304. The paired data are $[(X_1, Y_1), (X_2, Y_2), \ldots, (X_n, Y_n)]$, where X_i and Y_i are the before and after measurements, respectively, for the ith individual. For analysis, we reduce these data to the n individual changes,

$$C_i = Y_i - X_i, \quad i = 1, 2, \ldots, n.$$

When we assume that the X_i and Y_i are normal variables, it follows that the C_i are normal variables, and the methods of the section on page 381 apply directly to our data. That is,

$$Z = (\bar{C} - \mu_{co})/(\sigma_c/\sqrt{n}), \tag{10.24}$$

and the rejection regions, (10.3)-(10.5), can be used to test H_o: $\mu_c = \mu_{co}$ when σ_c^2 is known and

$$\overline{C} = \sum_{i=1}^{n} C_i/n.$$

Or, if σ_c^2 is unknown and

$$S_c^2 = \sum_{i=1}^{n} (C_i - \overline{C})^2/(n - 1),$$

$$t = (\overline{C} - \mu_{co})/(S_c/\sqrt{n}), \tag{10.25}$$

and the rejection regions, (10.7)-(10.9), can be used to test H_o: $\mu_c = \mu_{co}$.

EXAMPLE 8: The heart beat rate for 10 patients has been recorded before, X, and after, Y, being administered a drug to slow the rate. (See Example 7 on page 276.) The drug manufacturer claims that the average change produced by the drug is negative. Use $\overline{c} = -4.9$ and $s_c^2 = 4.99$ to test H_o: $\mu_c = 0$ against H_1: $\mu_c < 0$. (Use $\alpha = .05$.)

SOLUTION: To test H_o: $\mu_c = 0$ against H_1: $\mu_c < 0$ with $\alpha = .05$, we use

$$R: \{t | t \leq -t(9, .05) = -1.83\}.$$

Now

$$t = -4.9/(\sqrt{4.99}/\sqrt{10}) = -4.9/.71 = -6.9.$$

Since $t = -6.90 < -1.83$, we reject H_o and conclude that the average change, μ_c, produced by the drug is less than 0 (negative). That is, the manufacturer's claim is reasonable.

In those cases where we assume that the changes, $C_i = Y_i - X_i$, are continuous and symmetric but not necessarily normal, we can use the modified Wilcoxon statistic in Eq. (10.18). To apply that method to paired data we assign signed ranks to the changes, C_i, and proceed to calculate the test statistic, Y, as defined by Eq. (10.18). The appropriate rejection region is selected from (10.21)-(10.23) to correspond with the given alternative hypothesis, H_1. The null hypothesis, H_o, is that there is no change. Thus, H_o: $\mu_c = 0$ is the standard null hypothesis for the nonparametric paired sample test. The next example illustrates this method.

EXAMPLE 9: The achievement scores for 8 students have been recorded before, X, and after, Y, taking a brief refresher course in algebra. Is there any doubt that H_o: $\mu_c = 0$ when $\alpha = .05$?

Student	1	2	3	4	5	6	7	8
X	68	69	75	98	71	95	88	76
Y	72	70	81	95	72	98	80	79

SOLUTION: To obtain the signed ranks, R_i, we complete the following computations:

$C_i = Y_i - X_i$: 4, 1, 6, -3, 1, 3, -8, 3.

Ordered $|C_i|$: 1, 1, 3, 3, 3, 4, 6, 8.

$|R_i|$: 1.5, 1.5, 4, 4, 4, 6, 7, 8.

R_i: 1.5, 1.5, -4, 4, 4, 6, 7, -8.

Then to test H_o: $\mu_c = 0$ against $H_1 = \mu_c \neq 0$, we use

$$R: \{y \mid |y| > y(8, .025) = 2.16\}.$$

Now

$$\bar{R} = 12/8 = 1.5.$$

Using Eq. (10.19),

$$Z = 1.5/[(9)(17)/48]^{1/2}$$
$$= .84.$$

Using Eq. (10.18),

$$y = (.84/2)[1 + \sqrt{7/(8-.71)}]$$
$$= (.84/2)(1 + .98) = .83.$$

Since $|y| = .83 < 2.16$, we accept the hypothesis that the refresher course does not yield any change in achievement.

In a similar way, the Wilcoxon signed rank test, as described in Conover [1], can be used to test hypotheses about the average change in before-after experiments.

EXERCISES

1. A teacher of 3rd grade wants to test the hypothesis that during the school year 3rd-grade boys gain an average of 5 pounds. A sample of 16 boys shows $\overline{c} = 6.5$ and $s_c^2 = 9$. Test H_o: $\mu_c = 5$ against H_1: $\mu_c \neq 5$ with $\alpha = .10$.

2. Identify one or more testing situations for which paired data could be used for decision-making. If possible, select one such situation, and collect a set of data to test any hypothesis of interest.

3. In a study of body temperatures to see if they vary during the day, 10 subjects recorded their temperatures at 7:00 a.m., X, and 1:00 p.m., Y. Use the t-test and the data below to test H_o: $\mu_c = 0$ against H_1: $\mu_c \neq 0$ with $\alpha = .05$.

Subject	1	2	3	4	5	6	7	8	9	10
X	96.3	97.6	96.8	97.6	98.7	97.5	98.0	98.7	97.6	97.2
Y	98.2	97.4	98.8	99.0	98.9	98.6	99.1	98.2	97.5	99.1

4. Use the modified Wilcoxon procedure to test H_o: $\mu_c = 0$ against H_1: $\mu_c \neq 0$ with $\alpha = .05$, where μ_c is the average change in temperature between 7:00 a.m. and 1:00 p.m. Use the data in Exercise 3.

5. (a) Use the t-test with $\alpha = .05$ to analyze the data in Exercise 4 (on page 307) to test H_o: $\mu_c = 0$ against H_1: $\mu_c > 0$, where μ_c is the average gain in pounds for 3rd-grade boys during the school year.
   (b) Repeat part (a) using the modified Wilcoxon procedure.

6. (a) Use the data $(X_1, Y_1), \ldots, (X_{25}, Y_{25})$ from Experiment 2 at the beginning of this chapter and the t-test to see if the mean change for guesses, μ_c, is 0 when $C_i = X_i - Y_i$. That is, test H_o: $\mu_c = 0$ against H_1: $\mu_c \neq 0$ with $\alpha = .05$.
   (b) Repeat part (a) using the modified Wilcoxon procedure.

*10.4 TESTS OF H_o: $\sigma^2 = \sigma_o^2$

We now turn our attention to the problem of testing the hypothesis, H_o: $\sigma^2 = \sigma_o^2$. That is, we want to use sample data to either accept or reject a statement that the variance of a variable being studied is some value, σ_o^2. As for the section on page 381, the data are assumed to be measurements on a normal variable, X. There are no nonparametric procedures for testing the null hypothesis of this section. So our discussion is restricted to making decisions about variables that are approximately normally distributed.

When H_o: $\sigma^2 = \sigma_o^2$ is true,

$$Y = \Sigma(X_i - \overline{X})^2/\sigma_o^2 = (n - 1)s^2/\sigma_o^2 \qquad (10.26)$$

is distributed as a χ^2 variable with $(n - 1)$ d.f. Thus, Y satisfies the requirements of a test statistic. The α-level rejection regions for Y are

$$R: \quad \{y|y \geq \chi^2(n - 1, \alpha/2) \text{ or } y \leq \chi^2(n - 1, 1 - \alpha/2)\} \qquad (10.27)$$

for testing H_o against H_1: $\sigma^2 \neq \sigma_o^2$,

$$R: \quad \{y|y \geq \chi^2(n - 1, \alpha)\} \qquad (10.28)$$

for testing H_o against H_1: $\sigma^2 > \sigma_o^2$, and

$$R: \quad \{y|y \leq \chi^2(n - 1, 1 - \alpha)\} \qquad (10.29)$$

for testing H_o against H_1: $\sigma^2 < \sigma_o^2$.

The next example and the exercises that follow provide illustrations of the procedure used.

EXAMPLE 10: The net weights for 9 boxes of breakfast cereal (from Example 2) yield $s^2 = .25$. Is there any reason to doubt that H_o: $\sigma^2 = .16$? (Use H_1: $\sigma^2 \neq .16$ with $\alpha = .05$.)

SOLUTION: To test H_o: $\sigma^2 = .16$ against H_1: $\sigma^2 \neq .16$ with $\alpha = .05$, we use

$$R: \quad \{y|y \geq \chi^2(8, .025) = 17.53 \text{ or } y \leq \chi^2(8, .975) = 2.18\}.$$

Now

$$y = 8(.25)/.16 = 2.00/.16 = 12.5.$$

Since $2.18 < y = 12.5 < 17.53$, we accept H_o as a reasonable hypothesis.

We note that the tests provided in (10.27)-(10.29) can be used for testing H_o: $\sigma^2 = \sigma_{co}^2$ against appropriate alternative hypotheses, where σ_c^2 is the variance of the variable, C_i, denoting the change occurring in a paired data experiment. These tests are valid as long as C_i is approximately a normal random variable.

EXERCISES

1. Select a sample of 16 men on campus, and record their weights. Test H_o: $\sigma^2 = 225$ against H_1: $\sigma^2 \neq 225$ with $\alpha = .05$. (Refer to Exercise 1 on page 387.)
2. A sample of 25 tires shows $s = 2000$. Is there any reason to doubt that H_o: $\sigma^2 = 9{,}000{,}000$? Use H_1: $\sigma^2 \neq 9{,}000{,}000$ and $\alpha = .10$. (Refer to Exercise 2 on page 387.)
3. The manufacturer of an automatic sugar-bagging machine claims that the variance of bag weights, σ^2, is less than or equal to .01. Do the following data cast any doubt on that claim with $\alpha = .05$? 10.1, 9.8, 10.1, 9.9, 10.0, 9.7, 9.9, 10.0, 9.8. (Refer to Exercise 3 on page 387.)
4. A total of 16 randomly selected statistics exam papers shows $s = 8$. Is H_o: $\sigma^2 = 100$ acceptable when tested against H_1: $\sigma^2 \neq 100$ with $\alpha = .10$? (Refer to Exercise 4 on page 387.)
5. Use any data you have for Experiment 1 at the beginning of this chapter to test H_o: $\sigma^2 = 1$ against H_1: $\sigma^2 \neq 1$ with $\alpha = .10$.
6. Use any data you have for Experiment 3 at the beginning of this chapter to test H_o: $\sigma^2 = 225$ against H_1: $\sigma^2 \neq 225$ with $\alpha =$ .05.

10.5 SUMMARY

The contents of this chapter are concerned with inference about the location and dispersion of a random variable when a random sample of observations is available for analysis. The classical Z- and t-tests are presented for testing H_o: $\mu = \mu_o$ when it is reasonable to assume that $\overline{X}$ is approximately a normal variable. For use when the normality assumption cannot be made, a modification of the Wilcoxon signed ranks nonparametric procedure is described. These procedures are applied to data obtained from either observations on one variable or paired observations where the one variable of interest is the change occurring between measurements. Finally, a chi-square procedure for testing H_o: $\sigma^2 = \sigma_o^2$ is included.

Both the classical and nonparametric procedures of this chapter are described and illustrated for use in analyzing one random variable. That is, we use one sample to answer questions about the location and dispersion of the distribution of one random variable. Therefore, it is natural that the next step in our study is the development of statistical methods that allow us to simultaneously study the location and dispersion of two or more random variables. The development uses independent random samples of observations of each variable under study, and this begins in the next chapter.

10.6 REFERENCES

[1] Conover, W. J. 1971. *Practical nonparametric statistics*. New York: Wiley.

[2] Iman, R. L. 1974. Use of a t-statistic as an approximation of the exact distribution of the Wilcoxon signed ranks test statistic. *Communications in Statistics* 3: 795-806.

[3] Wilcoxon, F. 1945. Individual comparisons by ranking methods. *Biometrics* 1: 80-83.

10.7 REVIEW EXERCISES

1. The boiling temperature has been recorded before, X, and after, Y, a quantity of salt was added to 6 samples of water. The data are as follows (measurements being given in degrees Centigrade):

Sample	1	2	3	4	5	6
X	99.6	98.6	99.5	98.7	99.1	98.8
Y	97.3	95.7	96.1	96.2	97.0	95.8

(a) Use a paired t-test [see Eq. (10.24)] to test H_o: $\mu_c = 0$ with $\alpha = .05$.

(b) Use the modified Wilcoxon procedure to test H_o: $\mu_c = 0$ with $\alpha = .05$.

(c) Compare the analyses in (a) and (b).

2. Use the data for X in Exercise 6 on page 395 to test H_o: $\mu = 100$ by:

(a) the t-test; and

(b) the modified Wilcoxon test. (Use $\alpha = .05$.)

3. The weights of 16 professional football players show $\bar{x} = 230$ and $s = 20$.

(a) Test H_o: $\mu = 220$ against H_1: $\mu \neq 220$ with $\alpha = .05$.

(b) Test H_o: $\sigma^2 = 225$ against H_1: $\sigma^2 \neq 225$ with $\alpha = .10$.

CHAPTER 11

Analysis of Variance and Other Testing Procedures for Two or More Samples

11.1 INTRODUCTION

Just as there are reasons to test hypotheses about the location and dispersion of a single distribution by the methods used in Chapter 10, there are also situations that require techniques for comparing the locations and/or dispersions of two or more distributions. For example, an instructor of elementary statistics may wish to compare the average examination score for k different classes. Or an agronomist may wish to compare the average yield for k varieties of wheat. These are only two examples that cause experimenters to be interested in the simultaneous study of two or more distributions.

The present discussion is divided into two general parts. Section 11.2 is concerned with procedures for testing whether or not k distributions have equal location values. Section 11.3 then treats the problems of testing for equal locations and dispersions for two distributions. Although the results in Section 11.2 hold for two distributions, it is of practical importance to highlight the case for $k = 2$. In both cases, classical and nonparametric procedures are provided. The classical methods are appropriate when the data being studied are approximately normal variables, while the nonparametric procedures are appropriate when the normality assumption is not valid for the data.

EXPERIMENT: Practice sample data for use in this chapter can be collected as follows. Select 4 samples of 10($n = 10$) observations each from Table D3. Then add 2 to each element in the first sample, add 5 to each element in the second sample, add 4 to each element in the third sample, and add 3 to each element in the fourth sample. That procedure provides 4 independent random samples of size 10 from the normal distributions, $N(2, 1)$, $N(5, 1)$, $N(4, 1)$, and $N(3, 1)$, respectively. Notice that the 4 distributions being sampled have the same variance (namely, 1), but different means.

11.2 TESTS FOR EQUAL LOCATIONS

In this section, we suppose that there are k distributions (populations) under study. Furthermore, it is assumed that the means and variances of the distributions are $(\mu_1, \mu_2, \ldots, \mu_k)$ and $(\sigma_1^2, \sigma_2^2, \ldots, \sigma_k^2)$, respectively. We assume that the k distributions have equal dispersion, or, in symbols, we suppose that

$$\sigma_1^2 = \sigma_2^2 = \ldots = \sigma_k^2 = \sigma^2, \tag{11.1}$$

where σ^2 denotes the common variance. In the section on page 401, we further suppose that the k distributions are normal, but in the section on page 413 that assumption is relaxed as we present the Kruskal-Wallis nonparametric procedure.

The data are provided by a random sample from each distribution. These data are represented by using the following symbols:

x_{ij} denotes the ith observation from the jth distribution.

n_j denotes the number of observations from the jth distribution. (11.2)

By using the symbols in (11.2), we can indicate the sample data as presented in Table 11.1.

Frequently, the sample sizes are equal, and we use n to represent the common value. That is, $n_1 = n_2 = \ldots = n_k = n$ for many experiments. So in the development, we present the formulations needed for the general case of unequal sample sizes and then indicate the special results for equal sample sizes.

The presentation used here follows the format of Chapters 9 and 10. First, we present the null hypothesis being tested, as well as any appropriate alternative hypothesis. Second, the test statistic and rejection regions are indicated. Third, examples are given to demonstrate the procedures and illustrate necessary conclusions. The first method presented is the classical one-way *analysis of variance* (ANOVA).

TABLE 11.1
Sample data are given from k distributions.

Sample #	1	2	...	k
	x_{11}	x_{12}		x_{1k}
	x_{21}	x_{22}		x_{2k}
	$\vdots$	$\vdots$		$\vdots$
	$x_{n_1 1}$			
		$x_{n_2 2}$		$x_{n_k k}$

CLASSICAL ONE-WAY ANALYSIS OF VARIANCE: In addition to the assumption of equal variances made in Eq. (11.1), we suppose that the k populations (distributions) being sampled are normal for the analysis in this section. Under these assumptions, our objective is to use data as given in Table 11.1 to test H_0: $\mu_1 = \mu_2 = \ldots = \mu_k$ against H_1: Some of the μ_j are unequal. The technique to be used compares the variation *within* the k samples with the variation *among* the k samples. When the variation among the samples gets too large relative to the variation within the k samples, we reject H_0 and conclude that not all of the μ_j are equal. To be more explicit in the discussion, the definitions and results that follow are needed.

First, the *combined sample size*, m, the *over-all mean*, $\overline{x}$, and the *sample means*, $\overline{x}_j$, are defined by

$$m = \sum_{j=1}^{k} n_j, \quad \overline{x} = \sum_{j=1}^{k} \sum_{i=1}^{n_j} x_{ij}/m, \text{ and}$$

$$\overline{x}_j = \sum_{i=1}^{n_j} x_{ij}/n_j, \tag{11.3}$$

where x_{ij} and n_j are defined in (11.2). Second, we define the *total variation* in the experiment to be the *total sum of squared deviations* from the over-all mean, $\overline{x}$. Thus, we use SST to denote the *total sum of squares* and write

$$\text{SST} = \sum_{j=1}^{k} \sum_{i=1}^{n_j} (x_{ij} - \overline{x})^2. \tag{11.4}$$

Third, the *variation within* the sample is defined to be the *sum of squared deviations* from the respective sample means, $\overline{x}_j$. Using SSW to denote the *within sum of squares*, we write

$$SSW = \sum_{j=1}^{k} \sum_{i=1}^{n_j} (x_{ij} - \overline{x}_j)^2. \tag{11.5}$$

Finally, the *variation among* the samples represents the variation among the sample means, $\overline{x}_j$, with respect to the over-all mean, $\overline{x}$. We use SSA to denote the *among sum of squares* and write

$$SSA = \sum_{j=1}^{k} \sum_{i=1}^{n_j} (\overline{x}_j - \overline{x})^2 = \sum_{j=1}^{k} n_j (\overline{x}_j - \overline{x})^2. \tag{11.6}$$

It can be shown that SST = SSW + SSA, or

$$\Sigma\,\Sigma(x_{ij} - \overline{x})^2 = \Sigma\,\Sigma(x_{ij} - \overline{x}_j)^2 + \Sigma\, n_j(\overline{x}_j - \overline{x})^2. \tag{11.7}$$

Proof: $SST = \Sigma\ \Sigma(x_{ij} - \overline{x})^2$

$= \Sigma\ \Sigma[(x_{ij} - \overline{x}_j) + (\overline{x}_j - \overline{x})]^2$. Add and subtract $\overline{x}_j$.

$= \Sigma\ \Sigma[(x_{ij} - \overline{x}_j)^2 + 2(x_{ij} - \overline{x}_j)(\overline{x}_j - \overline{x}) + (\overline{x}_j - \overline{x})^2]$.

Expand this using the binomial theorem.

Now $\sum_j \sum_i (x_{ij} - \overline{x}_j)(\overline{x}_j - \overline{x}) =$

$\sum_j (\overline{x}_j - \overline{x}) \sum_{i=1}^{n_j} (x_{ij} - \overline{x}_j)$.

However, $\sum_i (x_{ij} - \overline{x}_j) = 0$, by the definition of $\overline{x}_j$.

Therefore, $\Sigma\ \Sigma\ (x_{ij} - \overline{x}_j)(\overline{x}_j - \overline{x}) = 0$.

Thus, $SST = \Sigma\ \Sigma\ (x_{ij} - \overline{x}_j)^2 + \Sigma\ \Sigma\ (\overline{x}_j - \overline{x})^2$

$= SSW + SSA$.

Use Eqs. (11.5) and (11.6).

EXAMPLE 1: Illustrate the additive relationship between the within (SSW) and among (SSA) components of variation and the total (SST) variation as stated in Eq. (11.7) by means of the following data:

Sample #	1	2	3
	3	2	7
	5	1	9
	4	4	
		1	

SOLUTION: For the data, we have $n_1 = 3$, $n_2 = 4$, and $n_3 = 2$; $m = 9$ and $k = 3$; $\overline{x} = 36/9 = 4$; and $\overline{x}_1 = 12/3 = 4$, $\overline{x}_2 = 8/4 = 2$, and $\overline{x}_3 = 16/2 = 8$. Thus,

$$\begin{aligned} SST &= (3-4)^2 + (5-4)^2 + (4-4)^2 + (2-4)^2 \\ &\quad + \dots + (9-4)^2 \\ &= 58, \end{aligned}$$

$$\begin{aligned} SSW &= (3-4)^2 + (5-4)^2 + (4-4)^2 + (2-2)^2 \\ &\quad + (1-2)^2 + (4-2)^2 + (1-2)^2 + (7-8)^2 \\ &\quad + (9-8)^2 \\ &= 10, \end{aligned}$$

$$\begin{aligned} SSA &= 3(4-4)^2 + 4(2-4)^2 + 2(8-4)^2 \\ &= 48. \end{aligned}$$

Therefore, the previous data satisfy Eq. (11.7).

Equation (11.7) is true in general as proved by the preceding algebraic arguments. However, arithmetic round-off errors may cause a slight discrepancy between SST and SSW + SSA for a particular example.

It can further be shown that the $(m - 1)$ degrees of freedom associated with the total sum of squares, SST, are partitioned into

$$(m - k) \text{ d.f. for SSW and } (k - 1) \text{ d.f. for SSA.} \qquad (11.8)$$

When a sum of squares is divided by its d.f., the result is called a *mean square*. Thus, the *mean square within samples* is

$$MSW = SSW/(m-k) = \Sigma\ \Sigma(x_{ij} - \bar{x}_j)^2/(m-k), \qquad (11.9)$$

and the *mean squares among samples* is

$$MSA = SSA/(k-1) = \Sigma\ n_j(\bar{x}_j - \bar{x})^2/(k-1). \qquad (11.10)$$

The analysis of variance test statistic is

$$F = MSA/MSW, \qquad (11.11)$$

where F is distributed as an F-statistic with (k - 1) and (m - k) d.f. for numerator and denominator, respectively. The critical values for the test statistic are obtained from Table E3. That is, the best α-level test for testing

$$H_o: \ \mu_1 = \mu_2 = \ldots = \mu_k$$

against

$$H_1: \text{ Some of the } \mu_j \text{ are unequal}$$

is defined by the rejection region

$$R: \ \{F|F > F(k-1, m-k, \alpha)\},$$

where $F(a, b, \alpha)$ is defined by Eq. (7.57) and is obtained from Table E3.

We remark that the rejection region contains only large values of F. To rationalize that property, notice that differences among the μ_j generally lead to differences among the $\bar{x}_j$. Furthermore, large differences among the $\bar{x}_j$ lead to large values of MSA. Thus, small values of F tend to support H_o, while large values of F tend to dispute H_o. This reasoning is further developed at the end of this section.

It is convenient to summarize the pertinent results for completing an ANOVA in table form such as that of Table 11.2.

To complete an ANOVA table as given in Table 11.2 for any set of samples, we first complete the d.f. column. That is accomplished by determining the total number of data points, m, and the number of samples, k. Second,

TABLE 11.2.
A one-way ANOVA table is shown for summarizing the data from k samples.

Source of variation	d.f.	Sum of squares	Mean squares	F
Among samples	$k-1$	$SSA = \Sigma\, n_j(\bar{x}_j - \bar{x})^2$	$MSA = SSA/(k-1)$	$F = MSA/MSW$
Within samples	$m-k$	$SSW = \Sigma\,\Sigma(x_{ij} - \bar{x}_j)^2$	$MSW = SSW/(m-k)$	
Total	$m-1$	$SST = \Sigma\,\Sigma(x_{ij} - \bar{x})^2$		

the sum of squares column is completed by using the following computation steps:

1. Calculate $T = \sum_{j=1}^{k} \sum_{i=1}^{n_j} x_{ij}^2$.

2. Calculate $A = \sum_{j=1}^{k} \left(\sum_{i=1}^{n_j} x_{ij} \right)^2 / n_j$.

3. Calculate $C = \left(\sum_{j=1}^{k} \sum_{j=1}^{n_j} x_{ij} \right)^2 / m$.

4. Calculate $SST = T - C$.

5. Calculate $SSA = A - C$.

6. Calculate $SSW = T - A$ (or $SST - SSA$).

It can be shown algebraically that the results produced by steps 1 - 6 are equivalent to the results given by Eqs. (11.4) - (11.6). Yet for most data, the sums of squares are more easily computed by steps 1-6. After the d.f. and sums of squares columns are available, the completion of the mean squares and F columns is direct. The next example illustrates the full procedure.

EXAMPLE 2: Use the data in Example 1 to illustrate the calculation method for completing an ANOVA. (Use $\alpha = .05$.)

SOLUTION: The null and alternative hypotheses are

$$H_o: \mu_1 = \mu_2 = \mu_3,$$

$$H_1: \text{Not all } \mu_j \text{ are equal.}$$

There are three samples of sizes 3, 4, and 2, respectively. Therefore, m = 9 and k = 3. So the rejection region is

$$R: \{F | F > F(2, 6, .05) = 5.14\}.$$

The data are as follows:

Sample #	1	2	3
	3	2	7
	5	1	9
	4	4	
		1	
Sample totals	12	8	16
Grand total			36

The calculation steps yield the following sums of squares for use in completing the ANOVA table.

1. $T = 3^2 + 5^2 + 4^2 + 2^2 + 1^2 + 4^2 + 1^2 + 7^2 + 9^2$

   $= 202.$

2. $A = (12)^2/3 + (8)^2/4 + (16)^2/2$

   $= 192.$

3. $C = (36)^2/9 = 144.$

4. $SST = 202 - 144 = 58.$

5. $SSA = 192 - 144 = 48.$

6. $SSW = 202 - 192 = 10.$

ANOVA table

Source	d.f.	S.S.	M.S.	F
Among samples	2	48	24	14.4
Within samples	6	10	10/6	
Total	8	58		

Since F = 14.4 > 5.14, we reject H_o and conclude that the three distributions being studied have different means.

When $n_1 = n_2 = \ldots = n_k$, we use n to denote the common sample size and write m = nk and

$$A = (1/n) \sum_{j=1}^{k} \left(\sum_{i=1}^{n} x_{ij} \right)^2 .$$

These simplifications offer some reduction in the arithmetic needed to calculate A in step 2. The next example gives an illustration of the full procedure for completing an ANOVA when the sample sizes are equal.

EXAMPLE 3: The following four samples have been obtained from Table D3 as outlined at the beginning of this chapter. Complete the ANOVA table to test H_o: $\mu_1 = \mu_2 = \mu_3 = \mu_4$, and draw appropriate conclusions. (Use $\alpha = .05$.) The data are given on the next page, following the Solution.

SOLUTION: The degrees of freedom are obtained from m = 4(10) = 40 and k = 4. Thus, k - 1 = 3, m - k = 40 - 4 = 36, and the rejection region is

$$R: \quad \{F \mid F > F(3, 36, .05) = 2.87\}.$$

The sum of squares calculation is as follows:

1. $T = (2.52)^2 + (2.23)^2 + \ldots + (3.10)^2 + (3.62)^2$

   $= 564.62.$

2. $A = [(21.44)^2 + (44.78)^2 + (44.37)^2$

   $+ (30.42)^2]/10$

   $= 535.90.$

3. $C = (141.01)^2/40 = 497.10$.

4. SST = 564.62 - 497.10 = 67.52.

5. SSA = 535.90 - 497.10 = 38.80.

6. SSW = 564.62 - 535.90 = 28.72.

Sample #	1	2	3	4	
	2.52	3.71	5.37	3.85	
	2.23	4.46	5.77	2.46	
	1.38	4.18	2.94	4.61	
	1.73	5.24	5.65	3.43	
	2.42	3.93	4.48	2.38	
	1.31	2.81	3.55	2.98	
	2.23	4.10	5.72	2.50	
	2.52	5.06	3.19	1.49	
	2.07	6.10	3.96	3.10	
	3.03	5.19	3.74	3.62	
Sample totals	21.44	44.78	44.37	30.42	
Grand total					141.01

ANOVA table

Source	d.f.	S.S.	M.S.	F
Among samples	3	38.80	12.93	16.16
Within samples	36	28.72	.80	
Total	39	67.52		

Since $F = 16.16 > 2.87$, we reject H_o and conclude that some of the 4 μ_j are unequal. Note that for this example we have made the correct decision because, by construction, $\mu_1 = 2$, $\mu_2 = 5$, $\mu_3 = 4$, and $\mu_5 = 3$. Of course, in any real-life experiment, we never know whether or not our decision is correct.

We conclude this section with a few general remarks about the ANOVA technique. First, Eq. (11.7) shows that the *total variation*, SST, in any set of samples is equal to the sum of the *within samples variation*, SSW, and the *among samples variation*, SSA. Thus, for any fixed value of SST, a decrease in either SSW or SSA must result in an increase of the other factor. For example, when all sample means, $\bar{x}_j$, are equal, SSA is 0. Or when all data within each sample are equal, but the elements in different samples vary, SSW is 0. These cases represent the two extremes and help to illustrate why SSA tends to be large when the samples are representing distributions with means that vary widely. Thus, as stated previously, it is reasonable that large values of SSA lead to the rejection of

$$H_o: \mu_1 = \mu_2 = \dots = \mu_k.$$

Second, we observe that MSA represents the variation among samples per d.f., and MSW represents the variation within samples per d.f. Whereas SSA and SSW are not directly comparable quantities because they are based on different d.f., MSA and MSW are expressions for the average variation per d.f. for the two sources of variation. Also Eq. (7.56) shows that E(F) (F=MSA/MSW) is $(m-k)/(m-k-2)$ when H_o is true. Furthermore, when H_o is false, we expect the magnitude of MSA to increase. Thus, it is logical that large values of F lead to the rejection of H_o. This reasoning leads to the choice of

$$R: \{F | F > F(k - 1, m - k, \alpha)\}$$

as the best test of H_o.

In conclusion, we see that the analysis of variance technique is a procedure that compares the variation among samples to the variation within samples (which is assumed to be equal for all distributions being studied) in order to decide whether or not

$$H_o: \mu_1 = \mu_2 = \dots = \mu_k$$

is an acceptable hypothesis. Furthermore, the analysis in this section assumes that the distributions being sampled are normal. We note also that the procedure is useful and gives good results as long as the data are approximately normal. However, if there is evidence that the data are not normal, then the Kruskal-Wallis nonparametric procedure presented in the next section provides a more appropriate analysis.

EXERCISES

1. Complete the analysis of variance for the following data to test H_o: $\mu_1 = \mu_2 = \mu_3$. (The data have been rounded off to integers to provide easier arithmetic.) Use $\alpha = .05$.

Sample #	1	2	3
	3	8	3
	5	7	5
	4	6	6
	2	7	4

2. Complete an analysis of variance for any data you collected for the experiment outlined at the beginning of this chapter. Choose any appropriate α.

3. Select 3 random samples of size 5 from Table D3, and complete an analysis of variance to test H_o: $\mu_1 = \mu_2 = \mu_3$. Compare your F-statistic with the F-statistic of your classmates, or repeat the experiment to obtain independent F-statistics. Are the F values close to 1.2?

4. The following data are samples of exam scores for 5 different statistics classes. Use $\alpha = .05$ to test H_o: $\mu_1 = \mu_2 = \mu_3 = \mu_4 = \mu_5$.

Sample #	1	2	3	4	5
	90	95	96	95	89
	87	93	89	93	88
	85	92	87	90	86
	84	90	85	89	85
	82	86	83	85	82
	74	82	82	81	69
	72	78	76		68
	68	74	71		64
		69			

5. The following data are adapted from King [1]. A study designed to analyze the effect of management's expectations on new innovations for production personnel used the output for 4 plants. The variable being measured is the average number of clothing patterns per day as recorded for the 12 months of a year. The first data set was observed in the year prior to changing the duties of production personnel and telling the managers of plants 1 and 3 that the changes were expected to increase production. The second set of data was observed following the changes in personnel duties, the telling of managers of plants 1 and 3 to expect increased production, and the telling of managers at plants 2 and 4 to expect no change in production. The variable, that is, the average daily production, is expressed in units of 1000 clothing patterns.

(1) Data for year prior to experiment:

Plant #	1	2	3	4
	18.1	18.3	18.2	18.4
	18.3	18.2	18.3	18.1
	18.2	18.1	18.0	18.4
	17.9	18.3	18.1	18.3
	18.3	18.2	17.9	17.9
	18.2	17.8	18.3	18.3
	18.3	18.3	17.9	18.1
	17.9	18.4	18.3	18.2
	18.2	17.7	18.4	18.4
	18.2	18.3	18.4	18.0
	18.1	18.1	18.1	18.5
	18.2	18.2	17.6	18.2

(2) Data for year following experiment:

Plant #	1	2	3	4
	18.5	18.3	18.6	18.3
	18.7	18.3	18.8	18.5
	18.9	18.4	19.0	18.3
	19.4	18.4	19.2	18.1
	19.4	18.4	19.2	18.1
	19.6	18.5	19.4	18.4
	19.5	18.6	19.8	18.2
	19.9	18.7	19.9	18.4
	19.9	18.6	19.9	18.4
	20.1	18.4	20.2	18.3
	20.2	18.0	20.2	18.3
	20.1	18.7	20.0	18.1

(a) Complete an ANOVA for data set 1 to test whether or not the 4 plants show equal daily production. (Use $\alpha = .05$.)

(b) Complete an ANOVA for data set 2 to test whether or not the 4 plants show equal daily production following the experiment. (Use $\alpha = .05$.)

(c) What conclusions do you draw from the results of parts (a) and (b)?

6. Five experimental plots for each of 3 new varieties of winter wheat are harvested to compare yields. Complete an ANOVA for the data, and draw appropriate conclusions. The yield is measured in bushels per plot.

Variety #	1	2	3
	22.4	20.1	23.9
	18.3	21.6	22.1
	22.5	20.5	20.5
	20.5	22.5	22.8
	21.0	18.9	23.2

Would you recommend one of the varieties over the other two?

KRUSKAL-WALLIS NONPARAMETRIC ANALYSIS OF VARIANCE: In those cases where we want to test whether or not k random samples are representing identical distributions (populations), but where we do not assume that the populations are normal, we can use the nonparametric procedure presented by Kruskal and Wallis [2]. That procedure is basically a test of location for the k populations being studied. The hypotheses are as follows:

H_o: The k samples are from identical distributions.
H_1: The k distributions being sampled have different means. (11.12)

The statement of H_1 above is used to emphasize that the Kruskal-Wallis analysis is sensitive to shifts in location (means) and is useful for our purposes. Yet the alternative hypothesis, H_1, is sometimes stated as follows. At least one of the distributions tends to contain larger values than at least one of the other distributions. Conover [3] presents a full discussion of the procedure and can be referenced for additional information.

The data for analysis are as presented in Table 11.1. We continue to use

$$m = \sum_{j=1}^{k} n_j$$

to denote the total number of data in the k samples. With that in mind, the first step in completing the analysis is to assign the ranks 1, 2, ..., m to the m data points. Rank 1 is given to the smallest observation, rank 2 is given to the next to smallest observation, and finally rank m is given to the largest observation. If R_{ij} denotes the rank assigned to the ith observation in the jth sample, x_{ij}, and if

$$A_j = \sum_{i=1}^{n_j} R_{ij}, \; j = 1, 2, \ldots, k \qquad (11.13)$$

is the sum of the ranks for the jth sample, then the test statistic is

$$H = [12/m(m + 1)] \sum_{j=1}^{k} (A_j^2/n_j) - 3(m + 1). \qquad (11.14)$$

If $k \leq 3$ and all $n_j \leq 5$, the exact α-level values of H are available in either the original paper by Kruskal and Wallis [2] or in the book by Conover [3]. Otherwise, and for the cases we consider, it has been shown that a chi-square distribution with (k - 1) d.f. provides a good approximation for the distribution of H. Thus, for our purposes, the α-level rejection region for the Kruskal-Wallis nonparametric analysis of variance is

$$R: \{H | H > \chi^2 (k - 1, \alpha)\}. \qquad (11.15)$$

In ranking the combined samples, if any ties are encountered, then the average of the ranks for which the tied data are competing is assigned to each tied data point.

We note that the Kruskal-Wallis procedure is equivalent to completing the classical analysis of variance as presented in the previous section on the ranks of the data. That is, we replace the actual data, which cannot be assumed to be normally distributed, by the ranks and then complete an ANOVA for the ranks. However, that analysis is equivalent to using the statistic H in Eq. (11.14). Example 4 on page 415 illustrates the procedure.

Sample #	1	2	3
	96	95	89
	89	93	88
	87	90	86
	85	89	85
	83	85	82
	82	81	69
	76		68
	71		64

EXAMPLE 4: The preceding data are the exam scores for a sample of students from 3 different statistics classes. Complete a Kruskal-Wallis analysis, and draw appropriate conclusions. (Use $\alpha = .05$.)

SOLUTION: The hypotheses are

H_o: Grades for the 3 classes have the same distribution.

H_1: The grade distributions have different means.

The rejection region is

$$R: \ \{H | H > \chi^2 (2, .05) = 5.99\}.$$

The ranks for the data are as follows:

Sample #	1	2	3
	22	21	17
	17	20	15
	14	19	13
	11	17	11
	9	11	7.5
	7.5	6	3
	5		2
	4		1
A_j	89.5	94	69.5

$$\sum_{j=1}^{3} A_j^2/n_j = (89.5)^2/8 + (94)^2/6 + (69.5)^2/8 = 3077.73.$$

Therefore,

$$H = [12/(22)(23)] \ (3077.73) - 3(23)$$

$$= 72.99 - 69$$

$$= 3.99.$$

Since $H = 3.99 < 5.99$, we accept H_o and conclude that the scores for the 3 classes have the same distribution.

EXERCISES

1. The ranks of scores for samples of size 7 selected from 4 classes of statistics are as follows:

Sample #	1	2	3	4
	15	18	2	26
	17	9	7	16
Ranks	21	12	4	28
	10	6	13	23
	8	11	3	19
	27	1	14	22
	25	5	20	24

Select an appropriate value of α, and state an H_o and H_1 to be tested by the Kruskal-Wallis procedure.

2. Apply the Kruskal-Wallis procedure (α = .05) to any data available from the experiment outlined at the beginning of this chapter.

3. Complete the Kruskal-Wallis analysis of variance for the data in Example 3. (Use α = .05.)

4. Table D3 was used to select 4 random samples of 5 observations each from the following distributions: (1) N(3, 1); (2) N(4, 1); (3) N(5, 1); and (4) N(6, 1). The data are as follows:

Sample #	1	2	3	4
	6.3	3.1	5.4	5.9
	2.7	5.3	4.8	5.6
	3.5	4.1	4.5	6.0
	1.7	4.3	4.6	5.3
	1.8	3.2	5.5	6.5

(a) Complete an ANOVA for the data. (Use α = .05.)
(b) Complete a Kruskal-Wallis analysis of variance for the data (α = .05.).
(c) Compare the results for parts (a) and (b).

5. Analyze the following yields of corn per plot produced by 5, 6, and 4 plots, respectively, for each of 3 varieties. (Use $\alpha = .10$.)

Sample #	1	2	3
	48	52	45
	56	56	47
	49	58	50
	51	54	46
	52	55	
		49	

11.3 TESTS FOR EQUALITY OF LOCATION AND DISPERSION WHEN STUDYING TWO SAMPLES

In this section, we consider the special analyses for studying changes in the location and dispersion of two distributions. The results in the section on page 401 for location hold when $k = 2$ and can be applied to two-population experiments, but the most frequent use of the ANOVA technique is for $k \geq 3$. Also, the results in the section on page 413 concern location and are for $k \geq 3$, but a similar analysis can be derived for $k = 2$. Thus, the purpose of this section is to emphasize analyses for two-sample experiments. For the most part, the location part of the presentation is merely a specialization of the test statistics used in Section 11.2 for experiments involving 3 or more distributions. Yet one basic difference between experiments studying exactly 2 populations and those studying more than 2 populations is the interest in one-sided alternative hypotheses. Thus, in this section some of the developments include testing procedures for both one- and two-sided alternative hypotheses.

The data for the 2 samples are denoted by

$$(X_{11}, X_{21}, \ldots, X_{n_1 1}) \quad \text{and} \quad (X_{12}, X_{22}, \ldots, X_{n_2 2}),$$

which are the first and second columns of Table 11.1.

The following experiments can be used to collect data for reference throughout the remainder of this section.

EXPERIMENT 1: Collect 2 random samples of size 10 from Table D3 so that $X_{i1} \sim N(5, 1)$ and $X_{i2} \sim N(6, 1)$.

EXPERIMENT 2: Collect 2 random samples of size 10 from Table D3 so that $X_{i1} \sim N(5, 1)$ and $X_{i2} \sim N(5, 4)$.

TWO-SAMPLE TESTS OF LOCATION: The next three subsections consider hypothesis-testing techniques that use two independent random samples to test one of the following pairs of hypotheses:

$$\begin{cases} H_o: & \mu_1 = \mu_2 \\ H_1: & \mu_1 \neq \mu_2 \end{cases} \qquad \begin{cases} H_o: & \mu_1 = \mu_2 \\ H_1: & \mu_1 > \mu_2 \end{cases} \qquad \begin{cases} H_o: & \mu_1 = \mu_2 \\ H_1: & \mu_1 < \mu_2 . \end{cases} \tag{11.16}$$

If we assume that σ_1^2 and σ_2^2 are known, then we obtain a procedure that uses a normally distributed test statistic. The second procedure is based on a specialization of Eq. (11.10) and the square root of F as defined in Eq. (11.11). The third procedure is the Mann-Whitney [4] nonparametric procedure, which is the two-sample counterpart for the Kruskal-Wallis procedure.

TWO-SAMPLE z-TESTS: When sampling two normal distributions for which σ_1^2 and σ_2^2 are known, the z-statistic in Eq. (7.45) is used to develop a testing procedure for the hypotheses in (11.16). In fact, the statistic

$$Z = (\bar{X}_1 - \bar{X}_2)/\sqrt{(\sigma_1^2/n_1 + \sigma_2^2/n_2)} \sim N(0, 1) \tag{11.17}$$

when H_o: $\mu_1 = \mu_2$ is true. Thus, the best α-level tests for testing $H_o : \mu_1 = \mu_2$ are

$$R: \quad \{z \mid |z| > z(\alpha/2)\} \tag{11.18}$$

when the alternative hypothesis is H_1: $\mu_1 \neq \mu_2$,

$$R: \quad \{z \mid z > z(\alpha)\} \tag{11.19}$$

when the alternative hypothesis is H_1: $\mu_1 > \mu_2$, and

$$R: \quad \{z \mid z < -z(\alpha)\} \tag{11.20}$$

when the alternative hypothesis is H_1: $\mu_1 < \mu_2$.

EXAMPLE 5: (See Example 15 on page 296.) A random sample of the weights (X_1) for 20 6th-grade boys exhibited $\bar{x}_1 = 100$ pounds, while a sample of the weights (X_2) of 25 6th-grade girls exhibited $\bar{x}_2 = 85$ pounds. If we assume $\sigma_1^2 = 200$ and $\sigma_2^2 = 150$, is there any reason to suspect that μ_1 is greater than μ_2?

SOLUTION: The hypotheses are

$$H_0: \mu_1 = \mu_2,$$
$$H_1: \mu_1 > \mu_2.$$

If we choose $\alpha = .05$, the rejection region is

$$R: \{z | z > z(.05) = 1.645\}.$$

The test statistic is

$$z = (100 - 85)/\sqrt{(200/20 + 150/25)} = 15/\sqrt{16}$$
$$= 3.75.$$

Since $z = 3.75 > 1.645$, we reject H_0 and conclude that the average weight of 6th-grade boys is greater than the average weight of 6th-grade girls.

As n_1 and n_2 increase, the sample variances, S_1^2 and S_2^2, converge to the population variances, σ_1^2 and σ_2^2, respectively. Thus, the test statistic Z in Eq. (11.17), with S_1^2 and S_2^2 replacing σ_1^2 and σ_2^2, is sometimes called the *large-sample test statistic* for testing $H_0 : \mu_1 = \mu_2$ when σ_1^2 and σ_2^2 are unknown and not assumed to be equal. As a rule of thumb, that approximate test should be good if both n_1 and n_2 are greater than or equal to 25. We note that there are no exact two-sample procedures when σ_1^2 and σ_2^2 are unknown and not assumed to be equal, but that for moderate- to large-size samples the approximate z-test is adequate for most practical purposes.

EXERCISES

1. Use any data you have collected for Experiment 1 in Section 11.3 to test H_0: $\mu_1 = \mu_2$ against H_1: $\mu_1 \neq \mu_2$ when $\sigma_1^2 = \sigma_2^2 = 1$. (Use $\alpha = .05$.)

2. Use any data you have collected for Experiment 2 in Section 11.3 to test H_0: $\mu_1 = \mu_2$ against H_1: $\mu_1 \neq \mu_2$ when $\sigma_1^2 = 1$ and $\sigma_2^2 = 4$. (Use $\alpha = .10$.)

3. Two brands of tires have been road-tested for mileage to test whether or not they have the same average life. The data are as follows:

Data	Brand #1	$n_1 = 40$, $\bar{x}_1 = 41{,}000$, $s_1^2 = 4{,}000{,}000$
	Brand #2	$n_2 = 30$, $\bar{x}_2 = 35{,}000$, $s_2^2 = 1{,}800{,}000$

Test H_o: $\mu_1 = \mu_2$ against H_1: $\mu_1 \neq \mu_2$ by using the approximate z-test defined by Eq. (11.17). Choose any appropriate α.

4. Thirty plots each for two new varieties of wheat are harvested to determine whether or not the average yields are equal. The data are as follows:

Data	Wheat #1	$n_1 = 30$, $\bar{x}_1 = 47$, $s_1^2 = 12$
	Wheat #2	$n_2 = 30$, $\bar{x}_1 = 44$, $s_2^2 = 18$

Test H_o: $\mu_1 = \mu_2$ against H_1: $\mu_1 \neq \mu_2$ by using the approximate z-test defined by Eq. (11.17). (Use $\alpha = .05$.)

TWO-SAMPLE t-TESTS: When k = 2 in the ANOVA procedure of the section beginning on page 401, we can rewrite the mean square among term, MSA, as follows:

$$\begin{aligned} MSA &= n_1(\bar{x}_1 - \bar{x})^2 + n_2(\bar{x}_2 - \bar{x})^2 \\ &= (\bar{x}_1 - \bar{x}_2)^2\,[n_1 n_2/(n_1 + n_2)]. \end{aligned} \tag{11.21}$$

Proof: When k = 2, $\bar{x} = (n_1\,\bar{x}_1 + n_2\,\bar{x}_2)/(n_1 + n_2)$.

Substitution into Eq. (11.21) gives

$$\begin{aligned} MSA &= n_1\,\{[n_2/(n_1 + n_2)](\bar{x}_1 - \bar{x}_2)\}^2 \\ &\quad + n_2\,\{[n_1/(n_1 + n_2)](\bar{x}_2 - \bar{x}_1)\}^2 \\ &= [(n_1\,n_2)/(n_1 + n_2)^2](\bar{x}_1 - \bar{x}_2)^2(n_2 + n_1) \\ &= [(n_1\,n_2)/(n_1 + n_2)](\bar{x}_1 - \bar{x}_2)^2. \end{aligned}$$

Furthermore, the mean square within samples in Eq. (11.9) reduces to

$$MSW = [\Sigma(x_{j1} - \overline{x}_1)^2 + \Sigma(x_{j2} - \overline{x}_2)^2]/(n_1 + n_2 - 2) \qquad (11.22)$$

when k = 2.

Thus, MSW is just the pooled estimator of the common variance, S_p^2, as defined in Eq. (7.12) and used in Eq. (7.50). That result is consistent with the ANOVA assumption that

$$\sigma_1^2 = \sigma_2^2 = \sigma^2.$$

Next we observe that

$$(n_1 + n_2)/n_1 n_2 = (1/n_1 + 1/n_2). \qquad (11.23)$$

Then by substitution into Eq. (11.11), it follows that, for k = 2,

$$F = (\overline{x}_1 - \overline{x}_2)^2/[MSW(1/n_1 + 1/n_2)]. \qquad (11.24)$$

By taking the square root of F and referring to Eq. (7.50), we find that

$$t = (\overline{x}_1 - \overline{x}_2)/\sqrt{MSW(1/n_1 + 1/n_2)} \sim t(n_1 + n_2 - 2). \qquad (11.25)$$

It can be shown in general that the square root of an F statistic with 1 and k_2 d.f. is a t-statistic with k_2 d.f. Thus, the result in Eq. (11.25) is one special case of that general result.

When using the test statistic, t, as defined in Eq. (11.25) for testing H_o: $\mu_1 = \mu_2$, the best α-level rejection regions are

$$R: \{t \mid |t| > t(n_1 + n_2 - 2, \alpha/2)\} \qquad (11.26)$$

when the alternative hypothesis is H_1: $\mu_1 \neq \mu_2$,

$$R: \{t \mid t > t(n_1 + n_2 - 2, \alpha)\} \qquad (11.27)$$

when the alternative hypothesis is H_1: $\mu_1 > \mu_2$, and

$$R: \{t \mid t < -t(n_1 + n_2 - 2, \alpha)\} \qquad (11.28)$$

when the alternative hypothesis is H_1: $\mu_1 < \mu_2$.

EXAMPLE 6: Suppose the following two samples of size 10 are such that $X_1 \sim N(\mu_1, 1)$ and $X_2 \sim N(\mu_2, 1)$.

Data	x_1	4.6, 4.7, 4.9, 4.5, 6.5, 3.6, 5.6, 5.3, 5.0, 5.6
	x_2	5.7, 6.0, 7.8, 5.0, 3.8, 5.9, 6.8, 5.1, 6.2, 5.3

Test H_o: $\mu_1 = \mu_2$ against H_1: $\mu_1 \neq \mu_2$ with $\alpha = .05$.

SOLUTION: The hypotheses are

$$H_o: \mu_1 = \mu_2,$$
$$H_1: \mu_1 \neq \mu_2.$$

The rejection region is

$$R: \{t \mid |t| > t(18, .025) = 2.10\}.$$

To calculate t we follow the calculation steps in the section on page 405 to obtain MSW. First,

$$T = (4.6)^2 + (4.7)^2 + \ldots + (6.2)^2 + (5.3)^2$$
$$= 600.89.$$

$$A = (4.6 + 4.7 + \ldots + 5.6)^2/10$$
$$+ (5.7 + 6.0 + \ldots + 5.3)^2/10$$
$$= (50.3)^2/10 + (57.3)^2/10$$
$$= 581.34.$$

Second,

$$MSW = (T - A)/(20 - 2) = 1.09.$$

Third,

$$\bar{x}_1 = 50.3/10 = 5.03,$$
$$\bar{x}_2 = 57.3/10 = 5.73.$$

Fourth,

$$t = (5.03 - 5.73)/\sqrt{(1.09)(1/10 + 1/10)}$$

$$= -.70/\sqrt{.218} = -.70/.467$$

$$= -1.50.$$

We accept H_0 because $|t| = 1.50 < 2.10$. Note that the preceding data were generated from Table D3 in Appendix D in accordance with Experiment 1 from Section 11.2. Thus, in this case, our test failed to detect the true difference of one unit between $\mu_1 = 5$ and $\mu_2 = 6$, and we committed a type II error. This does not mean that if we were to repeat the previous experiment that we would not reject H_0.

Notice that MSW can be calculated either as an ANOVA (see Example 6) or as S_p^2 in Eq. (7.12). When given the actual data, the ANOVA approach is usually best, but if either S_1^2 and S_2^2 or $\Sigma(x_{i1} - \overline{x}_1)^2$ and $\Sigma(x_{i2} - \overline{x}_2)^2$ are available, then Eq. (7.12) provides the better approach.

EXERCISES

1. Use any data from Experiment 1 at the beginning of Section 11.3 to test H_0: $\mu_1 = \mu_2$ against H_1: $\mu_1 \neq \mu_2$. Choose any appropriate α.

2. A random sample of the weights for 5 players from each of the 2 football teams playing in the Super Bowl yields the following data:

Data	x_1	226, 250, 220, 224, 250
	x_2	218, 180, 190, 255, 195

Is H_0: $\mu_1 = \mu_2$ an acceptable hypothesis? (Use $\alpha = .10$.)

3. A sample of grades from 2 different statistics classes gives the following data:

Data	x_1	95, 93, 92, 90, 86, 82, 78, 74, 69
	x_2	89, 88, 86, 85, 82, 69, 68, 64

Is it reasonable to suppose that the 2 classes have the same mean? Choose any appropriate α.

THE MANN-WHITNEY PROCEDURE: The two preceding sections contain the classical two-sample procedures for testing the equality of location for two distributions. So this and the following section present two nonparametric procedures which are useful for testing the equality of location for two distributions when the assumption of normality is not appropriate. The first of these procedures is called the Mann-Whitney [4] procedure and is a two-sample analogue of the Kruskal-Wallis analysis of variance procedure. We retain the assumption of equal dispersion for the two distributions under study.

To begin the analysis we combine the two samples and assign ranks, R_{ij}, to the combined samples. As for the Kruskal-Wallis procedure, R_{ij} denotes the rank of the ith observation in the jth sample, x_{ij}, when the smallest observation in the combined sample is given rank 1, the next smallest observation is given rank 2, ..., and the largest observation in the combined sample is given rank $(n_1 + n_2)$. Next, the test statistic is

$$T = \sum_{i=1}^{n_1} R_{i1} - [n_1(n_1 + 1)/2], \qquad (11.29)$$

where ΣR_{i1} is the sum of those ranks assigned to the observations in the first sample, $(x_{11}, \ldots, x_{n_1 1})$. Exact tables of critical values for T can be deduced from Verdooren [5] for $\alpha = .001, .005, .01, .025, .05, .10$; $n_1 = 2(1)25$; and $n_2 = 2(1)25$. Tables for $\alpha = .005, .01, .025, .05$, and for n_1 and n_2 ranging from 2 to 20 are given in Table F3. Thus, as long as both n_1 and n_2 are less than or equal to 20, the best α-level rejection regions for testing H_0: $\mu_1 = \mu_2$ are

$$R: \{T \mid T < T(n_1, n_2, \alpha/2) \text{ or } T > n_1 n_2 - T(n_1, n_2, \alpha/2)\} \qquad (11.30)$$

when the alternative hypothesis is H_1: $\mu_1 \neq \mu_2$,

$$R: \{T \mid T > n_1 n_2 - T(n_1, n_2, \alpha)\} \qquad (11.31)$$

when the alternative hypothesis is H_1: $\mu_1 > \mu_2$, and

$$R: \{T \mid T < T(n_1, n_2, \alpha)\} \qquad (11.32)$$

when the alternative hypothesis is H_1: $\mu_1 < \mu_2$. We note that Table F3 gives only the lower-tail critical values, $T(n_1, n_2, \alpha)$, while the upper-tail critical values are obtained by $n_1 n_2 - T(n_1, n_2, \alpha)$. Furthermore, observe that since we are dealing with a discrete variable, the rejection region does not include the critical points, $T(n_1, n_2, \alpha)$ and $n_1n_2 - T(n_1, n_2, \alpha)$.

EXAMPLE 7: Two types of 60-watt light bulbs are placed on test to determine their average life. The data are as follows:

Data	Bulb #1	749	952	873	898	1005
	Bulb #2	947	868	975	1010	969

Do the two types of bulbs exhibit the same average life? (Use $\alpha = .05$.)

SOLUTION: The hypotheses are

$$H_o: \mu_1 = \mu_2,$$
$$H_o: \mu_1 \neq \mu_2.$$

The rejection region is

$$R: \{T|T < T(5, 5, .025) = 3 \text{ or}$$
$$T > 25 - T(5, 5, .025) = 22\},$$

where $T(5, 5, .025)$ is taken from Table F3.

The ordered, combined sample and ranks are as follows:

X_1	749		873	898		952			1005	
X_2		868			947		969	975		1010
R_{ij}	1	2	3	4	5	6	7	8	9	10

Now

$$\Sigma R_{i1} = 1 + 3 + 4 + 6 + 9 = 23.$$

Therefore,

$$T = 23 - [5(6)/2] = 8.$$

Since $T = 8$ is neither < 3 nor > 22, we accept H_o and conclude that the two types of bulbs have the same average life.

If either n_1 or n_2 is greater than 20, a normal approximation for T can be used to obtain critical values. The approximate test is based on the fact that $E(T) = n_1n_2/2$ and $V(T) = n_1n_2(n_1 + n_2 + 1)/12$ are the mean and variance of T, respectively. We use those properties to define the test statistic,

$$Z = [T - (n_1n_2/2)]/\sqrt{n_1n_2(n_1 + n_2 + 1)/12}, \quad (11.33)$$

which, for moderate to large values of n_1 and n_2, is approximately an N(0, 1) variable. Thus, when either n_1 or n_2 is greater than 20, we use the α-level values of a normal variable, z(α) or z(α/2), to define the following rejection regions:

$$R: \{z \mid |z| > z(\alpha/2)\} \quad (11.34)$$

for testing H_0: $\mu_1 = \mu_2$ against H_1: $\mu_1 \neq \mu_2$,

$$R: \{z \mid z > z(\alpha)\} \quad (11.35)$$

for testing H_o: $\mu_1 = \mu_2$ against H_1: $\mu_1 > \mu_2$,

$$R: \{z \mid z < -z(\alpha)\} \quad (11.36)$$

for testing H_0: $\mu_1 = \mu_2$ against H_1: $\mu_1 < \mu_2$.

EXAMPLE 8: Fifteen light bulbs for one type of bulb and 25 of a second type of bulb are tested to determine their lifetime. Use the following data to see if the two types of bulbs have the same average lifetime. The ordered data and ranks are given after the Solution.

SOLUTION: The hypotheses are

$$H_o: \mu_1 = \mu_2,$$

$$H_1: \mu_1 \neq \mu_2.$$

The rejection region for α = .05 is

$$R: \{z \mid |z| > z(.025) = 1.96\}.$$

Now

$$\Sigma R_{i1} = 7 + 9 + 10 + 19.5 + 23$$
$$+ 31 + 32 + \ldots + 40$$
$$= 423.5.$$

Thus,

$$T = 423.5 - (15)(16)/2$$
$$= 303.5.$$

The value of the test statistic is

$$z = [303.5 - (15)(25)/2]/\sqrt{(15)(25)(15 + 25 + 1)/12}$$
$$= (303.5 - 187.5)/\sqrt{1281.25} = 116/35.8$$
$$= 3.24.$$

Since z = 3.24 > 1.96, we reject H_0 and conclude that the type 1 bulb has a longer life than the type 2 bulb.

X_1	X_2	Rank	X_1	X_2	Rank	X_1	X_2	Rank
	870	1		939	13.5		959	27
	879	2		941	15		960	28
	882	3		943	16		964	29.5
	909	4		944	17		964	29.5
	914	5		947	18	968		31
	915	6		950	19.5	987		32
917		7	950		19.5	990		33
	926	8		953	21.5	993		34
927		9		953	21.5	996		35
932		10	954		23	1000		36
	935	11		955	24	1001		37
	937	12		956	25.5	1005		38
	939	13.5		956	25.5	1110		39
						1111		40

EXERCISES

1. Use the Mann-Whitney procedure to test H_o: $\mu_1 = \mu_2$ against H_1: $\mu_1 \neq \mu_2$ for the normal data collected for Experiment 1 in Section 11.3. (Use $\alpha = .05$.)

2. To study the effect of fluctuating voltage on the life of light bulbs, 6 bulbs are placed on life test where the voltage varies during the test. Eight other bulbs are tested with constant voltage. Use the following data to determine the effect of varying the voltage.

Data	Changing voltage	879, 915, 955, 950, 956, 870
	Constant voltage	920, 1005, 996, 954, 987, 1111, 932, 950

3. A class of 29 students is given an examination with 150 possible points. Of the students, 8 are married and 21 are single. Do the following data support the claim of one married student that married students perform better in the classroom? (Use α = .05.)

Data	X_1	142, 117, 97, 149, 137, 138, 148, 73
	X_2	108, 119, 96, 145, 105, 140, 113, 76, 114, 83, 87, 110, 131, 120, 93, 75, 102, 118, 122, 70, 125

***TWO-SAMPLE TESTS OF DISPERSION:** The purpose of the next two subsections is to present procedures used for testing whether or not two distributions have the same dispersion or variability. The first of the procedures is the classical F-test which uses 2 samples of sizes n_1 and n_2 for the normal variables, X_1 and X_2, to test H_o: $\sigma_1^2 = \sigma_2^2$. As indicated in Section 11.3, the data are $(X_{11}, X_{21}, \ldots, X_{n_1 1})$ and $(X_{12}, X_{22}, \ldots, X_{n_2 2})$. The second of the procedures is a nonparametric method proposed by Siegel and Tukey [6]. This procedure is based on the Mann-Whitney statistic in the section on page 424 and differs from that location procedure only in the method for assigning ranks.

Procedures for testing whether or not k-distributions have equal variances are sometimes used as a pretest for ANOVA studies in which the assumption of equal variances is questioned. Thus, prior to beginning the development of the two-sample procedures, we note that Hartley [7] presents a method for testing the equality of variances in k-distributions.

***CLASSICAL F-TEST FOR H_o: $\sigma_1^2 = \sigma_2^2$:** In this section, we assume that two independent samples of sizes n_1 and n_2 are available from normal distributions having variances σ_1^2 and σ_2^2. We further suppose that S_1^2 and S_2^2 are the usual, unbiased estimators for σ_1^2 and σ_2^2, respectively. Now the null hypothesis being tested states that $\sigma_1^2 = \sigma_2^2$. Thus, Eq. (7.53) implies that the test statistics,

$$F_1 = S_1^2/S_2^2 \quad \text{and} \quad F_2 = S_2^2/S_1^2, \tag{11.37}$$

are $F(n_1 - 1, n_2 - 1)$ and $F(n_2 - 1, n_1 - 1)$ variables, respectively, when H_o is true. So Table E3 can be used to obtain the α-level values needed to build a testing procedure around F_1 and F_2. In fact, the best α-level rejection regions for testing H_o: $\sigma_1^2 = \sigma_2^2$ are

$$\begin{aligned} R: \quad &\{F_1 | F_1 > F(n_1 - 1, n_2 - 1, \alpha/2)\} \\ &\cup \{F_2 | F_2 > F(n_2 - 1, n_1 - 1, \alpha/2)\} \end{aligned} \tag{11.38}$$

for H_1: $\sigma_1^2 \neq \sigma_2^2$,

$$R: \quad \{F_1 | F_1 > F(n_1 - 1, n_2 - 1, \alpha)\} \tag{11.39}$$

for H_1: $\sigma_1^2 > \sigma_2^2$, and

$$R: \quad \{F_2 | F_2 > F(n_2 - 1, n_1 - 1, \alpha)\} \tag{11.40}$$

for H_1: $\sigma_1^2 < \sigma_2^2$. Notice that the rejection region for the two-sided test has two parts. That is, H_o is to be rejected either if S_1^2/S_2^2 is too large or if S_2^2/S_1^2 is too large. Furthermore, rejection of H_o because F_1 is too large implies that σ_1^2 is probably greater than σ_2^2, while rejection of H_o because F_2 is too large implies that σ_2^2 is probably greater than σ_1^2.

EXAMPLE 9: Two samples for Experiment 2 as described in Section 11.3 produce the following data:

Data	x_1	4.2, 6.0, 4.0, 5.9, 5.9, 4.3, 6.4, 5.7, 4.6, 4.4
	x_2	1.5, 5.3, 5.6, 4.1, 4.8, 6.0, 3.0, 8.1, 3.0, 7.2

Use the data to test H_o: $\sigma_1^2 = \sigma_2^2$ against H_1: $\sigma_1^2 \neq \sigma_2^2$.

SOLUTION: For the hypotheses, H_o: $\sigma_1^2 = \sigma_2^2$ and H_1: $\sigma_1^2 \neq \sigma_2^2$, the rejection region for $\alpha = .05$. is

$$R: \{F_1 \mid F_1 > F(9, 9, .025) = 4.03\}$$

$$\cup \{F_2 \mid F_2 > F(9, 9, .025) = 4.03\}.$$

(Note: Since $n_1 = n_2 = 10$, the two critical points are equal.)

By using Eq. (7.5) to calculate the variance for each sample, we find that $S_1^2 = .84$ and $S_2^2 = 4.09$. Thus,

$$F_1 = .84/4.09 = .21$$

$$F_2 = 4.09/.84 = 4.87.$$

Since $F_2 = 4.87 > 4.03$, we reject H_o and conclude that $\sigma_2^2 > \sigma_1^2$. Notice that for the two distributions being sampled, N(5, 1) and N(5, 4), we have made a correct decision.

EXERCISES

1. (a) Test H_o: $\sigma_1^2 = \sigma_2^2$ against H_1: $\sigma_1^2 \neq \sigma_2^2$ by using any data available for Experiment 1 in Section 11.3. Use $\alpha = .05$.
   (b) Repeat part (a) with data from Experiment 2 in Section 11.3. Use $\alpha = .05$.
   (c) If the test in part (a) were repeated 100 times, how many times would we expect to reject H_o? Can you make a similar statement for part (b)?

2. A random sample of the weights for 10 2nd-grade boys shows a variance of $s_1^2 = 16$. A similar sample of 16 9th-grade boys yields a variance of $s_2^2 = 80$. Are the weights of 9th graders more variable than the weights of 2nd graders? Choose any α you desire.

3. The weights of 25 boxes of cereal boxed by the old machine show $s_o^2 = .16$, while the weights for 20 boxes of cereal boxed by the new machine show $s_n^2 = .09$. Would you advise the management to consider the new machine as being superior to the old machine?

4. The grades for samples of exam scores from 2 statistics classes are as follows:

Data	x_1	95, 93, 90, 89, 85, 81
	x_2	89, 88, 86, 85, 82, 69, 69, 64

Is H_o: $\sigma_1^2 = \sigma_2^2$ an acceptable hypothesis when $\alpha = .05$ and H_1: $\sigma_1^2 \neq \sigma_2^2$ is the alternative hypothesis?

5. Refer to the pattern production data in Exercise 5 on page 411 to determine whether or not the variances of the monthly production values for plant #1 remain the same for the 2 years considered. The pertinent data are as follows:

Data	x_1	18.1, 18.3, 18.2, 17.9, 18.3, 18.2, 18.3, 17.9, 18.2, 18.2, 18.1, 18.2
	x_2	18.5, 18.7, 18.9, 19.4, 19.4, 19.6, 19.5, 19.9, 19.9, 20.1, 20.2, 20.1

***NONPARAMETRIC TESTS FOR DISPERSION:** If we assume that the two distributions being studied have equal location parameters but do not assume normal distributions, then the Siegel-Tukey [6] nonparametric procedure is one method for testing H_o: $\sigma_1^2 = \sigma_2^2$. That technique is a modification of the Mann-Whitney test for location. The change occurs in the assignment of ranks for the data. First, the combined sample is ordered from small to large. Second, the ranks are assigned by using rank 1 for the *smallest* value, rank 2 for the *largest* value, rank 3 for the next to *largest* value, rank 4 for the next to *smallest* value, rank 5 for the third *smallest* value, and so forth by alternating the assignment of two ranks at each end of the array until all values are ranked. The ranks for the combined, ordered sample when aligned small to large from left to right are

$$1, 4, 5, 8, 9, \ldots, (n_1 + n_2), \ldots, 7, 6, 3, 2. \qquad (11.41)$$

In case of ties, the average ranks are used. After the ranks in (11.41) are assigned to the data, the Siegel-Tukey procedure is used to calculate the Mann-Whitney statistic as presented in the section on page 424. Thus, the test statistic is given by Eq. (11.29) for n_1 and n_2 less than or equal to 20, and the critical values are obtained from Table F3. If either n_1 or n_2 is greater than 20, then the procedure uses the normal approximation for T with the test-statistic Z as defined in Eq. (11.33).

Since the size of the test statistic (either T or Z) depends on whether the sum of ranks for the X_1 variable is large or small, an inspection of (11.41) shows that the test statistic tends to be small (large) when X_1 has greater (less) dispersion than X_2.

When both $n_1 \leq 20$ and $n_2 \leq 20$, we use the test statistic T as given in Eq. (11.29) for testing the equality of two variances. For that case, the α-level rejection regions for testing H_0: $\sigma_1^2 = \sigma_2^2$ are

$$R: \{T | T < T(n_1, n_2, \alpha/2) \text{ or}$$

$$T > n_1 n_2 - T(n_1, n_2, \alpha/2)\} \quad (11.42)$$

when H_1: $\sigma_1^2 \neq \sigma_2^2$ is the alternative hypothesis,

$$R: \{T | T < T(n_1, n_2, \alpha)\} \quad (11.43)$$

when H_1: $\sigma_1^2 > \sigma_2^2$ is the alternative hypothesis, and

$$R: \{T | T > n_1 n_2 - T(n_1, n_2, \alpha)\} \quad (11.44)$$

when H_1: $\sigma_1^2 < \sigma_2^2$ is the alternative hypothesis. Similarly, when either $n_1 > 20$ or $n_2 > 20$, the test statistic is Z, as given in Eq. (11.33), and the α-level rejection regions are

$$R: \{z \,|\, |z| > z(\alpha/2)\}, \quad (11.45)$$

$$R: \{z | z < -z(\alpha)\}, \quad (11.46)$$

and

$$R: \{z | z > z(\alpha)\}, \quad (11.47)$$

when the alternative hypotheses are H_1: $\sigma_1^2 \neq \sigma_2^2$, H_1: $\sigma_1^2 > \sigma_2^2$, and H_1: $\sigma_1^2 < \sigma_2^2$, respectively.

EXAMPLE 10: Two types of movie light bulbs are being tested. Both are listed as 4-hour bulbs. Six bulbs of brand I and 8 bulbs of brand II are tested. The data are as follows:

Data	x_1	3.6, 4.0, 3.7, 3.4, 5.5, 4.7
	x_2	4.5, 4.2, 2.0, 2.8, 4.0, 6.0, 4.9, 5.0

Test H_0: $\sigma_1^2 = \sigma_2^2$ against H_1: $\sigma_1^2 \neq \sigma_2^2$ with $\alpha = .05$.

SOLUTION: Since $n_1 = 6 < 20$ and $n_2 = 8 < 20$, the rejection region from Table F3 for testing H_o: $\sigma_1^2 = \sigma_2^2$ against H_1: $\sigma_1^2 \neq \sigma_2^2$ is

$$R: \{T | T < T(6, 8, .025) = 9 \quad \text{or} \quad T > 48 - 9 = 39\}.$$

The combined sample and ranks are as follows:

x_1			3.4,	3.6,	3.7,	4.0,				4.7,			5.5	
x_2	2.0,	2.8,					4.0,	4.2,	4.5,		4.9,	5.0,		6.0
R_{ij}	1	4	5	8	9	12.5	12.5	14	11	10	7	6	3	2

$$\Sigma R_{i1} = 5 + 8 + 9 + 12.5 + 10 + 3 = 47.5.$$

Therefore, Eq. (11.29) gives

$$T = 47.5 - [6(7)/2]$$
$$= 26.5.$$

Since $9 \leq T = 26.5 \leq 39$, we accept H_o and conclude that the 2 samples have equal variances.

EXAMPLE 11: Suppose a study similar to that in Example 10 gives data (shown in the table on page 434) with the ranks as assigned. Is there any reason to question H_o: $\sigma_1^2 = \sigma_2^2$?

SOLUTION: Since $n_1 = 12$ and $n_2 = 21 > 20$, the .05-level rejection region for testing H_o: $\sigma_1^2 = \sigma_2^2$ against H_1: $\sigma_1^2 \neq \sigma_2^2$ is

$$R: \{z | |z| > 1.96\}.$$

Now

$$\Sigma R_{i1} = 749/3.$$

Using Eq. (11.33), we find

$$z = [171.67 - (12)(21)/2]/\sqrt{(12)(21)(34)/12}$$
$$= 45.67/\sqrt{714}$$
$$= 1.71.$$

Since z = 1.71 < 1.96, we accept H_0.

x_1	x_2	R_{ij}	x_1	x_2	R_{ij}	x_1	x_2	R_{ij}
	2.4	1	3.8		$23\frac{1}{3}$		5.1	22
	2.5	4		3.8	$23\frac{1}{3}$		5.2	19
	2.9	5		3.9	28		5.3	18
	3.1	8	4.0		29	5.5		15
	3.2	9	4.2		32	5.6		14
3.3		12		4.3	33	5.7		11
3.4		13	4.4		31		5.9	10
	3.5	16		4.5	30		6.4	7
	3.6	17		4.6	27		6.6	6
3.7		20	4.7		26		6.7	3
3.8		$23\frac{1}{3}$		4.9	23		7.0	2

(Note: The three 3.8 values are tied for ranks 21, 24, and 25, so each is ranked 23 1/3.)

There are other nonparametric procedures for testing the equality of dispersion for 2 samples. Conover [3] discusses those methods and provides recommendations for their use.

EXERCISES

1. Use the Siegel-Tukey procedure to test H_0: $\sigma_1^2 = \sigma_2^2$ against H_1: $\sigma_1^2 \neq \sigma_2^2$ for the data you have collected for Experiment 2 in Section 11.3.

2. In Example 7 on page 425, we accepted the hypothesis that the 2 types of bulbs have the same average life. Do those data indicate that the variances for the 2 distributions are equal? (Use $\alpha = .05$.)

Data	x_1	749	952	873	898	1005
	x_2	947	868	975	1010	969

3. Two varieties of corn produce the following yields per plot:

Data	x_1	48	56	49	51	52	
	x_2	52	56	58	54	55	49

Are the yield variances for the two varieties equal?

4. Samples from 2 normal distributions produce the following data: $n_1 = 21$, $n_2 = 25$, and

$$\sum_{1}^{21} R_{i1} = 407.$$

Is H_o: $\sigma_1^2 = \sigma_2^2$ favored over H_1: $\sigma_1^2 \neq \sigma_2^2$ if $\alpha = .05$?

11.4 OUTLINE OF TESTING PROCEDURES

The key results and test statistics for the testing procedures presented in Chapters 8 - 11 are outlined here for easy reference. The particular section in the text which discusses the main topics is indicated by page number in parentheses.

I. Basic definitions (page 316)

II. Steps of a statistical test (page 322)

III. Uniform models (page 323)

IV. Binomial models
   A. Exact tests for H_o: $p = p_o$ (page 334)
   B. Approximate chi-square tests for H_o: $p = p_o$ (page 344)

$$Y = (X - np_o)^2/np_o(1 - p_o) \sim \chi^2(1)$$

   C. Chi-square test for H_o: $p_1 = p_2 = \ldots = p_c$ for c binomial models (page 347)

$$Y = \sum_{1}^{c} (X_i - n_i\hat{p}_o)^2/n_i\hat{p}_o(1 - \hat{p}_o) \sim \chi^2_{(c-1)},$$

where $\hat{p}_o = \sum_{1}^{c} X_i / \sum_{1}^{c} n_i$

V. Multinomial models
Tests for H_o: $p_i = p_{io}$, $i = 1, 2, \ldots, k$ (page 349)

$$Y = \sum_1^k (X_i - np_{io})^2/np_{io} \sim \chi^2(k-1)$$

VI. Tests of independence for $r \times c$ contingency tables (page 353)

$$Y = \sum_1^r \sum_1^c (n_{ij} - E_{ij})^2/E_{ij} \sim \chi^2[(r-1)(c-1)],$$

where $E_{ij} = n_{i.}\ n_{.j}/n$

VII. Goodness of fit procedures (page 361)
A. Uniform models (page 362)
B. Binomial models (page 367)
C. Normal models (page 371)
The test statistic for these procedures is of the form

$$Y = \sum_1^k (f_i - np_i)^2/np_i \sim \chi^2(k-1-m),$$

where k is the number of categories used to form a distribution of the data, and m is the number of parameters estimated by using the data.

VIII. Tests of location
A. Techniques using one sample
1. Classical tests for H_o: $\mu = \mu_o$ (page 381)

a. $Z = (\overline{X} - \mu)/(\sigma/\sqrt{n}) \sim N(0,1)$

b. $t = (\overline{X} - \mu)/(S/\sqrt{n}) \sim t(n-1)$

2. Modified Wilcoxon nonparametric procedure (page 388)

*B. Paired sample tests (page 392)
C. Procedures using two or more samples (page 400)
1. Classical one-way analysis of variance for H_o: $\mu_1 = \mu_2 = \ldots = \mu_k$ (page 401)

$$F = [\Sigma n_j(\overline{x}_j - \overline{x})^2/(k-1)]/[\Sigma\Sigma(x_{ij} - \overline{x}_j)^2/(m-k)]$$

$$\sim F(k-1,m-k),$$

where m is the combined sample size.

2. Kruskal-Wallis nonparametric one-way analysis of variance (page 413)

$$H = [12/m(m+1)]\sum_{1}^{k} (A_j^2/n_j) - 3(m+1) \sim \chi^2(k-1)$$

3. Two-sample tests (page 418)

a. z-test (page 418)

$$Z = (\bar{X}_1 - \bar{X}_2)/\sqrt{(\sigma_1^2/n_1 + \sigma_2^2/n_2)} \sim N(0,1)$$

b. t-test (page 420)

$$t = (\bar{X}_1 - \bar{X}_2)/\sqrt{\Sigma\Sigma(X_{ij} - \bar{X}_j)^2/(n_1 + n_2 - 2)} \cdot$$

$$\sqrt{(1/n_1 + 1/n_2)} \sim t(n_1 + n_2 - 2)$$

c. Mann-Whitney test (page 424)

$$T = \Sigma R_{i1} - [n_1(n_1 + 1)/2]$$

*IX. Tests of dispersion

A. Classical chi-square test for H_o: $\sigma^2 = \sigma_o^2$ (page 395)

$$Y = (n-1)S^2/\sigma_o^2 \sim \chi^2(n-1)$$

B. Classical F-test for H_o: $\sigma_1^2 = \sigma_2^2$ (page 429)

$$F_1 = S_1^2/S_2^2 \quad \text{or} \quad F_2 = S_2^2/S_1^2$$

C. Siegel-Tukey nonparametric procedure for dispersion (page 431)

11.5 REFERENCES

[1] King, A. S. 1973. Organizational innovation and experimental effects: a conceptual schema for determining the effect of management innovations and expectations in differing organizational contents. *Social Science Quarterly* 54: 384 - 395.

[2] Kruskal, W. H., and Wallis W. A. 1952. Use of ranks on one-criterion variance analysis. *Journal of the American Statistical Assoc.* 47: 583 - 621.

[3] Conover, W. J. 1971. *Practical nonparametric statistics.* New York: Wiley.

[4] Mann, H. B., and Whitney, D. R. 1947. On a test of whether one of two random variables is stochastically larger than the other. *The Annals of Math. Statistics* 18: 50 - 60.

[5] Verdooren, L. R. 1963. Extended tables of critical values for Wilcoxon's test statistic. *Biometrika* 50: 177 - 186.

[6] Siegel, S., and Tukey, J. W. 1960. A nonparametric sum of ranks procedure for relative spread in unpaired samples. *Journal of the American Statistical Assoc.* 55: 429 - 444.

[7] Hartley, H. O. 1950. The maximum F-ratio as a short-cut test for heterogeneity of variance. *Biometrika* 37: 308 - 312.

11.6 REVIEW EXERCISES

1. Four gasolines are tested by running 6 mileage trials for each gasoline in a test car. A measured amount of gasoline is used for each trial, and the resulting mileage is recorded. The data are as follows:

Gasoline	1	2	3	4
	15.2	18.6	16.1	17.0
	16.4	18.2	17.1	15.3
	17.5	13.4	18.6	15.6
	16.2	11.6	17.5	15.3
	17.3	16.6	20.6	17.2
	14.9	16.4	18.1	20.6

Complete an ANOVA for the test data, and draw appropriate conclusions.

2. Complete a Kruskal-Wallis analysis for the data in Exercise 1, and compare the results with those for the previous ANOVA.

3. Use a 2-sample t-test procedure to test H_o: $\mu_1 = \mu_3$ against H_1: $\mu_1 < \mu_3$ for gasolines #1 and #3 in Exercise 1.

4. The following data show a random sample of the heights for 10 players from each of 2 professional football teams.

Data	x_1	73, 72, 75, 78, 77, 78, 72, 74, 72, 68
	x_2	74, 74, 74, 71, 76, 74, 73, 76, 75, 75

Use a t-test to test H_o: $\mu_1 = \mu_2$ against H_1: $\mu_1 \neq \mu_2$ with $\alpha = .05$.

*5. Use an F-test to test H_o: $\sigma_1^2 = \sigma_2^2$ against H_1: $\sigma_1^2 \neq \sigma_2^2$ for the data in Exercise 4. Choose any appropriate α.

6. The following exam scores represent random samples from 2 classes for the same exam.

Data	x_1	87, 90, 63, 55, 26, 58, 23, 47, 99
	x_2	60, 72, 20, 58, 54, 81, 37, 43, 60

Use a Mann-Whitney test to determine whether or not the level of performance is the same for both classes.

*7. Use an F-test to test H_o: $\sigma_1^2 = \sigma_2^2$ against H_1: $\sigma_1^2 \neq \sigma_2^2$ for the data in Exercise 6.

*8. Complete a Siegel-Tukey analysis for the score data in Exercise 6, and compare your results with the results of Exercise 7.

9. An exam in basic statistics has been taken by 32 students, and 22 of those students have returned to see their exam results, while 10 did not come to see their results. To determine if the following data tend to support the hypothesis that the better students want to see their papers, test H_o: $\mu_1 = \mu_2$ by any procedure that you think is appropriate.

Data	x_1	81, 72, 26, 90, 46, 20, 69, 64, 60, 98, 64, 70
		63, 25, 99, 55, 37, 92, 43, 75, 87, 52
	x_2	23, 81, 60, 33, 68, 47, 67, 88, 37, 58

10. Use the following experiment to test the ability of people to guess the length of a line. First, obtain 2 sheets of plain paper. Draw a vertical line 5.5 inches long on one sheet, and draw a horizontal line 5.5 inches long on the other sheet. Have 2 independent groups (10 or more) of people guess the lengths of the two lines. Record X_v for the vertical guesses by one group and X_h for the horizontal guesses of the second group. Use data to test H_o: $\mu_v = \mu_h$.

CHAPTER 12

Simple Linear Regression and Correlation

12.1 INTRODUCTION

Thus far in our study, most analyses have been made using measurements of *one* variable for a set of individuals. That is, some questions regarding the elements of a designated population were posed, and the answers to those questions were obtained by studying a sample of measurements relative to one variable. For example, a sample of student exam scores was studied to make inference about either the average score, μ, for all students or the variance of the scores, σ^2. However, when introducing the basic concepts in Chapter 1, we indicated that each element in a population can be the source for numerous measurements. As an example, any student can be the source for numerical measurements such as age, height, weight, exam score in elementary statistics, college cumulative grade point average, and college entrance exam score. Then the question or questions of interest may pertain to relationships among various variables. For example, the contingency tables in Chapter 9 are used to investigate independence relationships between categories of 2 variables. In a similar fashion, the procedures in the chapter investigate relationships between *2 variables* by studying 2 measurements from each population element selected for the sample.

The data in this chapter consist of n pairs of observations (random variables) denoted by (X_1, Y_1), (X_2, Y_2), ..., (X_n, Y_n). The notation indicates that both an X-measurement and a Y-measurement are obtained for each of the n elements selected for the sample. Some examples of variables, X and Y, are as follows:

- X = years of education; Y = lifetime earnings
- X = college entrance exam score; Y = cumulative grade point average
- X = height; Y = weight
- X = age; Y = height
- X = age; Y = weight
- X = pounds of fertilizer per acre; Y = yield per acre
- X = sales aptitude test score; Y = monthly sales
- X = course grade in algebra; Y = score on first statistics exam
- X = age of an automobile; Y = value of an automobile.

The techniques of this chapter are directed toward answering the following types of questions:

(1) Given an observed value of the variable X, what linear function of X provides a good estimate of Y?

(2) Are X and Y linearly dependent? That is, does a linear function, say $\hat{Y} = a + bX$, provide a good estimate of Y?

Thus, we are investigating the association between the 2 variables, X and Y. In one case, we use the data (n pairs of measurements) to determine a linear equation for predicting Y-values for given X-values. That type of study is called *regression analysis*. For the second case, we use the data to investigate whether or not X and Y tend to vary in a linear (straight line) fashion. *Correlation analysis* is the name for that type of study.

One or more of the following experiments can be used to collect data for use throughout this chapter.

EXPERIMENT 1: Obtain the heights (X) and weights (Y) of 15 men (or women).

EXPERIMENT 2: Obtain the ages (X) and heights (Y) of 15 men (or women).

EXPERIMENT 3: Secure the college entrance exam scores (X) and the cumulative grade point average (Y) for 15 students.

EXPERIMENT 4: Secure the cumulative grade point average (X) and the class average to the present time in this statistics course (Y) for each member of the class.

Note that in each experiment an X measurement and a Y measurement are obtained from *each* individual selected for the sample.

The presentation of Chapter 12 is separated into two major divisions and numerous subdivisions. Section 12.2 develops regression analysis. Correlation analysis is discussed in Section 12.3. The topics of estimation and hypothesis testing for both methods are presented within those sections.

12.2 SIMPLE LINEAR REGRESSION

Simple linear regression analysis uses a random sample of observations, say

$$[(x_1, y_1), (x_2, y_2), \ldots, (x_n, y_n)], \qquad (12.1)$$

to determine a linear function of one variable, say X, to predict a value for another variable, say Y. In that context, we refer to X as the *independent variable* and to Y as the *dependent variable*. Thus, our objective is to determine coefficients a and b so that

$$\hat{y} = a + bx \qquad (12.2)$$

is the best linear equation for predicting values of Y, say $\hat{y}$, for known values of X. The exact meaning of *best* is discussed in the section on page 446, after the assumptions for regression analysis are presented and illustrated.

There are three basic assumptions for simple linear regression analysis:

(1) The independent variable X is fixed. That is, we suppose that there is a given set of points, $(x_1, x_2, \ldots, x_n)$, for which we want to predict y-values.

(2) For each specified value of X, say x, there exists a normal distribution for the dependent variable, Y, such that the conditional mean of

Y given x, $\mu_{Y|x}$, is a linear function of x. In particular, we assume that

$$\mu_{Y|x} = A + Bx. \qquad (12.3)$$

(3) All of the normal distributions of Y for each value of x have the same variance. That is,

$$\sigma^2_{Y|x_1} = \sigma^2_{Y|x_2} = \dots = \sigma^2_{Y|x_n} = \sigma^2. \qquad (12.4)$$

We use σ^2 to denote the common variance and say that the distributions of Y are *homoscedastic* (have the same variances).

Those three assumptions can be graphically presented for 4 values of x as in Figure 12.1. Notice that the means of the 4 normal distributions lie on the line, $\mu_{Y|x} = A + Bx$. Furthermore, the 4 normal distributions are pictured as having the same variance.

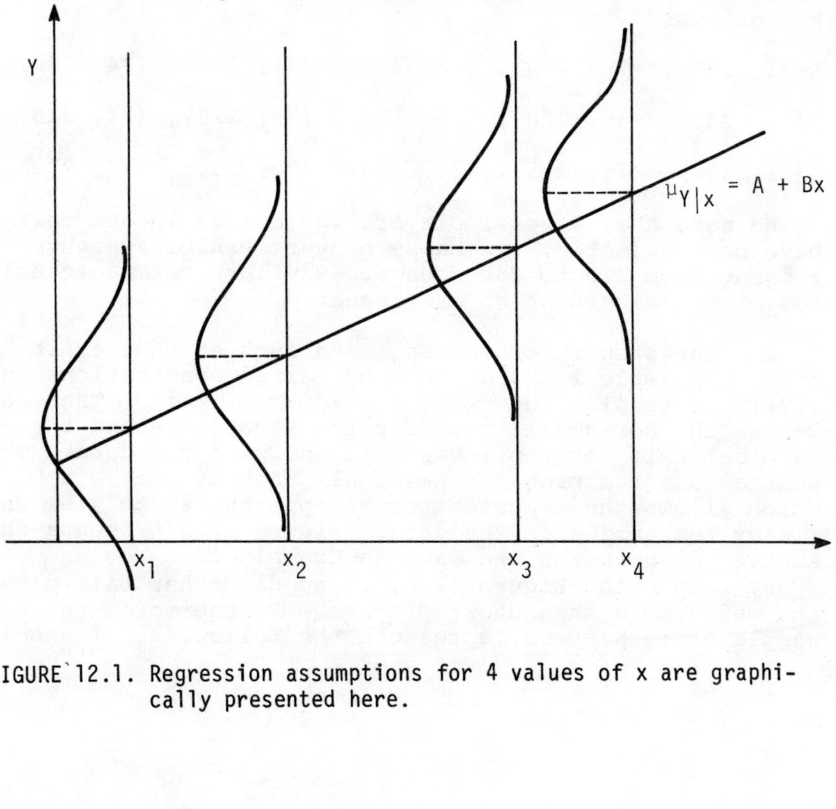

FIGURE 12.1. Regression assumptions for 4 values of x are graphically presented here.

When we sample a situation such as that shown in Figure 12.1, we expect to find several different values of Y paired with the same value of x. This is natural because we assume that for each x there is an infinity of possible Y values. For example, all men who are 6 feet tall are not the same weight. Actually, the set of men 6 feet tall exhibits a distribution of weights. That distribution of weights is assumed to be normal for the discussion of regression. Of course, the weights for men (or women) of any other given heights are also assumed to be normal. However, in any population, men (or women) of different heights are not segregated. So when selecting a random sample, we expect to observe a mixture of individuals who exhibit a variety of heights and weights. The following example illustrates the point.

EXAMPLE 1: The combined rosters of football players for the 2 teams playing in the Super Bowl have been sampled to determine whether or not the weights of professional football players are a linear function of their heights. By expressing the measurements as ordered pairs (X, Y), where X denotes the player's height in inches and Y denotes the player's weight in pounds, the sample data are as follows:

(71, 178), (76, 250), (74, 250), (71, 190), (74, 230),

(70, 205), (75, 220), (69, 170), (74, 225), (74, 215),

(71, 204), (76, 235), (74, 218), (78, 250), (77, 270).

We note that several players who are 74 inches tall have been selected, yet those players exhibit weights ranging from 215 to 250 pounds. Similar statements hold for other heights which are repeated.

A first step in organizing data such as that exhibited in Example 1 and denoted by paired observations in (12.1) is to plot the data in a *scatter diagram*. That is, we use the horizontal axis for the X-variable and the vertical axis for the Y-variable and plot the data. The scatter diagram provides a visual graph of the data which allows the experimenter an opportunity to view any linear tendencies in the data. Figure 12.2 presents the scatter diagram for the data in Example 1.

By inspecting Figure 12.2, it appears that tall players weigh more than short players. Furthermore, the sample of 15 players represents the following: 1 sample

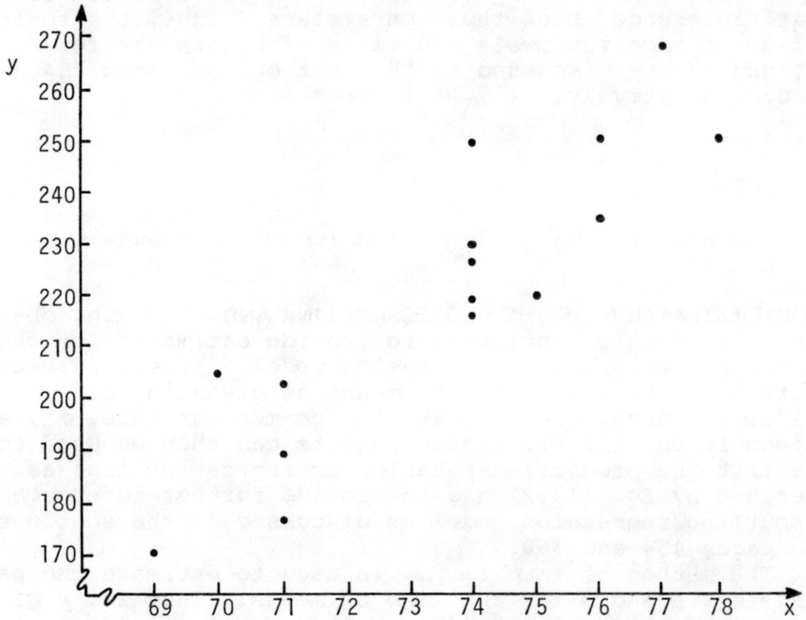

FIGURE 12.2. A scatter diagram is shown for the weights and heights of 15 professional football players, as presented in Example 1.

each from the distributions of weights for heights of 69, 70, 75, 77, and 78 inches; 2 samples from the weight distribution for 76 inches; 3 samples from the weight distribution for 71 inches; and 5 samples from the weight distribution for 74 inches. Also, a quick inspection of the scatter diagram indicates that the changes in weights for changes in heights could be modeled by a straight line. That is, when trying to select a best curve to describe the relationship between the weights and heights of the professional football players, a straight line appears to be a reasonable choice.

With those ideas in mind, we devote the next section to the development and presentation of methods for estimating the line of means, $\mu_{Y|x} = A + Bx$, in the linear regression model. In addition to obtaining point estimates for A and B in order to estimate the line of means, we also determine an estimate for the common variance, σ^2. After establishing point estimates for the 3 parameters, A, B, and σ^2, our efforts are directed toward inference about those parameters. Thus, the topics of confidence intervals and tests of hypotheses for A, B, and σ^2 are discussed in the sections on pages 454 and 460, respectively.

EXERCISE

1. Plot scatter diagrams for any data for the experiments in Section 12.1.

POINT ESTIMATION OF THE REGRESSION LINE AND σ^2: The objective of this section is to provide estimates for the three parameters in a regression model. First, we estimate A and B in the line of means as given in Eq. (12.3). Second, we estimate the common variance, σ^2, as given in Eq. (12.4). Those results can then be used to exhibit the prediction equation or regression line as defined by Eq. (12.2) and to provide further inference about the regression model as discussed in the sections on pages 454 and 460.

The method of *least squares* is used to estimate the parameters A and B for the line of means. That is, a given set of observations $[(x_1, y_1), (x_2, y_2), \ldots, (x_n, y_n)]$ is used to determine a and b so that the sum of squared deviations,

$$\sum_{i=1}^{n} (y_i - \hat{y}_i)^2 = \sum_{i=1}^{n} [y_i - (a + bx_i)]^2, \qquad (12.5)$$

is minimum. In Eq. (12.5), the choice of a and b is such that the sum of squared deviations between the *observed y_i-values* and the *predicted y_i-values*, $\hat{y}_i$, is minimized. The minimization property of a and b is our definition of best line. Note that in relation to a scatter diagram such as Figure 12.2, the sum of squared vertical deviations is minimized.

By using the calculus, it can be shown that the desired a and b must satisfy the following pair of simultaneous equations:

$$\begin{aligned} a \cdot n + b\,\Sigma x_i &= \Sigma y_i \\ a\,\Sigma x_i + b\,\Sigma x_i^2 &= \Sigma x_i y_i. \end{aligned} \qquad (12.6)$$

The simultaneous solution for the equations in (12.6) is

$$b = [\Sigma x_i y_i - (\Sigma x_i)(\Sigma y_i)/n]/[\Sigma x_i^2 - (\Sigma x_i)^2/n]$$
$$a = \overline{y} - b\overline{x} = (\Sigma y_i - b\ \Sigma x_i)/n. \tag{12.7}$$

Proof: Solving the first equation in (12.6) for a gives

$$a = (\Sigma y_i - b\ \Sigma x_i)/n.$$

Substituting that value into the second equation in (12.6) gives

$$[(\Sigma y_i - b\ \Sigma x_i)/n](\Sigma x_i) + b\ \Sigma x_i^2 = \Sigma x_i y_i.$$

Therefore,

$$b[\Sigma x_i^2 - (\Sigma x_i)^2/n] = \Sigma x_i y_i - (\Sigma x_i)(\Sigma y_i)/n.$$

Thus,

$$b = [\Sigma x_i y_i - (\Sigma x_i)(\Sigma y_i)/n]/[\Sigma x_i^2 - (\Sigma x_i)^2/n].$$

At this point, we take a brief aside to offer some rationale for b and relate it to our earlier study of variances. In Chapter 3, we used the variance of a sample to describe the variation of the observations (x_1, x_2, ..., x_n) from their mean, $\overline{x}$. Equations (3.18) and (3.19) defined

$$s^2 = \Sigma(x_i - \overline{x})^2/(n - 1) = [\Sigma x_i^2 - (\Sigma x_i)^2/n]/(n - 1).$$

When studying measurements on two variables, X and Y, by means of a bivariate sample, $\{(x_1, y_1), (x_2, y_2), \ldots, (x_n, y_n)\}$, we can still use the usual definitions to calculate the sample means, $\overline{x}$ and $\overline{y}$, and the sample variances, s_x^2 and s_y^2. However, we can also calculate an additional descriptive statistic called the *covariance of X and Y*. That statistic is defined by

$$\text{cov}(x, y) = \Sigma(x_i - \overline{x})(y_i - \overline{y})/(n - 1)$$
$$= \{\Sigma x_i y_i - [(\Sigma x_i)(\Sigma y_i)/n]\}/(n - 1). \tag{12.8}$$

Note that the numerator of cov(x, y) is the *product* of the deviations of x_i and y_i from the respective means, $\overline{x}$ and $\overline{y}$, rather than the *square* of a deviation for one variable in the variance. Furthermore, the terms x_i^2 and $(\Sigma x_i)^2$ in s^2 are replaced by $x_i y_i$ and $(\Sigma x_i)(\Sigma y_i)$, respectively, in cov(x, y). Thus, cov(x, y) is a measure of how the observation points (x_i, y_i) vary from the mean point $(\overline{x}, \overline{y})$ and have the name covariance. Notice that the covariance may either be negative or positive. In particular, cov(x, y) is *negative* when $(x_i - \overline{x})$ and $(y_i - \overline{y})$ have *different* signs for most values of i and *positive* when $(x_i - \overline{x})$ and $(y_i - \overline{y})$ have the *same* sign for most values of i.

The foregoing discussion shows that b is just the ratio of the covariance, cov(x, y), to the variance, v(x) $= s_x^2$. Thus the sign of b is the same as the sign of cov(x, y). Also, we see that the formulas in (12.7) may be written as

$$b = \Sigma(x_i - \overline{x})(y_i - \overline{y})/\Sigma(x_i - \overline{x})^2,$$

$$a = \overline{y} - b\,\overline{x}. \qquad (12.9)$$

If $\overline{x}$ and $\overline{y}$ are integers, then (12.9) provides a convenient way for completing the necessary calculations for obtaining a and b. However, when either $\overline{x}$ or $\overline{y}$ is not an integer, (12.7) is usually a more convenient method for computing the estimates, a and b. That convenience is equivalent to the convenience provided by (3.19) over (3.18) when calculating the variance. The following examples illustrate the preceding ideas and concepts.

EXAMPLE 2: To illustrate the basic ideas of least squares estimation of A and B, we consider the following 5 pairs of hypothetical data: (1, 1), (1, 3), (3, 3), (5, 3), (5, 5). Find the estimates, a and b.

SOLUTION: Method #1. Using (12.7) to calculate a and b, we need the following 4 totals: Σx_i, Σx_i^2, Σy_i, and $\Sigma x_i y_i$. One convenient method for obtaining these totals is to use a computation table in which the columns are defined for the 4 quantities: x_i, y_i, x_i^2, and $x_i y_i$, as shown in the table on the next page. By substituting the totals in the computation table and n = 5 into (12.7), it follows immediately that

$$b = [53 - (15)(15)/5]/[61 - (15)^2/5]$$

$$= 8/16 = .5.$$

$$a = [15 - .5\ (15)]/5 = 1.5.$$

x_i	y_i	x_i^2	$x_i y_i$
1	1	1	1
1	3	1	3
3	3	9	9
5	3	25	15
5	5	25	25
15	15	61	53

Method #2. Using (12.9) to calculate a and b, we need to calculate $\overline{x}$, $\overline{y}$, $\Sigma(x_i - \overline{x})(y_i - \overline{y})$, and $\Sigma(x_i - \overline{x})^2$. These computations can be completed in table form as follows:

x_i	y_i	$(x_i - \overline{x})$	$(y_i - \overline{y})$	$(x_i - \overline{x})^2$	$(x_i - \overline{x})(y_i - \overline{y})$
1	1	-2	-2	4	4
1	3	-2	0	4	0
3	3	0	0	0	0
5	3	2	0	4	0
5	5	2	2	4	4
15	15	0	0	16	8

Using the totals in the table with n = 5 gives $\overline{x}$ = 15/5 = 3 and $\overline{y}$ = 15/5 = 3. Substitution of the above values into (12.9) gives the desired estimates to be as follows:

$$b = 8/16 = .5.$$

$$a = 3 - (.5)(3) = 1.5.$$

Thus, the estimates provided by (12.7) and (12.9) yield identical results. Notice that for this example cov(x, y) = 8/4 = 2 and s_x^2 = 16/4 = 4. Thus,

$$b = \text{cov}(x, y)/s_x^2 = 2/4 = .5,$$

as for both methods previously described.

The use of a prediction equation and the optimal nature of a least-squares regression solution are illustrated by the next example.

EXAMPLE 3: Find the prediction equation for the data in Example 2. Find a predicted y-value for x=4. Graph the prediction equation $\hat{y} = a + bx$ and the equation $\tilde{y} = x$ on a scatter diagram for the data. Calculate $\Sigma(y_i - \hat{y}_i)^2$ and $\Sigma(y_i - \tilde{y})^2$.

SOLUTION: From Example 2, b = .5 and a = 1.5. Therefore, the prediction equation is

$$\hat{y} = 1.5 + .5x.$$

Thus, if x = 4, the predicted value of Y is

$$\hat{y} = 1.5 + .5(4) = 3.5.$$

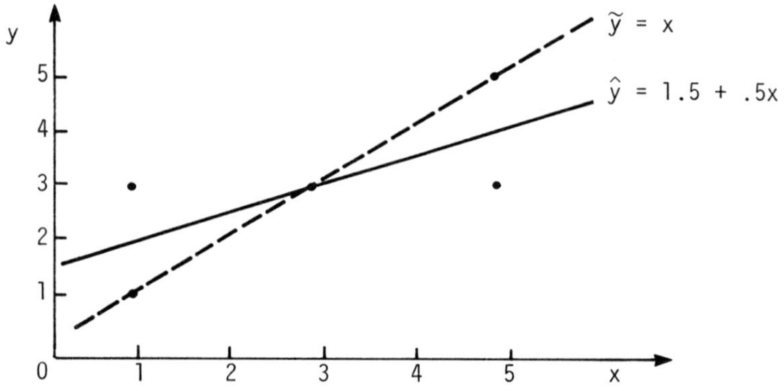

The figure above shows the scatter diagram and the lines, $\tilde{y} = x$ and $\hat{y} = 1.5 + .5x$.

x	y	$\hat{y}$	$\tilde{y}$	$(y - \hat{y})^2$	$(y - \tilde{y})^2$
1	1	2	1	1	0
1	3	2	1	1	4
3	3	3	3	0	0
5	3	4	5	1	4
5	5	4	5	1	0
				4	8

Note that the least-squares line, $\hat{y} = 1.5 + .5x$, passes through the middle of the scatter diagram and actually hits 1 of the 5 points, while the line, $\tilde{y} = x$, passes through 3 of the 5 data points. When judging closeness by the sum of squared deviations, however, the least-squares line shows $\Sigma(y - \hat{y})^2 = 4$, as opposed to $\Sigma(y - \tilde{y})^2 = 8$, for the other line. This illustrates the meaning of least in the term, least squares. In fact, the a and b determined by the method of least squares is such that the sum of squared deviations for any line other than $\hat{y} = a + bx$ is greater than $\Sigma(y - \hat{y})^2$.

The primary use for a least-squares regression line is the prediction of y-values for observed x-values. For example, a college admissions officer may use the entrance examination score (x) for a prospective student to predict a cumulative grade point average (y) for that student. Or a farmer may predict the yield (y) for a crop in terms of the amount of fertilizer (x) applied to the field. Or a businessman may predict next month's sales (y) in terms of either the economy (x) or last month's sales (x).

We observe that the predictions provided by a regression equation are predictions for the unknown mean values for the normal distributions associated with various values of x. That is, given the x-value associated with any regression problem, we predict the mean of the corresponding y-distribution as the y-value for the individual. Thus, it is true that the predicted value for both the mean, $\mu_{y|x}$, and the y-value for a specific x-value is $\hat{y} = a + bx$.

One further comment on the use of a regression equation concerns the range of x-values for which the equation is useful. The assumptions for a regression analysis state that there is a fixed set of x-values. Thus, it is to that initial reference set that the prediction equation is applicable. For example, the heights (y) of teenage boys (or girls) can be expressed as a linear function of their age (x). However, if a model derived from a set of (age, height) data for teenagers were applied to a young adult of age x = 25, the resulting prediction is likely to be ridiculous. Even more ridiculous is the possibility that the teenage model may yield a negative height for very young individuals (e.g., x = 2). Thus, when using regression models, we must remain ever aware of the range of x-values for which the prediction equation is useful.

We now conclude this section by providing an estimator for the common variance, σ^2, for the normal distributions present in a regression model. The usual estimator is

$$s^2_{y|x} = \sum_{i=1}^{n} (y_i - \hat{y}_i)^2/(n - 2). \qquad (12.10)$$

It can be shown that $s^2_{y|x}$ is an unbiased estimator of σ^2. Furthermore, it resembles the earlier sample variance estimator (7.5) of a variance σ^2, in that $\hat{y}_i$ is an estimator of the mean, $\mu_{y|x}$. Thus, the numerator of $s^2_{y|x}$ is the sum of squared deviations between a set of observations, y_i, and the predicted (estimated) means, $\hat{y}_i$.

The use of Eq. (12.10) to estimate σ^2 requires that the $\hat{y}_i$ values for all i be available, which in turn requires the evaluation of the prediction equation for each x_i. When n is large, that task can be a great burden, so it is expedient to find an alternative expression for $s^2_{y|x}$. Algebraic manipulation of Eq. (12.10) provides the following useful alternative definitions of $s^2_{y|x}$.

$$s^2_{y|x} = (\Sigma y_i^2 - a\,\Sigma y_i - b\,\Sigma x_i y_i)/(n - 2).$$

$$s^2_{y|x} = [\Sigma (y_i - \bar{y})^2 - b\,\Sigma (x_i - \bar{x})(y_i - \bar{y})]/(n - 2). \qquad (12.11)$$

The next example illustrates the use of both Eqs. (12.10) and (12.11) to estimate σ^2.

EXAMPLE 4: Use the hypothetical data in Example 2 to estimate σ^2.

SOLUTION: Method #1. To use Eq. (12.10) we refer to Example 3 and find that $\Sigma(y_i - \hat{y}_i)^2 = 4$. Since n = 5,

$$s^2_{y|x} = 4/3.$$

Method #2. To use (12.11) we refer to Example 2 to find a = 1.5, b = .5, $\Sigma y_i = 15$, and $\Sigma x_i y_i = 53$. Next

$$\Sigma y_i^2 = 1 + 9 + 9 + 9 + 25 = 53.$$

Thus,

$$s^2_{y|x} = [53 - (1.5)(15) - .5(53)]/3 = 4/3.$$

The trade-off between Eqs. (12.10) and 12.11) for calculating $s^2_{y|x}$ is between calculating $\hat{y}_i$ to obtain

$\Sigma(y_i - \hat{y}_i)^2$ and calculating Σy_i^2 to use with the available values for a, b, Σy_i, and $\Sigma x_i y_i$ to obtain $\Sigma y_i^2 - a\,\Sigma y_i - b\,\Sigma x_i y_i$. Frequently, the latter is the easier method. Furthermore, the results from (12.11) offer greater precision whenever the quantities a, b, and $\hat{y}_i$ involve round-off errors.

Following the presentation of some exercises to further illustrate the point estimation techniques for regression models, we shift our interest to the problems of finding confidence intervals for A, B, and σ^2.

EXERCISES

1. Find the sample regression line and the estimate of σ^2 for data sets that yield the following totals:
   (a) $\Sigma x_i = 10$, $\Sigma y_i = 20$, $\Sigma x_i^2 = 35$, $\Sigma y_i^2 = 76$, $\Sigma x_i y_i = 45$, $n = 10$.
   (b) $\Sigma x_i = 10$, $\Sigma y_i = 20$, $\Sigma x_i^2 = 35$, $\Sigma y_i^2 = 76$, $\Sigma x_i y_i = -5$, $n = 10$.
   (c) $\overline{x} = 5$, $\overline{y} = 10$, $\Sigma(x_i - \overline{x})^2 = 10$, $\Sigma(y_i - \overline{y})^2 = 40$, $\Sigma(x_i - \overline{x})(y_i - \overline{y}) = 16$, $n = 20$.

2. Consider the following data:

x	1	2	3	4	5
y	4	4	6	10	12

   (a) Find the cov(x, y).
   (b) Find the least-squares regression line.
   (c) Find $s^2_{y|x}$.
   (d) Find $\hat{y}$ for x = 3.
   (e) Plot the regression line and the data on a scatter diagram.

3. Consider the following data:

x	1	3	5	7
y	10	7	7	4

   (a) Find the cov(x, y).
   (b) Find the least-squares regression line.
   (c) Find $s^2_{y|x}$.
   (d) Find $\hat{y}$ for x = 4.
   (e) Plot the regression line and the data on a scatter diagram.

4. Find the regression line for the football players' heights and weights data in Example 1 on page 444. What weight is predicted for a player who stands 75 inches tall? Would the model be appropriate for basketball players?

5. A sample of 10 brand X used cars (1 - 4 years old) has been studied to determine the age (X) and the price (Y). The data shows that $\overline{x} = 2$ years, $\overline{y} = \$2000$, $\Sigma(x_i - \overline{x})^2 = 6$, $\Sigma(y_i - \overline{y})^2 = 6{,}000{,}000$, and $\Sigma(x_i - \overline{x})(y_i - \overline{y}) = -3600$. Find the regression line for the data. Predict the value of a brand X car which is 3 years old. Is the model appropriate for predicting the value of a 7-year-old brand X car?

6. The ages (X) and weights (Y) for 20 elementary school boys revealed that $\overline{x} = 9.5$ years, $\overline{y} = 75$ pounds, $\Sigma(x_i - \overline{x})^2 = 20$, $\Sigma(y_i - \overline{y})^2 = 400$, $\Sigma(x_i - \overline{x})(y_i - \overline{y}) = 80$. Find the prediction equation for the data, and predict the weight for an 11-year-old boy. Is the prediction equation appropriate for boys of age 16?

7. Find a least-squares prediction equation for any data available for the experiments in Section 12.1.

CONFIDENCE INTERVALS AND TESTS OF HYPOTHESES FOR THE PARAMETERS IN A REGRESSION ANALYSIS: The preceding discussion has centered on the determination of point estimators for A, B, and σ^2. From the estimators, a and b, we define $\hat{y} = a + bx$ to be an estimator for $\mu_{Y|x} = A + Bx$. Thus, the four values in any regression analysis for which inference is desired are A, B, σ^2, and $\mu_{Y|x}$. We present $(1 - \alpha)$-level two-sided confidence intervals for those four parameters in the section below. Then the section on page 460 provides tests of hypotheses for the parameters A, B, and σ^2.

We do not need the assumption that Y for given values of X is a normal variable in the point estimation techniques of the last section. However, that assumption is basic to all the statistical inference techniques in this section. In fact, the normality assumption provides that the estimators, $\hat{y}$, a, and b, have normal sampling distributions and that the estimator, $(n - 2)S^2_{Y|x}/\sigma^2$, has a chi-square sampling distribution. Thus, the inference techniques of this section are similar to those used in Chapters 7 and 11 to make inference about the mean and variance of a normal variable. The similarities appear throughout the development.

Since the pivotal quantities for obtaining the desired confidence intervals and the test statistics for testing the indicated hypotheses are similar, we present both forms of inference before giving any exercises.

*** INTERVAL ESTIMATIONS FOR $\mu_{Y|x}$, A, B, AND σ^2:** The results in this section are similar to the confidence intervals presented in Chapter 7. Basic to all of the results is the assumption that the y-values for each x-value are normally distributed. A factor that plays a dominant role in all the inference of both this section and of the section on page 460 is that all variances and standard deviations involved are a function of $S^2_{Y|x}$, the estimate of σ^2. Furthermore, the confidence intervals are developed from a pivotal quantity that has a known sampling distribution. Thus, the presentation for each parameter begins with the indication of a pivotal quantity and its sampling distribution. Then inversion techniques similar to those used to justify Eqs. (7.26), (7.34), and (7.40) are employed to produce the desired confidence intervals.

We begin the discussion by consideration of a *confidence interval for the conditional mean*, $\mu_{Y|x}$. Now the point estimate of $\mu_{Y|x}$ is $\hat{y} = a + bx$, and it can be shown that the expected value and variance of $\hat{y}$ are as follows:

$$E(\hat{y}) = \mu_{Y|x},$$
$$V(\hat{y}) = \sigma^2[1/n + (x - \overline{x})^2/\sum_1^n (x_i - \overline{x})^2]. \tag{12.12}$$

Since the y-values are assumed to be normal variables,

$$t = (\hat{y} - \mu_{Y|x})/S_{\hat{y}} \sim t(n - 2), \tag{12.13}$$

where

$$S_{\hat{y}}^2 = S_{y|x}^2 \; [1/n + (x - \overline{x})^2/\Sigma(x_i - \overline{x})^2]$$

is the estimator of $V(\hat{y})$ provided by using $S_{y|x}^2$ in place of σ^2 in (12.12). Thus, the techniques of Section 7.3 show that the two-sided $(1 - \alpha)$-level confidence interval for $\mu_{Y|x}$ is given by

$$CI_{1-\alpha}: \quad \hat{y} - t(n - 2, \alpha/2)\, S_{\hat{y}} \leq \mu_{Y|x} \leq \hat{y} + t(n - 2, \alpha/2)\, S_{\hat{y}}. \tag{12.14}$$

In using (12.14), we note that the t-statistic depends on $(n - 2)$ d.f., rather than on $(n - 1)$ for earlier confidence intervals. Next, observe that $V(\hat{y})$ increases as the x-value for which the CI is desired varies from $\overline{x}$. This property reflects the fact that we have estimated a line by estimating its y-intercept, A, and its slope, B. Thus, the line, $\hat{y} = a + bx$, may differ from the true line, $\mu_{Y|x} = A + Bx$, by having either the wrong intercept or the wrong slope or both. So a longer interval is needed to include the same confidence for the means, $\mu_{Y|x}$, of those normal distributions depending on an x which is far removed from $\overline{x}$. The $(x - \overline{x})^2$ factor in $V(\hat{y})$ provides the proper adjustment in the length of interval.

EXAMPLE 5: In Examples 2 - 4, we found a regression equation, $\hat{y} = 1.5 + .5x$, and an estimate of σ^2, $s^2_{Y|x} = 4/3$, for the 5 data points, (1, 1), (1, 3), (3, 3), (5, 3), and (5, 5). Use that information to find 95% confidence intervals for $\mu_{Y|3}$ and $\mu_{Y|5}$.

SOLUTION: From Example 2, $\Sigma(x_i - \bar{x})^2 = 16$. From Table E1, $t(3, .025) = 3.18$. When $x = 3$,

$$s^2_{\hat{Y}} = (4/3)\,[1/5 + (3 - 3)^2/16] = 4/15$$

and

$$\hat{y} = 1.5 + .5(3) = 3.$$

Therefore,

$$CI_{95}:\quad 3 - 3.18\sqrt{4/15} \leq \mu_{Y|3} \leq 3 + 3.18\sqrt{4/15}.$$

After simplification,

$$CI_{95}:\quad 1.36 \leq \mu_{Y|3} \leq 4.64.$$

Similarly, $x = 5$ gives

$$s^2_{\hat{Y}} = (4/3)[1/5 + (5 - 3)^2/16] = 3/5$$

and

$$\hat{y} = 1.5 + .5(5) = 4.$$

Therefore,

$$CI_{95}:\quad 4 - 3.18\sqrt{3/5} \leq \mu_{Y|5} \leq 4 + 3.18\sqrt{3/5}.$$

Following simplification, we have

$$CI_{95}:\quad 1.54 \leq \mu_{Y|5} \leq 6.46.$$

Notice that the 95% CI for $x = 5$ is 4.92 units long, while the 95% CI for $x = \bar{x} = 3$ is only 3.28 units long. This illustrates the fact that the CI associated with $x = \bar{x}$ is shorter than the CI associated with any other value of x.

The next technique is concerned with a *confidence interval for the parameter* A, which represents the y-intercept (point at which the regression intersects the y-axis) for the model, $\mu_{Y|x} = A + Bx$, and is estimated by a. It can be shown that, under the normality assumption, a is a normal variable with the following mean and variance:

$$E(a) = A.$$
$$V(a) = \sigma^2[1/n + \bar{x}^2/\Sigma(x_i - \bar{x})^2]. \tag{12.15}$$

Thus, the quantity,

$$t = (a - A)/S_a \sim t(n - 2), \tag{12.16}$$

where

$$S_a^2 = S_{y|x}^2 \, [1/n + \bar{x}^2/\Sigma(x_i - \bar{x})^2]$$

is the estimator of V(a) given by replacing σ^2 with $S_{y|x}^2$ in (12.15). The usual algebraic manipulation shows that

$$CI_{1-\alpha}: \quad a - t(n - 2, \alpha/2)\, S_a \leq A \leq a + t(n - 2, \alpha/2)\, S_a \tag{12.17}$$

is the desired two-sided $(1-\alpha)$-level confidence interval for A. One particular point of interest is whether or not 0 is a reasonable value for A. The CI given in (12.17) answers that question as follows: If 0 is in the CI, then 0 is a reasonable value for A. When 0 is not in the CI, we suspect that A is either strictly positive $[a - t(n - 2, \alpha/2)\, S_a > 0]$ or A is strictly negative $[a + t(n - 2, \alpha/2)\, S_a < 0]$. If we believe that $A = 0$, there is a simple ratio relationship between $\mu_{Y|x}$ and x. That is, $\mu_{Y|x} = Bx$. Thus, the means of Y for given values of x are a linear function of x which passes through the origin.

EXAMPLE 6: Continuing with the data used in Example 5, we can find a 95% confidence interval for A.

SOLUTION: From Example 5, $a = 1.5$, $s_{y|x}^2 = 4/3$, $\bar{x} = 3$, $n = 5$, $\Sigma(x_i - \bar{x})^2 = 16$, and $t(3, .025) = 3.18$. Therefore,

$$s_a^2 = (4/3)(1/5 + 3^2/16) = 61/60$$

and

$$CI_{95}: \quad 1.5 - 3.18\sqrt{61/60} \leq A \leq 1.5 + 3.18\sqrt{61/60}.$$

After some arithmetic simplification, we find that

$$CI_{95}: \quad -1.71 \leq A \leq 4.71.$$

Since the CI spans 0, it is possible that there exists a simple ratio relationship between $\mu_{Y|x}$ and x.

In turning our attention to *confidence intervals for the slope parameter*, *B*, we again find that the assumption of normally distributed values of Y for fixed x-values implies that b is normally distributed with

$$E(b) = B,$$
$$V(b) = \sigma^2/\Sigma(x_i - \bar{x})^2. \tag{12.18}$$

Thus we find that

$$t = (b - B)/S_b \sim t(n - 2), \tag{12.19}$$

where

$$S_b^2 = S_{Y|x}^2/\Sigma(x_i - \bar{x})^2$$

is the estimator of V(b). Replacing σ^2 with $S_{Y|x}^2$ in (12.18) then leads to the following $(1 - \alpha)$-level two-sided confidence interval for B:

$$CI_{1-\alpha}: \quad b - t(n - 2, \alpha/2)\, S_b \leq B \leq b + t(n - 2, \alpha/2)\, S_b. \tag{12.20}$$

As for A, one point of interest when doing inference about B is whether or not B = 0. If B = 0 is a possibility, it follows that $\mu_{Y|x} = A$. That is, the means of Y for the various x-values are equal and neither increase nor decrease as x changes. In that case, the knowledge of x does not provide a better estimator of $\mu_{Y|x}$.

EXAMPLE 7: Use the results available in Example 5 to find a 95% confidence interval for B.

SOLUTION: From Example 5, n = 5, b = .5, $s_{Y|x}^2 = 4/3$, $\Sigma(x_i - \bar{x})^2 = 16$, and $t(3, .025) = 3.18$. Therefore,

$$s_b^2 = (4/3)/16 = 1/12$$

and

$$CI_{95}: \; .5 - 3.18\sqrt{1/12} \le B \le .5 + 3.18\sqrt{1/12}.$$

After arithmetic simplification,

$$CI_{95}: \; -.42 \le B \le 1.42.$$

Since 0 is included in the interval, the small sample available does not provide any evidence that B is different from 0.

To obtain confidence intervals for σ^2, we use the estimator, $S^2_{y|x}$, to define the following χ^2 variable:

$$(n - 2)\, S^2_{y|x}/\sigma^2 = \Sigma(y_i - \hat{y})^2/\sigma^2 \sim \chi^2(n - 2). \quad (12.21)$$

Then the arguments are the same as those for (7.41) to show that the (1 - α)-level two-sided confidence interval is

$$CI_{1-\alpha}: \; \Sigma(y_i - \hat{y})^2/\chi^2(n - 2, \alpha/2) \le \sigma^2 \le \Sigma(y_i - \hat{y})^2/\chi^2(n - 2, 1 - \alpha/2). \quad (12.22)$$

The next example illustrates the necessary arithmetic.

EXAMPLE 8: Use the results of Example 3 to find a 95% confidence interval for σ^2 based on the 5 data points in Example 2.

SOLUTION: From Example 3, $\Sigma(y_i - \hat{y})^2 = 4$ and $n = 5$. Now Table E2 gives $\chi^2(3, .025) = 9.35$ and $\chi^2(3, .975) = .22$. Thus,

$$CI_{95}: \; 4/9.35 \le \sigma^2 \le 4/.22,$$

which is equivalent to

$$CI_{95}: \; .43 \le \sigma^2 \le 18.18.$$

Note that when only 5 data points are available for obtaining the interval for σ^2, the resulting interval is quite wide.

All of the confidence intervals presented and illustrated in this section are two-sided. In those cases where a one-sided interval is desired, the procedure is identical to that used in Chapter 7. We simply use one of the two limits given by the two-sided interval and replace $\alpha/2$ by α to determine the appropriate critical value. The other limit is set equal to its natural limit. That is, when we find $LCI_{1-\alpha}$ for one of the parameters, $\mu_{Y|x}$, A, B, and σ^2, the upper limit is set equal to ∞. Similarly, the lower limit is set equal to $-\infty$ for $\mu_{Y|x}$, A, and B and set equal to 0 for σ^2 when finding $UCI_{1-\alpha}$ for one of the parameters.

***TESTS OF HYPOTHESES FOR A, B, AND σ^2:** The t-variables in Eqs. (12.16) and (12.19) and the χ^2 variable in Eq. (12.21) provide natural test statistics for testing hypotheses about the parameters, A, B, and σ^2, respectively. The purpose of this section is to present and illustrate those testing techniques. An alternate method for testing H_o: B = 0 is presented in Section 12.3.

Just as the discussion in the last section was limited to two-sided confidence intervals, the discussion for tests of hypotheses is limited to two-sided alternative hypotheses. The techniques for one-sided tests are straightforward and identical to those in Chapter 10.

To test that A is some specified value, A_o, we state the following hypotheses:

$$H_o: \quad A = A_o$$

$$H_1: \quad A \neq A_o.$$

The test statistic is defined by

$$t = (a - A_o)/S_a, \qquad (12.23)$$

where S_a^2 is defined for Eq. (12.16). Since the test statistic t has a t-distribution with (n - 2) d.f. when H_o is true, the α-level rejection region for testing H_o against H_1 is

$$R: \quad \{t \mid |t| > t(n - 2, \alpha/2)\}. \qquad (12.24)$$

In a similar fashion, we state the following hypotheses for testing whether or not B is some specified value, B_o:

$$H_o: \quad B = B_o$$

$$H_1: \quad B \neq B_o.$$

The test statistic is

$$t = (b - B_o)/S_b, \tag{12.25}$$

where S_b^2 is defined for Eq. (12.19). Since t is distributed as a t-variable when H_o is true, the α-level rejection region for testing H_o against H_1 is

$$R: \ \{t \,|\, |t| > t(n - 2, \alpha/2)\}. \tag{12.26}$$

A common value of B_o is 0 as the hypothesis H_o: B = 0 is equivalent to saying that all of the Y-distributions have a common mean, i.e., $\mu_{Y|x} = A$. Thus, accepting H_o: B = 0 leads to the conclusion that the mean of Y does not depend on x.

EXAMPLE 9: We use the data in Example 2 to test the following pairs of hypotheses with $\alpha = .05$:

$$\text{(a)}\ H_o:\ A = 2 \qquad \text{(b)}\ H_o:\ B = 0$$

$$H_1:\ A \neq 2. \qquad H_1:\ B \neq 0.$$

SOLUTION: Example 2 gives a = 1.5 and b = .5. Example 6 gives $s_a^2 = 61/60$, and Example 7 gives $s_b^2 = 1/12$. Table E1 gives t(3, .025) = 3.18.
(a) To test H_o: A = 2 against H_1: $A \neq 2$ when $\alpha = .05$, we use the rejection region, R: $\{t \,|\, |t| > 3.18\}$, where $t = (a - 2)/s_a$. The above data give

$$t = (1.5 - 2)/\sqrt{61/60} = -.50.$$

Since $|t| = .50 < 3.18$, we accept H_o. Thus, A = 2 is an acceptable value for the true y-intercept.
(b) To test H_o: B = 0 against H_1: $B \neq 0$ when $\alpha = .05$, we use the rejection region, R: $\{t \,|\, |t| > 3.18\}$, where $t = (b - 0)/s_b$. The data above give

$$t = .5/\sqrt{1/12} = 1.73.$$

Since $|t| = 1.73 < 3.18$, we accept H_o and conclude that 0 is an acceptable value of the slope. That is, the data do not indicate that the regression line has a non-zero slope. So the means, $\mu_{Y|x}$, may be equal for the 3 values of x included in the 5 points.

The procedure for testing hypotheses about σ^2 is the same as that presented in Chapter 10, except that the test statistic now depends on (n - 2) d.f. In particular, we test H_o: $\sigma^2 = \sigma_o^2$ against H_1: $\sigma^2 \neq \sigma_o^2$ by using the test statistic,

$$Y = \Sigma(y_i - \hat{y}_i)^2/\sigma_o^2 = (n - 2)\ s_{y|x}^2/\sigma_o^2, \qquad (12.27)$$

and the two-sided $(1-\alpha)$-level rejection region,

$$R:\ \{y|y < \chi^2(n - 2,\ 1 - \alpha/2) \text{ or } y > \chi^2(n - 2,\ \alpha/2)\}. \qquad (12.28)$$

EXAMPLE 10: Use the data in Example 2 to test H_o: $\sigma^2 = 5$ against H_1: $\sigma^2 \neq 5$ when $\alpha = .05$.

SOLUTION: Examples 3 and 4 give $s_{y|x}^2 = 4/3$. Table E2 gives

$$\chi^2\ (3,\ .975) = .22 \text{ and } \chi^2\ (3,\ .025) = 9.35.$$

Thus, we test H_o against H_1 by using the rejection region,

$$R:\ \{y|y < .22 \text{ or } y > 9.35\}.$$

For the data being studied,

$$y = 4/5 = .8.$$

Since $y = .8$ is between .22 and 9.35, we accept H_o and conclude that 5 is a possible value for σ^2.

We notice that the examples in this and the preceding sections are hypothetical and are used to illustrate the necessary arithmetic for completing a regression analysis. The exercises below attempt to provide more realistic data to illustrate the uses of regression.

EXERCISES

1. A sample of elementary school boys are measured on their birthdays. The age (x) and the height (y) are recorded for each boy. The data are as follows:

x	5	10	8	7	9	8	10	12	5	6
y	42	55	50	46	53	48	57	60	36	40

   (a) Plot a scatter diagram for the data.
   (b) Find and plot the least-squares regression line.
   (c) What height do you predict for a 9 1/2-year-old boy?
   (d) Estimate σ^2.

(e) Find a 90% CI for $\mu_{Y|9.5}$.
(f) Find a 95% CI for A.
(g) Test H_o: $B = 0$ against H_1: $B \neq 0$ with $\alpha = .10$.

2. In a study of the ability of laboratory rats to complete a maze, some rats were allowed one trial, others two trials, and still others three trials. The number of trials (x) and the time in minutes (y) were recorded for each rat to obtain the following data:

x	1	1	2	2	2	3	3
y	17	13	10	12	8	3	7

(a) Plot the data on a scatter diagram.
(b) Find and plot the least-squares regression line.
(c) If a new rat is given two trials to run the maze, what required time do you predict for the second trial?
(d) Estimate σ^2 and find a 95% CI for σ^2.
(e) Test H_o: $B = 0$ against H_1: $B \neq 0$ with $\alpha = .01$.
(f) Find a 95% CI for the mean time required to complete the maze on the first trial.

3. Refer to Exercise 4 on page 453, and find a 95% CI for the mean weight of professional football players who are 76 inches tall. For the same data, test H_o: $B = 0$ with $\alpha = .05$.

4. Use any or all of the data sets you collected for the experiments outlined in Section 12.1 to answer the following exercises:
(a) Test H_o: $B = 0$ against H_1: $B \neq 0$ with any appropriate choice for α. Interpret your results in terms of a dependence relationship between x and y.
(b) Choose any x-value in the range of observed values, and compute a 95% CI for the mean of y for that value of x.

5. An experiment to study the effect of fertilizer on the yield of wheat is conducted by treating 4 one-quarter acre plots with the following amounts of fertilizer per plot: 10 pounds, 15 pounds, and 20 pounds. The corresponding yields (y) are observed to be as follows:

x	10	10	10	10	15	15	15	15	20	20	20	20
y	7	10	11	10	9	13	10	12	16	14	13	17

(a) Find the least-squares regression line, and plot it on a scatter diagram for the data.
(b) Find a 95% CI for the mean yield when the fertilizer rate is 18 pounds per plot.
(c) Test H_o: $B = 0$ against H_1: $B > 0$ with $\alpha = .05$.

6. A group of 10 salesmen is ranked (X) from 1 to 10 on the basis of an aptitude score and a personal interview. Their sales productions for the first year (Y) are recorded in units of \$10,000. The pertinent sums are as follows: $\Sigma x = 55$, $\Sigma x^2 = 385$, $\Sigma y = 500$, $\Sigma y^2 = 30{,}000$, $\Sigma xy = 3080$, and $n = 10$.
   (a) Find the least-squares regression line for the data.
   (b) If a new salesman is ranked 4 in a scale from 1 to 10, what do you predict for his first year's sales production?
   (c) Is H_0: $B = 0$ an acceptable hypothesis?

12.3 CORRELATION ANALYSIS

The regression methods of Section 12.2 were used to study linear relationships between a dependent variable, Y, and fixed values of an independent variable, X. Those investigations were conducted by studying the sample regression coefficient, b. In this section, we study the problem of linear relationships between two jointly distributed variables, X and Y, by means of correlation analysis. The basic assumptions for a correlation analysis differ from those of a regression analysis in that the x-values are fixed for regression, but for a correlation analysis, we assume that X and Y have a bivariate distribution. Thus, there is a distribution of y-values for each value of x, and there is a distribution of x-values for each value of y. For example, when measuring the heights and weights for a population of elementary school boys (or girls), we expect to find a distribution of weights (y) for any fixed height (x), and we also expect to find a distribution of heights (x) for any fixed weight (y). The purpose of correlation analysis is to determine the *strength of any linear association* between the two jointly distributed variables. If the variables X and Y tend to vary in a linear pattern (with either a negative slope or a positive slope), we say that they are *highly correlated.* If the variables exhibit no tendency to cluster along a line, we say that the variables are *uncorrelated.* Figure 12.3 depicts 3 scatter diagrams to illustrate a high positive correlation, a high negative correlation, and no correlation.

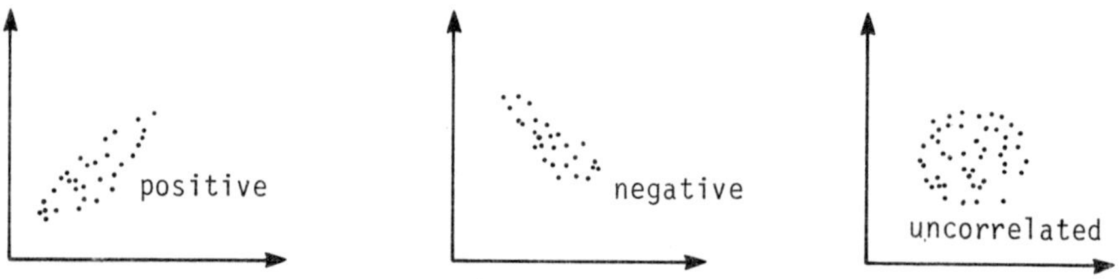

FIGURE 12.3. Scatter diagrams show positive, negative, and no correlation between X and Y.

The section below presents the usual Pearson product moment correlation coefficient for use in studying the linear relation between two variables, X and Y, which have a bivariate normal distribution. The section on page 475 presents a corresponding nonparametric rank correlation coefficient, proposed by Spearman, which uses the ranks of the data and is appropriate when the assumption of a bivariate normal model for the data is not valid.

PEARSON'S PRODUCT MOMENT CORRELATION COEFFICIENT: When we consider a random sample from a bivariate normal distribution, the sample points $[(x_1, y_1), (x_2, y_2), \ldots, (x_n, y_n)]$ have the joint density function,

$$f(x, y) = \frac{1}{\sigma_x \sigma_y \sqrt{(1 - \rho^2)}\ 2\pi} \qquad (12.29)$$

$$e^{-\frac{1}{2(1-\rho^2)}\left\{\left(\frac{x-\mu_x}{\sigma_x}\right)^2 + 2\rho\left(\frac{x-\mu_x}{\sigma_x}\right)\left(\frac{y-\mu_y}{\sigma_y}\right) + \left(\frac{y-\mu_y}{\sigma_y}\right)^2\right\}},$$

for any points (x, y) such that $-\infty < x < \infty$ and $-\infty < y < \infty$. The distribution depends on the following 5 parameters: μ_x is the mean of the variable, X; σ_x^2 is the variance of X; μ_y is the mean of the variable, Y; σ_y^2 is the variance of Y; and ρ is the correlation between the variables, X and Y. It can be shown that $-1 \leq \rho \leq 1$. *Furthermore, $\rho = 1$ means that y is a linear function of x with a positive slope; $\rho = -1$ means that y is a linear function of x with a negative slope; and $\rho = 0$ means that x and y are not linearly related.* Thus, ρ is a measure of the strength of linear association between the two variables, X and Y. If ρ is near either +1 or -1, we say that X and Y are highly correlated to indicate that there is a strong linear relationship between X and Y. When ρ is near 0, we say that X and Y are uncorrelated, meaning that there is no linear relationship between X and Y.

Now that we have illustrated the parameter, ρ, and have given some interpretations for special values, the next step is to present an estimator for ρ. That is, we want to use a random sample of n pairs of observations, say $\{(x_1, y_1), (x_2, y_2), \ldots, (x_n, y_n)\}$, from (12.29) to estimate ρ. The estimator we use is called the Pearson product moment correlation coefficient and is defined as

$$r = \frac{[\Sigma x_i y_i - (\Sigma x_i)(\Sigma y_i)/n]}{\{[\Sigma x_i^2 - (\Sigma x_i)^2/n][\Sigma y_i^2 - (\Sigma y_i)^2/n]\}^{\frac{1}{2}}}. \qquad (12.30)$$

It is easy to show that an alternate definition for r is as follows:

$$r = \Sigma(x_i - \bar{x})(y_i - \bar{y})/\{[\Sigma(x_i - \bar{x})^2][\Sigma(y_i - \bar{y})^2]\}^{\frac{1}{2}}$$

$$= \text{cov}(x, y)/(S_x\ S_y). \qquad (12.31)$$

Proof: The development of Eq. (3.19) shows that

$$\Sigma(x_i - \bar{x})^2 = \Sigma x_i^2 - (\Sigma x_i)^2/n.$$

Similarly,

$$\Sigma(y_i - \bar{y})^2 = \Sigma y_i^2 - (\Sigma y_i)^2/n.$$

Then, by multiplication and simplification,

$$\Sigma(x_i - \bar{x})(y_i - \bar{y}) = \Sigma(x_i y_i - \bar{x}y_i - x_i\bar{y} + \overline{xy})$$

$$= \Sigma x_i y_i - n\,\bar{x}\,\bar{y} - n\,\bar{x}\,\bar{y} + n\,\bar{x}\,\bar{y}$$

$$= \Sigma x_i y_i - (\Sigma x_i)(\Sigma y_i)/n.$$

Thus, Eqs. (12.30) and (12.31) are algebraically equivalent. That $r = \text{cov}(x, y)/S_x\ S_y$ follows directly from (12.8) and the definitions of S_x and S_y.

Before giving some examples, we note that the arithmetic for computing r in a correlation analysis requires the same quantities as required for computing and making inference about A and B in a regression analysis. That is, we require the following 6 quantities: Σx_i, Σy_i, Σx_i^2, Σy_i^2, $\Sigma x_i y_i$, and n.

EXAMPLE 11: The following three data sets illustrate both the calculation of r and the change in r as a reflection of the linear relation between X and Y.

(a)

x	1	2	3
y	3	5	7

(b)

x	1	1	2	3	3
y	10	8	7	4	6

(c)

x	1	2	3
y	3	6	3

SOLUTION: (a) For the given data, $\Sigma x_i = 6$, $\Sigma x_i^2 = 14$, $\Sigma y_i = 15$, $\Sigma y_i^2 = 83$, $\Sigma x_i y_i = 34$, and $n = 3$. Direct substitution into Eq. (12.30) gives

$$r = [34 - (6)(15)/3]/\{[14 - (6)^2/3][83 - (15)^2/3]\}^{\frac{1}{2}}$$

$$= 4/\sqrt{(2)(8)} = 1.$$

Since the three given points fall on the line, $y = 2x + 1$, we find $r = 1$. Thus, we can predict y as an *exact linear function* of x.

(b) For the given data, $\Sigma x_i = 10$, $\Sigma x_i^2 = 24$, $\Sigma y_i = 35$, $\Sigma y_i^2 = 265$, $\Sigma x_i y_i = 62$, and $n = 5$. Direct substitution into Eq. (12.30) gives

$$r = [62 - (10)(35)/5]/\{[24 - (10)^2/5][265 - (35)^2/5]\}^{\frac{1}{2}}$$

$$= -8/\sqrt{(4)(20)} = -.894.$$

Notice that the linear relationship between X and Y has a negative slope (i.e., X and Y vary inversely). Also, since r is near -1, there is a tendency for X and Y to vary linearly.

(c) For the given data, $\Sigma x_i = 6$, $\Sigma x_i^2 = 14$, $\Sigma y_i = 12$, $\Sigma y_i^2 = 54$, $\Sigma xy = 24$, and $n = 3$. Direct substitution into Eq.(12.30) gives

$$r = [24 - (6)(12)/3]/\{[14 - (6)^2/3][54 - (12)^2/3]\}^{\frac{1}{2}}$$

$$= 0/\sqrt{(2)(6)} = 0.$$

Thus, there is no better line than the horizontal line, $\hat{y} = \bar{y} = 4$, to represent the relationship between X and Y. Note that $r = 0$, or r near 0 means only that there is no linear relationship between X and Y. That is, there may exist a perfect (all points on a curve) curvilinear relationship between X and Y for data which gives r near 0. For these data the quadratic curve, $y = -3x^2 + 12x - 6$, provides an exact description for the data, but $r = 0$.

EXAMPLE 12: Find and interpret r for the 15 pairs of (height, weight) data for professional football players in Example 1. The necessary sums with X denoting height and Y denoting weight are as follows: $\Sigma x_i = 1104$, $\Sigma x_i^2 = 81{,}354$, $\Sigma y_i = 3310$, $\Sigma y_i^2 = 741{,}624$, $\Sigma x_i y_i = 244{,}554$, and $n = 15$.

SOLUTION: The totals in Eq. (12.30) give

$$r = [244{,}554 - (1104)(3310)/15]/$$
$$\{[81{,}354 - (1104)^2/15][741{,}624 - (3310)^2/15]\}^{\frac{1}{2}}$$
$$= 938/\sqrt{(99.6)(11{,}217.34)} = 938/1057 = .887.$$

Thus there is a strong linear relationship between the heights and weights of professional football players. Furthermore, the relationship is such that the heavier players tend to be taller. That fact is indicated by a positive value of r.

The problem of finding confidence intervals for ρ based on r is beyond the scope of this text in that there is no simple pivotal quantity which can be used for that purpose. Also, in that same way, the methods for testing H_o: $\rho = \rho_o$ for a general value, ρ_o, require more sophisticated mathematics than we assume for the present treatment. However, the hypothesis, H_o: $\rho = 0$, can be tested against H_1: $\rho \neq 0$ by an analysis of variance technique which is both useful and instructive. Furthermore, H_o: $\rho = 0$ is of primary interest in that its rejection leads to the conclusion that the variables, X and Y, are correlated (linearly dependent).

The basis for an ANOVA procedure as presented in Chapter 11 is the partitioning of a sum of squares into meaningful components that are independent estimates of the same parameter when the null hypothesis is true. In that spirit, we begin the development of a test for H_o: $\rho = 0$ by partitioning $\Sigma(y_i - \bar{y})^2$ into two components as follows:

$$\Sigma(y_i - \bar{y})^2 = \Sigma(y_i - \hat{y}_i)^2 + \Sigma(\hat{y}_i - \bar{y})^2. \quad (12.32)$$

For convenience, we adopt the following notation:

$$SST = \Sigma(y_i - \bar{y})^2; \qquad SSE = \Sigma(y_i - \hat{y}_i)^2;$$
$$SSR = \Sigma(\hat{y}_i - \bar{y})^2. \quad (12.33)$$

Proof: $\Sigma(y_i - \bar{y})^2 = \Sigma[(y_i - \hat{y}_i) + (\hat{y}_i - \bar{y})]^2$

$$= \Sigma(y_i - \hat{y}_i)^2 + \Sigma(\hat{y}_i - \bar{y})^2$$

$$+ 2\,\Sigma(y_i - \hat{y}_i)(\hat{y}_i - \bar{y}).$$

Now $\Sigma(y_i - \hat{y}_i)(\hat{y}_i - \bar{y})$

$= \Sigma[y_i - \bar{y} - b(x_i - \bar{x})][\bar{y} + b(x_i - \bar{x}) - \bar{y}]$

$= b\Sigma(y_i - \bar{y})(x_i - \bar{x}) - b^2\,\Sigma(x_i - \bar{x})^2$

$= 0$ by substitution of b from (12.9).

The above steps use $a = \bar{y} - b\bar{x}$ to write

$\hat{y} = a + bx_i = \bar{y} - b\bar{x} + bx_i = \bar{y} + b(x_i - \bar{x})$.

Then the definition of b shows that the final difference is 0.

Thus the total variation in Y is partitioned into two parts where SSE represents the *variation from the regression line* and SSR represents the *variation due to regression*. The variation due to regression denotes the change from $\bar{y}$, which is explained by the linear relationship between X and Y, while the variation from regression represents the *error variation* around the regression line. The parts of the partitioned sum of squares are illustrated in Figure 12.4. Notice that $(y_i - \bar{y}) = (y_i - \hat{y}_i) + (\hat{y}_i - \bar{y})$ for every y_i.

We observe that, for the calculation of $S^2_{y|x}$ in (12.10), the divisor, (n - 2), is used because SSE = $\Sigma(y_i - \hat{y}_i)^2$ is associated with (n - 2) d.f. Also, it can be shown that SSR = $\Sigma(\hat{y}_i - \bar{y})^2$ is associated with 1 d.f. These results can be summarized in an ANOVA table such as Table 11.2 for analysis. Before presenting the ANOVA table, however, we present some alternative formulas for calculating SSR and SSE. First,

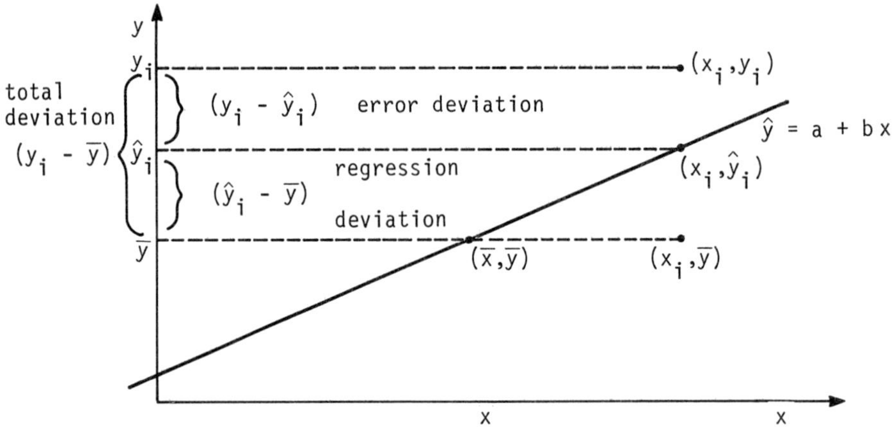

FIGURE 12.4. The partitioned sum of squares is shown here.

$$SSR = r^2 \Sigma (y_i - \bar{y})^2 = r^2 (SST). \qquad (12.34)$$

Second,

$$SSE = (1 - r^2) \Sigma (y_i - \bar{y})^2 = (1 - r^2)(SST). \qquad (12.35)$$

Proof: Since $\hat{y}_i = a + bx_i = (\bar{y} - b\bar{x}) + bx_i$

$$= \bar{y} + b(x_i - \bar{x}),$$

$$\Sigma(\hat{y}_i - \bar{y})^2 = \Sigma[b(x_i - \bar{x})]^2 = b^2 \Sigma(x_i - \bar{x})^2.$$

Now $b = \Sigma(x_i - \bar{x})(y_i - \bar{y})/\Sigma(x_i - \bar{x})^2$, so

$$\Sigma(\hat{y}_i - \bar{y})^2 = \{[\Sigma(x_i - \bar{x})(y_i - \bar{y})]^2/[\Sigma(x_i - \bar{x})^2]^2\}$$
$$\cdot \Sigma(x_i - \bar{x})^2[\Sigma(y_i - \bar{y})^2/\Sigma(y_i - \bar{y})^2].$$

By using Eq. (12.31),

$$SSR = \Sigma(\hat{y}_i - \bar{y})^2 = r^2 \Sigma(y_i - \bar{y})^2 = r^2 (SST).$$

By Eq. (12.32), it follows that

$$SSE = (1 - r^2) \Sigma(y_i - \bar{y})^2 = (1 - r^2)(SST).$$

We can now summarize the results in an ANOVA table such as given in Table 12.1. When H_o: $\rho = 0$ is a true hypothesis, it follows that the F-statistic as given in Table 12.1 is an F-variable with 1 and (n - 2) d.f. for its numerator and denominator, respectively. Thus, the best α-level rejection region for testing

$$H_o: \rho = 0 \quad \text{against} \quad H_1: \rho \neq 0$$

is

$$R: \{F | F > F(1, n-2, \alpha)\}. \tag{12.36}$$

TABLE 12.1.
An ANOVA table is given for testing
H_o: $\rho = 0$ against H_1: $\rho \neq 0$.

Source of variation	d.f.	Sum of squares	Mean squares	F
Regression	1	r^2 (SST)	$MSR = r^2 (SST)/1$	$F = \frac{MSR}{MSE}$
Error	n - 2	$(1 - r^2)$ (SST)	$MSE = \frac{(1 - r^2)(SST)}{(n-2)}$	$= \frac{(n-2)r^2}{(1-r^2)}$
Total	n - 1	$SST = \Sigma(y_i - \bar{y})^2$		

Observe that the final F-statistic depends only on r^2. So we do not need to complete the full table to test H_o: $\rho = 0$. That is, we may simply calculate

$$F = (n - 2)r^2/(1 - r^2) \tag{12.37}$$

and compare the value to the tabled critical value, $F(1, n - 2, \alpha)$. Yet there is useful information in Table 12.1. First, we observe that r^2 represents that portion of the total variation in Y which is explained by regression on X. In that sense, r^2 is called the *coefficient of determination.* Second, when r^2 is large relative to $(1 - r^2)/(n - 2)$, we know that a significant portion of the variation in Y is explained by regression on X. In reference to the hypothetical data in Example 11, linear regression in X explains 100%, 80%, and 0% of the variation in Y for parts (a), (b), and (c), respectively. Notice that the coefficient of determination is the same for positive and negative correlation coefficients of equal magnitude. Furthermore, note that the sign of r agrees with the sign of b and indicates whether the regression of Y on X has a positive or negative slope.

EXAMPLE 13: Use the analysis in Example 12 to test H_0: $\rho = 0$ against H_1: $\rho \neq 0$ for the height and weight data for professional football players. (Use $\alpha = .05$.)

SOLUTION: Example 12 gives $n = 15$ and $r = .887$. Thus, we use

$$R: \{F | F > F(1, 13, .05) = 4.67\}$$

to test H_0: $\rho = 0$ against H_1: $\rho \neq 0$ with $\alpha = .05$. Now $r^2 = .787$ and

$$F = 13(.787)/(1 - .787) = 48.$$

Since $F = 48 > 4.67$, we reject H_0 and conclude that there is a strong correlation between X (height) and Y (weight). In fact, 78.7% of the variation observed in weights is explained by linear regression on heights.

The preceding analysis could have been completed by using an ANOVA table to test H_0: $\rho = 0$. The following ANOVA table gives the actual partitioning of $\Sigma(y_i - \bar{y})^2 = 11{,}217.34$ as given in Example 12. In this table, we see that the total variation in weights, 11,217.34, is partitioned into 8828.05 for regression and 2389.29 for error about the regression line.

Source	d.f.	SS	MS	F
Regression	1	8828.05	8828.05	48
Error	13	2389.29	183.79	
Total	14	11,217.34		

The preceding discussion has used the regression of Y on X to compute the ANOVA table. Yet in a correlation model, we assume that X and Y have a bivariate normal distribution. Thus, there are two regression lines,

$$\mu_{Y|x} = A_1 + B_1 x \quad \text{and} \quad \mu_{X|y} = A_2 + B_2 y,$$

which represent the regression of Y on X and X on Y, respectively. The results for the regression of X on Y follow from the regression of Y on X (Section 12.2) by interchanging x_i and y_i in all the formulas. Since r is symmetric in x and y (i.e., we can interchange x_i and y_i

without changing the value of r), it follows that we could have used a partitioning of $(x_i - \bar{x})^2$ to obtain the F-statistic in Eq. (12.37). Note that the sum of squares and mean squares columns of an ANOVA table produced by partitioning the squared deviations for x would not agree with those columns for an ANOVA based on the squared deviations for y. However, the resulting F-statistics are equal.

Equation (12.25) defines a t-statistic, say

$$t_1 = b_1/S_{b_1},$$

for testing H_{01}: $B_1 = 0$ in the regression of Y on X. In a similar way, we can define a t-statistic, say

$$t_2 = b_2/S_{b_2},$$

for testing H_{01}: $B_2 = 0$ in the regression of X on Y. Now rejecting H_{01} leads to the conclusion that there is a linear relationship between Y and fixed values of X, while rejecting H_{02} implies that there is a linear relationship between X and fixed values of Y. Furthermore, rejecting H_0: $\rho = 0$ because F, as defined in Eq. (12.37), is too large indicates a linear association between the variables, X and Y. Thus, it seems that all three tests are investigating whether or not there is a linear association between X and Y. To show that in fact the tests are algebraically equivalent, we note (without proof) that, for t_1 and t_2 as defined above and for F as defined in Eq. (12.37),

$$F = t_1^2 = t_2^2. \tag{12.38}$$

Therefore, H_{01}: $B_1 = 0$, H_{02}: $B_2 = 0$, and H_0: $\rho = 0$ are either all accepted or all rejected when tested at the same level of α against a two-sided alternative hypothesis.

The foregoing discussion indicates that there are many computational similarities between regression and correlation. However, the choice of whether regression or correlation is appropriate for a given data set depends on what assumptions are made. A study which assumes that a normal variable Y is observed for fixed values of X requires regression of Y on X, while observation of a normal variable X for fixed values of Y implies regression of X on Y. If we assume that X and Y represent a random sample from a bivariate normal distribution, then correlation analysis is the appropriate choice.

EXERCISES

1. Find and interpret the correlation coefficient for any available data from the experiments in Section 12.1.

2. The heights (X) and weights (Y) for 7 elementary school girls are as follows:

x	38	40	52	45	58	40	61
y	38	42	70	60	75	45	82

   (a) Find and interpret r.
   (b) What percentage of the variability in weights of elementary school girls can be explained by regression on heights?
   (c) Test H_o: $\rho = 0$ against H_1: $\rho \neq 0$ with $\alpha = .05$.

3. Find and interpret r for the used car data in Exercise 5 on page 454. Are the assumptions of a correlation analysis met by the variables in this problem?

4. The college entrance exam percentile rank (X) and cumulative grade point average (Y) are recorded for 20 students. Those data give the following totals: $\Sigma x_i = 1500$, $\Sigma x_i^2 = 112{,}700$, $\Sigma y_i = 60$, $\Sigma y_i^2 = 188$, $\Sigma x_i y_i = 4536$.
   (a) Find r.
   (b) Find and interpret r^2.
   (c) Test H_o: $\rho = 0$ against H_1: $\rho \neq 0$ with $\alpha = .01$.

5. Find and interpret the correlation coefficient for the following data, where X denotes the percentage score for an algebra exam, and Y denotes the percentage score for a statistics exam: $\Sigma(x_i - \overline{x})^2 = 1000$, $\Sigma(y_i - \overline{y})^2 = 1000$, $\Sigma(x_i - \overline{x})(y_i - \overline{y}) = 750$, and $n = 10$. Is H_o: $\rho = 0$ an acceptable hypothesis?

6. (a) Find r for the following data:

x	1	3
y	3	5

   (b) Find r for the following data:

x	1	3
y	3	1

   (c) What are the possible values for r when the data consist of only two pairs of observations?
   (d) Use (x_1, y_1) and (x_2, y_2) to denote any two data points, and indicate the conditions that x_1, x_2, y_1, and y_2 must satisfy to exhibit each of the possible values of r.

(e) What are the implications of the preceding answers for a regression line based on two data points?

SPEARMAN'S RANK CORRELATION COEFFICIENT: Frequently we seek a measure of linear association between two variables when it is not reasonable to assume the bivariate normal model in Eq. (12.29). For such cases, we shift to a nonparametric statistic for purposes of making inference about the data. One such statistic is the Spearman rank correlation coefficient. That statistic is obtained by replacing the numerical measurements for X and Y with the ranks, 1, 2, ..., n. That is, the X-values are ranked from small (1) to large (n), and the Y-values are ranked from small (1) to large (n). After ranking the data, the Spearman rank correlation coefficient, r_s, is simply the Pearson product moment correlation coefficient, r, for the ranks. Thus, r_s may be calculated by Eq. (12.30) or (12.31), but it is easier to use the fact that the ranks for both variables take on values 1, 2, ..., n to derive a special formula for r_s.

Let $R(x_i)$ and $R(y_i)$ denote the ranks assigned to x_i and y_i, respectively. Then it can be shown that

$$r_s = 1 - (6 \sum_{i=1}^{n} d_i^2)/n(n^2 - 1), \qquad (12.39)$$

where $d_i = R(x_i) - R(y_i)$. Clearly, $r_s = 1$ when the ranks of X and Y are in perfect agreement because all the $d_i = 0$ in that case. It is also true (but not obvious) that $r_s = -1$ when the two rankings show exact opposite orderings. So r_s near either +1 or -1 indicates a strong agreement (linear association) between the ranks of X and Y, while r_s near 0 indicates a lack of linear association for the ranks.

EXAMPLE 14: Let $R(x_i)$ denote the pre-season ranking of the 10 teams in an athletic conference, and let $R(y_i)$ denote the finishing position for the season.

$R(x_i)$	1	2	3	4	5	6	7	8	9	10
$R(y_i)$	1	4	5	2	7	3	8	6	10	9

Find r_s.

SOLUTION: First, the d_i's are 0, -2, -2, 2, -2, 3, -1, 2, -1, and 1. Thus, $\Sigma d_i^2 = 32$. Second,

$$r_s = 1 - [6(32)/10(10^2 - 1)]$$

$$= 1 - (192/990) = .806.$$

Since r_s is positive, there is a tendency for the rankings to agree. That is, the actual finish is positively correlated with the pre-season predictions.

After calculating r_s as an indicator of association between two variables, it is only natural to ask for a method for testing whether or not the indicator is significantly different from 0. That is, we want to test the following H_o against the indicated H_1:

H_o: The variables X and Y are uncorrelated.
(12.40)
H_1: The variables X and Y are correlated.

The H_1 in (12.40) is a two-sided alternative hypothesis, and that is the one of more interest to us. However, we can also test H_o against one-sided alternative hypotheses as follows:

H_o: The variables X and Y are uncorrelated.
(12.41)
H_o: The variables X and Y are positively correlated.

H_o: The variables X and Y are uncorrelated.
(12.42)
H_1: The variables X and Y are negatively correlated.

Critical values of r_s, say $r_s(n,\alpha)$, as presented by Glasser and Winter [1] are given in Table F1 for use in testing the preceding hypotheses. These values are defined by

$$P[r_s > r_s(n, \alpha)] = \alpha.$$

The α-level rejection regions are

$$R: \{r_s \mid |r_s| > r_s(n, \alpha/2)\}, \qquad (12.43)$$

$$R: \{r_s \mid r_s > r_s(n, \alpha)\}, \qquad (12.44)$$

and

$$R: \{r_s \mid r_s < - r_s(n, \alpha)\}, \qquad (12.45)$$

for (12.40), (12.41), and (12.42), respectively.

EXAMPLE 15: Use the data in Example 14 to test whether or not pre-season predictions and actual finishes are correlated with $\alpha = .05$.

H_o: Pre-season prediction and actual finish are uncorrelated.

H_1: Pre-season prediction and actual finish are correlated.

SOLUTION: From Table F1, we find that $r_s(10, .025) = .64$. Thus, the rejection region is

$$R: \{r_s \mid |r_s| > .64\}.$$

Since Example 14 gives $r_s = .806$, we reject H_o and conclude that there is significant agreement between the pre-season predictions and the actual finish of the 10 teams.

When using r_s, the average rank technique, as illustrated in Example 6 of Chapter 10, is used to resolve ties. One advantage of r_s or any other rank correlation coefficient over the correlation coefficient, r, is that it is possible to rank subjects with respect to a variable without any measurements. For example, the heights of individuals can be ranked by means of a line-up, or the color of furs can be ranked (say from light to dark) without defining a color unit for measurement. The following exercises offer further illustrations and uses of r_s.

EXERCISES

1. Six candidates for local office are ranked in the primary election, $R(x_i)$, and again in the general election, $R(y_i)$, to give the following data:

$R(x_i)$	1	2	3	4	5	6
$R(y_i)$	4	3	5	2	1	6

   (a) Find r_s.
   (b) Is there any linear association between the ranks, $R(x_i)$ and $R(y_i)$?

2. A subsequent election to that in Exercise 1 shows the following:

$R(x_i)$	1	2	3	4	5	6
$R(y_i)$	2	3	1	4	6	5

(a) Find r_s.
(b) Use α = .05 to test the following H_o against the given H_1:

H_o: $R(x_i)$ and $R(y_i)$ are uncorrelated.

H_1: $R(x_i)$ and $R(y_i)$ are correlated.

3. Three different sports prediction groups provide independent rankings for the 8 football teams in an 8-team conference. Let $R(x_i)$, $R(y_i)$, and $R(z_i)$ denote the 3 sets of rankings. The data are as follows:

$R(x_i)$	1	2	3	4	5	6	7	8
$R(y_i)$	2	1	8	3	6	7	4	5
$R(z_i)$	1	2	4	3	7	8	5	6
Finish	1	2	6	3	8	4	6	6

(a) Which pre-season ranking gives the largest r_s when correlated with the actual finish?
(b) Find r_s for $R(x_i)$ and $R(y_i)$.
(c) Find r_s for $R(x_i)$ and $R(z_i)$.
(d) Find r_s for $R(y_i)$ and $R(z_i)$.

4. (a) Show that r_s = 1 for the following data:

$R(x_i)$	1	2	3	4
$R(y_i)$	1	2	3	4

(b) Show that r_s = -1 for the following data:

$R(x_i)$	1	2	3	4
$R(y_i)$	4	3	2	1

5. Rank the height and weight data in Example 1 to calculate r_s. Compare the value of r_s with r = .887 as given in Example 12 for the same data.

6. Calculate r_s for any data you have collected for the experiments described in Section 12.1. Compare r_s and r for each data set.

12.4 SOME CONCLUDING REMARKS

The early discussion in this chapter pertains to the problems of finding linear prediction equations for use in predicting values of a variable, Y, when given an observed value of an independent variable, X. There are two natural extensions of that work. First, there is the problem of determining a prediction equation which is a linear function of *more than one* independent variable. For example, a model for predicting the college performance (Y) of a student, which includes both a measure of high school performance (X_1) and a college entrance exam score (X_2), may provide a better prediction for Y than a model which uses X_2 alone. Topics of this type are called *multiple regression* problems and have wide application in fields such as education, economics, and agriculture. Second, there is the problem of using prediction models which are nonlinear functions of the independent variable or variables. These problems are discussed under such terms as *nonlinear regression*, *polynomial regression*, and *curvilinear regression*. Even though these topics are natural extensions of the simple linear regression of Section 12.2, they are not discussed in this text.

Corresponding to the topics that extend the regression results of this chapter, there are extensions of the concepts of correlation analysis to situations involving more than just pairs of observations. These topics include discussion of *multiple correlation* and *partial correlation*. Although these techniques are widely used in statistical analyses, we omit complete discussion of them because of the limitation of space.

Finally, we close with a word of caution regarding the use of least-squares prediction equations and correlation coefficients. Certainly it is folly to apply a prediction equation outside the range of a variable for which it was derived or to use it for variables other than the variable used in its development. For example, the prediction model for weights of professional football players is probably not appropriate for predicting weights of professional basketball players, women, children, men, or any number of other groups. Similarly, the model for predicting the heights of elementary school children in terms of their ages is not appropriate for predicting the heights of adults.

In turning to misuses of correlation coefficients, the most frequent error is to associate a cause-effect relationship with high correlations. One classic example is the claim that high teachers' salaries lead to high consumption of alcoholic beverages because the two variables show a high correlation. Another example is the claim that storks deliver babies because there is a high correlation between the number of storks on chimneys and the number of births in certain English communities. These examples serve only to emphasize that any two variables which vary linearly (either negatively or positively) show a high correlation. Further examples of interest may be found in books such as those by Huff [2] and by Wallis and Roberts [3]. Reading either or both of those books would be an excellent way to conclude a first study of statistics.

12.5 REFERENCES

[1] Glasser, G. S., and Winter, R. F. 1961. Critical values of the coefficient of rank correlation for testing the hypothesis of independence. *Biometrika* 48: 444 - 448.

[2] Huff, D. 1954. *How to lie with statistics.* New York: Norton.

[3] Wallis, W. A., and Roberts, H. V. 1965. *The nature of statistics.* New York: The Free Press.

12.6 REVIEW EXERCISES

1. The percentile ratings for two different achievement tests are recorded for a sample of 20 students. Let X denote the percentile for test 1 and Y denote the percentile for test 2. The following sums are obtained from the data: $\Sigma x_i = 1480$, $\Sigma x_i^2 = 110{,}800$, $\Sigma y_i = 1520$, $\Sigma y_i^2 = 116{,}960$, $\Sigma x_i y_i = 113{,}760$, and $n = 20$.
   (a) Find the least-squares regression line.
   (b) Find a point estimate for σ^2.
   (c) Find r.
   *(d) Test H_o: $B = 0$ against H_1: $B \neq 0$, with $\alpha = .05$.
   *(e) Test H_o: $\rho = 0$ against H_1: $\rho \neq 0$, with $\alpha = .05$.
   *(f) What percent of the variation in Y is explained by regression on X?
   *(g) Test H_o: $A = 0$ against H_1: $A \neq 0$, with $\alpha = .05$.
   *(h) Find a 95% confidence interval for B.
   *(i) Find a 95% confidence interval for $\mu_{Y|x}$ when $x = 80$.

2. Six brands of vanilla ice cream are ranked by two taste experts to provide the following data:

Expert 1	1	2	3	4	5	6
Expert 2	3	2	1	4	5	6

   (a) Find r_s.
   (b) Are the two sets of rankings correlated?

(c) Use Eq. (12.30) to calculate r where X denotes the measurement for expert 1 and Y denotes the measurement for expert 2. How do r and r_s compare?

3. The ages (X) and heights (Y) for a random sample of 10 participants in a recent Super Bowl football game are recorded as follows:

x	28	26	37	25	32	27	23	23	26	32
y	77	71	74	70	69	73	76	78	74	76

(a)-(h) Answer questions (a) -(h) for Exercise 1.
(i) Find r_s, and compare it with r as calculated for part (c).
(j) Are ages and heights of professional football players linearly dependent? Give reasons for your answer.

4. Find r_s for the data in Exercise 3, and test whether or not the rankings for ages and heights are correlated.

5. A fur dealer has ranked 10 pelts with respect to color (X) and with respect to price (Y). The data are as follows:

x	1	2	3	4	5	6	7	8	9	10
y	2	3	1	6	7	5	8	10	4	9

(a) Find r_s.
(b) Are the color and price rankings correlated?
(c) If the color ranks are from light (1) to dark (10), and if the price ranks are from low (1) to high (10), what are the interpretations of parts (a) and (b)?

APPENDIX A

Algebraic Results with Mathematical Tables

RULES OF EXPONENTS

1. When working with an exponential quantity, x^c, we refer to x as the *base* and c as the *exponent*. For example, 10^2 has a base, 10, and an exponent, 2.
2. If c is an integer,

$$x^c = \overbrace{x \cdot x \cdot \cdots \cdot x}^{c \text{ terms}}.$$

   For example, $10^3 = 10 \cdot 10 \cdot 10$.
3. If c is a rational number, say $c = m/n$, then

$$x^c = \sqrt[n]{x^m}.$$

   That is, $x^{m/n}$ is the nth root of x^m. For example,

$$4^{1/2} = \sqrt{4} = 2 \text{ and } 8^{1/3} = \sqrt[3]{8} = 2.$$

4. If c is a positive number, then $x^{-c} = 1/x^c$. For example, $4^{-2} = 1/4^2 = 1/16$.
5. To multiply quantities with the same base, we add exponents. That is,

$$x^c \cdot x^d = x^{c+d}.$$

For example,

$$10^2 \cdot 10^3 = 10^{2+3} = 10^5 = 100{,}000,$$

$$10^{-2} \cdot 10^3 = 10^{-2+3} = 10,$$

and

$$10^2 \cdot 10^{-3} = 10^{2-3} = 10^{-1} = .1.$$

6. To divide quantities with the same base, we use a negative exponent to express the denominator and apply the multiplication rule. That is,

$$x^c/x^d = x^c \cdot x^{-d} = x^{c-d}.$$

For example,

$$4^{15}/4^{10} = 4^{15} \cdot 4^{-10} = 4^5 = 1024$$

and

$$3^3/3^5 = 3^3 \cdot 3^{-5} = 3^{-2} = 1/9.$$

BINOMIAL THEOREM: If x and y are any two quantities and n is an integer, then

$$(x + y)^n = \sum_{r=0}^{n} \binom{n}{r} x^{n-r} y^r,$$

where $\binom{n}{r} = n!/r!(n - r)!$.

Proof: For n = 1, the left side of the equation is simply x + y, and the right side of the equation is

$$\binom{1}{0} x^{1-0} y^0 + \binom{1}{1} x^{1-1} y^1 = x + y.$$

Therefore, the result holds for n = 1. We use mathematical induction to prove the general result. Let the inductive hypothesis be that the result holds for n = s. That is, we assume that

$$(x + y)^s = \sum_{r=0}^{s} \binom{s}{r} x^{s-r} y^r.$$

Then

$$(x + y)^{s+1} = (x + y)^s (x + y)$$

$$= \sum_{r=0}^{s} \binom{s}{r} x^{s-r} y^r x + \sum_{r=0}^{s} \binom{s}{r} x^{s-r} y^r y$$

$$= x^{s+1} + \sum_{r=1}^{s} [\binom{s}{r} + \binom{s}{r-1}] x^{s+1-r} y^r + y^{s+1}.$$

Now

$$\binom{s}{r} + \binom{s}{r-1} = \frac{s!}{r!(s-r)!} + \frac{s!}{(r-1)!(s-r+1)!}$$

$$= \frac{s!}{r!(s+1-r)!} [(s+1-r) + r]$$

$$= \frac{(s+1)!}{r!(s+1-r)!} = \binom{s+1}{r} \text{ for } r = 1, 2, \ldots s.$$

Since

$$\binom{s+1}{0} = \binom{s+1}{s+1} = 1,$$

it follows that

$$(x + y)^{s+1} = \sum_{r=0}^{s+1} \binom{s+1}{r} x^{s+1-r} y^r.$$

Therefore, the result holds for all integer values of n.

For example,

$$(a + b)^3 = \binom{3}{0} a^3 b^0 + \binom{3}{1} a^2 b^1$$

$$+ \binom{3}{2} a^1 b^2 + \binom{3}{3} a^0 b^3$$

$$= a^3 + 3a^2b + 3\ ab^2 + b^3.$$

SPECIAL FUNCTIONS AND CONSTANTS:

1. Minimum function: min (x_i) = largest x_i in a set of observations, $(x_1, x_2, \ldots, x_n)$. For example, $\min(x_i) = 2$ when $(x_1, x_2, \ldots x_5) = (7, 6, 2, 5, 4)$.
2. Maximum function: max (x_i) largest x_i in a set of observations, $(x_1, x_2, \ldots, x_n)$. For example, max $(x_i) = 7$ when $(x_1, x_2, \ldots, x_5) = (7, 6, 2, 5, 4)$.
3. Two constants that occur in the expressions for probability distributions are as follows:

   π = 3.14159.... is the ratio between the circumference and diameter of a circle.

   e = 2.718.... is the base of the natural logarithms (a second set of logarithms which differs from those in Table A3 only with respect to the base).

SIMULTANEOUS EQUATIONS: When the conditions of a problem provide constraints on the values for two variables, say x and y, which can be exhibited by two equations, we call those equations *simultaneous equations*. Our objective is to determine those values for which the equations hold simultaneously. For example, a problem may require that

$$ax + by = c$$

$$dx + ey = f$$

when a, b, c, d, e, and f are known coefficients. Then it can be shown that

$$x = (ce - fb)/(ae - bd)$$

$$y = (af - dc)/(ae - bd),$$

as long as $(ae - bd) \neq 0$. That is, there is a unique solution as long as $(ae - bd) \neq 0$. When $(ae - bd) = 0$, either there is no solution (the lines are parallel), or there is an infinity of solutions (the two lines are coincident).

SQUARES, SQUARE ROOTS, AND RECIPROCALS: Table A1 provides the square, k^2; the square root, $\sqrt{k}$; and the reciprocal, $1/k$, for any positive number, k, which can be expressed with two significant digits. Approximations to the squares, square roots, and reciprocals for numbers having more than two significant digits can be obtained by interpolation. When k *is between* 1.0 and 9.9, Table A1 provides k^2, $\sqrt{k}$, and $1/k$ directly. When k *is not between* 1.0 and 9.9, the values in Table A1 must be multiplied by an adjustment factor to obtain k^2, $\sqrt{k}$, and $1/k$. The table below provides a partial listing of frequently used adjustment factors and establishes a pattern that can be extended in either direction from the row for $1 \leq k < 10$ to include either larger or smaller values of k.

This is a table of adjustment factors for squares, square roots, and reciprocals.

Number	Enter Table A1 at the row for n =	To find k^2, multiply n^2 by	To find $\sqrt{k}$, use	To find $1/k$, multiply $1/n$ by
$.001 \leq k < .01$	1000 k	.000001	$.01\sqrt{10n}$	1000
$.01 \leq k < .1$	100 k	.0001	$.1\sqrt{n}$	100
$.1 \leq k < 1$	10 k	.01	$.1\sqrt{10n}$	10
$1 \leq k < 10$	k	1	$\sqrt{n}$	1
$10 \leq k < 100$	.1 k	100	$\sqrt{10n}$	.1
$100 \leq k < 1000$	.01 k	10,000	$10\sqrt{n}$	.01
$1000 \leq k < 10000$	.001 k	1,000,000	$10\sqrt{10n}$	.001

EXAMPLES

1. Find the square, square root, and reciprocal for k = 6.9.

SOLUTION: By entering Table A1 at n = 6.9, we read the following results directly:

$$(6.9)^2 = 47.61.$$

$$\sqrt{6.9} = 2.6268.$$

$$1/6.9 = .1449.$$

2. Find the square, square root, and reciprocal for k = 69.

SOLUTION: By entering Table A1 at n = (.1)69 = 6.9 and using the adjustment factors for $10 \leq k < 100$, we obtain the following results:

$$(69)^2 = 100(47.61) = 4{,}761.$$

$$\sqrt{69} = 8.3066.$$

$$1/69 = .1(.1449) = .01449.$$

3. Find the square, square root, and reciprocal for k = .069.

SOLUTION: By entering Table A1 at n = 100(.069) = 6.9 and using the adjustment factors for $.01 \leq k < .1$, we obtain the following results:

$$(.069)^2 = .0001(47.61) = .004761.$$

$$\sqrt{.069} = .1(2.6268) = .26268.$$

$$1/.069 = 100(.1449) = 14.49.$$

TABLE A1.
Squares, square roots, and reciprocals are shown.

n	n^2	$\sqrt{n}$	$\sqrt{10n}$	1/n	n	n^2	$\sqrt{n}$	$\sqrt{10n}$	1/n
1.0	1.00	1.0000	3.1623	1.0000	5.5	30.25	2.3452	7.4162	.1818
1.1	1.21	1.0488	3.3166	.9091	5.6	31.36	2.3664	7.4833	.1786
1.2	1.44	1.0954	3.4641	.8333	5.7	32.49	2.3875	7.5498	.1754
1.3	1.69	1.1402	3.6056	.7692	5.8	33.64	2.4083	7.6158	.1724
1.4	1.96	1.1832	3.7417	.7143	5.9	34.81	2.4290	7.6811	.1695
1.5	2.25	1.2247	3.8730	.6667	6.0	36.00	2.4495	7.7460	.1667
1.6	2.56	1.2649	4.0000	.6250	6.1	37.21	2.4698	7.8102	.1639
1.7	2.89	1.3038	4.1231	.5882	6.2	38.44	2.4900	7.8740	.1613
1.8	3.24	1.3416	4.2426	.5556	6.3	39.69	2.5100	7.9373	.1587
1.9	3.61	1.3784	4.3589	.5263	6.4	40.96	2.5298	8.0000	.1563
2.0	4.00	1.4142	4.4721	.5000	6.5	42.25	2.5495	8.0623	.1538
2.1	4.41	1.4491	4.5826	.4762	6.6	43.56	2.5690	8.1240	.1515
2.2	4.84	1.4832	4.6904	.4545	6.7	44.89	2.5884	8.1854	.1493
2.3	5.29	1.5166	4.7958	.4348	6.8	46.24	2.6077	8.2462	.1471
2.4	5.76	1.5492	4.8990	.4167	6.9	47.61	2.6268	8.3066	.1449
2.5	6.25	1.5811	5.0000	.4000	7.0	49.00	2.6458	8.3666	.1429
2.6	6.76	1.6125	5.0990	.3846	7.1	50.41	2.6646	8.4261	.1408
2.7	7.29	1.6432	5.1962	.3704	7.2	51.84	2.6833	8.4853	.1389
2.8	7.84	1.6733	5.2915	.3571	7.3	53.29	2.7019	8.5440	.1370
2.9	8.41	1.7029	5.3852	.3448	7.4	54.76	2.7203	8.6023	.1351
3.0	9.00	1.7321	5.4772	.3333	7.5	56.25	2.7386	8.6603	.1333
3.1	9.61	1.7607	5.5678	.3226	7.6	57.76	2.7568	8.7178	.1316
3.2	10.24	1.7889	5.6569	.3125	7.7	59.29	2.7749	8.7750	.1299
3.3	10.89	1.8166	5.7446	.3030	7.8	60.84	2.7928	8.8318	.1282
3.4	11.56	1.8439	5.8310	.2941	7.9	62.41	2.8107	8.8882	.1266
3.5	12.25	1.8708	5.9161	.2857	8.0	64.00	2.8284	8.9443	.1250
3.6	12.96	1.8974	6.0000	.2778	8.1	65.61	2.8460	9.0000	.1235
3.7	13.69	1.9235	6.0828	.2703	8.2	67.24	2.8636	9.0554	.1220
3.8	14.44	1.9494	6.1644	.2632	8.3	68.89	2.8810	9.1104	.1205
3.9	15.21	1.9748	6.2450	.2564	8.4	70.56	2.8983	9.1652	.1190
4.0	16.00	2.0000	6.3246	.2500	8.5	72.25	2.9155	9.2195	.1176
4.1	16.81	2.0248	6.4031	.2439	8.6	73.96	2.9326	9.2736	.1163
4.2	17.64	2.0494	6.4807	.2381	8.7	75.69	2.9496	9.3274	.1149
4.3	18.49	2.0736	6.5574	.2326	8.8	77.44	2.9665	9.3808	.1136
4.4	19.36	2.0976	6.6332	.2273	8.9	79.21	2.9833	9.4340	.1124
4.5	20.25	2.1213	6.7082	.2222	9.0	81.00	3.0000	9.4868	.1111
4.6	21.16	2.1448	6.7823	.2174	9.1	82.81	3.0166	9.5394	.1099
4.7	22.09	2.1679	6.8557	.2128	9.2	84.64	3.0332	9.5917	.1087
4.8	23.04	2.1909	6.9282	.2083	9.3	86.49	3.0496	9.6437	.1075
4.9	24.01	2.2136	7.0000	.2041	9.4	88.36	3.0659	9.6954	.1064
5.0	25.00	2.2361	7.0711	.2000	9.5	90.25	3.0822	9.7468	.1053
5.1	26.01	2.2583	7.1414	.1961	9.6	92.16	3.0984	9.7980	.1042
5.2	27.04	2.2804	7.2111	.1923	9.7	94.09	3.1145	9.8489	.1031
5.3	28.09	2.3022	7.2801	.1887	9.8	96.04	3.1305	9.8995	.1020
5.4	29.16	2.3238	7.3485	.1852	9.9	98.01	3.1464	9.9499	.1010

TABLE A2.
10,000 random digits are given.

Row	Column									
	00-04	05-09	10-14	15-19	20-24	25-29	30-34	35-39	40-44	45-49
00	14567	85552	63853	17961	95084	32354	96169	33718	64958	50892
01	36824	35568	40619	98350	84776	08468	98987	77942	64417	01258
02	48900	57979	89058	18031	39305	31613	54515	45030	24392	26950
03	46377	55250	70452	00358	22379	11677	20124	17022	55416	67563
04	99569	23616	79207	69433	51929	57159	42992	17628	01909	05579
05	65345	48005	90776	03764	48299	56175	19444	69264	21318	41562
06	62685	72874	64078	06039	06460	29268	37455	05432	60388	88402
07	57645	34425	23110	10669	55700	63433	28070	58797	03706	40359
08	70592	52388	39164	28982	83526	61569	09246	95487	33097	33474
09	13419	88293	44208	15009	25049	68554	28052	44052	44053	40966
10	12740	35726	76990	24143	81214	36165	32114	14450	83705	57864
11	05807	03935	18124	07339	98422	95568	93536	34497	67176	04777
12	83890	99567	16880	63324	23202	50981	23860	30786	29604	02234
13	28839	50442	62899	88837	06418	77444	88478	24429	75614	19595
14	80854	91845	77097	00640	33546	36394	43648	88287	17003	21621
15	85034	41782	07226	47949	76206	22756	33109	88145	16493	27729
16	78011	46960	17172	85728	98812	18187	10364	60159	73815	08067
17	10871	44769	61159	53631	37693	64670	68027	12222	76049	41616
18	55324	22373	47264	87243	87210	33145	49849	12500	15861	18241
19	93167	54306	46616	89912	77698	78150	09093	36699	83993	64905
20	25540	09394	25969	96174	68343	29837	26782	01386	18335	27003
21	18264	04292	93596	48923	71088	72696	77368	93470	49886	80758
22	13300	47319	40713	47995	71042	20419	19704	69294	91842	52536
23	99251	27623	05575	42635	56760	62621	25377	05922	61869	34973
24	90930	55497	41510	09581	29655	81422	40167	92845	52748	05038
25	42894	43222	89277	22957	95941	11870	66776	22343	25057	81607
26	33323	66020	31493	09030	73037	57315	69159	39221	03973	47602
27	91556	58706	38481	58565	85325	88040	31258	40215	38149	98434
28	88500	36675	75255	77222	46214	07743	68342	50214	58369	89552
29	15746	76626	41044	79115	64866	90886	09919	72949	90117	92967
30	26982	80667	36643	96978	37258	29086	32138	73214	36842	25997
31	86107	40694	29077	19597	47354	60641	56363	49393	88726	99502
32	31380	53818	02429	84916	05182	82042	81413	37889	63254	14732
33	06517	30069	36146	79693	26532	73842	33009	31912	04356	36761
34	92733	37810	88295	16143	66372	40009	94054	07499	52245	82487
35	92784	51817	40287	37391	87504	59934	91844	84964	66511	92789
36	16248	91101	54434	48128	77297	80098	85432	84714	30304	58121
37	92327	88047	26298	67017	17570	65731	50411	68635	83439	76788
38	87723	82601	21916	31523	92740	94779	52310	01346	26695	97914
39	73285	42494	37715	62479	09549	14557	99077	38740	57543	62846

TABLE A2. 10,000 random digits are given (continued).

Row	Column									
	00-04	05-09	10-14	15-19	20-24	25-29	30-34	35-39	40-44	45-49
40	55612	62375	08569	71508	15531	48485	58276	69417	11601	08635
41	71472	82429	83717	94653	83133	53825	90228	08905	36381	92479
42	80733	00125	16601	36654	78878	34252	99935	24246	37527	09247
43	59758	11433	21493	46842	42401	46810	33129	02625	54898	53317
44	55184	96512	04756	04794	93871	55021	37845	31086	70370	41958
45	13949	45800	58454	15650	59114	56087	17633	35874	89566	74008
46	33379	38185	10575	47749	44415	63195	27469	26400	99267	27421
47	30143	19764	18947	42593	48443	28425	50934	39402	72932	70499
48	48676	85597	04610	10084	30936	31720	66760	48532	25212	94543
49	36125	52153	29274	71154	79705	73167	39137	06392	34673	82945
50	42282	68482	26348	77171	10328	08957	02976	44374	18895	04978
51	76191	78056	00708	14585	12763	72261	64394	69676	05834	67851
52	11621	98762	52353	25822	88410	90866	55788	07666	74126	53603
53	70470	38146	49203	69987	02622	15204	61740	49623	80932	65264
54	85206	99880	37987	99059	47990	44705	30903	95732	46984	43056
55	08489	35853	33627	27571	23054	89681	02614	87131	28058	34266
56	13617	37084	86996	35033	29238	85941	17976	41646	84816	81389
57	79163	02123	00772	61430	37663	93047	47879	02192	31704	32318
58	00177	87555	83577	85569	59477	15182	68026	67854	54570	33168
59	35618	55428	36614	35725	20796	72865	25118	62040	91780	89651
60	06484	96915	60006	07397	80692	19940	42504	20651	98461	15309
61	50948	68022	79803	57881	69707	21272	13533	45276	98797	92998
62	33205	48710	80823	22902	03063	65962	58754	89696	74092	35442
63	53865	63524	20834	56382	75178	02443	02431	63463	44734	78807
64	52668	92522	64618	92755	05936	89431	73781	69147	10682	89813
65	15473	15569	16195	22218	38649	02616	96957	56200	07943	91885
66	46693	59585	41711	47527	83715	56359	05619	38016	55207	05232
67	18421	87226	89705	90287	08677	09627	32075	47083	47209	11577
68	75275	22753	21715	77903	10418	76725	72310	64280	64411	54965
69	42944	66187	56771	50096	84238	96945	79606	67997	51135	48563
70	35485	34209	47818	26657	68697	18789	41819	77631	12984	87335
71	07972	03737	13879	35977	36828	01878	94569	81365	85093	60452
72	88559	68361	30832	18359	53951	33233	01038	94962	34057	43066
73	26065	72451	21354	52708	96152	65818	41448	93580	27206	02788
74	77604	35074	84409	38909	40945	92344	42332	38749	74737	67290
75	33822	66041	27296	26013	28999	43481	69008	35973	32600	85951
76	72536	97446	39738	63336	30129	79890	69309	71548	06161	64959
77	03712	80089	61756	77598	25471	21469	91242	77724	71575	20042
78	39216	24543	19154	33887	41635	56238	62788	47496	99696	10625
79	94955	53007	19703	72088	74670	32755	40267	84698	28812	06567

TABLE A2. 10,000 random digits are given (continued).

Row	Column									
	00-04	05-09	10-14	15-19	20-24	25-29	30-34	35-39	40-44	45-49
80	15781	48330	46238	90526	25192	36795	31699	85590	60249	40188
81	04963	09345	76495	53465	96853	40518	21800	86459	18193	98485
82	12734	23877	37484	33683	50137	61870	60109	51173	59875	09366
83	59739	28712	37403	53139	93639	36230	93545	80062	59674	36475
84	94315	97287	62046	16203	73659	54091	69366	46062	28706	44073
85	24305	92369	93042	11414	70986	38984	30288	46664	81755	31048
86	44330	14145	35024	95613	22140	38789	08996	87102	17965	54397
87	19215	87480	77310	09109	37021	84063	37096	38472	01637	00940
88	00558	58324	54089	16120	73350	49026	32074	63273	55614	91220
89	66844	25935	92949	30968	40113	88299	62017	00686	30084	38134
90	30975	12078	49903	53797	44502	40606	40536	53481	37992	86841
91	87013	82109	49178	46707	56107	66089	07401	52110	61346	43734
92	58894	89737	97815	75481	62087	57179	68172	66289	25037	98066
93	44885	77548	95326	11992	70659	57700	11641	83547	73810	85824
94	99755	27351	91668	60360	04773	35757	86751	20837	06147	96474
95	77603	81751	27499	92937	86186	01943	61852	85819	73515	77322
96	52106	38334	77272	00124	34059	96657	28763	04674	92116	09912
97	57570	83839	72650	32144	06629	94259	56355	64671	81706	84696
98	33399	26685	08546	01508	95977	94728	45274	76086	37721	15031
99	59499	01326	33657	30114	05128	94984	09798	13741	44045	89703

Row	Column									
	50-54	55-59	60-64	65-69	70-74	75-79	80-84	85-89	90-94	95-99
00	66694	98742	37331	28860	25534	57121	76247	43665	37565	87462
01	50538	67535	76834	28440	69715	11682	45268	94975	38053	25317
02	76515	28447	84318	88802	15165	94152	53921	11091	90764	32647
03	49431	16046	51029	30422	27989	78459	51302	69058	65636	46410
04	36457	16269	66055	91307	73169	32791	02217	16009	09588	66783
05	85683	36291	60524	68825	44092	36046	84804	12675	67563	14712
06	68967	16131	78845	65521	45325	62205	27389	03991	86164	15840
07	54750	85696	86376	96486	84024	21300	37338	77331	17292	70726
08	65597	24580	87669	68747	96551	32177	91519	58480	78237	95886
09	00080	48685	07976	50113	39777	20647	99505	32426	62506	12178
10	65643	58077	71380	93698	71074	61033	15955	52001	35759	63423
11	78004	50562	04221	08790	80186	19567	04088	52579	79519	80776
12	03705	97565	74810	45156	17998	52035	16291	79825	82783	58481
13	19321	80212	18781	46456	44925	74588	37837	33831	61185	72858
14	74498	92189	78789	27276	96673	14638	51015	86468	61176	21168
15	67682	04497	84997	00762	91589	55874	55208	34831	91433	80113
16	76964	58604	35708	14203	99118	32207	52581	88316	75870	03655
17	22132	95305	42951	53572	72593	49577	79966	00711	36231	65758
18	85227	46095	55640	55527	35451	98157	52408	20292	08203	66753
19	50649	55120	30272	50402	67640	92062	41697	63280	77096	32581

TABLE A2. 10,000 random digits are given (continued).

Row	Column									
	50-54	55-59	60-64	65-69	70-74	75-79	80-84	85-89	90-94	95-99
20	41170	51719	11244	62919	46426	66945	74680	86294	46017	85462
21	72515	02634	22664	17474	45043	47963	63757	43811	78822	83879
22	74081	79823	66994	21704	25621	92208	69919	98892	05159	05831
23	95812	38041	19788	63830	38928	24338	46278	24291	47106	96056
24	56038	86998	45020	95219	17127	88985	75600	86119	01354	28693
25	42441	65171	60118	53205	12424	38675	20215	92419	99679	12494
26	16326	92184	85456	75627	29799	65271	85604	04926	67616	25701
27	76208	50801	79924	32812	92661	20483	01449	05225	65113	94149
28	66831	27374	73413	28743	95701	09130	74483	36552	86079	78469
29	04513	58079	25051	80837	86428	55590	85692	49634	74818	42331
30	78936	70600	23149	26359	75939	60462	78189	84359	38561	23981
31	53172	76437	62838	01924	74354	94033	46046	27652	34039	03282
32	83117	78554	49614	94539	64073	35488	18529	04946	23028	78558
33	05789	92920	86403	70848	63349	15848	88230	41588	64958	84067
34	59925	09661	68287	66157	15734	45393	46584	03516	01501	27903
35	76425	08699	80552	00218	71954	17074	88136	82714	19469	27248
36	47464	25814	81035	84347	69426	26760	42925	42509	68984	45454
37	67720	56260	98762	96053	30360	58602	40298	32460	09810	81430
38	67481	00125	66797	92316	69290	11857	57097	12417	38161	96223
39	09729	36771	74460	83681	12093	51951	35290	27397	61271	10668
40	05152	58547	71789	38389	07515	41834	86899	68148	62789	33098
41	09530	32510	22074	57453	06291	42164	79454	00796	32215	88849
42	37909	24604	25129	46528	37496	26531	00182	44497	61183	85893
43	35968	57655	86486	45918	32447	05124	19134	00503	47587	18877
44	85957	22064	13203	24943	74566	98847	30013	19993	81059	25450
45	64302	01382	57103	00541	87359	82972	22120	26612	57740	94231
46	55491	99299	48343	23090	71070	35985	04678	54608	87443	27856
47	40133	13053	36380	93321	04975	18786	00444	04632	62258	71211
48	75199	84245	18120	76714	12784	09737	44031	58289	85012	55706
49	74395	44831	95401	61939	53561	87821	51959	65089	24365	63001
50	95146	62171	59386	25454	87799	97269	49416	40711	26930	72594
51	57906	04120	27685	15694	29724	20658	22785	59409	41614	73718
52	53531	47781	06794	35572	69166	83949	55050	23871	22029	48113
53	93260	86791	18053	46709	05762	21928	43666	68826	65759	57556
54	25410	30975	81676	26552	24875	13228	50796	25089	03977	90919
55	57091	10563	99684	94015	26336	38288	94774	00160	86918	01046
56	27741	53610	07058	49393	46475	62477	52457	86203	66561	86019
57	41034	36861	17316	57343	22523	52978	64897	32131	00166	59823
58	69999	35403	98188	19708	96605	82540	72725	19291	58941	87384
59	02382	13074	40761	12662	67575	73419	96497	17428	94676	80920

TABLE A2. 10,000 random digits are given (continued).

Row	Column									
	50-54	55-59	60-64	65-69	70-74	75-79	80-84	85-89	90-94	95-99
60	16609	75932	44535	73736	28577	90394	70578	15543	62210	75110
61	98142	42426	09049	92914	29018	67071	22598	49880	91943	85593
62	82429	99208	21780	91221	06538	09939	84259	11396	36418	05795
63	85505	61806	43803	74001	20934	66293	08628	03518	73010	19560
64	42402	15317	27372	47554	21502	64694	34959	05990	58477	83362
65	59884	11296	11139	73420	93994	40770	20822	07681	60518	03698
66	73613	34619	29289	08770	48368	92651	36294	90324	91510	33690
67	61951	69304	21123	05402	87172	84445	01666	51776	95935	02605
68	44257	42200	82300	16201	16734	62627	37425	75825	37002	79489
69	06880	71874	85580	05244	89963	61372	50118	77288	38419	44030
70	29075	34240	70248	03731	47465	91972	07074	38717	52313	62881
71	98322	36724	80251	54009	67374	55629	00044	34309	70082	09830
72	23747	08313	23693	88588	81215	19613	51894	09531	60268	46126
73	82351	22333	55854	63034	05311	74167	13908	06247	40090	46288
74	19142	79537	16266	58996	17275	27204	30899	45361	36811	34351
75	33456	48088	32587	67113	23365	54179	47194	66978	96037	92975
76	38622	18870	16883	13233	21888	64661	00230	67753	77191	63627
77	55173	53024	32048	11590	33811	36041	14442	00001	57753	53510
78	47088	06271	64268	14664	60810	39888	98323	22211	71405	67461
79	00836	98423	34503	10676	04009	93923	61900	94638	14169	63172
80	90080	93566	27329	50767	84801	70527	23323	20887	42776	99925
81	76846	76196	84396	06639	81097	23852	51387	86360	58097	44342
82	51555	12026	56238	77724	29936	80939	05770	58113	27651	07722
83	57699	85517	96093	66933	59376	08044	64808	66468	79839	08063
84	80372	32291	62233	00100	81576	81150	25755	88025	91555	03338
85	16813	40917	75619	96125	12194	29508	87697	81816	37254	53871
86	51074	49632	58531	49352	82731	78690	50363	20332	58246	70677
87	66512	21518	99263	59414	52863	23655	91471	95403	12743	23557
88	26905	61963	52200	73504	30185	56730	92971	09871	63266	08776
89	26284	03831	87709	66855	36474	58780	92752	71525	24917	67077
90	55412	01187	58509	57951	40555	87909	12419	30532	55922	83284
91	85969	34432	40185	33915	08685	54648	23349	56955	48514	91275
92	33290	12901	60072	55262	56688	12078	40029	37066	31516	67580
93	70536	51935	53692	60140	58156	65111	92151	33092	99065	19962
94	08670	12093	50916	54035	88765	03143	56497	46807	33499	03911
95	53560	13448	71914	95353	31927	50418	22454	55980	98087	99384
96	25782	09544	01479	43353	57183	68789	2-527	15826	15186	46187
97	59582	57836	80548	27617	76303	77749	56114	34648	16044	79526
98	58710	15963	99987	34208	49351	67047	76014	84383	17047	38097
99	39058	49127	88829	72468	29152	66797	28351	85146	49758	66817

COMMON LOGARITHMS: One special class of exponents is called *logarithms*. A logarithm is the exponent, b, to which a base, x, must be raised to give a number, n. In symbols,

$$\log_x n = b \text{ and } x^b = n$$

are equivalent statements. The base for the common logarithms in Table A3 is x = 10. Thus, the logarithm for any number, n, with respect to the base, 10, is the exponent to which 10 must be raised to provide n. Since our primary concern is with logarithms to base 10, we do not write the base each time. That is, we simply write that

$$\log 1 = 0 \quad \text{because } 10^0 = 1,$$

$$\log 10 = 1 \quad \text{because } 10^1 = 10,$$

$$\log 100 = 2 \quad \text{because } 10^2 = 100, \text{ etc.}$$

Thus, it follows that the logarithm of any number between 1 and 10 is a decimal between 0 and 1; the logarithm of any number between 10 and 100 is between 1 and 2, etc. So a logarithm consists of two parts. The integer part is called the *characteristic*, and the decimal part is called the *mantissa*. Table A3 provides the mantissa for the logarithms of selected numbers, while the characteristic is determined by either inspection or the following table.

Number, n	Characteristic
$1 \le n < 10$	0
$10 \le n < 100$	1
$100 \le n < 1{,}000$	2
$1{,}000 \le n < 10{,}000$	3
$10{,}000 \le n < 100{,}000$	4
$100{,}000 \le n < 1{,}000{,}000$	5

For example,

$$\log 1.11 = 0.0453,$$

$$\log 11.1 = 1.0453,$$

$$\log 111.0 = 2.0453,$$

$$\text{and } \log 1110.0 = 3.0453.$$

Since logarithms are exponents, the rules of exponents as presented above apply and provide the following rules:

1. The logarithm of a product is the sum of the logarithms. That is,

   $$\log (xy) = \log x + \log y.$$

   For example, $\log (2 \cdot 4) = \log 2 + \log 4$ or

   $$.9031 = .3010 + .6021.$$

2. The logarithm of a quotient is the difference of the logarithms. That is,

   $$\log (x/y) = \log x - \log y.$$

   For example, $\log (8/2) = \log 8 - \log 2$ or

   $$.6021 = .9031 - .3010.$$

3. $\log x^c = c \log x$.

   For example, $\log 9^3 = 3 \log 9$

   $$= 3(.9542) = 2.8626.$$

   Also $\log\sqrt{29} = \log (29)^{\frac{1}{2}} = \frac{1}{2} \log 29$

   $$= \tfrac{1}{2} (1.4624) = .7312.$$

4. Extending #1 to a product of n numbers gives

$$\log\left(\prod_{i=1}^{n} x_i\right) = \log (x_1 \cdot x_2 \cdot \ldots x_n) = \sum_{i=1}^{n} \log x_i.$$

There are two basic uses of Table A3. First, it is used to obtain the mantissa for the logarithm of a specific number. Second, it is used to obtain a number for a given logarithm. The two examples below illustrate those two uses of the table.

EXAMPLES

1. Find the logarithm of 175.

SOLUTION: First, the characteristic is 2 since $100 \leq 175 < 1000$. Second, the row for 1.7 and the column for 5 show the mantissa to be .2430. Thus,

$$\log 175 = 2.2430.$$

2. Find X so that log X = 3.6946.

SOLUTION: We begin by searching the body of Table A3 to find a mantissa of .6946. That search provides the 3 digits of X to be 495. Then the characteristic, 3, indicates that $1000 \leq X < 10{,}000$. Thus,

$$X = 4950.$$

Note: If the number, X, for which we want the logarithm has more than 3 significant digits, we use interpolation between the two closest tabled values to determine the logarithm. Similarly, interpolation is used when the mantissa of a number being determined does not appear in the table.

TABLE A3.
Common logarithms (logarithms to base 10) are given here.

	0	1	2	3	4	5	6	7	8	9
1.0	.0000	.0043	.0086	.0128	.0170	.0212	.0253	.0294	.0334	.0374
1.1	.0414	.0453	.0492	.0531	.0569	.0607	.0645	.0682	.0719	.0755
1.2	.0792	.0828	.0864	.0899	.0934	.0969	.1004	.1038	.1072	.1106
1.3	.1139	.1173	.1206	.1239	.1271	.1303	.1335	.1367	.1399	.1430
1.4	.1461	.1492	.1523	.1553	.1584	.1614	.1644	.1673	.1703	.1732
1.5	.1761	.1790	.1818	.1847	.1875	.1903	.1931	.1959	.1987	.2014
1.6	.2041	.2068	.2095	.2122	.2148	.2175	.2201	.2227	.2253	.2279
1.7	.2304	.2330	.2355	.2380	.2405	.2430	.2455	.2480	.2504	.2529
1.8	.2553	.2577	.2601	.2625	.2648	.2672	.2695	.2718	.2742	.2765
1.9	.2786	.2810	.2833	.2856	.2878	.2900	.2923	.2945	.2967	.2989
2.0	.3010	.3032	.3054	.3075	.3096	.3118	.3139	.3160	.3181	.3201
2.1	.3222	.3243	.3263	.3284	.3304	.3324	.3345	.3364	.3385	.3404
2.2	.3424	.3444	.3464	.3483	.3502	.3522	.3541	.3560	.3579	.3598
2.3	.3617	.3636	.3655	.3674	.3692	.3711	.3729	.3747	.3766	.3784
2.4	.3802	.3820	.3838	.3856	.3874	.3892	.3909	.3927	.3945	.3962
2.5	.3979	.3997	.4014	.4031	.4048	.4065	.4082	.4099	.4116	.4133
2.6	.4150	.4166	.4183	.4200	.4216	.4232	.4249	.4265	.4281	.4298
2.7	.4314	.4330	.4346	.4362	.4378	.4393	.4409	.4425	.4440	.4456
2.8	.4472	.4487	.4502	.4518	.4533	.4548	.4564	.4579	.4594	.4609
2.9	.4624	.4639	.4654	.4669	.4683	.4698	.4713	.4728	.4742	.4757
3.0	.4771	.4786	.4800	.4814	.4829	.4843	.4857	.4871	.4886	.4900
3.1	.4914	.4928	.4942	.4955	.4969	.4983	.4997	.5011	.5024	.5038
3.2	.5051	.5065	.5079	.5092	.5105	.5119	.5132	.5145	.5149	.5172
3.3	.5185	.5198	.5211	.5224	.5237	.5250	.5263	.5276	.5289	.5302
3.4	.5315	.5328	.5340	.5353	.5366	.5378	.5391	.5403	.5416	.5428
3.5	.5441	.5453	.5465	.5478	.5490	.5502	.5514	.5527	.5539	.5551
3.6	.5563	.5575	.5587	.5599	.5611	.5623	.5635	.5647	.5658	.5670
3.7	.5682	.5694	.5705	.5717	.5729	.5740	.5752	.5763	.5775	.5786
3.8	.5798	.5809	.5821	.5832	.5843	.5855	.5866	.5877	.5888	.5899
3.9	.5911	.5922	.5933	.5944	.5955	.5966	.5977	.5988	.5999	.6010
4.0	.6021	.6031	.6042	.6053	.6064	.6075	.6085	.6096	.6107	.6117
4.1	.6128	.6138	.6149	.6160	.6170	.6180	.6191	.6201	.6212	.6222
4.2	.6232	.6243	.6253	.6263	.6274	.6284	.6294	.6304	.6314	.6325
4.3	.6335	.6345	.6355	.6365	.6375	.6385	.6395	.6405	.6415	.6425
4.4	.6435	.6444	.6454	.6464	.6474	.6484	.6493	.6503	.6513	.6522
4.5	.6532	.6542	.6551	.6561	.6571	.6580	.6590	.6599	.6609	.6618
4.6	.6628	.6637	.6646	.6656	.6665	.6675	.6684	.6693	.6702	.6712
4.7	.6721	.6730	.6739	.6749	.6758	.6767	.6776	.6785	.6794	.6803
4.8	.6812	.6821	.6830	.6839	.6848	.6857	.6866	.6875	.6884	.6893
4.9	.6902	.6911	.6920	.6928	.6937	.6946	.6955	.6964	.6972	.6981
5.0	.6990	.6998	.7007	.7016	.7024	.7033	.7042	.7050	.7059	.7067
5.1	.7076	.7084	.7093	.7101	.7110	.7118	.7126	.7135	.7143	.7152
5.2	.7160	.7168	.7177	.7185	.7193	.7202	.7210	.7218	.7226	.7235
5.3	.7243	.7251	.7259	.7267	.7275	.7284	.7292	.7300	.7308	.7316
5.4	.7324	.7332	.7340	.7348	.7356	.7364	.7372	.7380	.7388	.7396

TABLE A3. Common logarithms (logarithms to base 10) are given here (continued).

	0	1	2	3	4	5	6	7	8	9
5.5	.7404	.7412	.7419	.7427	.7435	.7443	.7451	.7459	.7466	.7474
5.6	.7482	.7490	.7497	.7505	.7513	.7520	.7528	.7536	.7543	.7551
5.7	.7559	.7566	.7574	.7582	.7589	.7597	.7604	.7612	.7619	.7627
5.8	.7634	.7642	.7649	.7657	.7664	.7672	.7679	.7686	.7694	.7701
5.9	.7709	.7716	.7723	.7731	.7738	.7745	.7752	.7760	.7767	.7774
6.0	.7782	.7789	.7796	.7803	.7810	.7818	.7825	.7832	.7839	.7846
6.1	.7853	.7860	.7868	.7875	.7882	.7889	.7896	.7903	.7910	.7917
6.2	.7924	.7931	.7938	.7945	.7952	.7959	.7966	.7973	.7980	.7987
6.3	.7993	.8000	.8007	.8014	.8021	.8028	.8035	.8041	.8048	.8055
6.4	.8062	.8069	.8075	.8082	.8089	.8096	.8102	.8109	.8116	.8122
6.5	.8129	.8126	.8142	.8149	.8155	.8162	.8169	.8176	.8182	.8189
6.6	.8195	.8202	.8209	.8215	.8222	.8228	.8235	.8241	.8248	.8254
6.7	.8261	.8267	.8274	.8280	.8287	.8293	.8299	.8306	.8312	.8319
6.8	.8325	.8331	.8338	.8344	.8351	.8357	.8363	.8370	.8376	.8382
6.9	.8388	.8395	.8401	.8407	.8414	.8420	.8426	.8432	.8439	.8445
7.0	.8451	.8457	.8463	.8470	.8476	.8482	.8488	.8494	.8500	.8506
7.1	.8513	.8519	.8525	.8531	.8537	.8543	.8549	.8555	.8561	.8567
7.2	.8573	.8579	.8585	.8591	.8597	.8603	.8609	.8615	.8621	.8627
7.3	.8633	.8639	.8645	.8651	.8657	.8663	.8669	.8675	.8681	.8686
7.4	.8692	.8698	.8704	.8710	.8716	.8722	.8727	.8733	.8739	.8745
7.5	.8751	.8756	.8762	.8768	.8774	.8779	.8785	.8791	.8797	.8802
7.6	.8808	.8814	.8820	.8825	.8831	.8837	.8842	.8848	.8854	.8859
7.7	.8865	.8871	.8876	.8882	.8887	.8893	.8899	.8904	.8910	.8915
7.8	.8921	.8927	.8932	.8938	.8943	.8949	.8954	.8960	.8965	.8971
7.9	.8976	.8982	.8987	.8993	.8998	.9004	.9009	.9015	.9020	.9025
8.0	.9031	.9036	.9042	.9047	.9053	.9058	.9063	.9069	.9074	.9079
8.1	.9085	.9090	.9096	.9101	.9106	.9112	.9117	.9122	.9128	.9133
8.2	.9138	.9143	.9149	.9154	.9159	.9165	.9170	.9175	.9180	.9186
8.3	.9191	.9196	.9201	.9206	.9212	.9217	.9222	.9227	.9232	.9238
8.4	.9243	.9248	.9253	.9258	.9263	.9269	.9274	.9279	.9284	.9289
8.5	.9294	.9299	.9304	.9309	.9315	.9320	.9325	.9330	.9335	.9340
8.6	.9345	.9350	.9355	.9360	.9365	.9370	.9375	.9380	.9385	.9390
8.7	.9395	.9400	.9405	.9410	.9415	.9420	.9425	.9430	.9435	.9440
8.8	.9445	.9450	.9455	.9460	.9465	.9469	.9474	.9479	.9484	.9489
8.9	.9494	.9499	.9504	.9509	.9513	.9518	.9523	.9528	.9533	.9538
9.0	.9542	.9547	.9552	.9557	.9562	.9566	.9571	.9576	.9581	.9586
9.1	.9590	.9595	.9600	.9605	.9609	.9614	.9619	.9624	.9628	.9633
9.2	.9638	.9643	.9647	.9652	.9657	.9661	.9666	.9671	.9675	.9680
9.3	.9685	.9689	.9694	.9699	.9703	.9708	.9713	.9717	.9722	.9727
9.4	.9731	.9736	.9741	.9745	.9750	.9754	.9759	.9763	.9768	.9773
9.5	.9777	.9782	.9786	.9791	.9795	.9800	.9805	.9809	.9814	.9818
9.6	.9823	.9827	.9832	.9836	.9841	.9845	.9850	.9854	.9859	.9863
9.7	.9868	.9872	.9877	.9881	.9886	.9890	.9894	.9899	.9903	.9908
9.8	.9912	.9917	.9921	.9926	.9930	.9934	.9939	.9943	.9948	.9952
9.9	.9956	.9961	.9965	.9969	.9974	.9978	.9983	.9987	.9991	.9996

APPENDIX B

Tables for Use with Binomial Models

The tables in this appendix are of three types. Table B1 provides the point probabilities and cumulative probabilities for selected values of n and p. Table B2 presents the binomial coefficients or combinations of n items selected r at a time for selected values of n. Table B3 gives confidence intervals for the binomial parameter, p, for three values of α and five values of n.

POINT AND CUMULATIVE PROBABILITIES FOR BINOMIAL MODELS: The purpose of Table B1 is to present the point probabilities, P(x), and the cumulative probabilities, F(x), as defined in (4.15) and (4.17), respectively. That is,

$$P(x) = \binom{n}{x} p^x q^{n-x}, \quad x = 0, 1, 2, \ldots, n$$

and

$$F(x) = \sum_{t=o}^{x} \binom{n}{t} p^t q^{n-t}.$$

If p = 1/10, 1/8, 1/6, 1/4, 1/3, or 1/2, the values of P(x) and F(x) are available in the table for n = 4, 5, 8, 10, 16, 20, and 25. These values are useful in modeling binomial experiments that involve dice-rolling and coin-flipping. If p = .1, .2, .3, .4, or .5, the values of P(x) and F(x) are available in the table for n = 20, 30, and 50. Those distributions are used in tests of hypotheses in Chapter 8. When p = .6, 2/3, .7, 3/4, .8, 5/6, 7/8, or .9, the desired values of P(x) and F(x) are found by interchanging the roles of p and q. In general, for any binomial variable, X,

$P(X = a) = P(a)$ and $P(X \leq a) = F(a)$ when $p \leq 1/2$,

and

$P(X = a) = P(n - a)$ and $P(X \leq a) = 1 - F(n - a - 1)$

when $p > 1/2$.

Notice that all arguments for P and F are x-values.

EXAMPLES

1. If $X \sim b(4, 1/3)$, find $P(X = 3)$ and $P(X \leq 3)$.

SOLUTION: Enter Table B1 with $n = 4$ and $p = 1/3$ at the row for $x = 3$ to find

$$P(X = 3) = P(3) = .0988,$$

$$P(X \leq 3) = F(3) = .9877.$$

2. If $X \sim b(4, 2/3)$, find $P(X = 3)$ and $P(X \leq 3)$.

SOLUTION: To find P(3) we enter Table B1 with $n = 4$, $p = 1/3$, and $n - x = 3$ to find

$$P(X = 3) = P(1) = .3951.$$

To find F(3) we can either add the probabilities for $(n - x) = 0, 1, 2,$ and 3 to obtain

$$F(3) = .0123 + .0988 + .2963 + .3951 = .8025,$$

or we can use the results above to write

$$P(X \leq 3) = 1 - F(0) = 1 - .1975 = .8025.$$

3. If $X \sim b(50, .7)$, find $P(X = 40)$ and $P(X \leq 30)$.

SOLUTION: By entering Table B1 for $n = 50$ and $p = .3$, we find

$$P(X = 40) = P(10) = .0386.$$

Similarly,

$$P(X \leq 30) = 1 - F(19) = 1 - .9152 = .0848.$$

REFERENCES

Additional listings of binomial probability distributions are available in the following sources.

Natl. Bureau of Standards. 1949. *Tables of binomial probability distribution*. Applied Math. Series No. 6. Washington, D.C.: U.S. Govt. Printing Office.

Romig, H. G. 1953. *50-100 binomial tables*. New York: Wiley.

TABLE B1.
Point and cumulative probabilities are given for binomial models.

p =	1/10		1/8		1/6		1/4		1/3		1/2		
x	P(x)	F(x)	P(x)	F(x)	P(x)	F(x)	P(x)	F(x)	P(x)	F(x)	P(x)	F(x)	n - x
					n = 4								
0	.6561	.6561	.5862	.5862	.4823	.4823	.3164	.3164	.1975	.1975	.0625	.0625	4
1	.2916	.9477	.3350	.9211	.3858	.8681	.4219	.7383	.3951	.5926	.2500	.3125	3
2	.0486	.9963	.0718	.9929	.1157	.9838	.2109	.9492	.2963	.8889	.3750	.6875	2
3	.0036	.9999	.0068	.9998	.0154	.9992	.0469	.9961	.0988	.9877	.2500	.9375	1
4	.0001	1.0000	.0002	1.0000	.0008	1.0000	.0039	1.0000	.0123	1.0000	.0625	1.0000	0
					n = 5								
0	.5905	.5905	.5129	.5129	.4019	.4019	.2373	.2373	.1317	.1317	.0313	.0313	5
1	.3281	.9185	.3664	.8793	.4019	.8038	.3955	.6328	.3292	.4609	.1563	.1875	4
2	.0729	.9914	.1047	.9840	.1608	.9645	.2637	.8965	.3292	.7901	.3125	.5000	3
3	.0081	.9995	.0150	.9989	.0321	.9967	.0879	.9844	.1646	.9547	.3125	.8125	2
4	.0005	1.0000	.0011	1.0000	.0032	.9999	.0146	.9990	.0412	.9959	.1563	.9688	1
5	.0000	1.0000	.0000	1.0000	.0001	1.0000	.0010	1.0000	.0041	1.0000	.0313	1.0000	0
					n = 8								
0	.4305	.4305	.3436	.3436	.2326	.2326	.1001	.1001	.0390	.0390	.0039	.0039	8
1	.3826	.8131	.3927	.7363	.3721	.6047	.2670	.3671	.1561	.1951	.0313	.0352	7
2	.1488	.9619	.1964	.9327	.2605	.8652	.3115	.6785	.2731	.4682	.1094	.1445	6
3	.0331	.9950	.0561	.9888	.1042	.9693	.2076	.8862	.2731	.7414	.2187	.3633	5
4	.0046	.9996	.0100	.9988	.0260	.9954	.0865	.9727	.1707	.9121	.2734	.6367	4
5	.0004	1.0000	.0011	.9999	.0042	.9996	.0231	.9958	.0683	.9803	.2188	.8555	3
6	.0000	1.0000	.0001	1.0000	.0004	1.0000	.0038	.9996	.0171	.9974	.1094	.9648	2
7	.0000	1.0000	.0000	1.0000	.0000	1.0000	.0004	1.0000	.0024	.9998	.0313	.9961	1
8	.0000	1.0000	.0000	1.0000	.0000	1.0000	.0000	1.0000	.0002	1.0000	.0039	1.0000	0

TABLE B1. Point and cumulative probabilities are given for binomial models (continued).

	p = 1/10		1/8		1/6		1/4		1/3		1/2		
x	P(x)	F(x)	P(x)	F(x)	P(x)	F(x)	P(x)	F(x)	P(x)	F(x)	P(x)	F(x)	n - x
	n = 10												
0	.3487	.3487	.2631	.2631	.1615	.1615	.0563	.0563	.0173	.0173	.0010	.0010	10
1	.3874	.7361	.3758	.6389	.3230	.4845	.1877	.2440	.0867	.1040	.0098	.0107	9
2	.1937	.9298	.2416	.8805	.2907	.7752	.2816	.5256	.1951	.2991	.0439	.0547	8
3	.0574	.9872	.0920	.9725	.1551	.9303	.2503	.7759	.2601	.5593	.1172	.1719	7
4	.0111	.9984	.0230	.9956	.0543	.9845	.1460	.9219	.2276	.7869	.2051	.3770	6
5	.0015	.9999	.0039	.9995	.0130	.9976	.0584	.9803	.1366	.9234	.2461	.6231	5
6	.0001	1.0000	.0005	1.0000	.0022	.9998	.0162	.9965	.0569	.9803	.2051	.8281	4
7	.0000	1.0000	.0000	1.0000	.0002	1.0000	.0031	.9996	.0163	.9966	.1172	.9453	3
8	.0000	1.0000	.0000	1.0000	.0000	1.0000	.0004	1.0000	.0030	.9996	.0439	.9893	2
9	.0000	1.0000	.0000	1.0000	.0000	1.0000	.0000	1.0000	.0004	1.0000	.0097	.9990	1
10	.0000	1.0000	.0000	1.0000	.0000	1.0000	.0000	1.0000	.0000	1.0000	.0010	1.0000	0
	n = 16												
0	.1853	.1853	.1181	.1181	.0541	.0541	.0100	.0100	.0015	.0015	.0000	.0000	16
1	.3294	.5147	.2699	.3879	.1731	.2272	.0535	.0635	.0122	.0137	.0002	.0003	15
2	.2745	.7893	.2891	.6771	.2596	.4868	.1336	.1971	.0457	.0594	.0018	.0021	14
3	.1423	.9316	.1928	.8698	.2423	.7291	.2079	.4050	.1066	.1660	.0085	.0106	13
4	.0514	.9830	.0895	.9593	.1575	.8866	.2252	.6302	.1732	.3391	.0278	.0384	12
5	.0137	.9967	.0307	.9900	.0756	.9622	.1802	.8104	.2078	.5469	.0667	.1051	11
6	.0028	.9995	.0080	.9981	.0277	.9899	.1101	.9204	.1905	.7374	.1222	.2273	10
7	.0004	.9999	.0016	.9997	.0079	.9978	.0524	.9729	.1361	.8735	.1746	.4018	9
8	.0001	1.0000	.0003	1.0000	.0018	.9996	.0197	.9925	.0765	.9500	.1964	.5982	8
9	.0000	1.0000	.0000	1.0000	.0003	.9999	.0058	.9984	.0340	.9841	.1746	.7728	7
10	.0000	1.0000	.0000	1.0000	.0001	1.0000	.0013	.9997	.0119	.9960	.1222	.8949	6
11	.0000	1.0000	.0000	1.0000	.0000	1.0000	.0002	.9999	.0032	.9992	.0667	.9616	5
12	.0000	1.0000	.0000	1.0000	.0000	1.0000	.0001	1.0000	.0007	.9999	.0278	.9894	4
13	.0000	1.0000	.0000	1.0000	.0000	1.0000	.0000	1.0000	.0001	1.0000	.0085	.9979	3
14	.0000	1.0000	.0000	1.0000	.0000	1.0000	.0000	1.0000	.0000	1.0000	.0018	.9997	2
15	.0000	1.0000	.0000	1.0000	.0000	1.0000	.0000	1.0000	.0000	1.0000	.0002	1.0000	1
16	.0000	1.0000	.0000	1.0000	.0000	1.0000	.0000	1.0000	.0000	1.0000	.0000	1.0000	0

TABLE B1. Point and cumulative probabilities are given for binomial models (continued).

	p = 1/10		1/8		1/6		1/4		1/3		1/2		
x	P(x)	F(x)	P(x)	F(x)	P(x)	F(x)	P(x)	F(x)	P(x)	F(x)	P(x)	F(x)	n - x
	n = 20												
0	.1216	.1216	.0692	.0692	.0261	.0261	.0032	.0032	.0003	.0003	.0000	.0000	20
1	.2702	.3918	.1977	.2670	.1043	.1304	.0211	.0243	.0030	.0033	.0000	.0000	19
2	.2852	.6769	.2684	.5353	.1982	.3287	.0669	.0913	.0143	.0176	.0002	.0002	18
3	.1901	.8670	.2300	.7653	.2379	.5666	.1339	.2252	.0429	.0604	.0011	.0013	17
4	.0898	.9568	.1397	.9050	.2022	.7688	.1897	.4148	.0911	.1515	.0046	.0059	16
5	.0319	.9887	.0638	.9688	.1294	.8982	.2023	.6172	.1457	.2972	.0148	.0207	15
6	.0089	.9976	.0228	.9916	.0647	.9629	.1686	.7858	.1821	.4793	.0370	.0577	14
7	.0020	.9996	.0065	.9982	.0259	.9888	.1124	.8982	.1821	.6615	.0739	.1316	13
8	.0004	1.0000	.0015	.9997	.0084	.9972	.0609	.9591	.1480	.8095	.1201	.2517	12
9	.0000	1.0000	.0003	1.0000	.0022	.9994	.0271	.9862	.0987	.9081	.1602	.4119	11
10	.0000	1.0000	.0000	1.0000	.0005	.9999	.0099	.9961	.0543	.9624	.1762	.5881	10
11	.0000	1.0000	.0000	1.0000	.0001	1.0000	.0030	.9991	.0247	.9870	.1602	.7483	9
12	.0000	1.0000	.0000	1.0000	.0000	1.0000	.0007	.9998	.0093	.9963	.1201	.8684	8
13	.0000	1.0000	.0000	1.0000	.0000	1.0000	.0002	1.0000	.0028	.9991	.0739	.9423	7
14	.0000	1.0000	.0000	1.0000	.0000	1.0000	.0000	1.0000	.0007	.9998	.0370	.9793	6
15	.0000	1.0000	.0000	1.0000	.0000	1.0000	.0000	1.0000	.0002	1.0000	.0148	.9941	5
16	.0000	1.0000	.0000	1.0000	.0000	1.0000	.0000	1.0000	.0000	1.0000	.0046	.9987	4
17	.0000	1.0000	.0000	1.0000	.0000	1.0000	.0000	1.0000	.0000	1.0000	.0011	.9998	3
18	.0000	1.0000	.0000	1.0000	.0000	1.0000	.0000	1.0000	.0000	1.0000	.0002	1.0000	2
19	.0000	1.0000	.0000	1.0000	.0000	1.0000	.0000	1.0000	.0000	1.0000	.0000	1.0000	1
20	.0000	1.0000	.0000	1.0000	.0000	1.0000	.0000	1.0000	.0000	1.0000	.0000	1.0000	0

TABLE B1. Point and cumulative probabilities are given for binomial models (continued).

x	p = 1/10 P(x)	1/10 F(x)	1/8 P(x)	1/8 F(x)	1/6 P(x)	1/6 F(x)	1/4 P(x)	1/4 F(x)	1/3 P(x)	1/3 F(x)	1/2 P(x)	1/2 F(x)	n - x
	n = 25												
0	.0718	.0718	.0355	.0355	.0105	.0105	.0008	.0008	.0000	.0000	.0000	.0000	25
1	.1994	.2712	.1268	.1623	.0524	.0629	.0063	.0070	.0005	.0005	.0000	.0000	24
2	.2659	.5371	.2173	.3796	.1258	.1887	.0251	.0321	.0030	.0035	.0000	.0000	23
3	.2265	.7636	.2380	.6176	.1929	.3816	.0641	.0962	.0114	.0149	.0001	.0001	22
4	.1384	.9020	.1870	.8047	.2122	.5937	.1175	.2137	.0313	.0462	.0004	.0005	21
5	.0646	.9666	.1122	.9169	.1782	.7720	.1645	.3783	.0658	.1120	.0016	.0020	20
6	.0239	.9905	.0534	.9703	.1188	.8908	.1828	.5611	.1096	.2216	.0053	.0073	19
7	.0072	.9977	.0207	.9910	.0645	.9553	.1654	.7265	.1487	.3703	.0143	.0216	18
8	.0018	.9995	.0067	.9977	.0290	.9843	.1241	.8506	.1673	.5376	.0322	.0539	17
9	.0004	.9999	.0018	.9995	.0110	.9953	.0781	.9287	.1580	.6956	.0609	.1148	16
10	.0001	1.0000	.0004	.9999	.0035	.9988	.0417	.9703	.1264	.8220	.0974	.2122	15
11	.0000	1.0000	.0001	1.0000	.0010	.9997	.0189	.9893	.0862	.9082	.1328	.3450	14
12	.0000	1.0000	.0000	1.0000	.0002	.9999	.0074	.9966	.0503	.9585	.1550	.5000	13
13	.0000	1.0000	.0000	1.0000	.0001	1.0000	.0025	.9991	.0251	.9836	.1550	.6550	12
14	.0000	1.0000	.0000	1.0000	.0000	1.0000	.0007	.9998	.0108	.9944	.1328	.7878	11
15	.0000	1.0000	.0000	1.0000	.0000	1.0000	.0002	1.0000	.0040	.9984	.0974	.8852	10
16	.0000	1.0000	.0000	1.0000	.0000	1.0000	.0000	1.0000	.0012	.9996	.0609	.9461	9
17	.0000	1.0000	.0000	1.0000	.0000	1.0000	.0000	1.0000	.0003	.9999	.0322	.9784	8
18	.0000	1.0000	.0000	1.0000	.0000	1.0000	.0000	1.0000	.0001	1.0000	.0143	.9927	7
19	.0000	1.0000	.0000	1.0000	.0000	1.0000	.0000	1.0000	.0000	1.0000	.0053	.9980	6
20	.0000	1.0000	.0000	1.0000	.0000	1.0000	.0000	1.0000	.0000	1.0000	.0016	.9995	5
21	.0000	1.0000	.0000	1.0000	.0000	1.0000	.0000	1.0000	.0000	1.0000	.0004	.9999	4
22	.0000	1.0000	.0000	1.0000	.0000	1.0000	.0000	1.0000	.0000	1.0000	.0001	1.0000	3
23	.0000	1.0000	.0000	1.0000	.0000	1.0000	.0000	1.0000	.0000	1.0000	.0000	1.0000	2
24	.0000	1.0000	.0000	1.0000	.0000	1.0000	.0000	1.0000	.0000	1.0000	.0000	1.0000	1
25	.0000	1.0000	.0000	1.0000	.0000	1.0000	.0000	1.0000	.0000	1.0000	.0000	1.0000	0

TABLE B1. Point and cumulative probabilities are given for binomial models (continued).

p =	.1		.2		.3		.4		.5		
x	P(x)	F(x)	P(x)	F(x)	P(x)	F(x)	P(x)	F(x)	P(x)	F(x)	n - x
	n = 20										
0	.1216	.1216	.0115	.0115	.0008	.0008	.0000	.0000	.0000	.0000	20
1	.2702	.3918	.0576	.0692	.0068	.0076	.0005	.0005	.0000	.0000	19
2	.2852	.6769	.1369	.2061	.0279	.0355	.0031	.0036	.0002	.0002	18
3	.1901	.8670	.2054	.4115	.0716	.1071	.0124	.0160	.0011	.0013	17
4	.0898	.9568	.2182	.6297	.1304	.2375	.0350	.0510	.0046	.0059	16
5	.0319	.9887	.1746	.8042	.1787	.4164	.0746	.1256	.0148	.0207	15
6	.0089	.9976	.1091	.9133	.1916	.6080	.1244	.2500	.0370	.0577	14
7	.0020	.9996	.0546	.9679	.1643	.7723	.1659	.4159	.0739	.1316	13
8	.0004	1.0000	.0222	.9900	.1144	.8867	.1797	.5956	.1201	.2517	12
9	.0000	1.0000	.0074	.9974	.0654	.9520	.1597	.7553	.1602	.4119	11
10	.0000	1.0000	.0020	.9994	.0308	.9829	.1171	.8725	.1762	.5881	10
11	.0000	1.0000	.0005	.9999	.0120	.9949	.0710	.9435	.1602	.7483	9
12	.0000	1.0000	.0001	1.0000	.0039	.9987	.0355	.9790	.1201	.8684	8
13	.0000	1.0000	.0000	1.0000	.0010	.9997	.0146	.9935	.0739	.9423	7
14	.0000	1.0000	.0000	1.0000	.0002	.9999	.0049	.9984	.0370	.9793	6
15	.0000	1.0000	.0000	1.0000	.0001	1.0000	.0013	.9997	.0148	.9941	5
16	.0000	1.0000	.0000	1.0000	.0000	1.0000	.0003	1.0000	.0046	.9987	4
17	.0000	1.0000	.0000	1.0000	.0000	1.0000	.0000	1.0000	.0011	.9998	3
18	.0000	1.0000	.0000	1.0000	.0000	1.0000	.0000	1.0000	.0002	1.0000	2
19	.0000	1.0000	.0000	1.0000	.0000	1.0000	.0000	1.0000	.0000	1.0000	1
20	.0000	1.0000	.0000	1.0000	.0000	1.0000	.0000	1.0000	.0000	1.0000	0

TABLE B1. Point and cumulative probabilities are given for binomial models (continued).

	p = .1		.2		.3		.4		.5		
x	P(x)	F(x)	P(x)	F(x)	P(x)	F(x)	P(x)	F(x)	P(x)	F(x)	n - x
					n = 30						
0	.0424	.0424	.0012	.0012	.0000	.0000	.0000	.0000	.0000	.0000	30
1	.1413	.1837	.0093	.0105	.0003	.0003	.0000	.0000	.0000	.0000	29
2	.2277	.4114	.0337	.0442	.0018	.0021	.0000	.0000	.0000	.0000	28
3	.2361	.6474	.0785	.1227	.0072	.0093	.0003	.0003	.0000	.0000	27
4	.1771	.8245	.1325	.2552	.0208	.0301	.0012	.0015	.0000	.0000	26
5	.1023	.9268	.1723	.4275	.0464	.0766	.0041	.0056	.0001	.0001	25
6	.0474	.9742	.1795	.6070	.0829	.1595	.0115	.0172	.0006	.0007	24
7	.0180	.9922	.1538	.7608	.1219	.2814	.0263	.0435	.0019	.0026	23
8	.0058	.9980	.1106	.8714	.1501	.4315	.0505	.0940	.0055	.0081	22
9	.0016	.9996	.0676	.9389	.1573	.5888	.0823	.1763	.0133	.0214	21
10	.0003	.9999	.0355	.9744	.1416	.7304	.1152	.2915	.0280	.0494	20
11	.0001	1.0000	.0161	.9905	.1103	.8407	.1396	.4311	.0509	.1002	19
12	.0000	1.0000	.0064	.9969	.0749	.9155	.1474	.5785	.0806	.1808	18
13	.0000	1.0000	.0022	.9991	.0444	.9599	.1360	.7145	.1115	.2923	17
14	.0000	1.0000	.0007	.9998	.0231	.9831	.1101	.8246	.1354	.4277	16
15	.0000	1.0000	.0002	1.0000	.0106	.9936	.0783	.9029	.1445	.5722	15
16	.0000	1.0000	.0000	1.0000	.0042	.9978	.0489	.9519	.1354	.7076	14
17	.0000	1.0000	.0000	1.0000	.0015	.9993	.0269	.9788	.1115	.8192	13
18	.0000	1.0000	.0000	1.0000	.0005	.9998	.0129	.9917	.0806	.8998	12
19	.0000	1.0000	.0000	1.0000	.0002	1.0000	.0054	.9971	.0509	.9506	11
20	.0000	1.0000	.0000	1.0000	.0000	1.0000	.0020	.9991	.0280	.9786	10
21	.0000	1.0000	.0000	1.0000	.0000	1.0000	.0007	.9998	.0133	.9919	9
22	.0000	1.0000	.0000	1.0000	.0000	1.0000	.0002	1.0000	.0055	.9974	8
23	.0000	1.0000	.0000	1.0000	.0000	1.0000	.0000	1.0000	.0019	.9993	7
24	.0000	1.0000	.0000	1.0000	.0000	1.0000	.0000	1.0000	.0006	.9999	6
25	.0000	1.0000	.0000	1.0000	.0000	1.0000	.0000	1.0000	.0001	1.0000	5
26	.0000	1.0000	.0000	1.0000	.0000	1.0000	.0000	1.0000	.0000	1.0000	4
27	.0000	1.0000	.0000	1.0000	.0000	1.0000	.0000	1.0000	.0000	1.0000	3
28	.0000	1.0000	.0000	1.0000	.0000	1.0000	.0000	1.0000	.0000	1.0000	2
29	.0000	1.0000	.0000	1.0000	.0000	1.0000	.0000	1.0000	.0000	1.0000	1
30	.0000	1.0000	.0000	1.0000	.0000	1.0000	.0000	1.0000	.0000	1.0000	0

TABLE B1. Point and cumulative probabilities are given for binomial models (continued).

	p = .1		.2		.3		.4		.5		
x	P(x)	F(x)	P(x)	F(x)	P(x)	F(x)	P(x)	F(x)	P(x)	F(x)	n - x
	n = 50										
0	.0052	.0052	.0000	.0000	.0000	.0000	.0000	.0000	.0000	.0000	50
1	.0286	.0338	.0002	.0002	.0000	.0000	.0000	.0000	.0000	.0000	49
2	.0779	.1117	.0011	.0013	.0000	.0000	.0000	.0000	.0000	.0000	48
3	.1386	.2503	.0044	.0057	.0000	.0000	.0000	.0000	.0000	.0000	47
4	.1809	.4312	.0128	.0185	.0002	.0002	.0000	.0000	.0000	.0000	46
5	.1849	.6161	.0295	.0480	.0006	.0007	.0000	.0000	.0000	.0000	45
6	.1541	.7702	.0554	.1034	.0018	.0025	.0000	.0000	.0000	.0000	44
7	.1076	.8778	.0870	.1904	.0048	.0073	.0001	.0001	.0000	.0000	43
8	.0643	.9421	.1169	.3073	.0110	.0183	.0002	.0003	.0000	.0000	42
9	.0333	.9755	.1364	.4437	.0220	.0403	.0005	.0008	.0000	.0000	41
10	.0152	.9907	.1398	.5836	.0386	.0789	.0014	.0022	.0000	.0000	40
11	.0061	.9968	.1271	.7107	.0602	.1391	.0035	.0057	.0000	.0000	39
12	.0022	.9990	.1033	.8140	.0838	.2229	.0076	.0133	.0001	.0002	38
13	.0007	.9997	.0755	.8894	.1050	.3279	.0147	.0280	.0003	.0005	37
14	.0002	.9999	.0499	.9393	.1190	.4468	.0260	.0540	.0008	.0013	36
15	.0001	1.0000	.0299	.9692	.1224	.5692	.0415	.0955	.0020	.0033	35
16	.0000	1.0000	.0164	.9856	.1147	.6839	.0606	.1561	.0044	.0077	34
17	.0000	1.0000	.0081	.9937	.0983	.7822	.0808	.2369	.0087	.0164	33
18	.0000	1.0000	.0038	.9975	.0772	.8594	.0987	.3356	.0160	.0324	32
19	.0000	1.0000	.0016	.9991	.0558	.9152	.1107	.4465	.0270	.0595	31
20	.0000	1.0000	.0006	.9997	.0370	.9522	.1146	.5610	.0419	.1013	30
21	.0000	1.0000	.0002	.9999	.0227	.9749	.1091	.6701	.0598	.1611	29
22	.0000	1.0000	.0001	1.0000	.0128	.9877	.0959	.7660	.0788	.2399	28
23	.0000	1.0000	.0000	1.0000	.0067	.9944	.0778	.8438	.0960	.3359	27
24	.0000	1.0000	.0000	1.0000	.0032	.9976	.0584	.9022	.1080	.4439	26
25	.0000	1.0000	.0000	1.0000	.0014	.9991	.0405	.9427	.1123	.5562	25

TABLE B1. Point and cumulative probabilities are given for binomial models (continued).

	p = .1		.2		.3		.4		.5		
x	P(x)	F(x)	P(x)	F(x)	P(x)	F(x)	P(x)	F(x)	P(x)	F(x)	n - x
	n = 50										
26	.0000	1.0000	.0000	1.0000	.0006	.9997	.0259	.9686	.1080	.6641	24
27	.0000	1.0000	.0000	1.0000	.0002	.9999	.0154	.9840	.0960	.7601	23
28	.0000	1.0000	.0000	1.0000	.0001	1.0000	.0084	.9924	.0788	.8389	22
29	.0000	1.0000	.0000	1.0000	.0000	1.0000	.0043	.9966	.0598	.8987	21
30	.0000	1.0000	.0000	1.0000	.0000	1.0000	.0020	.9986	.0419	.9406	20
31	.0000	1.0000	.0000	1.0000	.0000	1.0000	.0009	.9995	.0270	.9676	19
32	.0000	1.0000	.0000	1.0000	.0000	1.0000	.0003	.9998	.0160	.9836	18
33	.0000	1.0000	.0000	1.0000	.0000	1.0000	.0001	.9999	.0087	.9923	17
34	.0000	1.0000	.0000	1.0000	.0000	1.0000	.0001	1.0000	.0044	.9967	16
35	.0000	1.0000	.0000	1.0000	.0000	1.0000	.0000	1.0000	.0020	.9987	15
36	.0000	1.0000	.0000	1.0000	.0000	1.0000	.0000	1.0000	.0008	.9995	14
37	.0000	1.0000	.0000	1.0000	.0000	1.0000	.0000	1.0000	.0003	.9999	13
38	.0000	1.0000	.0000	1.0000	.0000	1.0000	.0000	1.0000	.0001	1.0000	12
39	.0000	1.0000	.0000	1.0000	.0000	1.0000	.0000	1.0000	.0000	1.0000	11
40	.0000	1.0000	.0000	1.0000	.0000	1.0000	.0000	1.0000	.0000	1.0000	10
41	.0000	1.0000	.0000	1.0000	.0000	1.0000	.0000	1.0000	.0000	1.0000	9
42	.0000	1.0000	.0000	1.0000	.0000	1.0000	.0000	1.0000	.0000	1.0000	8
43	.0000	1.0000	.0000	1.0000	.0000	1.0000	.0000	1.0000	.0000	1.0000	7
44	.0000	1.0000	.0000	1.0000	.0000	1.0000	.0000	1.0000	.0000	1.0000	6
45	.0000	1.0000	.0000	1.0000	.0000	1.0000	.0000	1.0000	.0000	1.0000	5
46	.0000	1.0000	.0000	1.0000	.0000	1.0000	.0000	1.0000	.0000	1.0000	4
47	.0000	1.0000	.0000	1.0000	.0000	1.0000	.0000	1.0000	.0000	1.0000	3
48	.0000	1.0000	.0000	1.0000	.0000	1.0000	.0000	1.0000	.0000	1.0000	2
49	.0000	1.0000	.0000	1.0000	.0000	1.0000	.0000	1.0000	.0000	1.0000	1
50	.0000	1.0000	.0000	1.0000	.0000	1.0000	.0000	1.0000	.0000	1.0000	0

BINOMIAL COEFFICIENTS OR COMBINATIONS: The values for the number of combinations of n items selected r at a time or the binomial coefficients,

$$\binom{n}{r} = n!/r!(n - r)!, \quad r = 0, 1, 2, \ldots, n,$$

can be read directly from Table B2 for n = 1, 2, ..., 14. If $14 < n \leq 25$ and $r > 14$, the missing values can be obtained by using the symmetry,

$$\binom{n}{r} = \binom{n}{n - r}.$$

The table can be extended by using Pascal's triangle to show that

$$\binom{n}{r} = \binom{n - 1}{r - 1} + \binom{n - 1}{r}.$$

EXAMPLES

1. $\binom{15}{8} = ?$

SOLUTION: Enter row n = 15 and column r = 8 to find

$$\binom{15}{8} = 6435.$$

2. $\binom{24}{17} = ?$

SOLUTION: First we note that

$$\binom{24}{17} = \binom{24}{7}.$$

Then we use n = 24 and r = 7 to find

$$\binom{24}{17} = \binom{24}{7} = 346104.$$

3. Find $\binom{26}{4}$.

SOLUTION: By Pascal's triangle,

$$\binom{26}{4} = \binom{25}{3} + \binom{25}{4} = 2300 + 12650 = 14950.$$

REFERENCE

Additional coefficients are available in the following source.

Abramowitz, M., and Stegun, I.A., eds. 1964. *Handbook of mathematical functions*. Applied Math. Series No. 55. U.S. Dept. of Commerce, Natl. Bureau of Standards. Washington, D.C.: U.S. Govt. Printing Office, pp. 828-830.

TABLE B2.
Shown here are binomial coefficients $\binom{n}{r}$.

n	r=0	1	2	3	4	5	6	7	8	9	10	11	12	13	14
1	1	1													
2	1	2	1												
3	1	3	3	1											
4	1	4	6	4	1										
5	1	5	10	10	5	1									
6	1	6	15	20	15	6	1								
7	1	7	21	35	35	21	7	1							
8	1	8	28	56	70	56	28	8	1						
9	1	9	36	84	126	126	84	36	9	1					
10	1	10	45	120	210	252	210	120	45	10	1				
11	1	11	55	165	330	462	462	330	165	55	11	1			
12	1	12	66	220	495	792	924	792	495	220	66	12	1		
13	1	13	78	286	715	1287	1716	1716	1287	715	286	78	13	1	
14	1	14	91	364	1001	2002	3003	3432	3003	2002	1001	364	91	14	1
15	1	15	105	455	1365	3003	5005	6435	6435	5005	3003	1365	455	105	15
16	1	16	120	560	1820	4368	8008	11440	12870	11440	8008	4368	1820	560	120
17	1	17	136	680	2380	6188	12376	19448	24310	24310	19448	12376	6188	2380	680
18	1	18	153	816	3060	8568	18564	31824	43758	48620	43758	31824	18564	8568	3060
19	1	19	171	969	3876	11628	27132	50388	75582	92378	92378	75582	50388	27132	11628
20	1	20	190	1140	4845	15504	38760	77520	125970	167960	184756	167960	125970	77520	38760
21	1	21	210	1330	5985	20349	54264	116280	203490	293930	352716	352716	293930	203490	116280
22	1	22	231	1540	7315	26334	74613	170544	319770	497420	646646	705432	646646	497420	319770
23	1	23	253	1771	8855	33649	100947	245157	490314	817190	1144066	1352078	1352078	1144066	817190
24	1	24	276	2024	10626	42504	134596	346104	735471	1307504	1961256	2496144	2704156	2496144	1961256
25	1	25	300	2300	12650	53130	177100	480700	1081575	2042975	3268760	4457400	5200300	5200300	4457400
30	1	30	435	4060	27405	142506	593775	2035800	5852925	14307150	30045015	54627300	86493225	119759850	145422675

CONFIDENCE INTERVALS FOR p: The lower and upper confidence limits, L(x) and U(x), for p can be obtained from Tables B3, B4, and B5. If either n = 20 or 50 or $x/n \leq .5$ for n = 100, 200, or 500, the limits are read directly from the table exhibiting the specified α. When n = 100, 200, or 500 and $x/n > .5$, we enter the table with x/n replaced by 1 - x/n and subtract the given values of L(x) and U(x) from 100.

EXAMPLES

1. A sample from a binomial model having n = 20 shows x = 8. Find a 95% CI for p.

SOLUTION: Enter Table B4 for $\alpha = .05$ with n = 20 and x = 8 to find L(x) = 19 and U(x) = 64. Since those values are expressed as percentages, the desired confidence interval is

$$CI_{.95}: \ .19 \leq p \leq .64.$$

2. A binomial sample with n = 100 shows x = 64. Find a 90% CI for p.

SOLUTION: Enter Table B3 for $\alpha = .10$ with n = 100 and x/n = 1 - .64 = .36 to find L(x) = 28 and U(x) = 45. Since $x/n > .5$ for this exercise, the desired limits are L(x) = 100 - 45 = 55 and U(x) = 100 - 28 = 72. Thus,

$$CI_{.90}: \ .55 \leq p \leq .72.$$

3. Find $UCI_{.95}$ for p when n = 20 and x = 5.

SOLUTION: For an upper limit only with $\alpha = .05$, we enter Table B3 for $\alpha = .10$ and set L(x) = 0. Thus, n = 20, and x = 5 implies that

$$UCI_{.95}: \ p \leq .46.$$

REFERENCE

Additional values can be calculated by using Eq. (7.70).

TABLE B3.
Confidence intervals are given for p(α = .10).

x	n = 20		n = 50		x/n	n = 100		n = 200		n = 500	
	L(x)	U(x)	L(x)	U(x)		L(x)	U(x)	L(x)	U(x)	L(x)	U(x)
0	0	14	0	6	0.00	0	3	0	1	0	0+
1	0	22	0	9	0.01	0	5	0	3	0	2
2	2	28	1	12	0.02	0	6	1	5	1	3-
3	4	34	2	15	0.03	1	8	1	6	2	5
4	7	40	3	17	0.04	1	9	2	7	3	6
5	10	46	4	20	0.05	2	10	3	8	3	7
6	14	51	5	22	0.06	3	12	4	10	4	8
7	18	56	7	25	0.07	3	13	4	11	5	9
8	22	61	8	27	0.08	4	14	5	12	6	10
9	26	65	11	29	0.09	5	15	6	13	7	11
10	30	70	11	32	0.10	6	16	7	14	8	12
11	35	74	13	34	0.11	6	18	8	15	9	14
12	39	78	14	36	0.12	7	19	8	16	10	15
13	44	82	16	38	0.13	8	20	9	18	11	16
14	49	86	18	40	0.14	9	21	10	19	12	17
15	54	90	20	42	0.15	9	22	11	20	13	18
16	60	93	21	44	0.16	10	23	12	21	13	19
17	66	96	23	47	0.17	11	24	13	22	14	20
18	72	98	25	49	0.18	12	26	13	23	15	21
19	78	100	26	51	0.19	13	27	14	24	16	22
20	86	100	28	53	0.20	14	28	15	25	17	23
21			30	55	0.21	14	29	16	26	18	24
22			32	57	0.22	15	30	17	27	19	25
23			34	59	0.23	16	31	18	29	20	26
24			36	60	0.24	17	32	19	29	21	27
25			38	62	0.25	18	33	20	31	22	28
26			40	64	0.26	19	34	21	32	23	29
27			41	66	0.27	20	35	22	33	24	30
28			43	68	0.28	21	36	23	34	25	31
29			45	70	0.29	22	37	24	35	26	32
30			47	72	0.30	22	38	25	36	27	33
31			49	74	0.31	23	40	25	37	28	34
32			51	75	0.32	24	41	26	38	29	36
33			53	77	0.33	25	42	27	39	30	37
34			56	79	0.34	26	43	28	40	31	38
35			58	80	0.35	27	44	29	41	32	39
36			60	82	0.36	28	45	30	42	33	40
37			62	84	0.37	29	46	31	43	34	41
38			64	86	0.38	30	47	32	44	34	42
39			66	87	0.39	31	48	33	45	36	43
40			68	89	0.40	32	49	34	46	37	44
41			71	90	0.41	33	50	35	47	38	45
42			73	92	0.42	34	51	36	48	39	46
43			75	93	0.43	35	52	37	49	40	47
44			78	95	0.44	36	53	38	50	41	48

TABLE B3. Confidence intervals are given for p(α = .10) (continued).

x	n = 20		n = 50		x/n	n = 100		n = 200		n = 500	
	L(x)	U(x)	L(x)	U(x)		L(x)	U(x)	L(x)	U(x)	L(x)	U(x)
45			80	96	0.45	36	54	39	51	41	49
46			83	97	0.46	37	55	40	52	42	50
47			85	98	0.47	38	56	41	53	43	51
48			88	99	0.48	39	57	42	54	44	52
49			91	100	0.49	40	58	43	55	45	53
50			94	100	0.50	41	59	44	56	46	54

NOTE: Reprinted by permission from H. C. Fryer, *Experimental Statistics* (New York: Allyn and Bacon, 1966), p. 563. © 1966 by the author.

TABLE B4.
Confidence intervals are given for p(α = .05).

x	n = 20		n = 50		x/n	n = 100		n = 200		n = 500	
	L(x)	U(x)	L(x)	U(x)		L(x)	U(x)	L(x)	U(x)	L(x)	U(x)
0	0	17	0	7	0.00	0	4	0	2	0	0+
1	0	25	0	11	0.01	0	5	0	4	0	2
2	1	32	1	14	0.02	0	7	1	5	1	4
3	3	38	1	17	0.03	1	9	1	6	2	5
4	6	44	2	19	0.04	1	10	2	8	3	6
5	9	49	3	22	0.05	2	11	2	9	3	7
6	12	54	5	24	0.06	2	13	3	10	4	9
7	15	59	6	27	0.07	3	14	4	12	5	10
8	19	64	7	29	0.08	4	15	5	13	6	11
9	23	68	9	31	0.09	4	16	6	14	7	12
10	27	73	10	34	0.10	5	18	6	15	8	13
11	32	77	11	36	0.11	6	19	7	17	8	14
12	36	81	13	38	0.12	6	20	8	18	9	15
13	41	85	15	40	0.13	7	21	9	19	10	16
14	46	88	16	43	0.14	8	22	10	20	11	17
15	51	91	18	45	0.15	9	24	11	21	12	18
16	56	94	19	47	0.16	9	25	11	23	13	19
17	62	97	21	49	0.17	10	26	12	24	14	21
18	68	99	23	51	0.18	11	27	13	25	15	22
19	75	100	25	53	0.19	12	28	14	26	16	23
20	83	100	26	55	0.20	13	29	15	27	17	24
21			28	57	0.21	14	30	16	28	18	25
22			30	59	0.22	14	31	17	29	19	26
23			32	61	0.23	15	33	18	30	19	27
24			33	63	0.24	16	34	18	31	20	28
25			35	65	0.25	17	35	19	32	21	29

TABLE B4. Confidence intervals are given for p(α = .05) (continued).

x	n = 20		n =50		x/n	n = 100		n = 200		n = 500	
	L(x)	U(x)	L(x)	U(x)		L(x)	U(x)	L(x)	U(x)	L(x)	U(x)
26			37	67	0.26	18	36	20	33	22	30
27			39	68	0.27	18	37	21	34	23	31
28			41	70	0.28	19	38	22	35	24	32
29			43	72	0.29	20	39	23	36	25	33
30			45	74	0.30	21	40	24	37	26	34
31			47	75	0.31	22	41	25	38	27	35
32			49	77	0.32	23	42	26	39	28	36
33			51	79	0.33	24	43	27	40	29	37
34			53	81	0.34	25	44	28	41	30	38
35			55	82	0.35	26	45	29	42	31	39
36			57	84	0.36	26	46	30	43	32	40
37			60	85	0.37	27	47	31	44	33	41
38			62	87	0.38	28	48	32	45	34	42
39			64	89	0.39	29	49	33	46	35	43
40			66	90	0.40	30	50	34	47	36	44
41			69	91	0.41	31	51	35	48	37	45
42			71	93	0.42	32	52	36	49	28	46
43			73	94	0.43	33	53	37	50	39	47
44			76	95	0.44	34	54	38	51	40	48
45			78	97	0.45	35	55	39	52	41	49
46			81	99	0.46	36	57	40	53	42	50
47			83	99	0.47	36	58	41	54	43	51
48			86	100	0.48	37	59	42	55	44	52
49			89	100	0.49	38	60	43	56	45	53
50			93	100	0.50	39	61	43	57	46	54

NOTE: Reprinted by permission from H. C. Fryer, *Experimental Statistics* (New York: Allyn and Bacon, 1966), p. 564. © by the author.

TABLE B5.
Confidence intervals are given for p(α = .01).

x	n = 20 L(x)	n = 20 U(x)	n = 50 L(x)	n = 50 U(x)	x/n	n = 100 L(x)	n = 100 U(x)	n = 200 L(x)	n = 200 U(x)	n = 500 L(x)	n = 500 U(x)
0	0	23	0	10	0.00	0	5	0	3	0	1
1	0	32	0	14	0.01	0	7	0	5	0	3
2	1	39	0	17	0.02	0	9	1	6	1	4
3	2	45	1	20	0.03	0	10	1	8	1	6
4	4	51	1	23	0.04	1	12	2	9	2	7
5	6	56	2	26	0.05	1	13	3	10	3	8
6	8	61	3	28	0.06	2	15	3	12	4	9
7	11	66	4	31	0.07	2	16	4	13	4	11
8	15	70	5	33	0.08	3	17	5	14	5	12
9	18	74	7	36	0.09	3	19	5	15	6	13
10	22	78	8	38	0.10	4	20	6	17	7	14
11	26	82	9	40	0.11	4	21	7	18	8	15
12	30	85	10	43	0.12	5	23	8	19	9	16
13	34	89	12	45	0.13	6	24	9	20	9	17
14	39	92	13	47	0.14	7	25	9	21	10	19
15	44	94	15	49	0.15	7	26	10	22	11	20
16	49	96	16	51	0.16	8	27	11	24	12	21
17	55	98	18	53	0.17	9	29	12	25	13	22
18	61	99	19	55	0.18	9	30	13	26	14	23
19	68	100	21	57	0.19	10	31	13	27	15	24
20	77	100	23	59	0.20	11	32	14	28	16	25
21			24	61	0.21	12	33	15	29	17	26
22			26	63	0.22	12	34	16	30	17	27
23			28	65	0.23	13	35	17	31	18	28
24			30	67	0.24	14	36	18	32	19	29
25			32	68	0.25	15	37	19	33	20	30
26			33	70	0.26	16	39	20	35	21	31
27			35	72	0.27	17	40	20	36	22	32
28			37	74	0.28	17	41	21	37	23	33
29			39	76	0.29	18	42	22	38	24	35
30			41	77	0.30	19	43	23	39	25	36
31			43	79	0.31	20	44	24	40	26	37
32			45	81	0.32	21	45	25	41	27	38
33			46	82	0.33	22	46	26	42	28	39
34			48	84	0.34	22	47	27	43	29	40
35			50	85	0.35	23	48	28	44	30	41
36			53	87	0.36	24	49	29	45	31	42
37			55	88	0.37	25	50	30	46	32	43
38			57	90	0.38	26	51	31	47	33	44
39			59	91	0.39	27	52	32	48	34	45
40			61	92	0.40	28	53	33	49	34	46
41			64	93	0.41	28	54	33	50	35	47
42			66	95	0.42	29	55	34	51	36	48
43			69	96	0.43	30	56	35	52	37	49
44			71	97	0.44	31	57	36	53	38	50

TABLE B5. Confidence intervals are given for p(α = .01) (continued).

x	n = 20		n = 50		x/n	n = 100		n = 200		n = 500	
	L(x)	U(x)	L(x)	U(x)		L(x)	U(x)	L(x)	U(x)	L(x)	U(x)
45			74	98	0.45	32	58	37	54	39	51
46			76	99	0.46	33	59	38	55	40	52
47			80	99	0.47	34	60	39	56	41	53
48			83	100	0.48	35	61	40	57	42	54
49			86	100	0.49	36	62	41	58	43	55
50			90	100	0.50	37	63	41	59	44	56

NOTE: Reprinted by permission from H. C. Fryer, *Experimental Statistics* (New York: Allyn and Bacon, 1966), p. 565. © 1966 by the author.

APPENDIX C

Tables for Use with Poisson Models

POINT AND CUMULATIVE PROBABILITIES FOR POISSON MODELS: For selected values of λ, Table C1 gives the point probabilities,

$$P(x) = e^{-\lambda}\lambda^{x}/x!, \quad x = 0, 1, 2, \ldots,$$

and the cumulative probabilities,

$$F(x) = \sum_{t=0}^{x} P(t).$$

Although P(x) is defined for all possible integer values of x, it is 0 for practical purposes after F(x) reaches 1.0000. Thus, P(x) = .0000 and F(x) = 1.0000 for all the blanks in Table C1.

EXAMPLE: If X is a Poisson variable with $\lambda = 3$, find $P(X = 4)$ and $P(X \leq 4)$.

SOLUTION: Entering Table C1 with $\lambda = 3$ gives

$$P(x = 4) = P(4) = .1680,$$

$$P(X \leq 4) = F(4) = .8153.$$

SELECTED VALUES OF λ, e^{λ}, AND $e^{-\lambda}$: The values listed in Table C2 can be used to evaluate Poisson point probabilities, $P(x)$, when λ is different from the values in Table C1.

EXAMPLE: Find $P(X = 2)$ when X is a Poisson variable with $\lambda = 6$.

SOLUTION: $P(X = 2) = e^{-6}\, 6^2/2! = (.00248)(36)/2 = .04464$.

REFERENCE

Additional tables of the Poisson probability distribution are available in the following source.

Molina, E. C. 1942. *Poisson's exponential binomial limit.* New York: D. Van Nostrand.

TABLE C1.
Point and cumulative probabilities are given for Poisson models.

x	P(x)	F(x)	P(x)	F(x)	P(x)	F(x)	P(x)	F(x)	P(x)	F(x)	P(x)	F(x)
	λ = .1		.5		1.0		1.5		2.0		2.5	
0	.9048	.9048	.6065	.6065	.3679	.3679	.2231	.2231	.1353	.1353	.0821	.0821
1	.0905	.9953	.3033	.9098	.3679	.7358	.3347	.5578	.2707	.4060	.2052	.2873
2	.0045	.9998	.0758	.9856	.1839	.9197	.2510	.8088	.2707	.6767	.2565	.5438
3	.0002	1.0000	.0126	.9982	.0613	.9810	.1255	.9344	.1805	.8571	.2138	.7576
4			.0016	.9998	.0153	.9963	.0471	.9814	.0902	.9473	.1336	.8912
5			.0002	1.0000	.0031	.9994	.0141	.9955	.0361	.9834	.0668	.9580
6					.0005	.9999	.0035	.9991	.0120	.9955	.0278	.9858
7					.0001	1.0000	.0008	.9999	.0034	.9989	.0099	.9957
8							.0001	1.0000	.0009	.9998	.0031	.9988
9									.0002	1.0000	.0009	.9997
10											.0002	.9999
11											.0001	1.0000
	λ = 3.0		3.5		4.0		5.0		10.0		15.0	
0	.0498	.0498	.0302	.0302	.0183	.0183	.0067	.0067	.0000	.0000	.0000	.0000
1	.1494	.1992	.1057	.1359	.0733	.0916	.0337	.0404	.0005	.0005	.0000	.0000
2	.2240	.4232	.1850	.3209	.1465	.2381	.0842	.1247	.0023	.0028	.0000	.0000
3	.2240	.6472	.2158	.5366	.1954	.4335	.1404	.2650	.0076	.0104	.0002	.0002
4	.1680	.8153	.1888	.7254	.1954	.6288	.1755	.4405	.0189	.0293	.0006	.0009
5	.1008	.9161	.1322	.8576	.1563	.7851	.1755	.6160	.0378	.0671	.0019	.0028
6	.0504	.9665	.0771	.9347	.1042	.8893	.1462	.7622	.0631	.1301	.0048	.0076
7	.0216	.9881	.0386	.9733	.0595	.9489	.1044	.8666	.0901	.2202	.0104	.0180
8	.0081	.9962	.0169	.9901	.0298	.9786	.0653	.9319	.1126	.3328	.0194	.0374
9	.0027	.9989	.0066	.9967	.0132	.9919	.0363	.9682	.1251	.4579	.0324	.0699
10	.0008	.9997	.0023	.9990	.0053	.9972	.0181	.9863	.1251	.5830	.0486	.1185
11	.0002	.9999	.0007	.9997	.0019	.9991	.0082	.9946	.1137	.6968	.0663	.1848
12	.0001	1.0000	.0002	.9999	.0006	.9997	.0034	.9980	.0948	.7916	.0829	.2676

TABLE C1. Point and cumulative probabilities are given for Poisson models (continued).

x	P(x)	F(x)	P(x)	F(x)	P(x)	F(x)	P(x)	F(x)	P(x)	F(x)	P(x)	F(x)
	λ = 3.0		3.5		4.0		5.0		10.0		15.0	
13			.0001	1.0000	.0002	.9999	.0013	.9993	.0729	.8645	.0956	.3632
14					.0001	1.0000	.0005	.9998	.0521	.9165	.1024	.4657
15							.0002	1.0000	.0347	.9513	.1024	.5681
16									.0217	.9730	.0960	.6641
17									.0128	.9857	.0847	.7489
18									.0071	.9928	.0706	.8195
19									.0037	.9966	.0557	.8752
20									.0019	.9984	.0418	.9170
21									.0009	.9993	.0299	.9469
22									.0004	.9997	.0204	.9673
23									.0002	.9999	.0133	.9805
24									.0001	1.0000	.0083	.9888
25											.0050	.9938
26											.0029	.9967
27											.0016	.9983
28											.0009	.9992
29											.0004	.9996
30											.0002	.9998
31											.0001	.9999
32											.0001	1.0000

TABLE C2.
Selected values of λ, e^{λ}, and $e^{-\lambda}$ are shown.

λ	e^{λ}	$e^{-\lambda}$	λ	e^{λ}	$e^{-\lambda}$
.00	1.00000	1.00000	2.0	7.3891	.13534
.05	1.0513	.95123	2.1	8.1662	.12246
.10	1.1052	.90484	2.2	9.0250	.11080
.15	1.1618	.86071	2.3	9.9742	.10026
.20	1.2214	.81873	2.4	11.023	.09072
.25	1.2840	.77880	2.5	12.182	.08209
.30	1.3499	.74082	2.6	13.464	.07427
.35	1.4191	.70469	2.7	14.880	.06721
.40	1.4981	.67032	2.8	16.445	.06081
.45	1.5683	.63763	2.9	18.174	.05502
.50	1.6487	.60653	3.0	20.086	.04979
.55	1.7333	.57695	3.5	33.115	.03020
.60	1.8221	.54881	4.0	54.598	.01832
.65	1.9155	.52205	4.5	90.017	.01111
.70	2.0138	.49659	5.0	148.41	.00674
.75	2.1170	.47237	5.5	244.69	.00409
.80	2.2255	.44933	6.0	403.43	.00248
.85	2.3396	.42471	6.5	665.14	.00150
.90	2.4596	.40657	7.0	1096.6	.00091
.95	2.5857	.38674	7.5	1808.0	.00055
1.00	2.7183	.36788	8.0	2981.0	.00034
1.10	3.0042	.33287	8.5	4914.8	.00020
1.20	3.3201	.30119	9.0	8103.1	.00012
1.30	3.6693	.27253	9.5	13360.	.00007
1.40	4.0552	.24660	10.0	22026.	.00005
1.50	4.4817	.22313	10.5	36316.	.00003
1.60	4.9530	.20190	11.0	59874.	.00002
1.70	5.4739	.18268	11.5	98716.	.00001
1.80	6.0496	.16530	12.0	162755.	.00001
1.90	6.6859	.14957	13.0	442413.	.00000

APPENDIX D

Tables for Use with Normal Models

Three tables are included in this appendix. Table D1 provides the ordinates, $f(z)$, for $.00 \le z \le 3.99$; Table D2 gives the probabilities, $P(.00 \le Z \le z)$, for $.00 \le z \le 3.99$; and Table D3 presents 1,000 random normal deviates that can be sampled by using the random digits in Table A2.

VALUES OF THE ORDINATES, f(z), WHEN $Z \sim N(0, 1)$: If $Z \sim N(0, 1)$, then

$$f(z) = (1/\sqrt{2\pi})e^{-z^2/2}, \quad -\infty < z < \infty .$$

By symmetry, $f(-z) = f(z)$. So the table values are given only for $z \ge 0$. If $X \sim N(\mu, \sigma^2)$, the ordinate of the density of X at the value x, say $g(x)$, is given by the following formula:

$$g(x) = f[(x - \mu)/\sigma]/\sigma.$$

EXAMPLES

1. If $Z \sim N(0, 1)$, find $f(1.23)$ and $f(-.72)$.

SOLUTION: Enter Table D1 at $z = 1.23$ to find

$$f(1.23) = .18724.$$

Since $-.72 < 0$, we enter Table D1 at $z = .72$ and use the symmetry to write

$$f(-.72) = f(.72) = .30785.$$

2. If $X \sim N(4, 9)$, find $g(4)$, $g(10)$, and $g(1)$.

SOLUTION: By the previous solution,

$$\begin{aligned} g(4) &= f[(4 - 4)/3]/3 = f(0)/3 \\ &= .39894/3 = .13298, \\ g(10) &= f[(10 - 4)/3]/3 = f(2)/3 \\ &= .05399/3 = .01800, \text{ and} \\ g(1) &= f[(1 - 4)/3]/3 = f(-1)/3 \\ &= f(1)/3 = .24197/3 = .08066. \end{aligned}$$

REFERENCE

For more extensive tables of the normal distribution, see the following source.

Natl. Bureau of Standards. 1952. *A guide to tables of the normal probability integral*. Applied Math. Series No. 21. U.S. Dept. of Commerce. Washington, D.C.: U.S. Govt. Printing Office.

TABLE D1.
Values of the ordinates, $f(z)$, are given when $Z \sim N(0, 1)$.

z	.00	.01	.02	.03	.04	.05	.06	.07	.08	.09
.0	.39894	.39892	.39886	.39876	.39862	.39844	.39822	.39797	.39767	.39733
.1	.39695	.39654	.39608	.39559	.39505	.39448	.39387	.39322	.39253	.39181
.2	.39104	.39024	.38940	.38853	.38762	.38667	.38568	.38466	.38361	.38251
.3	.38139	.38023	.37903	.37780	.37654	.37524	.37391	.37255	.37115	.36973
.4	.36827	.36678	.36526	.36371	.36213	.36053	.35889	.35723	.35553	.35381
.5	.35207	.35029	.34849	.34667	.34482	.34294	.34105	.33912	.33718	.33521
.6	.33322	.33121	.32918	.32713	.32506	.32297	.32086	.31874	.31659	.31443
.7	.31225	.31006	.30785	.30562	.30339	.30114	.29887	.29659	.29431	.29200
.8	.28969	.28737	.28504	.28269	.28034	.27798	.27561	.27324	.27086	.26848
.9	.26609	.26369	.26129	.25888	.25647	.25406	.25164	.24923	.24681	.24439
1.0	.24197	.23955	.23713	.23471	.23230	.22988	.22747	.22506	.22265	.22025
1.1	.21785	.21546	.21307	.21069	.20831	.20594	.20357	.20121	.19886	.19652
1.2	.19419	.19186	.18954	.18724	.18494	.18265	.18037	.17810	.17585	.17360
1.3	.17137	.16915	.16684	.16474	.16256	.16038	.15822	.15608	.15395	.15183
1.4	.14973	.14764	.14556	.14350	.14146	.13943	.13741	.13542	.13344	.13147
1.5	.12952	.12758	.12566	.12376	.12188	.12001	.11816	.11632	.11450	.11270
1.6	.11092	.10915	.10741	.10567	.10396	.10226	.10059	.09893	.09728	.09566
1.7	.09405	.09246	.09089	.08933	.08780	.08628	.08478	.08329	.08183	.08038
1.8	.07895	.07754	.07614	.07477	.07341	.07206	.07074	.06943	.06814	.06687
1.9	.06562	.06438	.06316	.06195	.06077	.05959	.05844	.05730	.05618	.05508
2.0	.05399	.05292	.05186	.05082	.04980	.04879	.04780	.04682	.04586	.04491
2.1	.04398	.04307	.04217	.04128	.04041	.03955	.03871	.03788	.03706	.03626
2.2	.03547	.03470	.03394	.03319	.03246	.03174	.03103	.03034	.02965	.02898
2.3	.02833	.02768	.02705	.02643	.02582	.02522	.02463	.02406	.02349	.02294
2.4	.02239	.02186	.02134	.02083	.02033	.01984	.01936	.01889	.01842	.01797

TABLE D1. Values of the ordinates, f(z), are given when $Z \sim N(0, 1)$ (continued).

z	.00	.01	.02	.03	.04	.05	.06	.07	.08	.09
2.5	.01753	.01709	.01667	.01625	.01585	.01545	.01506	.01468	.01431	.01394
2.6	.01358	.01323	.01289	.01256	.01223	.01191	.01160	.01130	.01100	.01071
2.7	.01042	.01014	.00987	.00961	.00935	.00909	.00885	.00861	.00837	.00814
2.8	.00792	.00770	.00748	.00727	.00707	.00687	.00668	.00649	.00631	.00613
2.9	.00595	.00578	.00562	.00545	.00530	.00514	.00499	.00485	.00471	.00457
3.0	.00443	.00430	.00417	.00405	.00393	.00381	.00370	.00358	.00348	.00337
3.1	.00327	.00317	.00307	.00298	.00288	.00279	.00271	.00262	.00254	.00246
3.2	.00238	.00231	.00224	.00216	.00210	.00203	.00196	.00190	.00184	.00178
3.3	.00172	.00167	.00161	.00156	.00151	.00145	.00141	.00136	.00132	.00127
3.4	.00123	.00119	.00115	.00111	.00107	.00104	.00100	.00097	.00094	.00090
3.5	.00087	.00084	.00081	.00079	.00076	.00073	.00071	.00068	.00066	.00063
3.6	.00061	.00059	.00057	.00055	.00053	.00051	.00049	.00047	.00046	.00044
3.7	.00042	.00041	.00039	.00038	.00037	.00035	.00034	.00033	.00031	.00030
3.8	.00029	.00028	.00027	.00026	.00025	.00024	.00023	.00022	.00021	.00021
3.9	.00020	.00019	.00018	.00017	.00017	.00016	.00016	.00015	.00014	.00014

NOTE: This table is adapted from Table 1, pages 110-116, of E. S. Pearson and H. O. Hartley, *Biometrika Tables for Statisticians*, 3rd ed. (Cambridge: The University Press, 1966), with permission of Professor E. S. Pearson for the publishers.

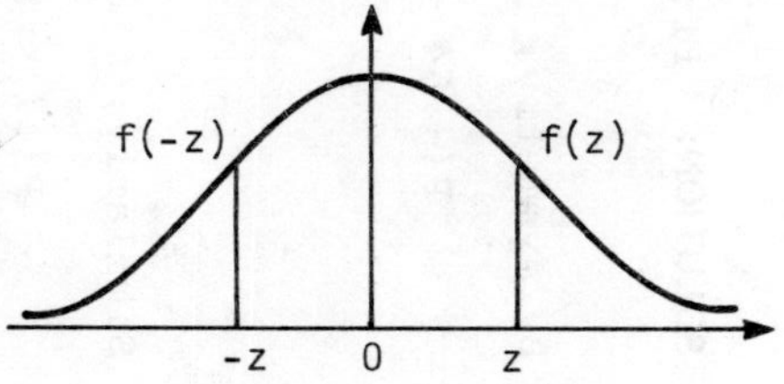

THE PROBABILITIES, $P(0 \leq Z \leq z)$, WHERE $Z \sim N(0, 1)$: Table D2 gives the probabilities, $P(0 \leq Z \leq z)$, for positive values of z between .00 and 3.99. By symmetry,

$$P(-z \leq Z \leq 0) = P(0 \leq Z \leq z)$$

for every $z > 0$. Those values, as well as

$$P(Z \leq 0) = P(Z \geq 0) = .5,$$

can be used to find probabilities associated with intervals for any normal variable, as illustrated by the examples in the section on page 199 and the examples below.

EXAMPLES

1. If $Z \sim N(0, 1)$, find $P(0 \leq Z \leq 1.05)$, $P(-.64 \leq Z \leq 2.14)$, and $P(.7 \leq Z \leq 1.67)$.

SOLUTION: From Table D2 with z = 1.05, we find

$$P(0 \leq Z \leq 1.05) = .35314.$$

By symmetry,

$$\begin{aligned} P(-.64 \leq Z \leq 2.14) &= P(0 \leq Z \leq .64) \\ &\quad + P(0 \leq Z \leq 2.14) \\ &= .23891 + .48386 \\ &= .72277. \end{aligned}$$

Similarly,

$$\begin{aligned} P(.7 \leq Z \leq 1.67) &= P(0 \leq Z \leq 1.67) \\ &\quad - P(0 \leq Z \leq .7) \\ &= .45254 - .25804 = .19450. \end{aligned}$$

2. If $X \sim N(5, 4)$, find $P(4 \leq X \leq 8)$.

SOLUTION: By using the Z-transformation, $Z = (X - \mu)/\sigma$,

$$\begin{aligned} P(4 \leq X \leq 8) &= P(-.5 \leq Z \leq 1.5) \\ &= P(0 \leq Z \leq .5) + P(0 \leq Z \leq 1.5) \\ &= .19146 + .43319 \\ &= .62465. \end{aligned}$$

REFERENCE

See the reference for Table D1.

TABLE D2.
The probabilities, $P(0 \le Z \le z)$, are given where $Z \sim N(0, 1)$.

z	.00	.01	.02	.03	.04	.05	.06	.07	.08	.09
.0	.00000	.00399	.00798	.01197	.01595	.01994	.02392	.02790	.03188	.03586
.1	.03983	.04380	.04776	.05172	.05567	.05962	.06356	.06749	.07142	.07535
.2	.07926	.08317	.08706	,09095	.09483	.09871	.10257	.10642	.11026	.11409
.3	.11791	.12172	.12552	.12930	.13307	.13683	.14058	.14431	.14803	.15173
.4	.15542	.15910	.16276	.16640	.17003	.17364	.17724	.18082	.18439	.18793
.5	.19146	.19497	.19847	.20194	.20540	.20884	.21226	.21566	.21904	.22240
.6	.22575	.22907	.23237	.23565	.23891	.24215	.24537	.24857	.25175	.25490
.7	.25804	.26115	.26424	.26730	.27035	.27337	.27637	.27935	.28230	.28524
.8	.28814	.29103	.29389	.29673	.29955	.30234	.30511	.30785	.31057	.31327
.9	.31594	.31859	.32121	.32382	.32639	.32894	.33147	.33398	.33646	.33891
1.0	.34134	.34375	.34614	.34850	.35083	.35314	.35543	.35769	.35993	.36214
1.1	.36433	.36650	.36864	.37076	.37286	.37493	.37698	.37900	.38100	.38298
1.2	.38493	.38686	.38877	.39065	.39251	.39435	.39617	.39796	.39973	.40147
1.3	.40320	.40490	.40658	.40824	.40988	.41149	.41309	.41466	.41621	.41774
1.4	.41924	.42073	.42220	.42364	.42507	.42647	.42786	.42922	.43056	.43189
1.5	.43319	.43448	.43574	.43699	.43822	.43943	.44062	.44179	.44295	.44408
1.6	.44520	.44630	.44738	.44845	.44950	.45053	.45154	.45254	.45352	.45449
1.7	.45543	.45637	.45728	.45818	.45907	.45994	.46080	.46164	.46246	.46327
1.8	.46407	.46485	.46562	.46638	.46712	.46784	.46856	.46926	.46995	.47062
1.9	.47128	.47193	.47257	.47320	.47381	.47441	.47500	.47558	.47615	.47670
2.0	.47725	.47778	.47831	.47882	.47932	.47982	.48030	.48077	.48124	.48169
2.1	.48214	.48257	.48300	.48341	.48386	.48422	.48461	.48500	.48537	.48574
2.2	.48610	.48645	.48679	.48713	.48745	.48778	.48809	.48840	.48870	.48899
2.3	.48928	.48956	.48983	.49010	.49036	.49061	.49086	.49111	.49134	.49158
2.4	.49180	.49202	.49224	.49245	.49266	.49286	.49305	.49324	.49343	.49361

TABLE D2. The probabilities, $P(0 \le Z \le z)$, are given where $Z \sim N(0, 1)$ (continued).

z	.00	.01	.02	.03	.04	.05	.06	.07	.08	.09
2.5	.49379	.49396	.49413	.49427	.49446	.49461	.49477	.49492	.49506	.49520
2.6	.49534	.49547	.49560	.49573	.49585	.49598	.49609	.49621	.49632	.49643
2.7	.49653	.49664	.49674	.49683	.49693	.49702	.49711	.49720	.49728	.49736
2.8	.49744	.49752	.49760	.49767	.49774	.49781	.49788	.49795	.49801	.49807
2.9	.49813	.49819	.49825	.49831	.49836	.49841	.49846	.49851	.49856	.49861
3.0	.49865	.49869	.49874	.49878	.49882	.49886	.49889	.49893	.49897	.49900
3.1	.49903	.49906	.49910	.49913	.49916	.49918	.49921	.49924	.49926	.49929
3.2	.49931	.49934	.49936	.49938	.49940	.49942	.49944	.49946	.49948	.49950
3.3	.49952	.49953	.49955	.49957	.49958	.49960	.49961	.49962	.49965	.49966
3.4	.49966	.49968	.49969	.49970	.49971	.49972	.49973	.49974	.49975	.49976
3.5	.49977	.49978	.49978	.49979	.49980	.49981	.49981	.49982	.49983	.49983
3.6	.49984	.49985	.49985	.49986	.49986	.49987	.49987	.49988	.49988	.49989
3.7	.49989	.49990	.49990	.49990	.49991	.49991	.49992	.49992	.49992	.49992
3.8	.49993	.49993	.49993	.49994	.49994	.49994	.49994	.49995	.49995	.49995
3.9	.49995	.49995	.49996	.49996	.49996	.49996	.49996	.49996	.49997	.49997

NOTE: This table is adapted from Table 1, pages 110-116, of E. S. Pearson and H. O. Hartley, *Biometrika Tables for Statisticians*, 3rd ed. (Cambridge: The University Press, 1966), with permission of Professor E. S. Pearson for the publishers.

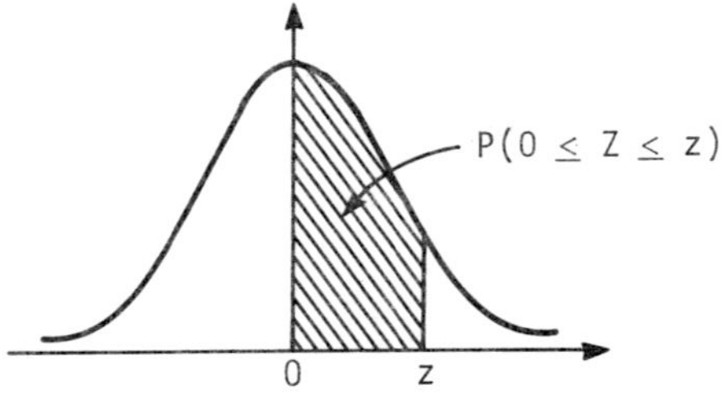

TABLE D3.
1,000 observations of a N(0, 1) variable are shown.

Row	Column 00	01	02	03	04	05	06	07	08	09
00	0.08	1.48	1.62	-0.10	-0.19	-1.51	0.71	1.05	1.05	-0.71
01	0.17	0.27	0.41	-0.10	0.29	-1.11	-1.32	0.68	0.01	-0.21
02	0.43	-0.32	-0.26	0.90	-1.03	-0.13	0.88	0.83	-0.68	1.03
03	-1.12	-1.38	0.08	-1.18	0.29	-1.25	-0.20	1.97	0.47	1.82
04	1.38	0.14	-1.24	-1.04	-1.00	-0.51	-0.97	-0.24	-2.00	-0.97
05	-1.49	-1.73	-1.36	-0.45	-1.37	1.85	-1.41	-1.48	-0.51	-0.09
06	-0.78	0.54	0.52	0.43	0.66	0.37	-1.36	-0.95	0.92	-0.66
07	-0.96	0.75	1.26	0.43	-0.12	0.59	-0.88	0.28	-0.33	0.05
08	0.12	-1.26	-1.18	-0.51	-0.67	-0.98	0.53	0.30	0.85	-1.48
09	1.03	-0.39	0.83	2.32	2.49	-0.49	0.56	1.35	-0.41	-0.34
10	-1.35	-0.50	-0.43	1.32	1.78	-1.47	1.35	0.24	-1.80	-0.40
11	0.27	0.62	-1.18	0.90	1.36	0.71	-0.85	0.43	-1.59	-1.23
12	-0.33	-1.48	-0.81	-1.51	-0.42	0.43	-0.88	0.82	-1.13	-0.47
13	-0.66	0.91	-0.86	1.81	0.63	0.06	-0.52	-0.77	-0.11	0.17
14	1.18	-0.08	-0.28	-0.42	0.89	-1.18	0.73	-2.45	-0.63	-0.32
15	0.43	-1.36	-0.78	0.50	0.07	0.94	1.45	0.39	0.15	0.15
16	2.02	-1.59	0.32	1.05	0.89	1.56	0.99	-0.34	-0.29	-0.37
17	-1.05	1.21	-0.74	1.65	0.63	0.18	1.24	-1.39	0.96	1.00
18	-1.34	-0.41	-0.31	0.47	-0.68	-0.86	0.17	0.30	-0.85	-0.62
19	-0.26	1.83	0.28	0.73	-0.90	-1.11	-0.09	-1.05	-1.42	0.04
20	0.05	0.42	0.19	-1.17	0.90	-1.08	-0.93	1.32	-0.14	1.93
21	0.33	0.30	-0.49	1.18	-0.63	-0.06	-0.59	-0.55	-0.10	1.25
22	-0.08	0.59	-0.28	-1.33	-0.70	0.61	0.89	0.40	0.17	-1.77
23	-0.50	-0.27	-0.97	-0.21	-0.54	0.75	0.78	-0.62	0.40	0.80
24	-0.63	2.74	0.86	-1.70	1.00	-1.61	-0.69	-1.28	-0.19	-0.59
25	-1.60	1.92	1.42	-1.51	-1.60	0.16	0.20	-1.83	-0.77	-1.56
26	-1.38	-0.98	0.19	-0.70	-0.05	-2.93	0.24	-0.61	0.27	-0.83
27	1.12	0.50	0.64	1.03	0.88	-1.67	1.64	-0.49	-0.74	-0.33
28	0.38	-0.72	1.46	-0.88	0.64	1.29	0.10	1.03	-1.62	0.51
29	0.84	1.11	-0.79	0.57	-0.07	-1.02	-1.32	-0.05	0.88	-1.22
30	-0.08	0.09	0.66	0.48	-1.92	-0.09	-1.22	-1.43	1.22	0.21
31	0.67	-0.36	1.80	-1.48	-2.18	0.92	0.12	-0.49	1.60	-0.55
32	-0.93	0.16	-0.10	-0.96	-0.14	-0.18	-1.10	1.40	0.67	-0.35
33	-0.73	-0.93	0.44	-0.54	-1.55	0.05	-0.59	-1.10	2.42	0.94
34	-0.05	1.67	0.90	0.08	0.17	0.55	-1.61	0.84	-0.13	-0.94
35	-0.53	-1.25	0.09	-0.37	-0.46	1.53	-0.37	-1.63	0.88	-0.30
36	-0.33	-0.27	0.86	0.42	0.51	-0.26	0.99	0.92	-0.96	0.86
37	0.82	-1.27	-0.96	2.19	0.56	0.55	-0.98	0.01	-1.78	0.55
38	0.74	1.40	-2.25	0.69	0.24	1.23	0.44	-1.04	-0.51	0.37
39	-0.48	0.94	0.25	-0.89	-0.25	-1.34	0.29	-1.42	0.77	0.65

TABLE D3. 1,000 observations of a N(0, 1) variable are shown (continued).

Row	Column 00	01	02	03	04	05	06	07	08	09
40	-2.05	-0.11	0.41	-1.13	-0.13	-0.32	-1.32	0.64	-0.51	-0.30
41	0.25	-0.69	2.05	0.30	-0.19	2.59	-0.27	-0.09	-1.01	-2.26
42	-1.29	1.42	-1.15	0.69	-0.61	-0.31	-1.23	1.31	1.03	-0.95
43	1.00	0.62	0.47	0.16	-0.32	1.07	-0.77	-0.80	-0.58	1.86
44	-1.62	-0.35	0.33	1.51	0.72	-1.78	-0.31	0.17	-1.31	0.38
45	0.44	1.41	0.99	-1.05	0.18	-0.44	-0.46	1.27	-0.19	-0.31
46	-0.33	-0.27	0.86	0.14	0.62	-0.15	-0.33	-0.24	-0.89	0.46
47	0.25	0.20	0.90	-1.15	0.74	0.34	-0.07	-0.26	1.47	0.18
48	-0.80	-1.31	-0.91	1.85	1.29	-0.43	0.26	-0.83	-0.70	-0.72
49	0.59	-1.17	0.06	0.08	1.27	-1.21	0.44	-0.44	-0.29	0.00

Row	Column 10	11	12	13	14	15	16	17	18	19
00	0.73	-0.37	0.25	0.87	1.28	1.61	-0.65	-0.66	0.68	-0.77
01	-0.90	-0.22	1.46	-0.38	0.88	-1.62	0.70	-0.67	0.45	0.30
02	0.09	0.59	-0.57	0.63	0.04	0.67	0.51	1.49	-0.26	0.13
03	0.14	-0.12	-0.17	-0.61	-2.08	0.88	-1.11	-0.29	0.51	-1.24
04	-0.42	1.14	-0.36	-0.94	1.19	-1.14	1.18	0.30	-0.45	1.62
05	0.15	0.84	0.13	0.03	0.44	-1.25	1.09	-1.27	-2 41	0.40
06	-0.49	0.63	-0.35	0.95	0.51	1.26	-0.15	-0.32	0.44	0.71
07	-1.32	-0.79	-1.82	-2.23	0.30	1.46	0.25	-0.27	-1.89	0.33
08	-0.20	-0.65	-1.34	-2.14	-0.07	1.26	1.64	-0.22	-0.56	-0.74
09	-0.60	0.41	0.08	-0.13	0.36	1.16	-0.53	0.79	0.67	-1.29
10	-0.25	-1.34	-0.32	-0.46	-0.61	0.52	1.80	0.50	0.75	0.09
11	0.35	0.33	0.48	-0.43	-1.41	-1.00	1.87	0.32	0.45	1.24
12	-1.28	-1.30	0.49	-0.57	1.66	-0.47	-0.19	-0.86	-0.93	0.44
13	-0.61	0.66	1.50	-1.38	0.49	-0.02	0.25	0.25	1.05	-1.70
14	0.12	-0.62	-0.11	0.56	-0.44	-0.51	1.27	0.32	1.44	-0.70
15	-1.46	0.03	-0.94	-0.68	-0.41	-0.21	0.28	0.45	-0.78	-0.09
16	0.02	0.26	-0.53	1.29	-0.45	-1.67	0.96	1.15	-1.83	-0.76
17	1.81	-0.64	-0.83	-0.06	0.88	-0.86	0.92	-1.92	-0.44	-0.01
18	-0.39	2.01	0.37	2.38	0.58	-0.31	0.64	-0.02	-0.35	0.17
19	0.20	-1.53	-1.23	-1.45	-0.31	-1.14	1.41	0.50	1.25	-1.00
20	0.56	0.09	0.60	0.40	0.38	0.07	-0.09	-0.66	-1.54	-0.30
21	0.37	-0.61	0.52	-0.46	0.45	-0.09	0.32	0.64	-1.56	0.67
22	-0.51	-1.61	1.61	0.87	0.59	-0.91	1.08	0.23	-0.53	-1.11
23	1.13	-0.21	-0.85	-1.83	-0.59	-0.12	0.51	-0.25	-0.45	0.87
24	-1.77	-0.93	-0.03	-1.87	0.24	0.18	-1.44	1.18	-0.24	1.05

TABLE D3. 1,000 observations of a N(0, 1) variable are shown (continued).

Row	Column 10	11	12	13	14	15	16	17	18	19
25	0.64	1.03	1.11	-0.14	0.40	0,57	-0.84	-0.80	-0.31	-1.77
26	-1.19	-0.92	0.25	2.07	1.53	0.04	-1.60	-2.01	1.75	1.77
27	-1.57	-0.49	-0.54	0.01	-0.43	0.14	1.47	0.93	1.05	-0.98
28	0.84	1.22	0.47	-0.80	-0.81	-2.33	-1.16	-1.17	-1.62	-0.01
29	0.87	0.83	-1.18	0.15	-0.42	0.51	-0.71	0.09	0.22	-1.37
30	-0.35	-0.37	-0.23	0.84	-1.16	-0.31	-0.39	-0.02	-0.40	-0.26
31	1.90	-0.01	-0.04	-1.80	-0.83	-.030	0.34	-0.83	-0.95	0.33
32	0.68	-1.38	0.49	-0.81	-0.59	-1.10	-1.51	1.26	0.83	0.84
33	0.39	-0.05	-1.07	-0.11	-1.34	0.62	0.31	-1.06	-0.90	1.64
34	-0.36	0.62	-0.81	0.46	0.47	0.91	0.64	-0.79	0.00	1.14
35	0.62	0.97	-0.18	1.56	0.18	0.03	-1.20	1.53	1.21	0.84
36	-1.30	-1.46	0.01	0.32	-0.35	1.00	1.64	0.26	-1.08	0.58
37	-0.09	-0.05	-1.66	-1.66	0.75	-0.46	0.10	-0.47	0.57	-1.12
38	-1.04	0.39	-0.05	-1.04	0.39	-0.13	-0.68	1.40	0.84	1.00
39	1.65	1.04	-0.13	-0.78	0.27	-0.91	-1.31	-1.21	-0.58	-0.42
40	0.63	1.31	-0.68	0.30	-1.11	-1.98	0.90	-0.63	-0.43	-0.16
41	-0.56	-0.94	0.12	-1.91	-0.50	0.22	-0.18	0.35	0.77	0.01
42	0.64	-0.82	-1.76	-0.10	0.78	0.16	-2.61	-0.25	-1.62	-2.85
43	-0.79	1.94	-0.35	-1.09	-0.16	-0.55	-1.61	1.32	1.87	0.43
44	1.58	-0.43	0.25	-0.69	-0.82	0.17	-0.73	-0.39	0.06	0.93
45	-0.63	0.07	-0.92	-0.66	0.50	-0.10	0.13	0.96	1.47	-3.19
46	-1.71	-1.22	-0.27	0.36	-2.13	-0.15	-0.68	1.54	-0.78	0.13
47	-1.14	0.19	-0.18	-0.58	-0.84	-1.05	-0.18	-1.87	-1.48	0.01
48	2.07	-0.15	1.32	0.43	0.10	2.24	-0.29	-1.17	-1.85	0.04
49	0.68	-1.37	-0.76	0.41	-0.40	-0.07	-1.00	-0.05	0.11	0.07

APPENDIX E

Tables of α-Level Values for Classical Statistical Procedures

The three tables in this appendix are used to obtain α-level values for the classical statistical procedures that use t-statistics, χ^2-statistics, and F-statistics.

α-LEVEL VALUES, t(k, α), FOR t-STATISTICS: Any variable, t, which has a t-distribution with k degrees of freedom, has upper-tail α-level values, $t(k, \alpha)$, defined by

$$P[t > t(k, \alpha)] = \alpha$$

and illustrated as follows:

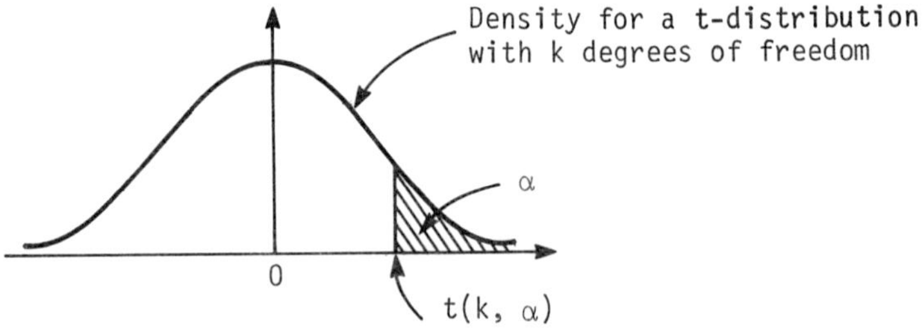

EXAMPLE: Find t(10, .05) and t(8, .95).

SOLUTION: Entering Table E1 with k = 10 and α = .05, we read directly

$$t(10, .05) = 1.812.$$

By symmetry we know that

$$t(8, .95) = -t(8, .05).$$

Thus,

$$t(8, .95) = -1.860.$$

TABLE E1.
α-level values t(k, α) are given for t-statistics.

k	α=.25	.20	.15	.10	.05	.025	.01	.005	.0005
1	1.000	1.376	1.963	3.078	6.314	12.706	31.821	63.657	636.619
2	.816	1.061	1.386	1.886	2.920	4.303	6.965	9.925	31.598
3	.765	.978	1.250	1.638	2.353	3.182	4.541	5.841	12.924
4	.741	.941	1.190	1.533	2.132	2.776	3.747	4.604	8.610
5	.727	.920	1.156	1.476	2.015	2.571	3.365	4.032	6.869
6	.718	.906	1.134	1.440	1.943	2.447	3.143	3.707	5.959
7	.711	.896	1.119	1.415	1.895	2.365	2.998	3.499	5.408
8	.706	.889	1.108	1.397	1.860	2.306	2.896	3.355	5.041
9	.703	.883	1.100	1.383	1.833	2.262	2.821	3.250	4.781
10	.700	.879	1.093	1.372	1.812	2.228	2.764	3.169	4.587
11	.697	.876	1.088	1.363	1.796	2.201	2.718	3.106	4.437
12	.695	.873	1.083	1.356	1.782	2.179	2.681	3.055	4.318
13	.694	.870	1.079	1.350	1.771	2.160	2.650	3.012	4.221
14	.692	.868	1.076	1.345	1.761	2.145	2.624	2.977	4.140
15	.691	.866	1.074	1.341	1.753	2.131	2.602	2.947	4.073
16	.690	.865	1.071	1.337	1.746	2.120	2.583	2.921	4.015
17	.689	.863	1.069	1.333	1.740	2.110	2.567	2.898	3.965
18	.688	.862	1.067	1.330	1.734	2.101	2.552	2.878	3.922
19	.688	.861	1.066	1.328	1.729	2.093	2.539	2.861	3.883
20	.687	.860	1.064	1.325	1.725	2.086	2.528	2.845	3.850
21	.686	.859	1.063	1.323	1.721	2.080	2.518	2.831	3.819
22	.686	.858	1.061	1.321	1.717	2.074	2.508	2.819	3.792
23	.685	.858	1.060	1.319	1.714	2.069	2.500	2.807	3.767
24	.685	.857	1.059	1.318	1.711	2.064	2.492	2.797	3.745
25	.684	.856	1.058	1.316	1.708	2.060	2.485	2.787	3.725
26	.684	.856	1.058	1.315	1.706	2.056	2.479	2.779	3.707
27	.684	.855	1.057	1.314	1.703	2.052	2.473	2.771	3.690
28	.683	.855	1.056	1.313	1.701	2.048	2.467	2.763	3.674
29	.683	.854	1.055	1.311	1.699	2.045	2.462	2.756	3.659
30	.683	.854	1.055	1.310	1.697	2.042	2.457	2.750	3.646
40	.681	.851	1.050	1.303	1.684	2.021	2.423	2.704	3.551
60	.679	.848	1.046	1.296	1.671	2.000	2.390	2.660	3.460
120	.677	.845	1.041	1.289	1.658	1.980	2.358	2.617	3.373
∞[z(α)]	.674	.842	1.036	1.282	1.645	1.960	2.326	2.576	3.291

NOTE: Table E1 is taken from Table III, page 46, of Fisher and Yates, *Statistical Tables for Biological, Agricultural and Medical Research.* (Edinburgh: Oliver and Boyd, 1967), by permission of the authors and publishers.

α-LEVEL VALUES, $\chi^2(d, \alpha)$, FOR χ^2-VARIABLES Since χ^2-variables have asymmetric distributions, we have $\chi^2(d, \alpha)$ for values of α near both 0 and 1 in Table E2. The α-level values are defined as follows and illustrated below. Suppose $Y \sim \chi^2(d)$. Then $\chi^2(d, \alpha)$ is that value of Y such that

$$P(Y > \chi^2(d, \alpha) = \alpha.$$

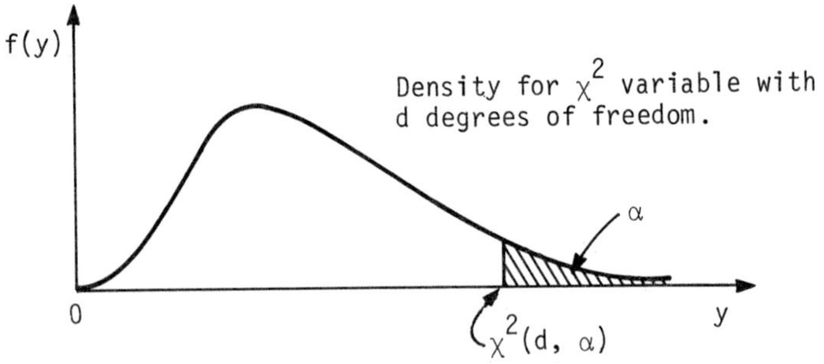

If $d > 30$, we use the following approximate α-level values:

$$\chi^2\ (d, \alpha) = [z(\alpha) + \sqrt{(2d - 1)}]^2/2,$$

where $z(\alpha)$ is the α-level value for a normal variable as given in the last line of Table E2.

EXAMPLE Find $\chi^2(7, .95)$, $\chi^2(10, .025)$, and $\chi^2(4], .025)$.

SOLUTION: Entering Table E2 at $d = 7$ and $\alpha = .95$, we have

$$\chi^2(7, .95) = 2.16735.$$

Likewise, $d = 10$ and $\alpha = .025$ gives

$$\chi^2(10, .025) = 20.4832.$$

By using the approximate formula above,

$$\begin{aligned} \chi^2(41, .025) &= (1.96 + \sqrt{82 - 1})^2/2 \\ &= (10.96)^2/2 \\ &= 120.1216/2 \\ &= 60.0608. \end{aligned}$$

TABLE E2.
α-level values, $\chi^2(d,\alpha)$ for a χ^2 variable.

d \ α	0·995	0·990	0·975	0·950	0·900	0·750	0·500
1	392704.10^{-10}	157088.10^{-9}	982069.10^{-9}	393214.10^{-8}	0·0157908	0·1015308	0·454936
2	0·0100251	0·0201007	0·0506356	0·102587	0·210721	0·575364	1·38629
3	0·0717218	0·114832	0·215795	0·351846	0·584374	1·212534	2·36597
4	0·206989	0·297109	0·484419	0·710723	1·063623	1·92256	3·35669
5	0·411742	0·554298	0·831212	1·145476	1·61031	2·67460	4·35146
6	0·675727	0·872090	1·23734	1·63538	2·20413	3·45460	5·34812
7	0·989256	1·239043	1·68987	2·16735	2·83311	4·25485	6·34581
8	1·34441	1·64650	2·17973	2·73264	3·48954	5·07064	7·34412
9	1·73493	2·08790	2·70039	3·32511	4·16816	5·89883	8·34283
10	2·15586	2·55821	3·24697	3·94030	4·86518	6·73720	9·34182
11	2·60322	3·05348	3·81575	4·57481	5·57778	7·58414	10·3410
12	3·07382	3·57057	4·40379	5·22603	6·30380	8·43842	11·3403
13	3·56503	4·10692	5·00875	5·89186	7·04150	9·29907	12·3398
14	4·07467	4·66043	5·62873	6·57063	7·78953	10·1653	13·3393
15	4·60092	5·22935	6·26214	7·26094	8·54676	11·0365	14·3389
16	5·14221	5·81221	6·90766	7·96165	9·31224	11·9122	15·3385
17	5·69722	6·40776	7·56419	8·67176	10·0852	12·7919	16·3382
18	6·26480	7·01491	8·23075	9·39046	10·8649	13·6753	17·3379
19	6·84397	7·63273	8·90652	10·1170	11·6509	14·5620	18·3377
20	7·43384	8·26040	9·59078	10·8508	12·4426	15·4518	19·3374
21	8·03365	8·89720	10·28293	11·5913	13·2396	16·3444	20·3372
22	8·64272	9·54249	10·9823	12·3380	14·0415	17·2396	21·3370
23	9·26043	10·19567	11·6886	13·0905	14·8480	18·1373	22·3369
24	9·88623	10·8564	12·4012	13·8484	15·6587	19·0373	23·3367
25	10·5197	11·5240	13·1197	14·6114	16·4734	19·9393	24·3366
26	11·1602	12·1981	13·8439	15·3792	17·2919	20·8434	25·3365
27	11·8076	12·8785	14·5734	16·1514	18·1139	21·7494	26·3363
28	12·4613	13·5647	15·3079	16·9279	18·9392	22·6572	27·3362
29	13·1211	14·2565	16·0471	17·7084	19·7677	23·5666	28·3361
30	13·7867	14·9535	16·7908	18·4927	20·5992	24·4776	29·3360
40	20·7065	22·1643	24·4330	26·5093	29·0505	33·6603	39·3353
50	27·9907	29·7067	32·3574	34·7643	37·6886	42·9421	49·3349
60	35·5345	37·4849	40·4817	43·1880	46·4589	52·2938	59·3347
70	43·2752	45·4417	48·7576	51·7393	55·3289	61·6983	69·3345
80	51·1719	53·5401	57·1532	60·3915	64·2778	71·1445	79·3343
90	59·1963	61·7541	65·6466	69·1260	73·2911	80·6247	89·3342
100	67·3276	70·0649	74·2219	77·9295	82·3581	90·1332	99·3341
z(α)	−2·5758	−2·3263	−1·9600	−1·6449	−1·2816	−0·6745	0·0000

Reproduced from Table 8 of E. S. Pearson and H. O. Hartley, *Biometrika Tables for Statisticians*, 3rd Edition, The University Press, Cambridge with the permission of Professor E. S. Pearson for the publishers.

TABLE E2. α-level values, $\chi^2(d,\alpha)$ for a χ^2 variable (continued).

d \ α	0·250	0·100	0·050	0·025	0·010	0·005	0·001
1	1·32330	2·70554	3·84146	5·02389	6·63490	7·87944	10·828
2	2·77259	4·60517	5·99146	7·37776	9·21034	10·5966	13·816
3	4·10834	6·25139	7·81473	9·34840	11·3449	12·8382	16·266
4	5·38527	7·77944	9·48773	11·1433	13·2767	14·8603	18·467
5	6·62568	9·23636	11·0705	12·8325	15·0863	16·7496	20·515
6	7·84080	10·6446	12·5916	14·4494	16·8119	18·5476	22·458
7	9·03715	12·0170	14·0671	16·0128	18·4753	20·2777	24·322
8	10·2189	13·3616	15·5073	17·5345	20·0902	21·9550	26·125
9	11·3888	14·6837	16·9190	19·0228	21·6660	23·5894	27·877
10	12·5489	15·9872	18·3070	20·4832	23·2093	25·1882	29·588
11	13·7007	17·2750	19·6751	21·9200	24·7250	26·7568	31·264
12	14·8454	18·5493	21·0261	23·3367	26·2170	28·2995	32·909
13	15·9839	19·8119	22·3620	24·7356	27·6882	29·8195	34·528
14	17·1169	21·0641	23·6848	26·1189	29·1412	31·3194	36·123
15	18·2451	22·3071	24·9958	27·4884	30·5779	32·8013	37·697
16	19·3689	23·5418	26·2962	28·8454	31·9999	34·2672	39·252
17	20·4887	24·7690	27·5871	30·1910	33·4087	35·7185	40·790
18	21·6049	25·9894	28·8693	31·5264	34·8053	37·1565	42·312
19	22·7178	27·2036	30·1435	32·8523	36·1909	38·5823	43·820
20	23·8277	28·4120	31·4104	34·1696	37·5662	39·9968	45·315
21	24·9348	29·6151	32·6706	35·4789	38·9322	41·4011	46·797
22	26·0393	30·8133	33·9244	36·7807	40·2894	42·7957	48·268
23	27·1413	32·0069	35·1725	38·0756	41·6384	44·1813	49·728
24	28·2412	33·1962	36·4150	39·3641	42·9798	45·5585	51·179
25	29·3389	34·3816	37·6525	40·6465	44·3141	46·9279	52·618
26	30·4346	35·5632	38·8851	41·9232	45·6417	48·2899	54·052
27	31·5284	36·7412	40·1133	43·1945	46·9629	49·6449	55·476
28	32·6205	37·9159	41·3371	44·4608	48·2782	50·9934	56·892
29	33·7109	39·0875	42·5570	45·7223	49·5879	52·3356	58·301
30	34·7997	40·2560	43·7730	46·9792	50·8922	53·6720	59·703
40	45·6160	51·8051	55·7585	59·3417	63·6907	66·7660	73·402
50	56·3336	63·1671	67·5048	71·4202	76·1539	79·4900	86·661
60	66·9815	74·3970	79·0819	83·2977	88·3794	91·9517	99·607
70	77·5767	85·5270	90·5312	95·0232	100·425	104·215	112·317
80	88·1303	96·5782	101·879	106·629	112·329	116·321	124·839
90	98·6499	107·565	113·145	118·136	124·116	128·299	137·208
100	109·141	118·498	124·342	129·561	135·807	140·169	149·449
z(α)	+0·6745	+1·2816	+1·6449	+1·9600	+2·3263	+2·5758	+3·0902

α-LEVEL VALUES, $F(k_1, k_2, \alpha)$, FOR F-VARIABLES: The tabled α-level values, $F(k_1, k_2, \alpha)$, given in Table E3, are defined as

$$P[F > F(k_1, k_2, \alpha)] = \alpha$$

and illustrated below:

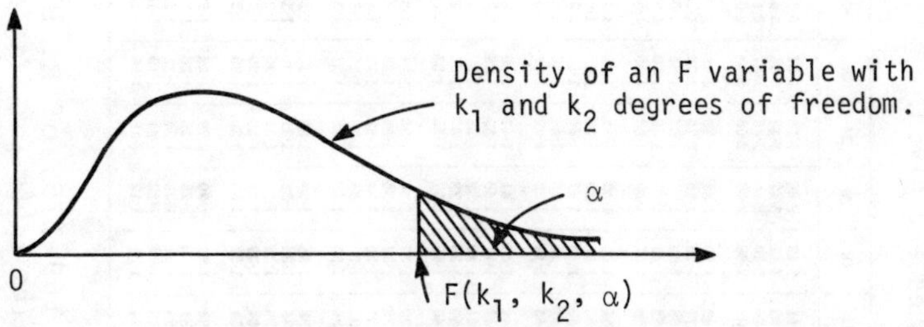

If $\alpha > .5$, we use the relationship,

$$F(k_1, k_2, \alpha) = 1/F(k_2, k_1, 1 - \alpha),$$

to obtain α-level values at the lower end of an F-distribution.

EXAMPLE: Find $F(2, 7, .05)$ and $F(4, 8, .75)$.

SOLUTION: Entering Table E3 with $\alpha = .05$, $k_1 = 2$, and $k_2 = 7$, we find that

$$F(2, 7, .05) = 4.74.$$

Since $.75 > .5$, we use the relationship above to write

$$F(4, 8, .75) = 1/F(8, 4, .25).$$

Then we use Table E3 with $\alpha = .25$, $k_1 = 8$, and $k_2 = 4$ to find

$$F(4, 8, .75) = 1/2.08 = .48.$$

TABLE E3.

$F(k_1, k_2, .25)$ values are given.

k_2 \ k_1	1	2	3	4	5	6	7	8	9	10	12	15	20	24	30	40	60	120	∞
1	5·83	7·50	8·20	8·58	8·82	8·98	9·10	9·19	9·26	9·32	9·41	9·49	9·58	9·63	9·67	9·71	9·76	9·80	9·85
2	2·57	3·00	3·15	3·23	3·28	3·31	3·34	3·35	3·37	3·38	3·39	3·41	3·43	3·43	3·44	3·45	3·46	3·47	3·48
3	2·02	2·28	2·36	2·39	2·41	2·42	2·43	2·44	2·44	2·44	2·45	2·46	2·46	2·46	2·47	2·47	2·47	2·47	2·47
4	1·81	2·00	2·05	2·06	2·07	2·08	2·08	2·08	2·08	2·08	2·08	2·08	2·08	2·08	2·08	2·08	2·08	2·08	2·08
5	1·69	1·85	1·88	1·89	1·89	1·89	1·89	1·89	1·89	1·89	1·89	1·89	1·88	1·88	1·88	1·88	1·87	1·87	1·87
6	1·62	1·76	1·78	1·79	1·79	1·78	1·78	1·78	1·77	1·77	1·77	1·76	1·76	1·75	1·75	1·75	1·74	1·74	1·74
7	1·57	1·70	1·72	1·72	1·71	1·71	1·70	1·70	1·69	1·69	1·68	1·68	1·67	1·67	1·66	1·66	1·65	1·65	1·65
8	1·54	1·66	1·67	1·66	1·66	1·65	1·64	1·64	1·63	1·63	1·62	1·62	1·61	1·60	1·60	1·59	1·59	1·58	1·58
9	1·51	1·62	1·63	1·63	1·62	1·61	1·60	1·60	1·59	1·59	1·58	1·57	1·56	1·56	1·55	1·54	1·54	1·53	1·53
10	1·49	1·60	1·60	1·59	1·59	1·58	1·57	1·56	1·56	1·55	1·54	1·53	1·52	1·52	1·51	1·51	1·50	1·49	1·48
11	1·47	1·58	1·58	1·57	1·56	1·55	1·54	1·53	1·53	1·52	1·51	1·50	1·49	1·49	1·48	1·47	1·47	1·46	1·45
12	1·46	1·56	1·56	1·55	1·54	1·53	1·52	1·51	1·51	1·50	1·49	1·48	1·47	1·46	1·45	1·45	1·44	1·43	1·42
13	1·45	1·55	1·55	1·53	1·52	1·51	1·50	1·49	1·49	1·48	1·47	1·46	1·45	1·44	1·43	1·42	1·42	1·41	1·40
14	1·44	1·53	1·53	1·52	1·51	1·50	1·49	1·48	1·47	1·46	1·45	1·44	1·43	1·42	1·41	1·41	1·40	1·39	1·38
15	1·43	1·52	1·52	1·51	1·49	1·48	1·47	1·46	1·46	1·45	1·44	1·43	1·41	1·41	1·40	1·39	1·38	1·37	1·36
16	1·42	1·51	1·51	1·50	1·48	1·47	1·46	1·45	1·44	1·44	1·43	1·41	1·40	1·39	1·38	1·37	1·36	1·35	1·34
17	1·42	1·51	1·50	1·49	1·47	1·46	1·45	1·44	1·43	1·43	1·41	1·40	1·39	1·38	1·37	1·36	1·35	1·34	1·33
18	1·41	1·50	1·49	1·48	1·46	1·45	1·44	1·43	1·42	1·42	1·40	1·39	1·38	1·37	1·36	1·35	1·34	1·33	1·32
19	1·41	1·49	1·49	1·47	1·46	1·44	1·43	1·42	1·41	1·41	1·40	1·38	1·37	1·36	1·35	1·34	1·33	1·32	1·30
20	1·40	1·49	1·48	1·47	1·45	1·44	1·43	1·42	1·41	1·40	1·39	1·37	1·36	1·35	1·34	1·33	1·32	1·31	1·29
21	1·40	1·48	1·48	1·46	1·44	1·43	1·42	1·41	1·40	1·39	1·38	1·37	1·35	1·34	1·33	1·32	1·31	1·30	1·28
22	1·40	1·48	1·47	1·45	1·44	1·42	1·41	1·40	1·39	1·39	1·37	1·36	1·34	1·33	1·32	1·31	1·30	1·29	1·28
23	1·39	1·47	1·47	1·45	1·43	1·42	1·41	1·40	1·39	1·38	1·37	1·35	1·34	1·33	1·32	1·31	1·30	1·28	1·27
24	1·39	1·47	1·46	1·44	1·43	1·41	1·40	1·39	1·38	1·38	1·36	1·35	1·33	1·32	1·31	1·30	1·29	1·28	1·26
25	1·39	1·47	1·46	1·44	1·42	1·41	1·40	1·39	1·38	1·37	1·36	1·34	1·33	1·32	1·31	1·29	1·28	1·27	1·25
26	1·38	1·46	1·45	1·44	1·42	1·41	1·39	1·38	1·37	1·37	1·35	1·34	1·32	1·31	1·30	1·29	1·28	1·26	1·25
27	1·38	1·46	1·45	1·43	1·42	1·40	1·39	1·38	1·37	1·36	1·35	1·33	1·32	1·31	1·30	1·28	1·27	1·26	1·24
28	1·38	1·46	1·45	1·43	1·41	1·40	1·39	1·38	1·37	1·36	1·34	1·33	1·31	1·30	1·29	1·28	1·27	1·25	1·24
29	1·38	1·45	1·45	1·43	1·41	1·40	1·38	1·37	1·36	1·35	1·34	1·32	1·31	1·30	1·29	1·27	1·26	1·25	1·23
30	1·38	1·45	1·44	1·42	1·41	1·39	1·38	1·37	1·36	1·35	1·34	1·32	1·30	1·29	1·28	1·27	1·26	1·24	1·23
40	1·36	1·44	1·42	1·40	1·39	1·37	1·36	1·35	1·34	1·33	1·31	1·30	1·28	1·26	1·25	1·24	1·22	1·21	1·19
60	1·35	1·42	1·41	1·38	1·37	1·35	1·33	1·32	1·31	1·30	1·29	1·27	1·25	1·24	1·22	1·21	1·19	1·17	1·15
120	1·34	1·40	1·39	1·37	1·35	1·33	1·31	1·30	1·29	1·28	1·26	1·24	1·22	1·21	1·19	1·18	1·16	1·13	1·10
∞	1·32	1·39	1·37	1·35	1·33	1·31	1·29	1·28	1·27	1·25	1·24	1·22	1·19	1·18	1·16	1·14	1·12	1·08	1·00

NOTE: Reproduced from Table 18, pages 169-175, of E. S. Pearson and H. O. Hartley *Biometrika Table for Statisticians*, 3rd Ed. (Cambridge: The University Press, with permission of Professor E. S. Pearson for the publishers.

TABLE E3. F(k_1, k_2, .10) values are given (continued).

k_2 \ k_1	1	2	3	4	5	6	7	8	9	10	12	15	20	24	30	40	60	120	∞
1	39·86	49·50	53·59	55·83	57·24	58·20	58·91	59·44	59·86	60·19	60·71	61·22	61·74	62·00	62·26	62·53	62·79	63·06	63·33
2	8·53	9·00	9·16	9·24	9·29	9·33	9·35	9·37	9·38	9·39	9·41	9·42	9·44	9·45	9·46	9·47	9·47	9·48	9·49
3	5·54	5·46	5·39	5·34	5·31	5·28	5·27	5·25	5·24	5·23	5·22	5·20	5·18	5·18	5·17	5·16	5·15	5·14	5·13
4	4·54	4·32	4·19	4·11	4·05	4·01	3·98	3·95	3·94	3·92	3·90	3·87	3·84	3·83	3·82	3·80	3·79	3·78	3·76
5	4·06	3·78	3·62	3·52	3·45	3·40	3·37	3·34	3·32	3·30	3·27	3·24	3·21	3·19	3·17	3·16	3·14	3·12	3·10
6	3·78	3·46	3·29	3·18	3·11	3·05	3·01	2·98	2·96	2·94	2·90	2·87	2·84	2·82	2·80	2·78	2·76	2·74	2·72
7	3·59	3·26	3·07	2·96	2·88	2·83	2·78	2·75	2·72	2·70	2·67	2·63	2·59	2·58	2·56	2·54	2·51	2·49	2·47
8	3·46	3·11	2·92	2·81	2·73	2·67	2·62	2·59	2·56	2·54	2·50	2·46	2·42	2·40	2·38	2·36	2·34	2·32	2·29
9	3·36	3·01	2·81	2·69	2·61	2·55	2·51	2·47	2·44	2·42	2·38	2·34	2·30	2·28	2·25	2·23	2·21	2·18	2·16
10	3·29	2·92	2·73	2·61	2·52	2·46	2·41	2·38	2·35	2·32	2·28	2·24	2·20	2·18	2·16	2·13	2·11	2·08	2·06
11	3·23	2·86	2·66	2·54	2·45	2·39	2·34	2·30	2·27	2·25	2·21	2·17	2·12	2·10	2·08	2·05	2·03	2·00	1·97
12	3·18	2·81	2·61	2·48	2·39	2·33	2·28	2·24	2·21	2·19	2·15	2·10	2·06	2·04	2·01	1·99	1·96	1·93	1·90
13	3·14	2·76	2·56	2·43	2·35	2·28	2·23	2·20	2·16	2·14	2·10	2·05	2·01	1·98	1·96	1·93	1·90	1·88	1·85
14	3·10	2·73	2·52	2·39	2·31	2·24	2·19	2·15	2·12	2·10	2·05	2·01	1·96	1·94	1·91	1·89	1·86	1·83	1·80
15	3·07	2·70	2·49	2·36	2·27	2·21	2·16	2·12	2·09	2·06	2·02	1·97	1·92	1·90	1·87	1·85	1·82	1·79	1·76
16	3·05	2·67	2·46	2·33	2·24	2·18	2·13	2·09	2·06	2·03	1·99	1·94	1·89	1·87	1·84	1·81	1·78	1·75	1·72
17	3·03	2·64	2·44	2·31	2·22	2·15	2·10	2·06	2·03	2·00	1·96	1·91	1·86	1·84	1·81	1·78	1·75	1·72	1·69
18	3·01	2·62	2·42	2·29	2·20	2·13	2·08	2·04	2·00	1·98	1·93	1·89	1·84	1·81	1·78	1·75	1·72	1·69	1·66
19	2·99	2·61	2·40	2·27	2·18	2·11	2·06	2·02	1·98	1·96	1·91	1·86	1·81	1·79	1·76	1·73	1·70	1·67	1·63
20	2·97	2·59	2·38	2·25	2·16	2·09	2·04	2·00	1·96	1·94	1·89	1·84	1·79	1·77	1·74	1·71	1·68	1·64	1·61
21	2·96	2·57	2·36	2·23	2·14	2·08	2·02	1·98	1·95	1·92	1·87	1·83	1·78	1·75	1·72	1·69	1·66	1·62	1·59
22	2·95	2·56	2·35	2·22	2·13	2·06	2·01	1·97	1·93	1·90	1·86	1·81	1·76	1·73	1·70	1·67	1·64	1·60	1·57
23	2·94	2·55	2·34	2·21	2·11	2·05	1·99	1·95	1·92	1·89	1·84	1·80	1·74	1·72	1·69	1·66	1·62	1·59	1·55
24	2·93	2·54	2·33	2·19	2·10	2·04	1·98	1·94	1·91	1·88	1·83	1·78	1·73	1·70	1·67	1·64	1·61	1·57	1·53
25	2·92	2·53	2·32	2·18	2·09	2·02	1·97	1·93	1·89	1·87	1·82	1·77	1·72	1·69	1·66	1·63	1·59	1·56	1·52
26	2·91	2·52	2·31	2·17	2·08	2·01	1·96	1·92	1·88	1·86	1·81	1·76	1·71	1·68	1·65	1·61	1·58	1·54	1·50
27	2·90	2·51	2·30	2·17	2·07	2·00	1·95	1·91	1·87	1·85	1·80	1·75	1·70	1·67	1·64	1·60	1·57	1·53	1·49
28	2·89	2·50	2·29	2·16	2·06	2·00	1·94	1·90	1·87	1·84	1·79	1·74	1·69	1·66	1·63	1·59	1·56	1·52	1·48
29	2·89	2·50	2·28	2·15	2·06	1·99	1·93	1·89	1·86	1·83	1·78	1·73	1·68	1·65	1·62	1·58	1·55	1·51	1·47
30	2·88	2·49	2·28	2·14	2·05	1·98	1·93	1·88	1·85	1·82	1·77	1·72	1·67	1·64	1·61	1·57	1·54	1·50	1·46
40	2·84	2·44	2·23	2·09	2·00	1·93	1·87	1·83	1·79	1·76	1·71	1·66	1·61	1·57	1·54	1·51	1·47	1·42	1·38
60	2·79	2·39	2·18	2·04	1·95	1·87	1·82	1·77	1·74	1·71	1·66	1·60	1·54	1·51	1·48	1·44	1·40	1·35	1·29
120	2·75	2·35	2·13	1·99	1·90	1·82	1·77	1·72	1·68	1·65	1·60	1·55	1·48	1·45	1·41	1·37	1·32	1·26	1·19
∞	2·71	2·30	2·08	1·94	1·85	1·77	1·72	1·67	1·63	1·60	1·55	1·49	1·42	1·38	1·34	1·30	1·24	1·17	1·00

TABLE E3. $F(k_1, k_2, .05)$ values are given (continued).

k_2 \ k_1	1	2	3	4	5	6	7	8	9	10	12	15	20	24	30	40	60	120	∞
1	161·4	199·5	215·7	224·6	230·2	234·0	236·8	238·9	240·5	241·9	243·9	245·9	248·0	249·1	250·1	251·1	252·2	253·3	254·3
2	18·51	19·00	19·16	19·25	19·30	19·33	19·35	19·37	19·38	19·40	19·41	19·43	19·45	19·45	19·46	19·47	19·48	19·49	19·50
3	10·13	9·55	9·28	9·12	9·01	8·94	8·89	8·85	8·81	8·79	8·74	8·70	8·66	8·64	8·62	8·59	8·57	8·55	8·53
4	7·71	6·94	6·59	6·39	6·26	6·16	6·09	6·04	6·00	5·96	5·91	5·86	5·80	5·77	5·75	5·72	5·69	5·66	5·63
5	6·61	5·79	5·41	5·19	5·05	4·95	4·88	4·82	4·77	4·74	4·68	4·62	4·56	4·53	4·50	4·46	4·43	4·40	4·36
6	5·99	5·14	4·76	4·53	4·39	4·28	4·21	4·15	4·10	4·06	4·00	3·94	3·87	3·84	3·81	3·77	3·74	3·70	3·67
7	5·59	4·74	4·35	4·12	3·97	3·87	3·79	3·73	3·68	3·64	3·57	3·51	3·44	3·41	3·38	3·34	3·30	3·27	3·23
8	5·32	4·46	4·07	3·84	3·69	3·58	3·50	3·44	3·39	3·35	3·28	3·22	3·15	3·12	3·08	3·04	3·01	2·97	2·93
9	5·12	4·26	3·86	3·63	3·48	3·37	3·29	3·23	3·18	3·14	3·07	3·01	2·94	2·90	2·86	2·83	2·79	2·75	2·71
10	4·96	4·10	3·71	3·48	3·33	3·22	3·14	3·07	3·02	2·98	2·91	2·85	2·77	2·74	2·70	2·66	2·62	2·58	2·54
11	4·84	3·98	3·59	3·36	3·20	3·09	3·01	2·95	2·90	2·85	2·79	2·72	2·65	2·61	2·57	2·53	2·49	2·45	2·40
12	4·75	3·89	3·49	3·26	3·11	3·00	2·91	2·85	2·80	2·75	2·69	2·62	2·54	2·51	2·47	2·43	2·38	2·34	2·30
13	4·67	3·81	3·41	3·18	3·03	2·92	2·83	2·77	2·71	2·67	2·60	2·53	2·46	2·42	2·38	2·34	2·30	2·25	2·21
14	4·60	3·74	3·34	3·11	2·96	2·85	2·76	2·70	2·65	2·60	2·53	2·46	2·39	2·35	2·31	2·27	2·22	2·18	2·13
15	4·54	3·68	3·29	3·06	2·90	2·79	2·71	2·64	2·59	2·54	2·48	2·40	2·33	2·29	2·25	2·20	2·16	2·11	2·07
16	4·49	3·63	3·24	3·01	2·85	2·74	2·66	2·59	2·54	2·49	2·42	2·35	2·28	2·24	2·19	2·15	2·11	2·06	2·01
17	4·45	3·59	3·20	2·96	2·81	2·70	2·61	2·55	2·49	2·45	2·38	2·31	2·23	2·19	2·15	2·10	2·06	2·01	1·96
18	4·41	3·55	3·16	2·93	2·77	2·66	2·58	2·51	2·46	2·41	2·34	2·27	2·19	2·15	2·11	2·06	2·02	1·97	1·92
19	4·38	3·52	3·13	2·90	2·74	2·63	2·54	2·48	2·42	2·38	2·31	2·23	2·16	2·11	2·07	2·03	1·98	1·93	1·88
20	4·35	3·49	3·10	2·87	2·71	2·60	2·51	2·45	2·39	2·35	2·28	2·20	2·12	2·08	2·04	1·99	1·95	1·90	1·84
21	4·32	3·47	3·07	2·84	2·68	2·57	2·49	2·42	2·37	2·32	2·25	2·18	2·10	2·05	2·01	1·96	1·92	1·87	1·81
22	4·30	3·44	3·05	2·82	2·66	2·55	2·46	2·40	2·34	2·30	2·23	2·15	2·07	2·03	1·98	1·94	1·89	1·84	1·78
23	4·28	3·42	3·03	2·80	2·64	2·53	2·44	2·37	2·32	2·27	2·20	2·13	2·05	2·01	1·96	1·91	1·86	1·81	1·76
24	4·26	3·40	3·01	2·78	2·62	2·51	2·42	2·36	2·30	2·25	2·18	2·11	2·03	1·98	1·94	1·89	1·84	1·79	1·73
25	4·24	3·39	2·99	2·76	2·60	2·49	2·40	2·34	2·28	2·24	2·16	2·09	2·01	1·96	1·92	1·87	1·82	1·77	1·71
26	4·23	3·37	2·98	2·74	2·59	2·47	2·39	2·32	2·27	2·22	2·15	2·07	1·99	1·95	1·90	1·85	1·80	1·75	1·69
27	4·21	3·35	2·96	2·73	2·57	2·46	2·37	2·31	2·25	2·20	2·13	2·06	1·97	1·93	1·88	1·84	1·79	1·73	1·67
28	4·20	3·34	2·95	2·71	2·56	2·45	2·36	2·29	2·24	2·19	2·12	2·04	1·96	1·91	1·87	1·82	1·77	1·71	1·65
29	4·18	3·33	2·93	2·70	2·55	2·43	2·35	2·28	2·22	2·18	2·10	2·03	1·94	1·90	1·85	1·81	1·75	1·70	1·64
30	4·17	3·32	2·92	2·69	2·53	2·42	2·33	2·27	2·21	2·16	2·09	2·01	1·93	1·89	1·84	1·79	1·74	1·68	1·62
40	4·08	3·23	2·84	2·61	2·45	2·34	2·25	2·18	2·12	2·08	2·00	1·92	1·84	1·79	1·74	1·69	1·64	1·58	1·51
60	4·00	3·15	2·76	2·53	2·37	2·25	2·17	2·10	2·04	1·99	1·92	1·84	1·75	1·70	1·65	1·59	1·53	1·47	1·39
120	3·92	3·07	2·68	2·45	2·29	2·17	2·09	2·02	1·96	1·91	1·83	1·75	1·66	1·61	1·55	1·50	1·43	1·35	1·25
∞	3·84	3·00	2·60	2·37	2·21	2·10	2·01	1·94	1·88	1·83	1·75	1·67	1·57	1·52	1·46	1·39	1·32	1·22	1·00

TABLE E3. $F(k_1, k_2, .025)$ values are given (continued).

$k_2 \backslash k_1$	1	2	3	4	5	6	7	8	9	10	12	15	20	24	30	40	60	120	∞
1	647·8	799·5	864·2	899·6	921·8	937·1	948·2	956·7	963·3	968·6	976·7	984·9	993·1	997·2	1001	1006	1010	1014	1018
2	38·51	39·00	39·17	39·25	39·30	39·33	39·36	39·37	39·39	39·40	39·41	39·43	39·45	39·46	39·46	39·47	39·48	39·49	39·50
3	17·44	16·04	15·44	15·10	14·88	14·73	14·62	14·54	14·47	14·42	14·34	14·25	14·17	14·12	14·08	14·04	13·99	13·95	13·90
4	12·22	10·65	9·98	9·60	9·36	9·20	9·07	8·98	8·90	8·84	8·75	8·66	8·56	8·51	8·46	8·41	8·36	8·31	8·26
5	10·01	8·43	7·76	7·39	7·15	6·98	6·85	6·76	6·68	6·62	6·52	6·43	6·33	6·28	6·23	6·18	6·12	6·07	6·02
6	8·81	7·26	6·60	6·23	5·99	5·82	5·70	5·60	5·52	5·46	5·37	5·27	5·17	5·12	5·07	5·01	4·96	4·90	4·85
7	8·07	6·54	5·89	5·52	5·29	5·12	4·99	4·90	4·82	4·76	4·67	4·57	4·47	4·42	4·36	4·31	4·25	4·20	4·14
8	7·57	6·06	5·42	5·05	4·82	4·65	4·53	4·43	4·36	4·30	4·20	4·10	4·00	3·95	3·89	3·84	3·78	3·73	3·67
9	7·21	5·71	5·08	4·72	4·48	4·32	4·20	4·10	4·03	3·96	3·87	3·77	3·67	3·61	3·56	3·51	3·45	3·39	3·33
10	6·94	5·46	4·83	4·47	4·24	4·07	3·95	3·85	3·78	3·72	3·62	3·52	3·42	3·37	3·31	3·26	3·20	3·14	3·08
11	6·72	5·26	4·63	4·28	4·04	3·88	3·76	3·66	3·59	3·53	3·43	3·33	3·23	3·17	3·12	3·06	3·00	2·94	2·88
12	6·55	5·10	4·47	4·12	3·89	3·73	3·61	3·51	3·44	3·37	3·28	3·18	3·07	3·02	2·96	2·91	2·85	2·79	2·72
13	6·41	4·97	4·35	4·00	3·77	3·60	3·48	3·39	3·31	3·25	3·15	3·05	2·95	2·89	2·84	2·78	2·72	2·66	2·60
14	6·30	4·86	4·24	3·89	3·66	3·50	3·38	3·29	3·21	3·15	3·05	2·95	2·84	2·79	2·73	2·67	2·61	2·55	2·49
15	6·20	4·77	4·15	3·80	3·58	3·41	3·29	3·20	3·12	3·06	2·96	2·86	2·76	2·70	2·64	2·59	2·52	2·46	2·40
16	6·12	4·69	4·08	3·73	3·50	3·34	3·22	3·12	3·05	2·99	2·89	2·79	2·68	2·63	2·57	2·51	2·45	2·38	2·32
17	6·04	4·62	4·01	3·66	3·44	3·28	3·16	3·06	2·98	2·92	2·82	2·72	2·62	2·56	2·50	2·44	2·38	2·32	2·25
18	5·98	4·56	3·95	3·61	3·38	3·22	3·10	3·01	2·93	2·87	2·77	2·67	2·56	2·50	2·44	2·38	2·32	2·26	2·19
19	5·92	4·51	3·90	3·56	3·33	3·17	3·05	2·96	2·88	2·82	2·72	2·62	2·51	2·45	2·39	2·33	2·27	2·20	2·13
20	5·87	4·46	3·86	3·51	3·29	3·13	3·01	2·91	2·84	2·77	2·68	2·57	2·46	2·41	2·35	2·29	2·22	2·16	2·09
21	5·83	4·42	3·82	3·48	3·25	3·09	2·97	2·87	2·80	2·73	2·64	2·53	2·42	2·37	2·31	2·25	2·18	2·11	2·04
22	5·79	4·38	3·78	3·44	3·22	3·05	2·93	2·84	2·76	2·70	2·60	2·50	2·39	2·33	2·27	2·21	2·14	2·08	2·00
23	5·75	4·35	3·75	3·41	3·18	3·02	2·90	2·81	2·73	2·67	2·57	2·47	2·36	2·30	2·24	2·18	2·11	2·04	1·97
24	5·72	4·32	3·72	3·38	3·15	2·99	2·87	2·78	2·70	2·64	2·54	2·44	2·33	2·27	2·21	2·15	2·08	2·01	1·94
25	5·69	4·29	3·69	3·35	3·13	2·97	2·85	2·75	2·68	2·61	2·51	2·41	2·30	2·24	2·18	2·12	2·05	1·98	1·91
26	5·66	4·27	3·67	3·33	3·10	2·94	2·82	2·73	2·65	2·59	2·49	2·39	2·28	2·22	2·16	2·09	2·03	1·95	1·88
27	5·63	4·24	3·65	3·31	3·08	2·92	2·80	2·71	2·63	2·57	2·47	2·36	2·25	2·19	2·13	2·07	2·00	1·93	1·85
28	5·61	4·22	3·63	3·29	3·06	2·90	2·78	2·69	2·61	2·55	2·45	2·34	2·23	2·17	2·11	2·05	1·98	1·91	1·83
29	5·59	4·20	3·61	3·27	3·04	2·88	2·76	2·67	2·59	2·53	2·43	2·32	2·21	2·15	2·09	2·03	1·96	1·89	1·81
30	5·57	4·18	3·59	3·25	3·03	2·87	2·75	2·65	2·57	2·51	2·41	2·31	2·20	2·14	2·07	2·01	1·94	1·87	1·79
40	5·42	4·05	3·46	3·13	2·90	2·74	2·62	2·53	2·45	2·39	2·29	2·18	2·07	2·01	1·94	1·88	1·80	1·72	1·64
60	5·29	3·93	3·34	3·01	2·79	2·63	2·51	2·41	2·33	2·27	2·17	2·06	1·94	1·88	1·82	1·74	1·67	1·58	1·48
120	5·15	3·80	3·23	2·89	2·67	2·52	2·39	2·30	2·22	2·16	2·05	1·94	1·82	1·76	1·69	1·61	1·53	1·43	1·31
∞	5·02	3·69	3·12	2·79	2·57	2·41	2·29	2·19	2·11	2·05	1·94	1·83	1·71	1·64	1·57	1·48	1·39	1·27	1·00

TABLE E3. $F(k_1, k_2, .01)$ values are given (continued).

$k_2 \backslash k_1$	1	2	3	4	5	6	7	8	9	10	12	15	20	24	30	40	60	120	∞
1	4052	4999·5	5403	5625	5764	5859	5928	5981	6022	6056	6106	6157	6209	6235	6261	6287	6313	6339	6366
2	98·50	99·00	99·17	99·25	99·30	99·33	99·36	99·37	99·39	99·40	99·42	99·43	99·45	99·46	99·47	99·47	99·48	99·49	99·50
3	34·12	30·82	29·46	28·71	28·24	27·91	27·67	27·49	27·35	27·23	27·05	26·87	26·69	26·60	26·50	26·41	26·32	26·22	26·13
4	21·20	18·00	16·69	15·98	15·52	15·21	14·98	14·80	14·66	14·55	14·37	14·20	14·02	13·93	13·84	13·75	13·65	13·56	13·46
5	16·26	13·27	12·06	11·39	10·97	10·67	10·46	10·29	10·16	10·05	9·89	9·72	9·55	9·47	9·38	9·29	9·20	9·11	9·02
6	13·75	10·92	9·78	9·15	8·75	8·47	8·26	8·10	7·98	7·87	7·72	7·56	7·40	7·31	7·23	7·14	7·06	6·97	6·88
7	12·25	9·55	8·45	7·85	7·46	7·19	6·99	6·84	6·72	6·62	6·47	6·31	6·16	6·07	5·99	5·91	5·82	5·74	5·65
8	11·26	8·65	7·59	7·01	6·63	6·37	6·18	6·03	5·91	5·81	5·67	5·52	5·36	5·28	5·20	5·12	5·03	4·95	4·86
9	10·56	8·02	6·99	6·42	6·06	5·80	5·61	5·47	5·35	5·26	5·11	4·96	4·81	4·73	4·65	4·57	4·48	4·40	4·31
10	10·04	7·56	6·55	5·99	5·64	5·39	5·20	5·06	4·94	4·85	4·71	4·56	4·41	4·33	4·25	4·17	4·08	4·00	3·91
11	9·65	7·21	6·22	5·67	5·32	5·07	4·89	4·74	4·63	4·54	4·40	4·25	4·10	4·02	3·94	3·86	3·78	3·69	3·60
12	9·33	6·93	5·95	5·41	5·06	4·82	4·64	4·50	4·39	4·30	4·16	4·01	3·86	3·78	3·70	3·62	3·54	3·45	3·36
13	9·07	6·70	5·74	5·21	4·86	4·62	4·44	4·30	4·19	4·10	3·96	3·82	3·66	3·59	3·51	3·43	3·34	3·25	3·17
14	8·86	6·51	5·56	5·04	4·69	4·46	4·28	4·14	4·03	3·94	3·80	3·66	3·51	3·43	3·35	3·27	3·18	3·09	3·00
15	8·68	6·36	5·42	4·89	4·56	4·32	4·14	4·00	3·89	3·80	3·67	3·52	3·37	3·29	3·21	3·13	3·05	2·96	2·87
16	8·53	6·23	5·29	4·77	4·44	4·20	4·03	3·89	3·78	3·69	3·55	3·41	3·26	3·18	3·10	3·02	2·93	2·84	2·75
17	8·40	6·11	5·18	4·67	4·34	4·10	3·93	3·79	3·68	3·59	3·46	3·31	3·16	3·08	3·00	2·92	2·83	2·75	2·65
18	8·29	6·01	5·09	4·58	4·25	4·01	3·84	3·71	3·60	3·51	3·37	3·23	3·08	3·00	2·92	2·84	2·75	2·66	2·57
19	8·18	5·93	5·01	4·50	4·17	3·94	3·77	3·63	3·52	3·43	3·30	3·15	3·00	2·92	2·84	2·76	2·67	2·58	2·49
20	8·10	5·85	4·94	4·43	4·10	3·87	3·70	3·56	3·46	3·37	3·23	3·09	2·94	2·86	2·78	2·69	2·61	2·52	2·42
21	8·02	5·78	4·87	4·37	4·04	3·81	3·64	3·51	3·40	3·31	3·17	3·03	2·88	2·80	2·72	2·64	2·55	2·46	2·36
22	7·95	5·72	4·82	4·31	3·99	3·76	3·59	3·45	3·35	3·26	3·12	2·98	2·83	2·75	2·67	2·58	2·50	2·40	2·31
23	7·88	5·66	4·76	4·26	3·94	3·71	3·54	3·41	3·30	3·21	3·07	2·93	2·78	2·70	2·62	2·54	2·45	2·35	2·26
24	7·82	5·61	4·72	4·22	3·90	3·67	3·50	3·36	3·26	3·17	3·03	2·89	2·74	2·66	2·58	2·49	2·40	2·31	2·21
25	7·77	5·57	4·68	4·18	3·85	3·63	3·46	3·32	3·22	3·13	2·99	2·85	2·70	2·62	2·54	2·45	2·36	2·27	2·17
26	7·72	5·53	4·64	4·14	3·82	3·59	3·42	3·29	3·18	3·09	2·96	2·81	2·66	2·58	2·50	2·42	2·33	2·23	2·13
27	7·68	5·49	4·60	4·11	3·78	3·56	3·39	3·26	3·15	3·06	2·93	2·78	2·63	2·55	2·47	2·38	2·29	2·20	2·10
28	7·64	5·45	4·57	4·07	3·75	3·53	3·36	3·23	3·12	3·03	2·90	2·75	2·60	2·52	2·44	2·35	2·26	2·17	2·06
29	7·60	5·42	4·54	4·04	3·73	3·50	3·33	3·20	3·09	3·00	2·87	2·73	2·57	2·49	2·41	2·33	2·23	2·14	2·03
30	7·56	5·39	4·51	4·02	3·70	3·47	3·30	3·17	3·07	2·98	2·84	2·70	2·55	2·47	2·39	2·30	2·21	2·11	2·01
40	7·31	5·18	4·31	3·83	3·51	3·29	3·12	2·99	2·89	2·80	2·66	2·52	2·37	2·29	2·20	2·11	2·02	1·92	1·80
60	7·08	4·98	4·13	3·65	3·34	3·12	2·95	2·82	2·72	2·63	2·50	2·35	2·20	2·12	2·03	1·94	1·84	1·73	1·60
120	6·85	4·79	3·95	3·48	3·17	2·96	2·79	2·66	2·56	2·47	2·34	2·19	2·03	1·95	1·86	1·76	1·66	1·53	1·38
∞	6·63	4·61	3·78	3·32	3·02	2·80	2·64	2·51	2·41	2·32	2·18	2·04	1·88	1·79	1·70	1·59	1·47	1·32	1·00

APPENDIX F

Tables of α-Level Values for Nonparametric Statistical Procedures

The three tables in this appendix present critical values for three nonparametric statistical procedures.

α-LEVEL VALUES, $r_s(n, \alpha)$, FOR THE RANK CORRELATION COEFFICIENT, r_s: The critical values, $r_s(n, \alpha)$, are defined by

$$P[r_s > r_s(n, \alpha)] \leq \alpha,$$

and symmetry gives

$$r_s(n, \alpha) = -r_s(n, 1 - \alpha) \text{ when } \alpha > .5.$$

EXAMPLE: Find $r_s(10, .05)$ and $r(15, .95)$.

SOLUTION: With $n = 10$ and $\alpha = .05$, Table F1 gives

$$r_s(10, .05) = .5515.$$

Using the symmetry relation above, as well as $n = 15$ and $\alpha = .05$, we have

$$r_s(15, .95) = -r_s(15, .05) = -.4429.$$

TABLE F1.
α-level values, $r_S(n, \alpha)$, are given for the rank correlation coefficient, r_S.

n	α =.10	.05	.025	.01	.005	.001
4	.8000	.8000				
5	.7000	.8000	.9000	.9000		
6	.6000	.7714	.8286	.8857	.9429	
7	.5357	.6786	.7450	.8571	.8929	.9643
8	.5000	.6190	.7143	.8095	.8571	.9286
9	.4667	.5833	.6833	.7667	.8167	.9000
10	.4424	.5515	.6364	.7333	.7818	.8667
11	.4182	.5273	.6091	.7000	.7545	.8455
12	.3986	.4965	.5804	.6713	.7273	.8182
13	.3791	.4780	.5549	.6429	.6978	.7912
14	.3626	.4593	.5341	.6220	.6747	.7670
15	.3500	.4429	.5179	.6000	.6536	.7464
16	.3382	.4265	.5000	.5824	.6324	.7265
17	.3260	.4118	.4853	.5637	.6152	.7083
18	.3148	.3994	.4716	.5480	.5975	.6904
19	.3070	.3895	.4579	.5333	.5825	.6737
20	.2977	.3789	.4451	.5203	.5684	.6586
21	.2909	.3688	.4351	.5078	.5545	.6455
22	.2829	.3597	.4241	.4963	.5426	.6318
23	.2767	.3518	.4150	.4852	.5306	.6186
24	.2704	.3435	.4061	.4748	.5200	.6070
25	.2646	.3362	.3977	.4654	.5100	.5962
26	.2588	.3299	.3894	.4564	.5002	.5856
27	.2540	.3236	.3822	.4481	.4915	.5757
28	.2490	.3175	.3749	.4401	.4828	.5660
29	.2443	.3113	.3685	.4320	.4744	.5567
30	.2400	.3059	.3620	.4251	.4665	.5479

NOTE: Adapted with minor corrections from Table 3, pages 444-484, G. J. Glasser and R. F. Winter, Critical values of the coefficient of rank correlation for testing the hypothesis of independence, *Biometrika:* 48 (1961), with permission of the authors and the publishers.

α-LEVEL VALUES, $y(k, \alpha) = [t(k-1, \alpha) + t(\infty, \alpha)]/2$: The critical values, $y(k, \alpha)$, as presented in Table F2 are for the test statistic, Y, as defined in Eq. (10.18). That is,

$$P[Y > y(k, \alpha)] = \alpha.$$

The symmetry of the t- and normal distributions implies that

$$y(k, \alpha) = -y(k, 1 - \alpha) \text{ for } \alpha > .5.$$

EXAMPLE: Find y(10, .05) and y(10, .975).

SOLUTION: Using k = 10 and α = .05 in Table F2 gives

$$y(10, .05) = 1.7390.$$

Similarly, using n = 10 and α = .025, as well as the symmetry above, gives

$$y(10, .975) = -y(10, .025) = -2.1111.$$

TABLE F2.
α-level values, $y(k, \alpha) = [t(k-1, \alpha) + t(\infty, \alpha)]/2$, are given.

k	α=.05	α=.025	α=.01	α=.005	k	α=.05	α=.025	α=.01	α=.005
5	1.8883	------	------	------	28	1.6741	2.0059	2.3995	2.6732
6	1.8299	2.2653	------	------	29	1.6730	2.0042	2.3967	2.6695
7	1.7940	2.2034	2.7345	------	30	1.6720	2.0026	2.3941	2.6661
8	1.7697	2.1623	2.6621	3.0376	31	1.6711	2.0011	2.3918	2.6629
9	1.7522	2.1330	2.6114	2.9656	32	1.6702	1.9997	2.3895	2.6599
10	1.7390	2.1111	2.5738	2.9128	33	1.6694	1.9984	2.3875	2.6571
11	1.7287	2.0940	2.5450	2.8725	34	1.6686	1.9972	2.3855	2.6545
12	1.7204	2.0805	2.5222	2.8408	35	1.6679	1.9961	2.3837	2.6521
13	1.7136	2.0694	2.5036	2.8151	36	1.6672	1.9950	2.3820	2.6498
14	1.7079	2.0602	2.4883	2.7940	37	1.6666	1.9940	2.3804	2.6476
15	1.7031	2.0524	2.4754	2.7763	38	1.6660	1.9931	2.3788	2.6456
16	1.6990	2.0457	2.4644	2.7612	39	1.6654	1.9922	2.3774	2.6437
17	1.6954	2.0399	2.4549	2.7483	40	1.6649	1.9913	2.3760	2.6418
18	1.6922	2.0349	2.4466	2.7370	41	1.6644	1.9905	2.3748	2.6401
19	1.6895	2.0304	2.4393	2.7271	42	1.6639	1.9897	2.3735	2.6385
20	1.6870	2.0265	2.4329	2.7183	43	1.6634	1.9890	2.3724	2.6369
21	1.6848	2.0230	2.4271	2.7105	44	1.6630	1.9883	2.3713	2.6354
22	1.6828	2.0198	2.4220	2.7036	45	1.6625	1.9877	2.3702	2.6340
23	1.6810	2.0169	2.4173	2.6973	46	1.6621	1.9870	2.3692	2.6327
24	1.6794	2.0143	2.4131	2.6915	47	1.6618	2.9864	2.3682	2.6314
25	1.6779	2.0119	2.4092	2.6863	48	1.6614	1.9858	2.3673	2.6302
26	1.6765	2.0097	2.4057	2.6816	49	1.6610	1.9853	2.3664	2.6290
27	1.6752	2.0077	2.4024	2.6772	50	1.6607	1.9848	2.3656	2.6279

NOTE: Reprinted from R. L. Iman, Use of a t-statistic as an approximation of the exact distribution of the Wilcoxon signed ranks test statistic, in *Comunications in statistics* (New York: Marcel Dekker, Inc., 1974), vol. 3, no. 8, p. 805, by permission of the publisher.

α-LEVEL VALUES, $T(n_1, n_2, \alpha)$, FOR THE MANN-WHITNEY STATISTIC, T: The α-level values, $T(n_1, n_2, \alpha)$, as presented in Table F3 are defined as

$$P[T < T(n_1, n_2, \alpha)] \leq \alpha,$$

where T is the Mann-Whitney statistic defined in (11.30). The values in the table are lower-tail α-level points, and the upper-tail values are given by

$$T(n_1, n_2, \alpha) = n_1 n_2 - T(n_1, n_2, 1 - \alpha) \text{ when } \alpha > .5.$$

EXAMPLE: Find T(5, 4, .05) and T(6, 8, .95).

SOLUTION: Entering Table F3 with $n_1 = 5$, $n_2 = 4$, and $\alpha = .05$ gives

$$T(5, 4, .05) = 3.$$

Using the relationship above, we have

$$T(6, 8, .95) = 48 - T(6, 8, .05).$$

Table F3 with $n_1 = 6$, $n_2 = 8$, and $\alpha = .05$ provides

$$T(6, 8, .95) = 48 - 11 = 37.$$

TABLE F3.
α-level values, $T(n_1, n_2, \alpha)$, are given for the Mann-Whitney statistic, T.

n_1	α	n_2= 2	3	4	5	6	7	8	9	10	11	12	13	14	15	16	17	18	19	20
2	.050	0	0	0	1	1	1	2	2	2	2	3	3	4	4	4	4	5	5	5
	.025	0	0	0	0	0	0	1	1	1	1	2	2	2	2	2	3	3	3	3
	.010	0	0	0	0	0	0	0	0	0	0	0	1	1	1	1	1	1	2	2
	.005	0	0	0	0	0	0	0	0	0	0	0	0	0	0	0	0	0	1	1
3	.050	0	1	1	2	3	3	4	5	5	6	6	7	8	8	9	10	10	11	12
	.025	0	0	0	1	2	2	3	3	4	4	5	5	6	6	7	7	8	8	9
	.010	0	0	0	0	0	1	1	2	2	2	3	3	3	4	4	5	5	5	6
	.005	0	0	0	0	0	0	0	1	1	1	2	2	2	3	3	3	3	4	4
4	.050	0	1	2	3	4	5	6	7	8	9	10	11	12	13	15	16	17	18	19
	.025	0	0	1	2	3	4	5	5	6	7	8	9	10	11	12	12	13	14	15
	.010	0	0	0	1	2	2	3	4	4	5	6	6	7	8	8	9	10	10	11
	.005	0	0	0	0	1	1	2	2	3	3	4	4	5	6	6	7	7	8	9
5	.050	1	2	3	5	6	7	9	10	12	13	14	16	17	19	20	21	23	24	26
	.025	0	1	2	3	4	6	7	8	9	10	12	13	14	15	16	18	19	20	21
	.010	0	0	1	2	3	4	5	6	7	8	9	10	11	12	13	14	15	16	17
	.005	0	0	0	1	2	2	3	4	5	6	7	8	8	9	10	11	12	13	14
6	.050	1	3	4	6	8	9	11	13	15	17	18	20	22	24	26	27	29	31	33
	.025	0	2	3	4	6	7	9	11	12	14	15	17	18	20	22	23	25	26	28
	.010	0	0	2	3	4	5	7	8	9	10	12	13	14	16	17	19	20	21	23
	.005	0	0	1	2	3	4	5	6	7	8	10	11	12	13	14	16	17	18	19
7	.050	1	3	5	7	9	12	14	16	18	20	22	25	27	29	31	34	36	38	40
	.025	0	2	4	6	7	9	11	13	15	17	19	21	23	25	27	29	31	33	35
	.010	0	1	2	4	5	7	8	10	12	13	15	17	18	20	22	24	25	27	29
	.005	0	0	1	2	4	5	7	8	10	11	13	14	16	17	19	20	22	23	25
8	.050	2	4	6	9	11	14	16	19	21	24	27	29	32	34	37	40	42	45	48
	.025	1	3	5	7	9	11	14	16	18	20	23	25	27	30	32	35	37	39	42
	.010	0	1	3	5	7	8	10	12	14	16	18	21	23	25	27	29	31	33	35
	.005	0	0	2	3	5	7	8	10	12	14	16	18	19	21	23	25	27	29	31
9	.050	2	5	7	10	13	16	19	22	25	28	31	34	37	40	43	46	49	52	55
	.025	1	3	5	8	11	13	16	18	21	24	27	29	32	35	38	40	43	46	49
	.010	0	2	4	6	8	10	12	15	17	19	22	24	27	29	32	34	37	39	41
	.005	0	1	2	4	6	8	10	12	14	17	19	21	23	25	28	30	32	34	37
10	.050	2	5	8	12	15	18	21	25	28	32	35	38	42	45	49	52	56	59	63
	.025	1	4	6	9	12	15	18	21	24	27	30	34	37	40	43	46	49	53	56
	.010	0	2	4	7	9	12	14	17	20	23	25	28	31	34	37	39	42	45	48
	.005	0	1	3	5	7	10	12	14	17	19	22	25	27	30	32	35	38	40	43

TABLE F3. α-level values, $T(n_1, n_2, \alpha)$, are given for the Mann-Whitney statistic, T (continued).

n_1	α	n_2= 2	3	4	5	6	7	8	9	10	11	12	13	14	15	16	17	18	19	20
11	.050	2	6	9	13	17	20	24	28	32	35	39	43	47	51	55	58	62	66	70
	.025	1	4	7	10	14	17	20	24	27	31	34	38	41	45	48	52	56	59	63
	.010	0	2	5	8	10	13	16	19	23	26	29	32	35	38	42	45	48	51	54
	.005	0	1	3	6	8	11	14	17	19	22	25	28	31	34	37	40	43	46	49
12	.050	3	6	10	14	18	22	27	31	35	39	43	48	52	56	61	65	69	73	78
	.025	2	5	8	12	15	19	23	27	30	34	38	42	46	50	54	58	62	66	70
	.010	0	3	6	9	12	15	18	22	25	29	32	36	39	43	47	50	54	57	61
	.005	0	2	4	7	10	13	16	19	22	25	28	32	35	38	42	45	48	52	55
13	.050	3	7	11	16	20	25	29	34	38	43	48	52	57	62	66	71	76	81	85
	.025	2	5	9	13	17	21	25	29	34	38	42	46	51	55	60	64	68	73	77
	.010	1	3	6	10	13	17	21	24	28	32	36	40	44	48	52	56	60	64	68
	.005	0	2	4	8	11	14	18	21	25	28	32	35	39	43	46	50	54	58	61
14	.050	4	8	12	17	22	27	32	37	42	47	52	57	62	67	72	78	83	88	93
	.025	2	6	10	14	18	23	27	32	37	41	46	51	56	60	65	70	75	79	84
	.010	1	3	7	11	14	18	23	27	31	35	39	44	48	52	57	61	66	70	74
	.005	0	2	5	8	12	16	19	23	27	31	35	39	43	47	51	55	59	64	68
15	.050	4	8	13	19	24	29	34	40	45	51	56	62	67	73	78	84	89	95	101
	.025	2	6	11	15	20	25	30	35	40	45	50	55	60	65	71	76	81	86	91
	.010	1	4	8	12	16	20	25	29	34	38	43	48	52	57	62	67	71	76	81
	.005	0	3	6	9	13	17	21	25	30	34	38	43	47	52	56	61	65	70	74
16	.050	4	9	15	20	26	31	37	43	49	55	61	66	72	78	84	90	96	102	108
	.025	2	7	12	16	22	27	32	38	43	48	54	60	65	71	76	82	87	93	99
	.010	1	4	8	13	17	22	27	32	37	42	47	52	57	62	67	72	77	83	88
	.005	0	3	6	10	14	19	23	28	32	37	42	46	51	56	61	66	71	75	80
17	.050	4	10	16	21	27	34	40	46	52	58	65	71	78	84	90	97	103	110	116
	.025	3	7	12	18	23	29	35	40	46	52	58	64	70	76	82	88	94	100	106
	.010	1	5	9	14	19	24	29	34	39	45	50	56	61	67	72	78	83	89	94
	.005	0	3	7	11	16	20	25	30	35	40	45	50	55	61	66	71	76	82	87
18	.050	5	10	17	23	29	36	42	49	56	62	69	76	83	89	96	103	110	117	124
	.025	3	8	13	19	25	31	37	43	49	56	62	68	75	81	87	94	100	107	113
	.010	1	5	10	15	20	25	31	37	42	48	54	60	66	71	77	83	89	95	101
	.005	0	3	7	12	17	22	27	32	38	43	48	54	59	65	71	76	82	88	93
19	.050	5	11	18	24	31	38	45	52	59	66	73	81	88	95	102	110	117	124	131
	.025	3	8	14	20	26	33	39	46	53	59	66	73	79	86	93	100	107	114	120
	.010	2	5	10	16	21	27	33	39	45	51	57	64	70	76	83	89	95	102	108
	.005	1	4	8	13	18	23	29	34	40	46	52	58	64	70	75	82	88	94	100

TABLE F3. α-level values, $T(n_1, n_2, \alpha)$, are given for the Mann-Whitney statistic, T (continued).

n_1	α	n_2= 2	3	4	5	6	7	8	9	10	11	12	13	14	15	16	17	18	19	20
20	.050	5	12	19	26	33	40	48	55	63	70	78	85	93	101	108	116	124	131	139
	.025	3	9	15	21	28	35	42	49	56	63	70	77	84	91	99	106	113	120	128
	.010	2	6	11	17	23	29	35	42	48	54	61	68	74	81	88	94	101	108	115
	.005	1	4	9	14	19	25	31	37	43	49	55	61	68	74	80	87	93	100	106

NOTE: Derived from Table 1 of L. R. Verdooren, Extended tables of critical values for Wilcoxon's test statistic, *Biometrika*: 50 (1963), pp. 179-182, with permission of the author and the publishers.

Index